COMPUTER-INTEGRATED DESIGN AND MANUFACTURING

McGraw-Hill Series in Industrial Engineering and Management Science

Consulting Editors:

Kenneth E. Case, *Department of Industrial Engineering and Management, Oklahoma State University*

Philip M. Wolfe, *Department of Industrial and Management Systems Engineering, Arizona State University*

Barish and Kaplan: Economic Analysis: For Engineering and Managerial Decision Making
Bedworth, Henderson, and Wolfe: Computer-Integrated Design and Manufacturing
Black: The Design of the Factory with a Future
Blank: Statistical Procedures for Engineering, Management, and Science
Denton: Safety Management: Improving Performance
Dervitsiotis: Operations Management
Hicks: Introduction to Industrial Engineering and Management Science
Huchingson: New Horizons for Human Factors in Design
Law and Kelton: Simulation Modeling and Analysis
Leherer: White-Collar Productivity
Moen, Nolan, and Provost: Improving Quality through Planned Experimentation
Niebel, Draper, and Wysk: Modern Manufacturing Process Engineering
Polk: Methods Analysis and Work Measurement
Riggs and West: Engineering Economics
Riggs and West: Essentials of Engineering Economics
Taguchi, Elsayed, and Hsiang: Quality Engineering in Production Systems
Wu and Coppins: Linear Programming and Extensions

COMPUTER-INTEGRATED DESIGN AND MANUFACTURING

David D. Bedworth
Department of Industrial and Management System Engineering

Mark R. Henderson
Department of Mechanical and Aerospace Engineering

Philip M. Wolfe
Department of Industrial and Management System Engineering
Arizona State University

McGraw-Hill, Inc.
New York   St. Louis   San Francisco   Auckland   Bogotá   Caracas   Hamburg
Lisbon   London   Madrid   Mexico   Milan   Montreal   New Delhi   Paris
San Juan   São Paulo   Singapore   Sydney   Tokyo   Toronto

COMPUTER-INTEGRATED DESIGN AND MANUFACTURING

234567890 DOH DOH 90987654321

ISBN 0-07-004204-7

This book was set in Times Roman by Bi-Comp, Inc.
The editors were Eric M. Munson and Margery Luhrs;
the production supervisor was Richard A. Ausburn.
The cover was designed by Edward Butler.
R. R. Donnelley & Sons Company was printer and binder.

Library of Congress Cataloging-in-Publication Data

Bedworth, David D.
    Computer-integrated design and manufacturing / David D. Bedworth,
Mark R. Henderson, Philip M. Wolfe.
        p.   cm. — (McGraw-Hill series in industrial engineering and
management science)
    Includes index.
    ISBN 0-07-004204-7
    1. CAD/CAM systems.   I. Henderson, Mark Richard, (date).
II. Wolfe, Philip M.   III. Title.   IV. Series.
TS155.6.B439   1991
670′.285—dc20   90-26510

ABOUT THE AUTHORS

David D. Bedworth (M.S.I.E., Ph.D., Industrial Engineering, Purdue University) is Professor of Engineering at Arizona State University. He has specialized in computer control and computer-aided manufacturing research and teaching for more than 25 years. In addition to publishing more than 60 research papers, he has co-authored a text on integrative production control and has received the Institute of Industrial Engineers Book-of-the-Year award for his text, *Industrial Systems*. He was also co-recipient of the American Society of Engineering Education's Wickenden Paper-of-the Year award for "CAD/CAM Education at Arizona State University." He is ASU's liaison for IBM's CIM in Higher Education program and is director of the associated laboratory.

Mark R. Henderson (M.S., Ph.D., Mechanical Engineering, Purdue University) is Associate Professor of Mechanical and Aerospace Engineering at Arizona State University, where he is co-founder of the CAE and Knowledge-Based Systems Laboratory. His research interests are in feature modeling and feature recognition, solid modeling, computer graphics, and knowledge-based systems. He is the author of many papers in the area of computer-integrated design and manufacturing and has received numerous research awards, including the NSF Presidential Young Investigator award. He has consulted in the area of automated manufacturing with Digital Equipment Corp., Boeing, General Motors, Sandia National Labs, and is currently a computer graphics instructor for Learning Tree International, Inc.

Philip M. Wolfe (M.S.E., Ph.D., Industrial Engineering, Arizona State University) is Professor and Chair in the Department of Industrial and Management Systems Engineering at Arizona State University. He has also held the position of Professor at Oklahoma State University. Formerly he was at the Garrett Turbine Engine Division of Allied Signal, where he served as Manager of Manufacturing Systems Engineering, and he has been employed by the Semiconductor Division of Motorola as a Senior Operations Research Analyst in the Information Systems group. He is active in the Institute of Industrial Engineers, currently serving as group vice president of chapter operations, and is one the board of trustees. His work, teaching, and research interests have related to computer applications in manufacturing, and he has authored two books and more than 30 technical articles.

v

for Ginny, Sue, and Laura;
and for our children, Alex, Ben, Erin, Jaree, Jennifer, Margaret, and Scot

The happiest moments of my life
have been the few I have passed at home
in the bosom of my family.

THOMAS JEFFERSON

CONTENTS

PREFACE   xvii

1 INTRODUCTION   1
   1.1 The Evolution of CAD/CAM   3
        1.1.1 The Evolution of CAD   3
        1.1.2 The Evolution of CAM   4
   1.2 The Scope of Computer-Integrated Manufacturing   10
   1.3 Operations Flow within CAD/CAM   13
   1.4 Topics of Subsequent Chapters   17
   1.5 Summary   20
        References   20

2 GEOMETRIC MODELING   21
   2.1 Geometry—The Language of CAE and CIM   21
   2.2 Key Definitions   24
   2.3 Geometric Modeling Techniques   25
        2.3.1 Multiple-View Two-Dimensional Input   26
        2.3.2 Wire Frame Geometry   27
        2.3.3 Surface Models   28
   2.4 Geometric Entities: Curves and Surfaces   30
        2.4.1 Points, Lines, Surfaces, and Solids   30
        2.4.2 Polygon (or Tesselated) Modeling   33
        2.4.3 Cubic Curves   36
        2.4.4 Bicubic Surfaces   42
   2.5 Solid Modelers   44
        2.5.1 Solid Modeling Example   45
        2.5.2 Solid Model Construction Techniques   46
        2.5.3 The Euler Formula   52
        2.5.4 Solid Modeler Storage Data Bases   54
   2.6 Choosing a Solid Modeler for CAD/CAM Integration   61
   2.7 Feature Recognition: A Case Study   62
        2.7.1 Feature-Based Design Using CSG Construction   62
        2.7.2 Using a BREP for Part Interpretation   65
   2.8 Data Transfer Standards   66
   2.9 Conclusion   67
        Exercises   67
        References   69

3	COMPUTER-AIDED DESIGN	71
3.1	Key Definitions	72
3.2	The Design Process	74
	3.2.1 Applying the Computer to Engineering Design	76
3.3	CAD Hardware	78
	3.3.1 Computers	78
	3.3.2 Input/Output Devices	80
	3.3.3 Output Devices	89
	3.3.4 Storing an Image	92
3.4	CAD Geometry	95
	3.4.1 Computer-Aided Drafting	96
3.5	Computer Graphics and the Part Model	98
	3.5.1 Interactive Graphics	98
	3.5.2 Graphics in CAD	100
	3.5.3 Two-Dimensional Graphics	101
	3.5.4 Two-Dimensional Transformations	103
	3.5.5 Three-Dimensional Graphics	108
	3.5.6 Three-Dimensional Transformations	109
	3.5.7 Composite Transformations in Three Dimensions	110
	3.5.8 Projection	112
	3.5.9 Realistic Image Generation	113
3.6	Analysis	116
	3.6.1 A Design Analysis Example	117
3.7	Integrated CAD	128
	Exercises	131
	References and Suggested Reading	132
4	CONCURRENT ENGINEERING	134
4.1	Key Definitions	137
4.2	Driving Forces behind Concurrent Engineering	138
4.3	The Meaning of Concurrent Engineering	141
4.4	Schemes for Concurrent Engineering	145
	4.4.1 Axiomatic Design	146
	4.4.2 DFM Guidelines	148
	4.4.3 Design Science	149
	4.4.4 Design for Assembly	150
	4.4.5 The Taguchi Method for Robust Design	158
	4.4.6 Manufacturing Process Design Rules	163
	4.4.7 Computer-Aided DFM	163
	4.4.8 Group Technology	167
	4.4.9 Failure-Mode and Effects Analysis	168
	4.4.10 Value-Engineering	169
4.5	Summary of Concurrent Engineering Tools	169
4.6	Conclusion	173
	Exercises	174
	References and Suggested Reading	175

5   GROUP TECHNOLOGY   177
    5.1   Key Definitions   178
    5.2   Background   180
          5.2.1   History of Group Technology   180
          5.2.2   The Role of Group Technology in CAD/CAM
                  Integration   182
    5.3   Methods for Developing Part Families   183
    5.4   Classification and Coding   186
          5.4.1   Hierarchical Code   187
          5.4.2   Attribute Code   189
          5.4.3   Hybrid Code   189
          5.4.4   Selecting a Coding System   190
          5.4.5   Developing Your Own Coding System   192
    5.5   Examples of Coding Systems   193
          5.5.1   DCLASS Coding System   196
          5.5.2   MICLASS Coding System   200
    5.6   Facility Design Using Group Technology   205
    5.7   Cell Example   211
    5.8   Economic Modeling in a Group Technology Environment   214
          5.8.1   Production Planning Cost Model   214
          5.8.2   Group Tooling Economic Analysis   218
    5.9   Economics of Group Technology   221
          5.9.1   Benefits in Design   221
          5.9.2   Benefits in Manufacturing   223
          5.9.3   Benefits to Management   225
          5.9.4   Group Technology Advantages/Disadvantages
                  Summarized   226
    5.10 Summary   227
          Exercises   228
          References and Suggested Reading   231

6   PROCESS PLANNING   233
    6.1   Key Definitions   235
    6.2   The Role of Process Planning in CAD/CAM Integration   236
    6.3   Approaches to Process Planning   237
          6.3.1   Manual Approach   238
          6.3.2   Variant Approach   238
          6.3.3   Generative Approach   239
    6.4   Example Process Planning Systems   242
          6.4.1   CAM-I Automated Process Planning (CAPP)   242
          6.4.2   DCLASS   249
          6.4.3   Computer Managed Process Planning (CMPP)   262
    6.5   Tolerance Charts   271
    6.6   Criteria for Selecting a CAPP System   275
    6.7   Research in CAPP   277
          6.7.1   Product Definition Data Standard   277

    6.7.2 *Part Feature Recognition*   281
    6.7.3 *Artificial Intelligence in Process Planning* 288
  6.8 Summary   289
    Exercises   290
    References and Suggested Reading   291

7 INTEGRATIVE MANUFACTURING PLANNING
AND CONTROL   295
  7.1 Key Definitions   299
  7.2 The Role of Integrative Manufacturing in CAD/CAM
    Integration   300
  7.3 Overview of Manufacturing Engineering 303
  7.4 Overview of Production Control 307
    7.4.1 *Forecasting*   309
    7.4.2 *Master Production Schedule* 312
    7.4.3 *Rough-Cut Capacity Planning* 314
    7.4.4 *Material Requirements Planning* 315
    7.4.5 *Capacity Planning*   317
    7.4.6 *Order Release*   322
    7.4.7 *Shop-Floor Control*   323
    7.4.8 *Quality Assurance*   328
    7.4.9 *Manufacturing Planning and Control Systems* 336
  7.5 Cellular Manufacturing   339
    7.5.1 *Overview*   339
    7.5.2 *Hierarchical Manufacturing Control Model* 341
  7.6 JIT Manufacturing Philosophy   346
  7.7 Integration of CAD/CAM Requires MRP II 348
  7.8 Summary   349
    Exercises   349
    References and Suggested Reading   352

8 MANUFACTURING CONTROL—COMPUTER CONTROL 354
  8.1 The Role of Computer Control in CAD/CAM Integration 357
  8.2 Key Definitions   359
  8.3 Background of Computer Control 361
    8.3.1 *Number Systems*   362
    8.3.2 *Data Handling with Microcomputers* 368
  8.4 Some Computer Control Programming Concepts 374
    8.4.1 *Digital Input*   374
    8.4.2 *Digital Output*   376
    8.4.3 *Analog Input*   377
    8.4.4 *Analog Output*   378
    8.4.5 *A Final Example*   380
  8.5 Timing, Interrupts, and Multitasking 381
    8.5.1 *Timing*   381

     8.5.2 Priority Interrupts   382
     8.5.3 Real-Time, Multitasking Operating Systems   384
  8.6 Programmable Controllers   388
     8.6.1 The Future of PLCs   401
  8.7 Summary   403
     Exercises   404
     References and Suggested Reading   408

9 MANUFACTURING CONTROL—NUMERICAL CONTROL   **409**
  9.1 The Role of NC in Integration   410
  9.2 Key Definitions   414
  9.3 NC Operation and Equipment   415
     9.3.1 The Analyst's Role in NC   416
     9.3.2 The Operator's Role in NC   416
     9.3.3 Axes in NC Operations   418
     9.3.4 Types of NC Systems   418
     9.3.5 NC/CNC/DNC   420
     9.3.6 Some Equipment Examples   423
  9.4 NC Programming   425
     9.4.1 Part Definition (Geometry)   426
     9.4.2 The Machining Plan   440
     9.4.3 Machining Specifications   451
     9.4.4 APT Contouring Example   454
  9.5 Computer Numerical Control (CNC)   457
     9.5.1 Canned Cycles   458
     9.5.2 Further CNC Possibilities   461
  9.6 Distributed Numerical Control (DNC)   461
     9.6.1 CLDATA   462
     9.6.2 Retrofitting to DNC   464
  9.7 Controls in NC   464
     9.7.1 Open-Loop and Closed-Loop Control   464
     9.7.2 Point-to-Point Control   466
     9.7.3 Contouring Control   466
     9.7.4 Adaptive Control   466
     9.7.5 Interpolation   467
  9.8 Concluding Comments   467
     9.8.1 When to Use NC   468
     9.8.2 Advantages of NC   468
     Exercises   469
     References and Suggested Reading   475

10 ROBOTICS   **477**
  10.1 The Role of Robotics in CAD/CAM Integration   479
  10.2 Key Definitions   481
  10.3 Characterization of Robots   482
     10.3.1 What Is a Robot?   482

	10.3.2 *Robot Motions*	484
	10.3.3 *Robot Drive Power*	487
	10.3.4 *Types of Robots*	489
10.4	Robot Motions	490
	10.4.1 *Introduction*	490
	10.4.2 *Kinematic Link Chains*	491
	10.4.3 *Link Geometries*	492
	10.4.4 *Frame of Reference*	493
	10.4.5 *Orientation*	494
	10.4.6 *Changing Frames of Reference*	495
	10.4.7 *Forward Transformation (Six Degrees of Freedom)*	497
	10.4.8 *Solving for Joint Angles*	498
10.5	Workspace Descriptions	499
10.6	Applications	502
	10.6.1 *Welding*	502
	10.6.2 *Spray Painting*	505
	10.6.3 *Materials Handling*	505
	10.6.4 *Assembly*	506
	10.6.5 *Inspection*	512
10.7	Robot Accuracies and Repeatabilities	516
10.8	Economic Justification of Robots	518
	10.8.1 *Life-Cycle Costs*	518
	10.8.2 *A Computer Program for Robot Evaluation*	520
	10.8.3 *Indirect Savings*	526
10.9	Robot Programming Languages	529
	10.9.1 *VAL II*	529
	10.9.2 *AML/X Programming Language*	541
	10.9.3 *Off-Line Programming and Simulation*	546
10.10	Summary	549
	Exercises	550
	References and Suggested Reading	552
11	**MEASUREMENT, ANALYSIS, AND ACTUATION**	554
11.1	Integrative Role in CAD/CAM	555
11.2	Key Definitions	556
11.3	Sensing and Measuring	557
	11.3.1 *Object Detection*	558
	11.3.2 *Object Identification*	563
	11.3.3 *Measurement of Conditions*	569
	11.3.4 *Machine Tool Sensing*	571
	11.3.5 *Robot Sensing*	574
11.4	Analysis	576
	11.4.1 *PID Control*	577
	11.4.2 *Analysis for Control by Programmable Controllers and Computers*	582

11.5	Actuation	585
	11.5.1 On/Off Switches	585
	11.5.2 Alarms and Annunciators	587
	11.5.3 Motor Drives	588
	11.5.4 Hydraulic and Pneumatic Actuation	592
11.6	Computer Communication with Sensors and Actuators—SCADA	593
11.7	Summary	595
	Exercises	595
	References and Suggested Reading	597
12	**COMPUTER-INTEGRATED MANUFACTURING**	**599**
12.1	A Definition of CIM	599
12.2	Key Definitions	601
12.3	Technology Issues	602
	12.3.1 The One-Model Concept	603
	12.3.2 Configuration Management	605
	12.3.3 Data Base Management Systems	607
	12.3.4 Networking	610
	12.3.5 Distributed Data Base Systems	614
	12.3.6 Management of Technology	614
	12.3.7 Other Emerging Issues	615
12.4	Fundamentals of Networking	616
	12.4.1 Networking Concepts	616
	12.4.2 OSI Fundamentals	620
	12.4.3 MAP/TOP Fundamentals	622
	12.4.4 Example MAP/TOP Network	625
12.5	Developing a Successful CIM Strategy	625
	12.5.1 Guidelines	625
	12.5.2 CIM Example	629
12.6	Chapter and Text Summary	638
	Exercises	638
	References and Suggested Reading	638
	INDEXES	
	Name Index	641
	Subject Index	645

PREFACE

It is hard to envisage technical fields that are more exciting than computer-aided design (CAD) and computer-aided manufacturing (CAM). These are fields that are extremely challenging to engineers of many disciplines, including electrical, industrial, mechanical, and manufacturing engineers. In fact, it is apparent that all engineering fields are concerned with either CAD or CAM. For example, the civil engineer tackles structural design with the aid of CAD packages, and the chemical engineer optimizes operating characteristics of processes by using a computer. However, it is to the application of CAD and CAM to discrete-item manufacturing that this text is primarily addressed—the manufacture of such items as automobiles, refrigerators, electronic circuit boards, and computers. This is the manufacturing category that is the concern of the four categories of engineers cited earlier.

Not only is CAD/CAM an exciting *technical* field, it is of paramount importance to the economic well-being of any nation in these times of great competition in the manufacturing arena. The major key to improving productivity in manufacturing lies in the judicious automation of design and manufacturing processes to ensure reduced design-to-prototype lead times, fewer problems with engineering design change implementations, flexible manufacturing capabilities, maximum production rates commensurate with reduced production costs, and many other needed production attributes. However, automation of computer-aided design and computer-aided manufacturing as separate entities does not lead to optimum solutions of these manufacturing attributes. The engineering design function has to be *integrated* with the manufacturing operation to ensure that the design engineer has a knowledge of the capability of manufacturing to implement specific designs into finished product. Similarly, the manufacturing engineer needs to know the requirements of the design in a clear and legible manner. This leads to the theme and title of this book. The theme will be to introduce the concepts of CAD and CAM to those persons involved with specifying, designing, and implementing CAD/CAM systems for discrete-item manufacturing. Further, the orientation will be to give state-of-the-art material relating to the possible integration of the two computerized areas. Thus the somewhat grandiose title of *Computer-Integrated Design and Manufacturing*. Not all aspects of design and manufacturing will be shown to be integrated. Rather, potential for integration and a route toward achieving that integration will be addressed.

The text is geared to the senior and beginning graduate student who is interested in the CAD and/or CAM fields. It is suggested that no mechanical engineering student should take a design course without an overview of the effects of design on manufacturing, as well as the effects of manufacturing on design. Similarly, the engineer who is concerned primarily with manufacturing needs to understand the design function so that effective dialogues can occur between the two groups responsible for these functions. Therefore, a mechanical engineering course should cover thoroughly the topics in the first four chapters, relating to CAD/CAM introduction, geometric modeling, engineering design, and concurrent engineering. Chapters 8 through 11, covering software and hardware characteristics relating to manufacturing control, are also logical in-depth chapters. A survey of the remaining four chapters would round out the introductory knowledge required. An industrial or manufacturing engineering course would probably survey the first three design chapters and include detailed coverage of the remaining manufacturing-oriented chapters. In either case, a one-semester course is feasible, as has been proved over many years of development at Arizona State University.

The 12 chapters follow what the authors feel is a logical progression from design through implementation. Chapter 1 introduces the field of CAD/CAM to the reader, while Chapter 2 covers geometric modeling to provide a firm foundation for Chapter 3, which presents a methodology for computer-aided design. The effects of concurrent engineering, including design for the complete product life cycle, are discussed in Chapter 4. The interface between design and manufacturing is handled through computer-aided process planning (CAPP). A key to effective CAPP is group technology, whereby families of parts are grouped for production on families of processing equipment. Chapter 5 presents this foundation material, and Chapter 6 discusses the salient points of computer-aided process planning. The heart of operating any manufacturing system, automated or not, is the production planning and control function, the topic of Chapter 7. Chapters 8 through 11 are concerned with the hardware and software aspects of controlling a manufacturing system. Chapter 8 covers computer control and discusses approaches to having a computer control process. Equipment control through numerical control and programmable controllers is covered in Chapter 9, while Chapter 10 evaluates the role of robotics in the CAD/CAM environment. Chapter 11 presents some of the instrumentation and analysis considerations involved in supporting the control activities. Finally, the concluding chapter considers the very important topic of CAD/CAM implementation.

It is impossible to thank all of those who have helped with the development of this book. First of all, thanks have to go to Anne Duffy, our initial editor at McGraw-Hill; John Corrigan, our editor after Anne moved to higher responsibilities; finally, Eric Munson, who had the difficult chore of giving the final push that got us to finish the book. We would also like to express our appreciation to our peers at Arizona State University and Garrett Turbine Engine Company. Without their aid and support this text would never have been completed. Special appreciation has to be given to the following individuals and companies who contributed specific thoughts and materials: Sue Baelen (Automatix); Dick Barnard (Crouzet Controls); Steve Belmont (Burr-Brown), Dennis Bennett, and Frank Charubin (ESAB North America, Inc.); John Bosch (Sheffield Measurement Division,

Cross & Trecker); Jo Brannan (Intelmatic Corp.); Ding-Yu Chan, Prasad Gavankar, Ed Stahlmann, and Rajendra Tapadia (Arizona State University); Keith Draper (Autotech Corp.); Lorraine A. Gorski (American Solenoid, Inc.); Richard Howard (Roberts Corp.); Steven Longren (Compumotor Corp.); James MacLaren (Westinghouse-Electric Corp.); Leonard Sharman (Micro-Relle); Terry Taylor and Randy Wiemer (Garrett Turbine Engine Co.); Larry Tucker (Hyde Park Electronics.); Pat Webster (Warner and Swasey Co.); and Albert White (IBM Corp.). We also thank Elinor Lindenberger, who typed part of the manuscript.

McGraw-Hill and the authors would like to thank the following reviewers for their many helpful comments and suggestions: Douglas E. Abbott, University of Massachusetts at Amherst; Han Bao, North Carolina State University; Jon F. Botsford, Texas A&M University; Gary L. Kinzel, Ohio State University; Richard J. Linn, Iowa State University; Peter O'Grady, North Carolina State University; Albert P. Pisano, University of California, Berkeley; Clark Radcliffe, Michigan State University; Harold J. Steudel, University of Wisconsin, Madison; and Eric Teicholz, Graphic Systems, Inc.

Finally, a debt of gratitude has to be tendered to our wives and families. Without the support of Ginny, Sue, and Laura, and the understanding of our children, this book could not have been written. Our first priority has always been to our respective families, but our attitudes while writing this book may not have shown it to those who are most dear to us.

David D. Bedworth
Mark R. Henderson
Philip M. Wolfe

CHAPTER

1

INTRODUCTION

The success of tomorrow's industrial plant will depend on its ability to gather, share, and utilize information.

J. T. O'Rourke (quoted in Ref. 1)

Given the fact that machine tool automation really began only with the development of numerical control in the 1950s—less than 40 years ago—it is amazing that there exist today manufacturing plants that are almost completely automated. Granted, these installations produce relatively few varieties of product, but it is clear that the physical components for a sophisticated automated enterprise are available.

Before beginning to discuss these capabilities, we must identify what we will mean by "manufacturing plants" in this book, since there are several categories of manufacturing (or production) that could be considered. We will consider manufacturing in three broad areas: (1) continuous-process production, (2) mass production, and (3) job-shop production. In an even broader classification, the latter two categories could be considered as *discrete-item production*.

- *Continuous-process production:* Product that flows in a *continuous* stream falls into this manufacturing category. Typical processes include petroleum, cement, steel rolling, petrochemical, and paper production, as well as many others. An obvious attribute of these processes is that the production equipment is utilized for a relatively small group of similar products. Combining this attribute with the continuous-flow characteristic allowed investment in automated equipment that realized benefits well before such advantages were seen in the other two categories of production. This text will *not* be oriented toward continuous-process production, though many of the topics will be as germane to this category as to the other two.

- *Mass production:* This entails the production of discrete units of production at very high rates of speed. *Discrete-item* production is used for such goods as automobiles, television sets, refrigerators, electronic components, and so on. Mass production attempts to emulate the characteristics of continuous-flow production for discrete products that are produced in high volumes with relatively small variations in product. As a result, mass production has historically realized many benefits from mechanization and automation.

- *Job-shop production:* A manufacturing facility that produces a large number of different discrete items that require different sequences through the production equipment is usually called a job shop. Because of the large number of different products and demands for those products, scheduling and routing problems are enormous. As a result, automation has at best been restricted to individual components of the job shop (sometimes referred to as *islands of automation*), and there have been few attempts to automate the entire facility by integrating the islands of automation.

Now we can clarify what we meant earlier when we said that the physical components for an automated manufacturing system exist although little automation of complete manufacturing facilities has actually been accomplished. First, we do not include continuous-flow processes, which comprise a relatively small percentage of manufacturing. We do include, however, mass production of discrete items, where *segments* of the production line are often quite automated, but not the entire line. For example, spot welding and painting in an automobile line may be completely automated using robotics, but much assembly work is accomplished by human labor. Similarly, job-shop facilities have long used automated machine tools, but only rarely has transfer of work in progress among these machines been handled automatically. Other than some physical equipment that might be needed to completely automate a specific discrete-item facility, a major problem is the one alluded to in the introductory quotation for this chapter: The software needed to integrate information has not evolved to the level of the available physical hardware.

In line with this observation, a major component of automated information that needs to be made available to the manufacturing operation in order to allow plant automation/integration must come from *product design*. However, information on manufacturing capability is frequently not available to the product design department, leading to what has often been called the ''wall'' between design and manufacturing. Manufacturing is more concerned with process design than product design.

The objective of this text is to present a state-of-the-art review of the hardware and software capabilities and needs required to make a job-shop manufacturing facility as automated as a continuous-flow enterprise. There are three components to this review, with two of the components, computer-aided design (CAD) and computer-aided manufacturing (CAM), being optimized into the third component, which has frequently been called computer-integrated manufacturing (CIM).

Before examining the potential scope of CIM, it will be of benefit to look briefly at the historical evolution of computer-aided design and computer-aided manufacturing. It has been wisely said that those who do not learn from the past will not be able to plan for the future.

1.1 THE EVOLUTION OF CAD/CAM

It is fun to attempt to trace those events that led to a major scientific discovery. In fact, a popular recent television series traced such *connections* for many such discoveries. The digital computer's roots, for example, were traced back to the Stone Age and counting on notched sticks. We will not attempt such an interesting task as showing that type of evolution but will limit ourselves to proven stepping stones.

1.1.1 The Evolution of CAD

The roots of CAM have been more clearly delineated in recent years than have those of CAD. Logically, CAD's development has come through the evolution of computer graphics and computer-aided drawing and drafting, often called CADD. Perhaps a moment should be taken now to define what these terms mean:

Computer graphics refers to the use of a computer (when the word "computer" is used in this text, it will be assumed to refer to a *digital* computer as contrasted to an *analog* computer) to assist in the generation of pictorial representations. These may range from business applications generating pie and bar charts to complex art representations that simulate paintings by the old masters.

Computer-aided drawing and drafting uses the computer to assist in the generation of blueprint-type data. This is usually in the form of two-dimensional representations of a part with associated dimensional data as well as other manufacturing information.

Computer-aided design goes far beyond CADD, allowing for *analysis* as well as graphical representation. For example, an automobile suspension system may be designed using CAD with testing under specific road conditions. Animation can be utilized to show the damping effects of the suspension design given the road conditions. The design may be interactively improved based on these results. As a result, CAD programs frequently incorporate complex engineering analysis routines. Further, CAD is not limited to the design of manufactured products. For example, an architectural layout for an office building might be considered as CADD if no analysis is included. However, if the computer package includes the capability of insuring that federal standards are maintained as regards maximum distance to restrooms, human factor characteristics of the office layout, and so on, then CAD is realized.

Besant [4] credits work by Ivan Sutherland at the Massachusetts Institute of Technology as being the first major step in the evolution of computer graphics and thus the evolution of computer-aided design. This work, which produced SKETCHPAD, was accomplished in 1963. A cathode-ray oscilloscope was driven by a Lincoln TX2 computer to allow graphical information to be displayed on a screen. Pictures could be drawn on the screen and then manipulated using a light pen. Besant calls this the inception of *interactive* graphics and Ivan Sutherland the father of computer graphics.

Machover [9] shows that at the end of the 1960s there were fewer than 200 CAD work stations in the United States, primarily in the aerospace and automotive industries. The next decade saw this rise to some 12,000 work stations, with industrial application much wider than the previous decade's narrow spectrum. A major impetus for this proliferation was the advent of the minicomputer and the development of display units that were much better for representing graphical information than the original oscilloscopes. Extremely complex computer-aided design software packages can now be resident in microcomputers, realizing significant cost advantages that allow engineering students the possibility of having their own package in their own home. Figure 1.1 shows such a microcomputer work station.

1.1.2   The Evolution of CAM

The roots for the automation of a *complete* factory could be shown to have come from many different sources, though it has been claimed that a mechanical flour mill patented by Oliver Evans of Philadelphia in 1795 was the world's first automatic factory [2]. This, of course, falls in the *continuous*-production classification.

The roots for automation of discrete-item production are firmly embedded in the mass production concepts developed by Henry Ford early in the twentieth century. The rate of production (cycle time) for an automobile might be such that a finished automobile comes off the line every minute and a half. The length of time required for actual assembly of an automobile is, of course, much longer. Assembly tasks are combined so that groups of tasks can be accomplished in less than the cycle time. Such groupings are often called *stations*, and many stations (assemblers, robots, etc.) are required along the line. This grouping of tasks is analogous to *group technology*, a technique that will be seen later, in Chapter 5, as being the key to realizing many of computer-aided manufacturing's objectives for *job-shop* production.

Logically, the digital computer is the key to computer-aided manufacturing as well as computer-aided design. Digital computers have been used to *control* manufacturing functions for more than 30 years. For example, it was reported in

FIGURE 1.1 Typical engineering work station. (Courtesy of Texas Instruments.)

1973 that a diverse grouping of computer control applications included [2]:

- Automobile traffic control
- Product testing and quality control
- Foundry control
- Numerical control equipment interface
- Space engineering research
- Neurological and biomedical research
- Television program/commercial switching
- Nuclear reactor control and monitoring
- Railroad freight-car monitoring
- Deep-sea data logging
- Cement plant blending and control

- Utility plant startup and control
- Hot-strip roughing—mill control
- Oxygen furnace control
- Blast furnace applications
- Nylon plant process control
- Ethylene plant control
- Oil refinery cat-cracker operation

There is an obvious reason for a majority of the applications falling into the process industry category. Any form of control requires information to be gathered from what is being controlled—say, the process; the information gathered has to be analyzed to determine if corrective action is needed; the process has to be adjusted if corrective action is needed. This three-stage control sequence comprises sensing→analysis→actuation. Automatic sensing and automatic actuation both require costly and complex instrumentation. Analysis requires a thorough mathematical knowledge of process characteristics. The types of manufacturing that had a good range of instrumentation available as well as a good knowledge of the underlying mathematics of the process fell in the continuous category. As an example, oil refineries and petrochemical plants already had independent units automated a great deal. As a result, it was relatively easy to integrate a digital computer into process control to allow integration of locally controlled segments of the plant as well as to allow a greater degree of computer ''decision making'' to be incorporated into the control procedures.

Discrete-item manufacturing, on the other hand, did not have the advantages of widely automated entities, and so the move to factory automation was quite slow in comparison to that in the process industries. The evolution of automated discrete-part manufacturing, starting with the Ford line, includes many interesting developments. A few of these are given below [2]. The dates of significant developments are in many cases blurred, and the authors concede that variation by up to five years might be possible in certain cases.

1909: *Ford production line.* As mentioned earlier, the Ford concepts have had far-reaching implications as regards feasibility for the automation of a production line for a particular product through the concept of division of labor. The original conveyorized line set the pace for mass production concepts.

1923: *Automatic transfer machines.* These were introduced at the Morris Engine factory in England. Transfer equipment, used for indexing parts down a production line, was the key to a completely mechanized production line, such as is used for the manufacture of engine blocks or transmission housings.

1952: *Numerical control (NC).* Machining operations are now widely controlled by numerical control operations whereby tool positioning is accomplished through computer commands. In 1952, these commands resulted in the generation of a punched paper tape that, when read by the machine tool's controller, caused tool positioning to effect the cutting desired. There is no doubt that today's computer-aided manufacturing would not be a reality without this far-reaching development. In fact, the APT compiler developed for NC can be argued to be a precursor to CAD as well as CAM. Robot developments are a natural extension of numerical control.

1959: *Control digital computer.* The first widely publicized application of control using a digital computer occurred at a Texaco refinery located in Port Arthur, Texas, where a catalytic cracking unit was optimized using a linear programming algorithm.

1960: *Robot implementation.* The first "Unimate," based on numerical control principles, was introduced in 1960. A Unimate robot was actually installed at the Ford Motor Company to tend a die-casting machine [7]. These innovations were the precursors for the later widespread use of robots in manufacturing processes.

1965: *Production-line computer control.* International Business Machines developed a computer-controlled production line for manufacturing circuit boards (then called solid logic technology boards). The line was designed to use a control computer to supply numerical control data for inserters and testers.

1970: *Multiple-machine computer control.* The Japanese National Railways placed seven lathes under the simultaneous control of one computer, with three of the lathes supposedly containing memories in which the programs were stored. This brings up the concepts of *direct numerical control (DNC)* and *computer numerical control (CNC).* The use of one mainframe computer to control several machines in a time-shared fashion evolved in the mid- to late 1960s and was originally called *direct numerical control.* DNC rapidly fell into disfavor, however, because of the high cost of mainframe computers and the fact that computer failure meant that all the machines went down. *Computer numerical control (CNC)* was developed in the early 1970s and proved so important that it is given its own itemization in this historical list.

1970–1972: *Computer numerical control.* The advent of the minicomputer meant that a machine tool could have its own computer memory. As a result, controlling programs could be stored in memory and did not have to be reloaded each time a part was to be produced. Also, certain "canned" routines could be used to accomplish many repetitive functions such as peck drilling and tapping. The memory capability led to the advent of distributed numerical control.

1975–1980: *Distributed numerical control.* The use of a mainframe computer to download numerical control programs to the desired machine tool's memory, distributing the work across machines depending on which applicable machine is available, is a concept that is key to today's computer-aided manufacturing advances in manufacturing cell control and flexible manufacturing systems.

1980s: *Manufacturing cells.* A reduction of the combinations involved in job-shop control can be achieved by determining families of parts that can be produced on a subset of the equipment available in the job shop. The determination of families and applicable equipment is most often done by a technique called *group technology*. Using group technology, a cell control computer can download NC programs and also effect materials handling between machines, frequently through a robot transfer. A typical cell configuration and control schema is shown in Figure 1.2.

1980s: *Flexible manufacturing systems*. The idea of using a set of machines to make a relatively wide variety of products, with automatic movement of products through any sequence of the machines, including testing, is the heart of a flexible manufacturing system. The combination of manufacturing cells and flexible manufacturing systems leads to the possibility of computer-integrated manufacturing.

The evolution of CAD and CAM has taken place over a relatively short time span. The technological capabilities are awesome. Machines have long had the capability of changing their own tools, thus allowing a diverse grouping of parts to be produced automatically. Figure 1.3 shows a robot sequencing tooling in a machine center's tool magazine to allow correct production of a complex part. Such machine centers are often the heart of flexible manufacturing systems. Cameras have been developed that can fit in the tool magazine and be treated like a

FIGURE 1.2 Typical cell functions and control. (From Ref. 8, courtesy of *Journal of Manufacturing Systems*, Society of Manufacturing Engineers.)

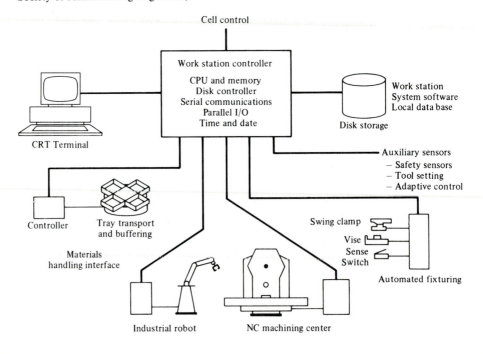

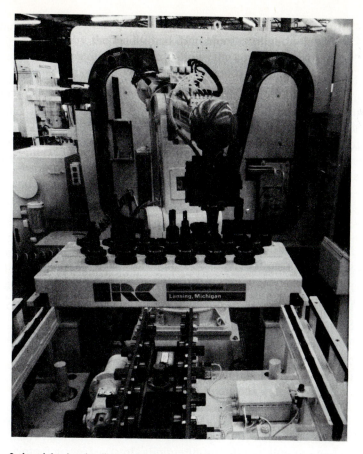

FIGURE 1.3   Industrial robot loading tooling for a machine tool. (Courtesy of The Roberts Company, Cross and Trecker Corporation.)

tool, being called to the part as desired to allow automatic verification of machining characteristics and allowing computer correction when these characteristics are not verified. Similarly, robots equipped with vision sensing equipment can determine part imperfections much faster and more positively than can a human operator. Tooling and components can be stored in and retrieved from computer-controlled automatic storage and retrieval facilities, called, not surprisingly, AS/RS facilities.

The *integration* of all such equipment requires the networking of information, both design and manufacturing. Further, to optimize the cost/profit situation automatically requires other functions to be brought into the picture, such as cost accounting. This integration philosophy is generally called *computer-integrated manufacturing*.

1.2   THE SCOPE OF COMPUTER-INTEGRATED MANUFACTURING

When all of the activities that comprise the modern manufacturing plant are considered as a whole, it is well-nigh mind-boggling even to think that a large portion might be automated, let alone trying to envisage automation of the whole. The systems approach is a technique that allows a large, complex system with interacting components to be analyzed and improved. Anyone charged with directing the automation of a complex system is advised to start by applying a technique similar to the traditional systems approach.

The steps involved in the systems approach generally follow those given by Bedworth and Bailey [3].

1. Determine the objectives of the system.
2. Structure the system (macro view) and set definable system boundaries.
3. Determine the significant components that make up the system.
4. Perform a detailed study on the components in light of the overall system.
5. Synthesize the analyzed components into the system.
6. Test the system according to some performance criterion (evolving from the original objectives).
7. Improve by cycling through steps 2 through 6 as needed.

No task, however small, should be tackled without knowledge of the task *objective*. This is the key ingredient which, when lacking, causes members of the same team to pull in different directions. In considering factory automation there could be many possible objectives. One might be to improve the performance of a specific process. Boundary conditions would then be limited to that process (as well as other processes that might be affected by increased output, such as material supply or assembly after production). Another objective might be to minimize costs in a segment of the operation, while a third might be profit maximization; obviously, it is rare that such multiple objectives can *all* be optimized, even though politicians seem to think so when it comes close to election day. When considering moving to a computer-integrated manufacturing operation, the objective would probably be related to being competitive, a problem that manufacturing plants are having at the micro level and a situation that is almost catastrophic for the United States at the macro level.

Setting system boundaries for a CIM project might at first appear to be concerned only with the engineering design and actual manufacture of the products. While the integration of these two components is a major task that is not satisfied in most facilities, CIM goes beyond these activities. Figure 1.4 shows graphically and dramatically what is involved in computer-integrated manufacturing.

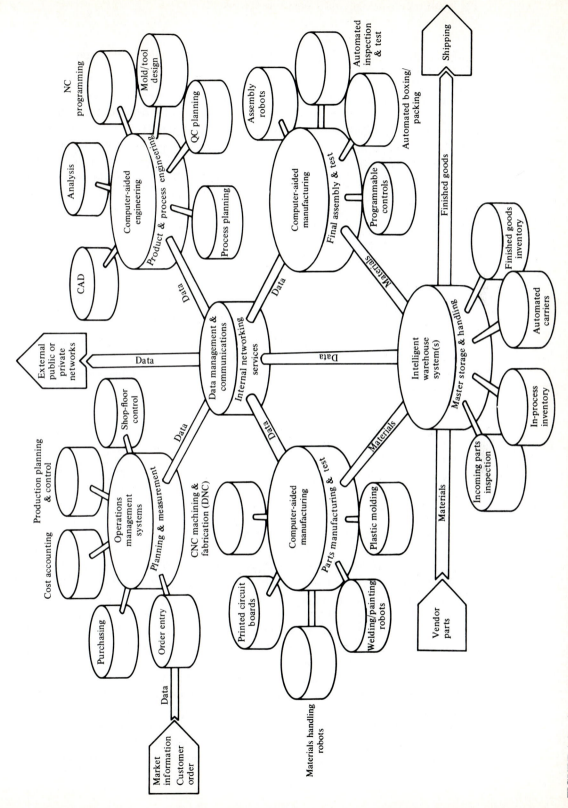

FIGURE 1.4 Structure of computer-integrated manufacturing. (From Ref. 6, courtesy of *Production Engineering*.)

As far as the plant itself is concerned, there are five major components:

1. *Computer-aided engineering (CAE)*. CAE encompasses CAD, NC programming, tool and fixture design, quality control (QC) planning, and process planning. This latter area is the glue that ties CAD and CAM together and when automated is called computer-aided process planning (CAPP).

2. *Operations management*. Operations management governs the acquisition of all materials needed for product manufacture; since a cost-effective system is required, it is mandatory to include cost accounting. Production planning and control is required to ensure that parts are routed in an efficient manner to keep equipment busy as well as to ensure that customer needs are satisfied. Shop-floor control is needed to ensure that production status data is available to production planning and control as well as to allow planned sequencing to be effected on the shop floor.

3. *Computer-aided manufacturing—manufacturing and test*. Parts have to be manufactured and tested. A major point is made in Figure 1.4 that should be amplified at this point. Many people think that computer-aided manufacturing is for *chip-cutting* parts only—those parts that are machined in some fashion. This is not the case, as cannot be stressed too forcefully. Electronic manufacturing has to follow the same CAM practices as discussed in this text if competitiveness is to be maintained or achieved. Assembly of parts into a circuit board, for example, is no different in concept from assembly of a turbine engine!

4. *Computer-aided manufacturing—assembly and test*. Assembly, inspection, and packing are the remaining ingredients of CAM.

5. *Intelligent warehousing*. The automatic storage and retrieval of materials, components, and finished goods is the fifth component of CIM. This includes not only incoming materials and finished products but also the temporary storage of work in progress.

Finally, to allow all the components to work as an integrated system, we have to integrate the five components with a network system as exemplified by the *data management* and *communications* component shown in Figure 1.4. Standards have been evolved for this function, but the *information* flow in CIM is still a major problem to be handled effectively.

Now, continuing with the *systems approach* comments, it is necessary to break the system into components for initial analysis. In fact, this overall systems approach would be applied to the six major components just discussed, including the component of data management and communications. Analyzing the entire system as a whole would initially be far too complex and so, as just mentioned, individual component analysis has to be accomplished. However, component analysis has to be done *in light of the entire system*. We have all heard horror stories about the plant that installed one computer system for design and a different one for manufacturing. Trying to integrate the design and manufacturing data in such a system becomes a major problem, frequently an insurmountable one.

After the components have been analyzed, integration into the system is mandated. With a CIM system, this is accomplished through a networking scheme with a smart manufacturing data base that can handle design, manufacturing, and planning. A *hierarchical* computer structure should be employed that allows this manufacturing data base to be resident in the computer at the top of the hierarchy, with cell control computers below the mainframe computer, and finally equipment-controlling and data-gathering computers at the bottom of the hierarchy.

We have not yet *fully* delineated the CIM system. Table 1.1 shows typical functions that fall under the umbrella of the components discussed so far. Every one of these functions can itself be a major undertaking. No text can tackle all of the functions shown in Table 1.1; this text, however, will consider the major functions that relate to CAD, CAM, and the production operating system.

1.3 OPERATIONS FLOW WITHIN CAD/CAM

Before giving a chapter-by-chapter synopsis of the material to be presented in this book, we will show briefly the operational flow of functions needed to process an item through a manufacturing facility. This operations flow within the CAD/CAM environment can be diagramed as shown in Figure 1.5. In the following discussion, sequence numbers refer to the box numbers in the figure.

1, 2. All planning has to be in terms of known customer orders and sales forecasts. If expected demands are not known and/or estimated, the enterprise will be operating in a vacuum.

3. Management decisions based on expected orders lead to long-term order requirements that need to be satisfied by either production or by subcontracting to outside sources (vendors).

4. In order to plan how parts can be produced, a relatively long-term evaluation of facility requirements is needed. For example, are enough machines of a particular capability available, will material be available, can we accomplish our needs with the current workforce, and so on. The aggregate planning function determines what product quantities should be produced in what time periods to satisfy the long-term requirements. The result of this activity is called the *master production schedule* or *master schedule*. It is a schedule for *final product*, not for the components that go into the final product.

5. The master schedule is affected by current status conditions, so feedback loops come from many sources—including problems that might occur with deliveries from vendors, trouble on the shop floor, analysis that reveals demands cannot be satisfied due to capacity problems, lack of vendors, and so on.

6. The material requirements planning (MRP) function takes current inventory levels for *all* components needed for the final products (a plant might have 20,000 part numbers and perhaps 100 final products for which master schedules have been determined) as well as the components' bills-of-materials and lead time information [obtained from design (7) and process planning (8)] and evolves component master schedules for all components needed by the demand requirements

TABLE 1.1
The scope of computer-integrated manufacturing

	Business		Production					Design		
	Resource management	**Economic accounting**	**Production planning**	**Part planning**	**Production control**	**Part processing**	**Document preparation**	**Test**	**Synthesis and analysis**
Industry	Trend analysis Resource availability Economic indicators		Capacity and delivery planning R&D	Machining technology data base R&D		R&D Testing	Standards Design CAD interface Parts data base		Design standards
Corporation	Trend analysis Facility planning Strategic planning Merger/acquisition Synergistic product Production levels Data management	Projections Simulation	Scheduling Facility planning Material requirement planning R&D	Machining technology data base Group technology R&D	Data management	R&D Testing	Parts data base Bills-of-materials GT/part classification Data management	Test data base Field report data base Computer-aided engineering	Computer-aided engineering Producibility analysis Design standards

	Plant layout	Cost tracking	Material requirement planning	Machining technology data base	Inventory	R&D	Parts data base	Computer-aided engineering	Computer-aided engineering
Plant	Plant layout Inventory Scheduling Manpower utilization Make/buy decision Data management	Cost tracking Customer billing Customer order Normal accounting Make/buy/economic order quantity Cost estimating Process justification	Material requirement planning Bill-of-materials Time standards Scheduling Make/buy decision Facility planning Capacity planning Plant layout Manpower utilization GT/operation sequence	Machining technology data base GT/plan retrieve Computer-assisted process planning R&D Part program Cost estimating	Routing/scheduling Material handling QC/QA Maintenance Purchase/receive Data management Standard methods	Testing	Computer-aided design and drafting Bill-of-materials GT/part classification Tool/fixture design and coding Data management	Testing	Design analysis System modeling Producibility analysis GT/design retrieval Design standards
Cell	Job sequencing Inventory Data management	Job tracking Economic data collection	Line balance Machine loading	Machining technology data base Computer-assisted process planning Part program NC verification	Material handling Routing/scheduling QC/QA Inspection Standard methods Inventory Data management	Automatic assembly Adaptive control Robotics Data collection	Process instructions	Data acquisition	GT/design retrieval Design standards
Work station		Economic data collection		Computer-assisted process planning	Maintenance/diagnostics	NC DNC CNC Adaptive control Automatic inspection Sensors Diagnostics Data collection	Process instructions	Data acquisition	

All of the various functions included under the umbrella of computer-integrated manufacturing are indicated in this matrix presented by Edward J. Adlard, supervisor of manufacturing software systems, Metcut Research Associates, Inc., at SME's Autofact 4 Conference. The matrix represents the various functions involved in CIM and shows where these functions impact the manufacturing organization, which itself consists of business, production, and design functions. The division along the left-hand side of the matrix represents the various levels of the overall manufacturing environment—industry, corporate, plant, cell, and work station. And within the matrix itself are the various systems and functions involved in a computerized manufacturing environment.

Source: From Ref. 5, courtesy of *Production Engineering.*

15

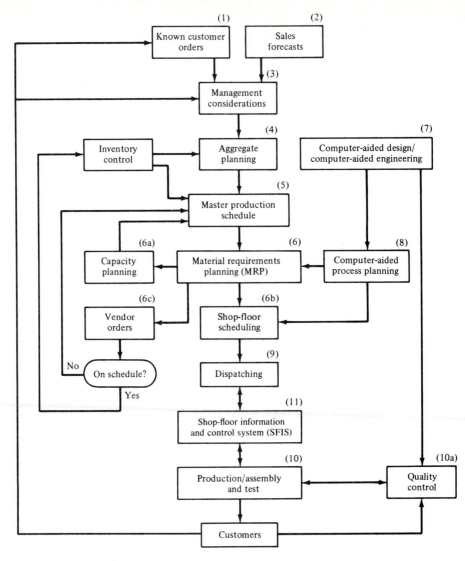

FIGURE 1.5   Operational flow for CIM.

agreed upon. MRP does not take into account whether manufacturing has sufficient capacity to handle the job releases, and so capacity planning (6a) evaluates
shop loading in terms of the requirements and feeds back to the master schedule
for corrective action if problems occur. A further function of MRP based on such
analysis is determining whether components should be produced in-house (6b) or
subcontracted to outside vendors (6c).

7. Computer-aided design, as discussed in Section 1.1, is the function that is
first accomplished after a need for a product has been determined, and so the

sequence in which it is discussed in this section is not the same as if we were just talking about the cycle from customer inception through design, manufacturing, assembly and testing, and back to the customer. The design engineer does not talk in the same terms as the manufacturing engineer. For example, geometric design deals with lines, splines, circles, and arcs whereas manufacturing talks about pockets, chamfers, holes, and so on. Process planning accomplishes the language transition from design to manufacturing, among other things.

8. Process planning evolves the sequence of operations required to manufacture a part, the times required to accomplish the operations, the machines and tooling required, and evaluates tolerance stacking problems that accrue from multiple cuts and/or multiple components that comprise a part [10]. The process planning function can ensure the profitability or nonprofitability of a part being manufactured because of the myriad ways in which a part can be produced. The process planning information feeds into the MRP analysis as well as into shop-floor scheduling (6b), so that detailed schedules can be evolved for machines, tooling, fixtures, people, material handling, testers, and materials. All have to come together at the right time or havoc will ensue.

9. Dispatching is the function of releasing all required items needed to perform an operation on a part so that part production may be accomplished at the time planned by the scheduling function.

10. Production and assembly is accomplished with NC-related equipment. Sequencing is accomplished through local control computers and/or programmable controllers.

11. Finally, the shop-floor information system is responsible for getting the required information down to the processing-equipment local controllers and sequencing controllers as well as capturing real-time status data from the equipment and parts so that the feedback loops can effect corrections or normal continuation of operation as required.

Subsequent chapters will provide in-depth coverage of the functions included within the CAD/CAM operational flow pattern just presented.

1.4 TOPICS OF SUBSEQUENT CHAPTERS

The structure of the remaining chapters follows (1) engineering design considerations, (2) design to manufacturing considerations, (3) manufacturing techniques to allow simplification and potential automation of the manufacturing process, (4) the basic characteristics for operating the manufacturing function, (5) the hardware characteristics of manufacturing control, and (6) implementation strategies for computer-integrated manufacturing. An attempt has been made to steer away from manufacturer-specific equipment, since the fields of CAD and CAM are changing at an almost unbelievable rate. Where material is so specific within the chapters, a justification will be given.

Chapter 2: Geometric Modeling. An understanding of geometric modeling is the key to understanding engineering design. Further, solid modeling is a major key to the integration of computer-aided design and computer-aided manufacturing. Both topics are covered. The material is presented in such a manner as to give the reader a clear and concise introduction to geometric modeling, one that the manufacturing engineer will be able to follow as easily as the design engineer.

Chapter 3: Computer-Aided Design. This chapter presents the design process and ways to automate this process. It is quite apparent that CAD is a key to an integrated CAD/CAM information data base, which itself is a major key to the feasibility for computer-integrated manufacturing. As with Chapter 2, this chapter gives a straightforward introduction to computer-aided design for all analysts and managers involved in the design/manufacturing operation. Topics include two-dimensional and three-dimensional computer graphics, feature modeling, and engineering analysis techniques.

Chapter 4: Concurrent Engineering. To decrease the time from product inception through prototype manufacture it is necessary to instill a knowledge of processing capabilities into the design process. Concurrent engineering stresses design of the product and production processes together with design of assembly, quality control, and field service—that is, the complete product life cycle. There are many ways that products can be designed to allow simplification of product manufacture. Further, concurrent engineering is often mandatory if the product is to be manufactured at all. This chapter discusses the product life cycle and considerations in designing for manufacturability, inspectability, functionality, and other "-ilities." A possibility for automating this aspect through a combined design/manufacturing data base is also apparent.

Chapter 5: Group Technology. In order to obtain a tractable subgrouping of a complex manufacturing system in order to allow automated groups of equipment and parts to be meshed together, it is beneficial to use a group technology technique. The premise of group technology is that a single solution can be evolved efficiently for a group of similar problems. This chapter introduces a variety of group technology schemes and discusses benefits to manufacturing of diverse products and design of a facility to manufacture those products, as well as benefits to design itself. Of particular interest to the design engineer is the fact that it is feasible to evolve a group technology code directly from a geometrically modeled computer-aided design.

Chapter 6: Process Planning. Once the parts have been determined for which CAM is to be applied—say, through group technology—the method by which these parts are to be manufactured has to be evolved. The determination procedure is called process planning and requires a very skilled individual to realize efficient and cost-effective plans. As a result, not only is this function the glue which ties design to manufacturing, it is also a very labor-intensive operation. Automating this function, a very difficult process in itself, is called computer-aided process planning (CAPP). The automatic generation of complex parts from design is possible through CAPP.

Chapter 7: Integrative Manufacturing Planning and Control. The key to a manufacturing facility's *running* in an efficient manner is the production planning and control function. Scheduling, inventory control, and the shop-floor control system are integrated through this function. Manufacturing resources planning (MRP II) will be seen as a means for tieing the manufacturing needs to the purchasing department. Chapter 7 has as a major objective an introduction to this important area so that all persons involved with CIM understand the operating characteristics of the manufacturing system. Also, because we are talking about automated manufacturing systems, it will be necessary to introduce the concept of automated production planning and control systems.

Chapter 8: Manufacturing Control—Computer Control. Computer-aided manufacturing obviously requires computer control of equipment. Chapter 8 introduces the concepts of how computers can control manufacturing processes, including the data input and output characteristics of control computers. A subset of control computers which is discussed is the area of programmable logic controllers (PLCs). PLCs are widely used in industrial applications for sequencing and other forms of local control. Thus a programmable controller might be linked to a cell controller through a network but have prime responsibility for relieving the cell controller from the minute-by-minute functions that could degrade the cell controller's operation.

Chapter 9: Manufacturing Control—Numerical Control. While shop-floor equipment in the CAM environment is controlled by control computers and programmable controllers, the information needed to actually cause positioning to achieve part satisfaction of design requirements has to be evolved. A numerical control system is the way to achieve this requirement. Most computer-aided design systems have a numerical control compiler ancillary to the design system. As a result, it is frequently possible to generate cutter location data directly from the design information through this compiler. The compiler most common to design packages is APT, *A*utomatically *P*rogramed *T*ools. As a result, the APT language will be used as the exemplary language in this chapter for numerical control.

Chapter 10: Robotics. While there have been many denigrating remarks made concerning the field of robotics and many plants have been automated without the use of a single robot, there is no doubt that robots will be a key component for many CAM situations. Chapter 10 gives an overview of the robotics field, including application considerations as well as economic considerations.

Chapter 11: Measurement, Analysis, and Actuation. All control requires status data to be gathered, analysis of that data, and actuation based on the analysis. Chapter 11 presents some of the instrumentation and techniques available to allow equipment to be controlled and shop-floor information to be gathered and disseminated. This field is changing extremely quickly, and so only a brief introduction to capabilities can be given.

Chapter 12: Computer-Integrated Manufacturing. Economic disasters have been caused by lack of planning for automated systems. This final chapter suggests ways to alleviate such disasters and exemplifies these ways with specific cases.

1.5 SUMMARY

The importance of computer-aided design and computer-aided manufacturing in improving productivity to allow manufacturing enterprises to maintain a competitive edge, in this era where no company or nation can be assured of a dominant position in a specific market, cannot be overstressed. It is paramount that discrete-item manufacturing adopt a posture of flexible automation leading to the possibility of single-digit parameters: batch sizes of 1, machine setup times of less than 10 minutes, defects down to 0. However, competitiveness will require manufacturing to go beyond these parameters. The time between engineering design of a product and the manufacture of the prototype unit also has to be decreased significantly. Integration of the design and manufacturing functions has to be achieved through human communication channels and through a common design/manufacturing data base. Both of these achievements will allow design requirements to be transmitted promptly and legibly to manufacturing. Just as important, the feasibility of manufacturing capabilities can be transmitted to the design engineer during the design phase to allow design for manufacturability to be realized.

In order for the entire process to be accomplished in as economical a manner as possible, cost accounting and purchasing have to be brought into the picture and information from these groups used in key design and manufacturing decisions. Integration of the *whole* enterprise information structure through networking techniques is classified as computer-integrated manufacturing. The objective of this book is to give the reader a knowledge of the design and manufacturing techniques that will allow computer-integrated manufacturing to be realized. It is difficult to envisage a more exciting engineering opportunity.

REFERENCES AND SUGGESTED READING

1. Bairstow, Jeffrey: "GM's Automotion Protocol," *High Technology,* October 1986.
2. Bedworth, David D.: *Industrial Systems,* Ronald Press, New York, 1973.
3. Bedworth, David D., and James E. Bailey: *Integrated Production Control Systems,* 2d ed., John Wiley, New York, 1987.
4. Besant, C. B.: *Computer-Aided Design and Manufacture,* Halsted Press, John Wiley, New York, 1980.
5. "Computer-Integrated Manufacturing—From Vision to Reality," *Production Engineering,* November 1983.
6. Curtin, Frank T.: "Justifying Factory Automation Systems," *Production Engineering,* May 1984.
7. Groover, M. P.: *Industrial Robotics,* McGraw-Hill, New York, 1986.
8. Jones, A. T., and C. R. McLean: "A Proposed Hierarchical Control Model for Automated Manufacturing Systems," *Journal of Manufacturing Systems,* vol. 5, no. 1, 1986.
9. Machover, Carl: "Introduction," in *The CAD/CAM Handbook,* Carl Machover and Robert E. Blauth, Eds., Computervision, Bedford, Mass., 1980.
10. Niebel, B. W., A. B. Draper, and R. A. Wysk: *Modern Manufacturing Process Engineering,* McGraw-Hill, New York, 1989.

CHAPTER 2

GEOMETRIC MODELING

In many design instances it is far easier, cheaper and safer to experiment with a model than with a real entity.

M. F. Hordeski [12]

In this chapter we will investigate the place of geometric modeling in CAD/CAM integration. Imagine, if you will, a CAD/CAM software package designed to automate the sequence of processes from design through manufacturing. A certain core of information must be available to the software at all times. Some of this information is the description of the part in terms of geometry. An exception is the microelectronics industry, in which geometry is replaced by circuit design primitives such as transistors and resistors. This chapter begins an odyssey in which we will trace the information from its input by the designer through its use in the design, planning, and manufacturing of mechanical parts. Geometric modeling will be the starting point of this discussion because it is the primary input from the part designer and because a mechanical part cannot be described without it. Several types of geometric definitions will be discussed together with the important concern of how to store and use geometry in a truly integrated environment.

2.1 GEOMETRY—THE LANGUAGE OF CAE AND CIM

Geometry is the language of design and manufacturing, at least for discrete mechanical parts. Geometry is also the driving force behind most manufacturing. It could also be said that material properties and/or inventory control, etc., are driving forces, but the fact is that without the geometry of the part to be made, the rest is worthless. This statement emphasizes the importance of geometry in integrated mechanical CAD/CAM.

In general, the CIM process can be represented by the block diagram shown in Figure 2.1 [19]. Selected CAD processes are shown in the *product design* block and a subset of the CAM processes is shown in the *production planning* block. All manufacturing and design data is stored in the *data management* system. Central to these modules is the *kernel,* the description of the part or parts to be made.

Note that three types of part representation are given. "2-D figure" represents an engineering drawing or blueprint which typically shows three orthogonal views of the part, either with or without hidden lines. One view is insufficient to ascertain the three-dimensionality of the part. Even three views are occasionally insufficient, however, and the draftsperson may have to add additional views of sections of the part that need more explanation. Because a 2-D drawing is limited in the information it can convey, a 3-D solid part description is preferred.

A part to be manufactured is defined first in terms of its geometry and second in terms of other parameters, such as material type and functionality. Geometry includes dimensions, tolerances, surface finish, definitions of surfaces and edges, and, in some cases, the fit between two mating parts. Traditionally, the designer has communicated this information to the manufacturing engineer through blueprints or engineering drawings which, in the ideal case, define the part unambiguously. Draftspersons learn very early that there are correct and incorrect ways to dimension a drawing. Over- or underdimensioning will either cause the part to be made wrong or the manufacturing engineer to request additional clarification.

FIGURE 2.1   Conceptual structure of an integrated CAD/CAM system. (From Ref. 19).

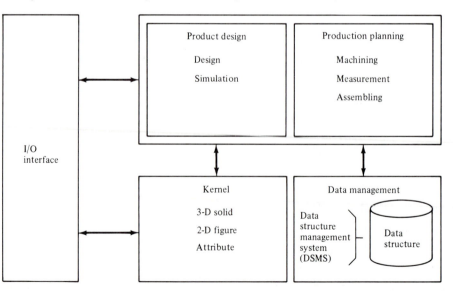

Hopefully, the latter occurs, most likely during the process planning stage where the manufacturing engineer plans in detail how to make this part the designer has suggested.

Geometry may be very simple, as in the case of the part in Figure 2.2. This simple block could appear in its final, totally specified form as the drawing in Figure 2.2. Notice the myriad of geometric specifications required to make this part. And this is a simple part; more complex parts may require many sheets of drawings, each filled with dimensions, tolerances, material specifications, and even manufacturing requirements such as welding, grinding, etc.

It is not important at this point to understand all of the notations on the drawing, but it is important to understand that CAD/CAM integration requires that all of this information appearing on a standard engineering drawing be stored in a concise, organized format within a computer data base. Today's CAD/CAM systems do not store or utilize the totality of information found on engineering drawings, but future, more powerful CAD/CAM systems will require the storage of at least all of the blueprint, and possibly even more information, such as descriptions and classifications of form features. Form features are manufacturable entities such as holes, slots, pockets, threads, knurls, and keyways which can be defined as sets of part faces and edges [14].

FIGURE 2.2   Object geometry.

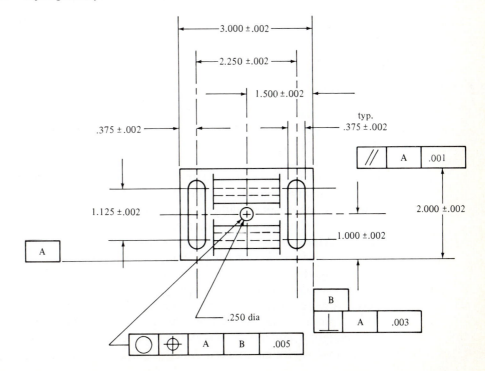

Geometric modeling, according to Mortenson [17], has three purposes in CAD/CAM: (1) *part representation*, which mandates a complete definition of the part for manufacturing and other applications; (2) *design*, which allows the user to input a geometric specification and manipulate it as one would a set of Tinkertoys or a piece of clay; and (3) *rendering,* an important application of geometric modeling which uses the geometry to paint a realistic picture of the object on a computer graphics output device. Each of these aspects has distinctly different requirements. *Part representation* requires completeness and a robust model. The model should allow the user to perform analyses and queries as though the part were real. *Design* requires flexibility and ease of use in changing geometries to suit the functionality or esthetics. *Rendering* implies a concise data base and fast data access for real-time part display and animation. Usually these three views are in conflict; a trade-off is required between concise storage and fast access. It is the unending goal to develop a geometric modeling scheme which is complete, fast, concise, and easy to massage.

This chapter will take us through the computer definition of a part from simple concepts such as wire frame modeling to more complete representations such as sculptured surfaces and solid models. Along the way we will examine the implementation of each of these methods, advantages and disadvantages, and finally where these techniques are used specifically in the integration of CAD and CAM. Before we begin, it is worth reemphasizing the importance of geometric modeling in CAD/CAM and CIM. The geometric model is the only *picture* of the object that can be used to automatically perform mathematical analyses, process planning, quality control inspection routines, and even premarket versions of sales brochures. This model, therefore, is the key to the successful automation of subsequent CAD/CAM integration procedures.

2.2 KEY DEFINITIONS

Boolean operator   Named after Alfred Boole, three operators for combining sets of objects. The operators are union (+), difference (−), and intersection (∗).

BREP   Boundary representation of a part. The description of a solid by enumerating all faces, edges, and vertices, in both geometry and topology.

CSG   Constructive solid geometry. The description of a solid as a tree structure of primitive volumes and the boolean operations for combining the volumes.

Curve   A line in space, either straight or curved.

Data base   A collection of data stored in a computer medium.

Data structure   A schema for organizing data in a computer data base.

Edge   A curve which serves to bound one side of a face.

Explicit equation   An equation defining a geometric entity in the form $y = f(x, z)$. In two dimensions the form may be $y = f(x)$.

Face   A finite region of an infinite surface used to define a side of a solid object.

Form feature   A set of faces or other geometric entities which together form a pattern useful in part analysis; e.g., a set of faces and edges defining a hole implies a drilling process.

Geometric modeling A technique of using computational geometry to define geometric objects. A subset is *solid geometric modeling,* in which only solid objects are defined.

Geometry The branch of mathematics dealing with the measurement of lines, angles, surfaces, and solids.

Implicit equation An equation defining a geometric entity in the form $f(x, y, z) = 0$. In two dimensions the form may be $f(x, y) = 0$.

Loop A closed path of edges and vertices in a surface which defines a finite face.

Parametric equation A set of equations defining a geometric entity in the form $x = f(a)$, $y = f(a)$, $z = f(a)$, where a is the parameter which usually varies from 0 to 1.

Polygon An n-sided, closed region of two-dimensional space defined by a loop of edges and vertices.

Polyhedron An n-sided, closed region of three-dimensional space defined by a set of polygons.

Process plan The plan used to process a part from raw material to finished goods. Usually this is a sequence of operations involving various machine tools, cutter geometries, and cutting parameters to achieve the desired dimensions.

Quadric surface A surface definable by a second-order equation in x, y, and z. The four natural quadrics are the plane, cylinder, cone, and sphere.

Surface A continuous sheet, either planar or sculptured, usually defined by a set of equations in x, y, and z.

Surface modeler A mathematical technique for modeling objects by specifying their vertices, edges, and surfaces, usually including only geometry and not topology.

Three-space (or 3-space) Three-dimensional space.

Topology The branch of mathematics dealing with the connectivity of geometric elements, not including the geometry of those elements.

Vector A mathematical entity with direction and magnitude, but no specific location. In this chapter a vector is a triplet of x, y, z coordinate values measured from the origin to the x, y, z point.

Vertex A point in two- or three-dimensional space which bounds one end of a line segment.

Wire frame Definition of an object by specifying merely the edges and vertices of the part, not including the surfaces.

2.3 GEOMETRIC MODELING TECHNIQUES

Traditionally, a designer defines an object by a two-dimensional drawing, usually in three views: a plan view and front and side elevations (Figure 2.3). The three-dimensionality is derived from simultaneous examination of the three views, pos-

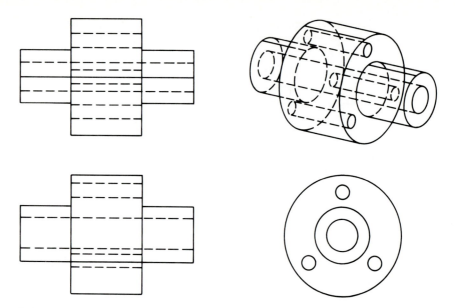

FIGURE 2.3 Orthographic drawing.

sibly with other oblique views. The questions to be asked pertaining to CAD/CAM and geometric modeling are:

1. How can a designer input this geometry into the computer?
2. Does this geometry represent a realizable, complete, unambiguous part?
3. Does the internal representation allow sufficient querying about important geometric aspects of the part for manufacturing and other applications?

2.3.1 Multiple-View Two-Dimensional Input

Looking back to the example in Figure 2.2, a box with a hole can be described by its plan and elevation views. A designer can sketch these views on a computer system using a tablet or a mouse or other locating device. The first step may be to define the vertices. These can be stored as (x, y) pairs in the elevation view and as (x, z) pairs in the plan view (Figure 2.4a). Next, the designer may connect the vertices with lines or edges. The entrance and bottom to the hole are circles, so the designer must specify a circular edge in these locations. Alternatively, the circles may be defined as many straight line segments. The final drawing may look like Figure 2.4b. This is the most elementary part representation. The computer data base contains two views, which are stored separately.

The designer may wish to query the data base about the part. An example question may be: What are the three coordinates (x, y, z) of the lower left corner? This information is not contained in the data base. Nowhere did the designer store

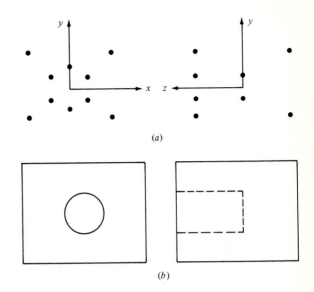

(a)

(b)

FIGURE 2.4   Constructing two views.

the relationships among the points in the two views. The lower left corner point is defined in two places. The answer to this query is not straightforward. The part exists not as a three-dimensional part, but as multiple two-dimensional views which are not necessarily related. A geometric modeler should represent the three-dimensionality of an object and be able to answer questions such as the one above.

2.3.2 Wire Frame Geometry

Many CAD/drafting systems use the 3-D wire frame method to define object geometry. In this method, the user enters 3-D vertices as (x, y, z) triplets. Joining the vertices creates a 3-D object called a wire frame. This representation contains only points and lines (usually straight lines). Now, as opposed to the case of multiple 2-D views above, the vertices and edges do have three-dimensionality. The user can inquire about the coordinates of the vertices and the length, in three dimensions, of the edges. Wire frames still are not complete, however. Surface geometries are not present. In the case of the block with the hole, the wire frame appears as in Figure 2.5a. The model contains only the points and edges. If the sides were sculptured to bulge outward or inward, the model would have no knowledge of it. Wire frame models lack the surface definition which may be important for later manufacturing processing and other analyses. Figure 2.5b shows a more complex and confusing wire frame object. A ramification of the lack of surface information arises when attempting to draw a realistic image of the part. For example, it is not possible to render a wire frame of an object with hidden surfaces removed, since there are no surfaces. In addition, wire frames, because they are merely a "connect-the-dots" construction technique, can be used to

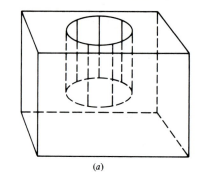

(a)

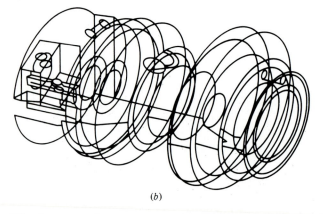

(b)

FIGURE 2.5   Wire frame examples.

represent nonsense objects (Figure 2.6). Even when the object appears to be realizable, as in Figure 2.7, it is apparent that the object can be ambiguous. The wire frame part shown in Figure 2.7a could represent any of the three solids in Figure 2.7b–2.7d.

Wire frames, then, lack the surface information needed to store a robust part model.

2.3.3 Surface Models

In an attempt to solve the drawback of wire frames, some modeling schemes add information about the surfaces in the part. In these systems, the user enters the vertices and edges as above, but in an ordered manner, outlining or bounding one face at a time. In our block-with-a-hole example, the user may first define the base of the block by specifying the vertices, then connect them in order to bound the bottom surface. The user can then select the type of surface to be fitted to these edges. Possible choices in this example are plane and sculptured surfaces. For faces with curved and straight-edge combinations, other surface types, such as cylinders or spheres, may fit. A sculptured (or irregular) surface can be made to fit

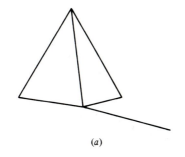

(a)

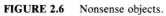

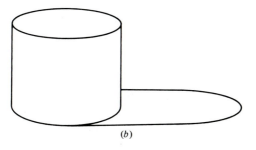

(b)

FIGURE 2.6   Nonsense objects.

FIGURE 2.7   Wire frame ambiguity.

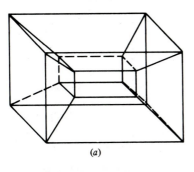

(a)

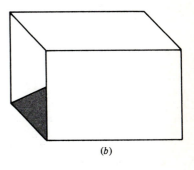

(b)

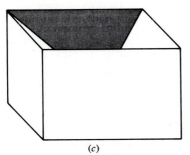

(c)

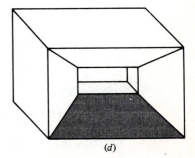

(d)

all legally connected vertex-edge loops. Sculptured surfaces are discussed later in this chapter.

A surface modeler, although it is a major advance over wire frame modelers, also has disadvantages. The modeler contains definitions of surfaces, edges, and vertices, but does not store the topology, or connectivity, of these entities. For this and other reasons, the modeler has no concept of the part "inside" and "outside." It does not store the face normal vectors and, therefore, cannot distinguish the side of the face pointing toward the air from the side pointing toward the material. Because of this, a surface modeler cannot calculate the volume of the part. Nor can it compute other mass properties such as moments of inertia and principal axes. In fact, a surface modeler cannot guarantee that the designer has described a realizable object. It may be a collection of surfaces which do not define a physical part, as in the case where the surfaces may not be connected. A complete part description would have sufficient information to answer any question one could ask about a physical solid object, including a guarantee of solidity.

2.4 GEOMETRIC ENTITIES: CURVES AND SURFACES

An example of a surface-modeled part is given in Figure 2.8. Note that, in this simple case, all faces are defined as planar polygons. A polygon is an ordered sequence of edges and vertices which form a closed region. The region need not be planar, but for the cases mentioned in this chapter we will consider only planar polygons. There are two reasons for this: Planar polygons are the most commonly used polygonal representation in geometric modeling because they are computationally simple and because they convey the principles of polygonal modeling without becoming mired in complex mathematics.

The primary uses of polygons in geometric modeling are to simplify the rendering algorithms for shaded image generation. Some geometric modelers (called *tesselated modelers*), rather than representing the exact shape of the surfaces of an object, approximate the object with many small tesselations or planar polygons. For example, an exact modeler (Figure 2.9a) may represent a finite cylinder as one cylindrical surface and two planes (top and bottom). Or, alternatively, a tesselated modeler would represent the cylinder as a collection of n rectangular planes arranged around the cylinder axis and $2n$ triangular planes, n for the top and n for the bottom, as shown in Figure 2.9b. The advantages of tesselated modeling will be discussed in Section 2.4.2; suffice it to say at this point that the storage of planar polygons is an important aspect of geometric modeling. The principles involved can demonstrate certain techniques which are common to all geometric model storage formats.

2.4.1 Points, Lines, Surfaces, and Solids

Construction of a surface model from polygons requires an understanding of the terms point, line, surface, and solid. These are the building blocks from which all objects are made, including other solids from combinations of solid objects.

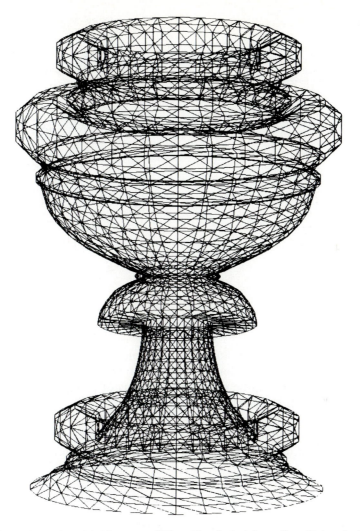

FIGURE 2.8   Polygonal model. (Courtesy of Diane Hansford, Arizona State University.)

A *point*, or *vertex*, is a zero-dimensional entity defining a point in 3-space and which can be described as a triplet of numbers: (x y z).

A *line* is a one-dimensional entity (sometimes called a curve or track) which can be drawn along one straight or curved axis. A line is infinite in length. A *line segment* is defined by a line and two endpoints. A line can be defined in several ways, depending on its type: straight, circular arc, ellipse, parabola, hyperbola, helix, or general nth-order curve. For example, a straight line (Figure 2.10) can be defined either by two points on the line [($x1$ $y1$ $z1$) and ($x2$ $y2$ $z2$)] or by a point on the line and a vector orientation [($x1$ $y1$ $z1$) and (i j k)], where (i j k) are direction

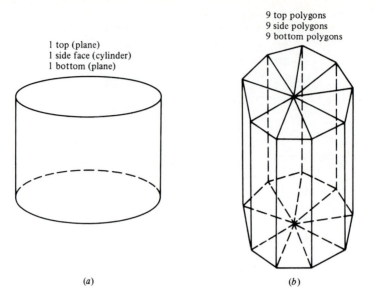

FIGURE 2.9   Tesselated modelers.

cosines of a unit vector along the direction of the line. The reader is advised to consult any text on vector algebra for an explanation of vectors and direction cosines. A circle can be expressed in terms of a center point, axis vector, and radius: $[(x\ y\ z),\ (i\ j\ k),\ r]$.

A *surface,* or *face,* is a two-dimensional entity because it can be drawn in a plane or deformed plane. It can be defined as a plane, cylinder, sphere, cone, etc., or, in general, as an nth-order surface. The four surfaces listed above constitute

FIGURE 2.10   Line representation formats: (*a*) two-point format; (*b*) point and vector.

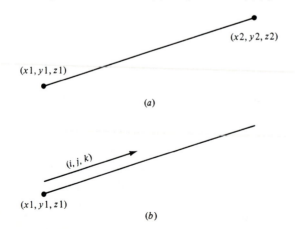

the four natural quadric surfaces; that is, they can be defined by a second-order equation. Higher-order surfaces will be described later.

A *solid* requires a full three dimensions to describe its collection of joined faces. We will return to the ways to store a solid object in Section 2.5.

2.4.2 Polygon (or Tesselated) Modeling

In order to learn the principles of polygon modeling, we shall start with the definition of a cube. A cube has 8 vertices, 12 edges, and 6 faces, as shown in Figure 2.11. A surface modeler stores descriptions of all faces, edges, and vertices (FEV) in terms of their geometry and a limited amount of connectivity information. The FEV can be labeled as shown in Figure 2.11. Faces *A*, *B*, *C*, *D*, *E*, and *F* are all rectangular polygons. Edges *a*, *b*, *c*, *d*, *e*, *f*, *g*, *h*, *i*, *j*, *k*, and *l* are all straight lines. Vertices 1, 2, 3, 4, 5, 6, 7, and 8 can be described by the (*x y z*) triplets.

Recall that the storage of an object in a computer data structure has the goals of conciseness, completeness, and ease and speed of accessing the relationships among the entities. As might be predicted, certain data structure formats are better than others. As a start, consider the data structure in Figure 2.12, called a *vertex list* structure. The only entities stored are the vertices and polygon names. Note that we do not need to store the edge definitions explicitly, because all the edges are straight and are drawn between the specified vertices.

It is possible to draw a picture of the cube by starting at the first polygon name and following the pointer to the list of vertices which connect to form that polygon. Notice that polygon closure is implied in this data structure. It is understood that the first and last vertices in the lists are connected to close the polygon.

The suitability of this and any data structure can be evaluated by determining (1) its efficiency in extracting the answer to the queries we expect the geometric model to be able to answer and (2) its speed in performing operations such as

FIGURE 2.11   A polygonal cube.

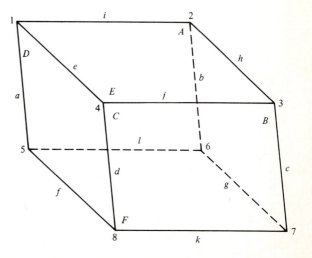

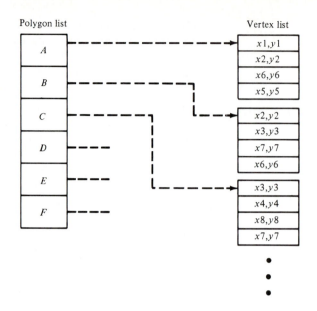

FIGURE 2.12   Vertex list.

drawing a picture of the object and editing the geometry. Included in the efficiency is the compactness of the storage size.

Consider the question, "Is face *A* adjacent to face *B*?" The answer to this question is obvious to the viewer of the object, but not so obvious to the viewer of the data structure alone. Answering this question requires an algorithm which we will call "Adjacent_Faces(FACE1 , FACE2)" (we will return to this relationship later in solid modeling) and which returns either true or false. The adjacency relationship can be proven or disproven by calling the procedure "Adjacent_Faces(A,B)." Instead of writing the code for this procedure, we can "walk" through the algorithm logically. The first step is to locate the polygon names *A* and *B*. From this point, the algorithm must check all vertex pairs in polygon *A* to see if they also exist in the polygon *B* definition. Note that the vertex pairs ($x1$, $y1$) and ($x2$, $y2$)—that is, the edge drawn from vertex 1 to vertex 2—appear in both of the polygon definitions, but in opposite directions.

The length of the search for the common edge, in this case, is proportional to the number of vertex pairs in polygon *A* times the number of vertex pairs in polygon *B*, or, for rectangles, of the order 16 comparisons (32 if both x and y values are considered).

Further, we should consider the speed with which the data structure can be used to draw a picture of the object. All polygons can be drawn by starting at the polygon label and connecting the vertices to which each polygon pointer points. Note that all edges common to two or more polygons are drawn more than once, one time for each polygon of which they are a member.

As a last test of the data structure we will examine what happens when the designer decides to modify a vertex. The vertex list structure contains multiple instances of each vertex which is a member of multiple polygons. Changing vertex $V1$, for example, would require searching through the data structure to find all occurrences of vertex $V1$. Clearly, a better structure would list each vertex only once, to make editing more efficient.

It is obvious that the vertex list is not the best way to organize geometric polygonal data. A better data structure would arrange the data such that, in the three applications above (finding adjacent faces, drawing polygons, and editing vertices), common edges are found more readily and each edge is drawn only once per picture. The data structure should be optimized for the particular applications it will encounter.

An alternative data structure is the *modified vertex list* as shown in Figure 2.13. This structure has the advantage that each vertex coordinate set is listed only once. Editing the geometry is more straightforward than before. Note, however, that the Adjacent_Face routine must still search for a common edge or pair of vertices among the polygons. Also, each edge will still be drawn twice when the polygons are plotted.

As a final alternative (there are many others, though), consider the *edge list* as shown in Figure 2.14. In this format the entrance to the data structure is the edge list. Vertices are listed only once, and polygons have multiple pointers to their edges. Upon examination of the three algorithms above, we see that the Adjacent_Face routine is now much easier. The procedure must first find the first polygon in the pair and follow the pointer to its edges and onto the adjacent polygons. Also, when drawing the polygons, each edge now is drawn only once by

FIGURE 2.13   Modified vertex list.

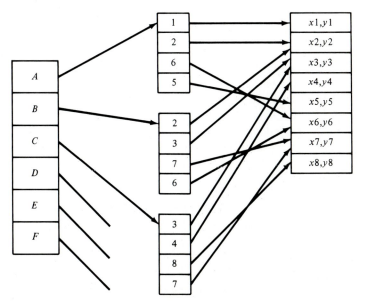

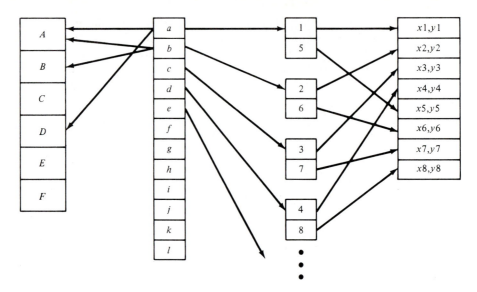

FIGURE 2.14   Edge list.

just drawing edges, not polygons. Finally, editing vertices requires altering only the single instance of the edited vertex. This data structure is more efficient for these three operations than the vertex list and modified vertex list. A formal evaluation of the three data structures can be carried out by measuring the order of complexity of each, that is, the worst-case search times for each application. Formal data structure evaluation techniques can be found in any good text on data structures.

2.4.3 Cubic Curves

Let us now extend geometric definition into the realm of nonplanar curves and surfaces. It is here that we can investigate the richest set of geometric entities. Also, if the geometric model is to be used for manufacturing applications (remember, the goal is CAD/CAM integration), then more complex curves and surfaces must be addressed.

Planar curves can be described by first- and second-order equations. For example, a straight line has the relation

$$y = m * x + b$$

where m is the slope and b is the y intercept. This is called an *explicit equation* because it is of the form $y = f(x)$. An example of a second-order curve is a circle,

$$(x - a)^2 + (y - b)^2 - r^2 = 0$$

where a and b are the coordinates of the center of the circle and r is the radius. This is called an *implicit equation* because it is of the form $f(x, y) = 0$. Other types of planar curves are ellipses, parabolas, and hyperbolas. The simplest form of a general nonplanar curve is the cubic equation. Traditionally, cubic curves and surfaces are represented by *parametric equations,* as opposed to the explicit and implicit forms mentioned above. A parametric set of cubic equations is of the form

$$x(t) = a_x t^3 + b_x t^2 + c_x t + d_x$$

$$y(t) = a_y t^3 + b_y t^2 + c_y t + d_y$$

$$z(t) = a_z t^3 + b_z t^2 + c_z t + d_z$$

where the a's, b's, c's, and d's are constant "shape" coefficients and t is the parameter which is varied to plot points on the curve. This form is called the *algebraic form* of the parametric equation. Usually t is defined such that

$$0 \leq t \leq 1$$

As t moves from 0 to 1, the curve is traced from beginning to end. This formulation defines cubic curve segments.

In order to specify the shape of the cubic curve, 12 coefficients must be determined. This is equivalent to defining four sets of (x, y, z) points in space. These four points which affect the shape of the curve are called *control points*. For a thorough treatment of cubic curves, the reader is referred to Ref. 7. For our purposes of applying geometric modeling to CAD/CAM integration, it will suffice to take a less thorough approach and learn by example. Therefore, we will proceed to describe the three most commonly used forms of cubic curve formulations: Hermite, Bezier, and B-spline. Understanding these methods is important because the geometric model input by the designer and stored as a geometric data base is typically the only source of part definition in a CAD data base. And, as has been stated earlier, the CAD data base content is a key to automated CAD/CAM integration.

HERMITE CURVES. *Hermite curves* are defined by specifying the two curve endpoints, p_0 and p_1, and the slopes at the two endpoints, p_0' and p_1', as shown in Figure 2.15. The coefficients a, b, c, and d can be found by setting $t = 0$ and $t = 1$ for both the parametric equations and their first derivatives as follows:

$$p(0) = a(0)^3 + b(0)^2 + c(0) + d$$

$$p(1) = a(1)^3 + b(1)^2 + c(1) + d$$

$$p'(0) = 3a(0)^2 + 2b(0) + c$$

$$p'(1) = 3a(1)^2 + 2b(1) + c$$

where p is x or y or z and a, b, c, and d are the x, y, or z coefficients. Here we have 12 equations (four each for x, y, and z) and 12 unknowns (a, b, c, and d for each x, y, and z). The p and p' values are given as data. Once the coefficients are known, the curve is calculated by varying t between 0 and 1. An example may help to clarify the procedure.

FIGURE 2.15   Hermite cubic curve.

Example: Hermite Curve Formulation. Consider that we want to draw a curve through the points $p(0) = (0,0,0)$ and $p(1) = (6,10,8)$ with end slopes of $p'(0) = (0,0,10)$ and $p'(1) = (-10,0,0)$. The slopes are given as direction vectors with no position. A $(0,0,10)$ vector is in the z direction, and $(-10,0,0)$ is in the negative x direction. The four equations for the two endpoints and slopes for all the x values are

$$x(0) = 0 = d_x$$

$$x(1) = 6 = a_x + b_x + c_x + d_x$$

$$x'(0) = 0 = c_x$$

$$x'(1) = -10 = 3a_x + 2b_x + c_x$$

Similar equations are written for the y and z values. Solving for the x coefficients, we find that

$$
\begin{array}{lll}
a_x = -22 & a_y = -20 & a_z = -6 \\
b_x = 28 & b_y = 30 & b_z = 4 \\
c_x = 0 & c_y = 0 & c_z = 10 \\
d_x = 0 & d_y = 0 & d_z = 0
\end{array}
$$

Therefore, the resulting parametric equations are

$$x(t) = -22t^3 + 28t^2$$

$$y(t) = -20t^3 + 30t^2$$

$$z(t) = -6t^3 + 4t^2 + 10t$$

Letting t vary from 0 to 1 generates the curve in Figure 2.16 shown in the x–z plane. Longer slope vectors have the effect of extending the slope for a longer distance along the curve and, in effect, straightening the curve near the endpoints. This example, together with calculations you make in experimenting, should demonstrate the flexibility of cubic curves.

An alternative form of the Hermite, and, in fact, any of the cubic curves, replaces the a, b, c, and d coefficients with the values of the control points. This

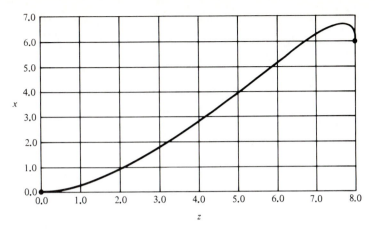

FIGURE 2.16 Cubic hermite curve.

Hermite form, called the *geometric form* of the cubic curve equation, appears as

$$p(t) = F_1(t)p(0) + F_2(t)p(1) + F_3(t)p'(0) + F_4(t)p'(1)$$

where p is either x, y, or z and p' is the x, y, or z slope magnitude. The F_i are called the *blending functions* and determine the total effect of the four control points at a specific t value. The blending functions are

$$F_1(t) = 2t^3 - 3t^2 + 1$$
$$F_2(t) = -2t^3 + 3t^2$$
$$F_3(t) = t^3 - 2t^2 + t$$
$$F_4(t) = t^3 - t^2$$

Substitution of values of t from 0 to 1 will demonstrate the effect of each blending function. The higher the blending function value for a certain value of t, the more that control point or slope affects the curve at that t value.

BEZIER CURVES. An alternative to the Hermite curve is the *Bezier curve*, developed by P. Bezier for use, together with a surface design package, in the PolySurf CAD system for Renault Automobile Company in France in the 1960s [3]. Because specifying slopes can be somewhat difficult and an art, Bezier surmised that selecting four control points might be more convenient than choosing two control points and two control slopes. The Bezier control points are defined at locations $p(0), p(\frac{1}{3}), p(\frac{2}{3}), p(1)$, where $t = 0, \frac{1}{3}, \frac{2}{3}$, and 1, as shown in Figure 2.17a. By defining (x, y, z) coordinates at these four locations, the 12 equations can be solved for the 12 coefficients. The form of the Bezier parametric equations is

$$p(t) = (1 - t)^3 p_0 + 3t(1 - t)^2 p_1 + 3t^2(1 - t)p_2 + t^3 p_3$$

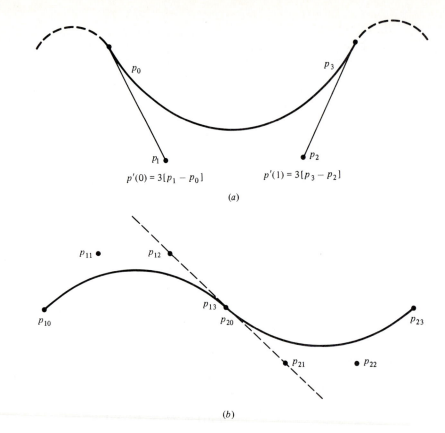

FIGURE 2.17   (*a*) Bezier curve; (*b*) composite Bezier curves.

where p_0, p_1, p_2, and p_3 are the coordinates of the four control points, and p stands for x or y or z. By substituting in the four values of $t = 0, \frac{1}{3}, \frac{2}{3}$, and 1, it can be seen that the curve passes through points p_0 and p_3, but not through the middle control points. The $\frac{1}{3}$ and $\frac{2}{3}$ points serve to shape the curve without lying on the curve. The slopes at the endpoints can be found by connecting each endpoint with its adjacent control point as shown.

A feature of the Bezier method, convenient for extended curve drawing, is that Bezier segments can be connected. By making $p(0)$ of one curve equal to $p(1)$ of another curve, one can establish zero-order continuity. Matching the slopes gives first-order continuity. Slopes may be matched by making the end control points collinear with the adjacent control points on each curve (Figure 2.17*b*). Connected Bezier curves can generate long, multicurvature lines while still possessing local curvature control established by intermediate control points.

B-SPLINE CURVES. The third type of curve is a variation on the two previous formulations. In the case of the *B-spline*, four control points are established, but the curve is not required to go through any of them (Figure 2.18*a*). B-splines were suggested by the traditional draftsman's spline, a flexible metal strip which, laid edge-on on a drafting board, allows the construction of free-form curves. Weights, called "ducks," are placed on the board to hold the strip in place. The strip does not pass through any of the duck centers, as is the case with the B-spline curve. The ducks are analogous to the control points. They affect the curvature but do not lie on the curve.

Multiple B-spline curves can be connected by overlapping their control points. Figure 2.18*b* shows three spliced B-splines. Control points $p0$, $p1$, $p2$, and $p3$ control the first B-spline. Points $p1$, $p2$, $p3$, and $p4$ control the second segment, and points $p2$, $p3$, $p4$, and $p5$ control the third segment. Overlapping the control points provides zero-order continuity. First- and second-order continuity are provided by the mathematics of the curve equations. B-spline curves have the advantage of higher-order continuity and, hence, look smoother. They also possess the advantage that the control points have more local control of the curvature. This allows the designer to change a small section of the composite curve without affecting the overall curvature. Automobile designers routinely use B-splines in the design of automotive sheet metal.

FIGURE 2.18 (*a*) A B-spline curve; (*b*) a three-segment B-spline curve.

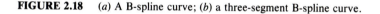

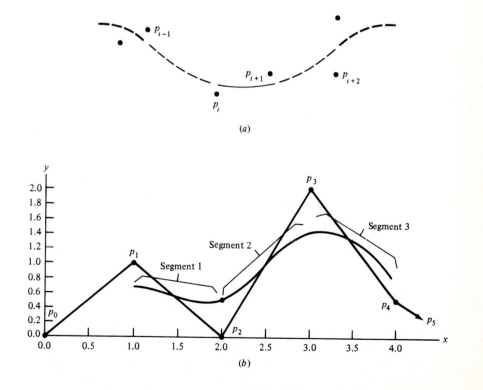

STORAGE OF CUBIC CURVES. The preceding three curve types all use four control data points to specify the curve. Therefore, storage of the curve requires only the storage of 12 numbers: the x, y, and z values for each of the control triplets, whether points, slopes, or combinations.

2.4.4   Bicubic Surfaces

Bicubic surfaces offer the same advantages as cubic curves: flexibility in sculpturing and convenience in storage of the surface description. Bicubic surfaces use the same concept as cubic curves; that is, each surface is defined by a certain number of control points, surface tangents (slopes), or a combination thereof. Before introducing the Hermite, Bezier, and B-spline surfaces, let us look at how a cubic surface is constructed.

Consider sculpting a surface from cubic curves. A skeleton could be fashioned from a criss-cross grid of cubic curves as shown in Figure 2.19. Each of the four border curves is a cubic curve, and the four intersect in pairs at each corner of the network. These border curves define the boundary, but more curves are needed to define the curvature of the interior. Adding two more curves in each direction gives the interior shape. There are now eight cubic curves, four in one direction and four in the cross direction. This is called a *bicubic surface*. Two parameters (s and t) are required, one in each of the cross directions. The general form of the bicubic equation is

$$p(s,t) = a_{33}s^3t^3 + a_{32}s^3t^2 + a_{31}s^3t + a_{30}s^3 + a_{23}s^2t^3 + a_{22}s^2t^2 + \ldots$$

where p is the x, y, or z coordinate and the a_{ij} terms are the coefficients for the x, y, or z equation. Altogether there are 16 coefficients for which to solve. To make a

FIGURE 2.19   Networked bicubic curves.

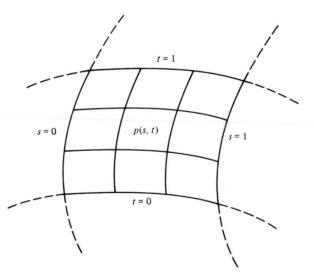

web, the curves must intersect along the edges, at the corners, and even in the interior at 16 points. Sixteen points can be defined to create these eight cubic curves. Alternatively, one could define these curves by specifying the four corner points and several slopes.

HERMITE SURFACES. The Hermite form of bicubic surfaces, like the Hermite curve, uses a combination of control points and slopes (Figure 2.20). The 16 surface data points include the four corner points (p_{00}, p_{01}, p_{11}, and p_{10}), the four s slopes at the corners, and the four t slopes at the corners (dp/ds and dp/dt). The remaining four triplets are called the twist vectors ($d^2p/ds\,dt$). A *twist vector* defines the interior of the surface by telling how the s and t slopes change along the boundaries. The twist vectors are equal in both the s and t directions at each corner. Surfaces can be sculpted as out of clay by changing these 16 definition points. Storage of a Hermite patch involves 16 (x, y, z) triplets, or 48 numbers.

BEZIER SURFACES. Bezier uses 16 control points as he used four control points for the Bezier curve. The network developed above can be defined as a network of straight lines connected at 16 control points (Figure 2.21). These control points define the eight Bezier curves, which, in turn, define the boundary and interior of the surface. Storage of a Bezier surface involves 16 control points, or 48 numbers.

Composite surfaces can be created by making the corner points of mating surface patches equal. In addition, the slopes of the surfaces must be equal at the junction. This is accomplished by constructing the boundary points and the first control point on either side of the boundary to be collinear, just as in the case of composite Bezier curves.

FIGURE 2.20   Hermite surface.

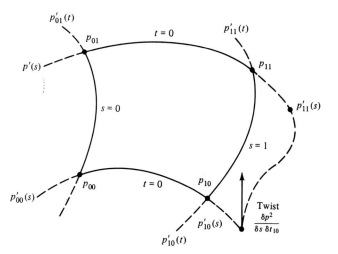

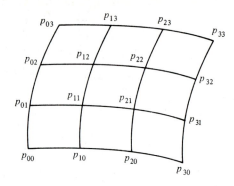

FIGURE 2.21   Bezier surface.

B-SPLINE SURFACES. B-spline surfaces also use 16 control points. As was the case for B-spline curves, however, the surface does not go through any of the points. Composite surfaces are created by overlapping the control points as was done for composite B-spline curves. B-spline composite surfaces possess first- and second-order continuity and, therefore, are smoother than Bezier or Hermite surfaces.

Before leaving the subject of surfaces, it should be noted that most surfaces in manufacturing are natural quadrics. It is possible to generate planes and quadric surfaces using bicubic patches. Some geometric modelers use *only* the bicubic formulation to achieve consistency and uniformity in representation. Manufacturing applications programs, though, may have a tough time decoding these over-specified forms of quadrics. This is especially true if the application is looking for cylindrical surfaces, for example. The only way to tell if a bicubic patch is really a cylinder is to try to match the bicubic equations to equations of a cylinder. This may not be easy, especially considering round-off and truncation errors encountered routinely in computer calculations.

The surfaces created by surface modelers can be very impressive and can define such difficult-to-manufacture items as turbine blades and airfoils. However, even though surface modelers are fairly sophisticated in their capabilities, they do have shortcomings: no guarantee of solidity, and lack of ability to perform mass property calculations, to name two.

NURBS. The trend in CAD geometry is to represent all surfaces as nonuniform rational B-splines (NURBS). NURBS mathematics has the advantage of better curvature control and because of the rational coefficients, alleviation of round-off errors in the storage of geometry.

2.5   SOLID MODELERS

The solution to the drawbacks of multiview orthographic drawings, wire frames, and surface modelers arose with the advent of solid geometric modeling schemes. The first solid modeling systems appeared in the early 1970s, and by the end of that decade a considerable body of knowledge was available in the literature. The primary research focus was on representation methods and algorithms for solids made up of either plane or simply curved surfaces (cylinders, cones, etc.). Braid is credited with the first attempt at solid modeling using the BUILD modeler [5]. Written in ALGOL, this software system allows the user to construct solid objects from primitive building blocks. The object can be displayed on a graphics device and analyzed for mass properties such as volume and surface area. These applications, though, are not as important as the major feature of BUILD and a characteristic of all solid modelers: a guarantee that the object constructed is a realizable solid! It is not possible in a true solid modeler to construct dangling edges or faces or, in fact, any object which, in theory, could not be made.

Solid modelers were one of the solutions manufacturing had required in order to fully automate the design/manufacturing process. Until this time, all geometry was verified by hand from engineering drawings. Numerical control had surfaced in the early 1950s at MIT; robotics was showing promise; but the missing link was a robust technique to store solid object information in the computer. Once stored, solid data could be used to plan robotic assembly, automated NC path calculation, and many other manufacturing-related tasks.

Since their inception, solid modelers have diversified and improved, especially in the area of user interface. Dozens of solid modelers exist, some as CAD/CAM commercial products and others as research vehicles (especially in academia) for continued solid modeler improvement. The remaining sections of this chapter will be devoted to the study of solid modeler techniques, including model storage, object construction techniques, and the use of solid models to integrate CAD/CAM.

2.5.1 Solid Modeling Example

An example solid modeling session may help lead into the topics to come. The goal of this example is to design and store the part shown in Figure 2.22. The user has chosen to build the part by describing the outer-turned profile cross section, generating a solid from the cross section, and adding the rectangular block at the end. Solid modelers can typically create much more complex objects than this, but this example will point out some of the important issues in model construction and storage.

The cross section of the round portion can be input by specifying the coordinates of the vertices in the cross section, similar to the procedure used by a draftsperson to draw a 2-D orthographic picture (Figure 2.23). The actual coordinate specification depends on the user interface. One method is to indicate the points on a graphics tablet. The vertices are connected in order from point to point, using a LINE command. Only half of the cross section must be defined, since the part is symmetric (Figure 2.23a). When all vertices have been connected, a SPIN command generates a solid volume by extruding the cross section 360° around the part axis (Figure 2.23b). The box on the end may be generated using a CREATE BLOCK command with parameters which specify the length,

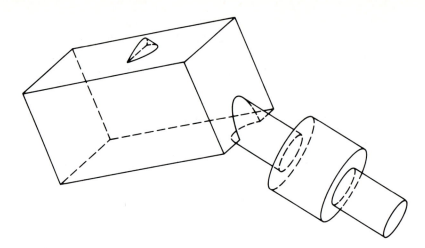

FIGURE 2.22   A sample object to construct.

width, height, location, and orientation of the box. By executing a UNITE command (Figure 2.23c), the box and round solids are joined and become one solid. Now a hidden surface view may be drawn and mass properties such as volume and moments of inertia calculated. In addition, the data structure of the solid model can be copied to a file for use in subsequent analysis and manufacturing programs.

Several questions should have occurred to you while following this example. First and foremost, how does the program guarantee that the final object is a valid solid? How do the SPIN and CREATE commands work? What other construction techniques are possible? What is the format of the final data structure of the part? How is this data structure usable in other applications (for example, manufacturing)?

We will approach the answers to these questions in a traditional manner. First, we will take a look at various ways to construct a model. Next, the data structure contents will be discussed. Then it will be appropriate to examine how the solidity is guaranteed, and finally, the solid modeler applications and use in CAD/CAM will be reviewed.

2.5.2   Solid Model Construction Techniques

At least six methods exist for constructing a 3-D object within a CAD solid modeling system:

- Pure primitive instancing (PPI)
- Spatial occupancy enumeration (SOE)
- Cell decomposition (CD)
- Sweeping (S)

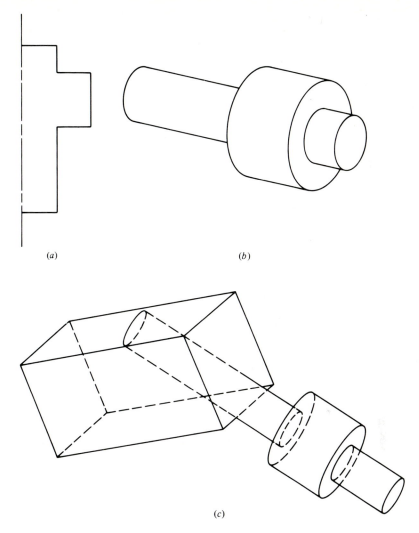

FIGURE 2.23 Creating a part: (*a*) the profile; (*b*) the swept solid; (*c*) overlayed parts.

- Constructive solid geometry (CSG)
- Boundary representation (BREP)

PPI involves recalling already-stored descriptions of primitive solids, such as blocks, spheres, and cylinders, and applying a scaling transform to the primitive. Figure 2.24 shows three primitives which have been instanced to yield objects which are larger, smaller, and distorted due to the applied transformations. The limitations are obvious. A complex object cannot be created unless it can be transformed from a primitive volume.

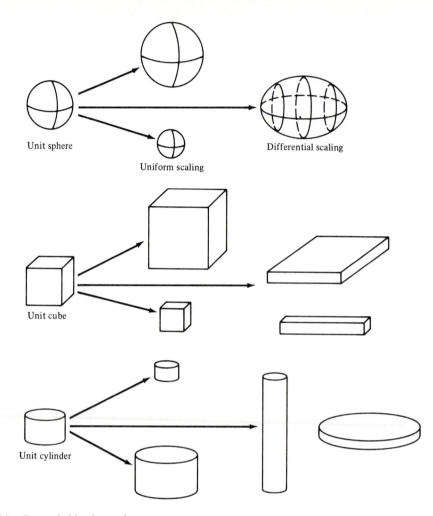

FIGURE 2.24 Pure primitive instancing.

The SOE method subdivides 3-D space into small volumes and classifies these volumes as either empty, full, or partially full of the solid. CD is similar to SOE, but it starts with the object, not with 3-D space, and subdivides that object into small volumes, all of which are full of the part (Figure 2.25). Usually the volumes are cubes. Therefore, the CD model for a cube is just a large cube with the same dimensions as the object. Irregular objects require more subcubes to account for the curved edges and surfaces. The cells are as small as the resolution of the model.

The last three construction techniques are considered the most significant [21]. Sweeping involves the notion that a polygon or polyhedron moving through space sweeps out a volume (a solid) described by the polygon or polyhedron and

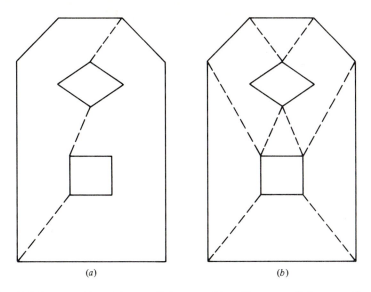

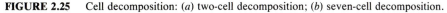

FIGURE 2.25   Cell decomposition: (*a*) two-cell decomposition; (*b*) seven-cell decomposition.

the trajectory (Figure 2.26) [17]. Sweeping was used in the example above to generate the turned section. Sweeping can also be performed along a trajectory, straight (as in an extrusion), or curved (as would be used in generating ductwork).

The CSG technique uses boolean combinations of primitive solids to build a part. The boolean operations are addition (+), subtraction (−), and intersection (∗), as illustrated in two and three dimensions in Figure 2.27. The primitives can be "pure," as in the case where a cube and a cylinder are joined. They can also be "super" primitives; that is, the primitives themselves may be combinations of solids. For example, imagine creating an airplane. The jet engine could be approximated by a cylinder with smaller cylinders subtracted from the inside to give the intake and exhaust ducts. This engine then could become a "super" primitive and be united with (or added to) the wing "super" primitive.

CSG is a popular method because adding and subtracting elementary volumes, called primitives, simulates the natural design process. These primitives may be readily translatable into machining operations. For example, drilling a hole can be interpreted as subtracting a cylinder.

By using just the four natural quadrics as these primitives (cones, spheres, planes, and cylinders), most batch manufacturing parts can be generated [14]. One method used to define these primitive solids for CSG is through explicit surface equations. Alternatively, these primitives can be generated implicitly from combinations of half-spaces. A half-space is the region on one side of a plane or other surface. An example of this type of construction is the building of a cube, of side length 4, from the intersection of the six half-spaces represented by all points with *x* coordinates less than the plane $x = 2$ and greater than the plane $x = -2$, with *y*

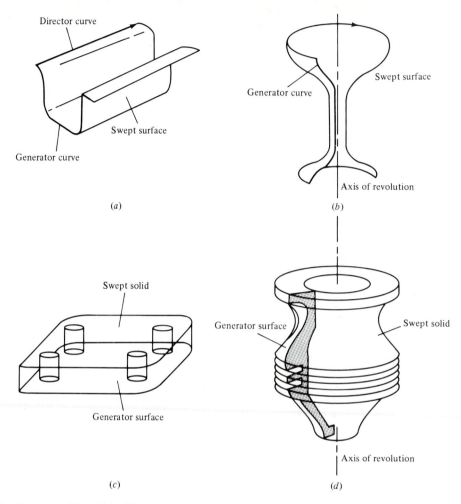

FIGURE 2.26 Sweeping. (From Ref. 17.)

coordinates less than the plane $y = 2$ and greater than the plane $y = -2$, and with z coordinates less than the plane $z = 2$ and greater than the plane $z = -2$. The half-space construction can utilize both planes and other infinite surfaces.

Keep in mind that the object itself exists only in the computer data base. The use of this data base is to answer our questions concerning the solid and maybe to draw a picture of the part for us. Therefore, the information present in the data structure must be sufficient at least to draw the edges and faces of the object. The CSG technique, then, must also have coresident with it a *boundary evaluator,* a piece of software which reads the CSG tree structure and calculates all new edges and faces resulting from the boolean operations on the primitives. A boundary evaluator is necessary in order to generate pictures or other representations involving the part faces, edges, and vertices.

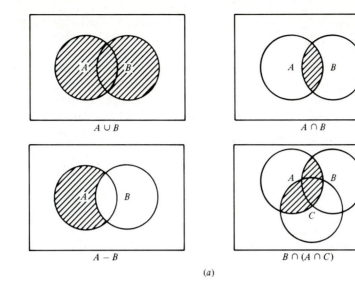

$A \cup B$

$A \cap B$

$A - B$

$B \cap (A \cap C)$

(a)

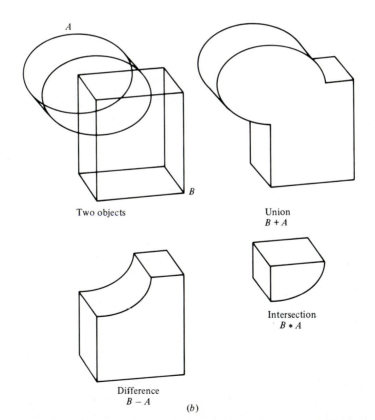

A

B

Two objects

Union
$B + A$

Difference
$B - A$

Intersection
$B * A$

(b)

FIGURE 2.27   (a) Two-dimensional boolean operations; (b) three-dimensional boolean operations.

BREP construction consists of entering all bounding edges for all surfaces. This is similar to copying an engineering drawing into the computer, line by line, surface by surface, with one important qualification: *The lines must be entered and surfaces oriented in such a way that they create valid volumes.* A cube, for example, would be entered one edge at a time, first outlining a face, then outlining adjoining faces, until a solid is created. The order of entry is important to avoid dangling edges and to ensure object validity. Solid modelers typically test for object validity, so the user is warned of inconsistencies [15].

2.5.3 The Euler Formula

A solid modeler guarantees that the object created is, in fact, a real solid object. There can be no Escher-type objects or nonsolids as shown in Figure 2.6. Most modelers use a derivation of the mathematical formula and procedures developed by Leonhard Euler (1707–1783) to test validity. The Euler formula defines a solid in terms of the number of faces (F), edges (E), and vertices (V) present on the object, where

$$F - E + V = 2$$

Satisfaction of the Euler formula is necessary but not a sufficient condition for solid validity. An object which satisfies this formula is called a *simple polyhedron*, a closed surface which has no passageways or holes in either the object or the faces. The formula can be extended to include *multiple polyhedra* with passageways and faces containing interior edge loops (face holes):

$$V - E + F - H + 2P - 2S = 0$$

where H is the number of holes in faces (interior edge loops), P is the number of passageways through the object, and S is the number of connected face sets, or shells, under consideration. As an example part, consider the object in Figure 2.28. Counting the $V, E, F, H, P,$ and S parameters in Euler's formula, a table can be completed as shown.

The formula works for this example. As an exercise, sketch some simple (or complex) parts and try the formula yourself. Remember that you are allowed to add edge loops in a face either to create a hole in the face or to create another interior face. You can also add multiple vertices between the original vertices, but each time you add or kill a vertex, edge, or face, be sure to recalculate the new number of vertices, edges, and faces. If one of the parameters in the formula, such as the number of vertices, changes, then another parameter (or parameters) must change to keep the equation balanced.

The Euler formula is worth more than just ensuring that an object is a valid polyhedron. It has also been used to define the permissible steps in constructing a solid. This set of *Euler operators* consists of the allowable operations which can be made during a piece-by-piece construction of a solid. The operators are listed in Table 2.1. The operators must be followed to guarantee that the object is a valid polyhedron. As an example, consider the construction of a tetrahedron. Figure

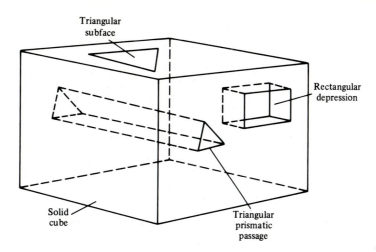

	F	$-$	E	$+$	V	$-$	H	$=$	2	$(S - P)$
Cube	6		12		8		0		1	0
Triangular passage	3		9		6		2		0	1
Rectangular depression	5		12		8		1		0	0
Triangular subface	1		3		3		1		0	0
Total	15		36		25		4		1	1

FIGURE 2.28   Euler's equation.

TABLE 2.1
Euler operators

	V 1	E −1	F 1	H −1	$2S$ −2	$2P$ 2
MEV	1	1	0	0	0	0
MFE	0	1	1	0	0	0
MBFV	1	0	1	0	1	0
MRB	0	0	0	0	1	1
ME − KH	0	1	0	−1	0	0
KEV	−1	−1	0	0	0	0
KFE	0	−1	−1	0	0	0
KBFV	−1	0	−1	0	−1	0
KRB	0	0	0	0	−1	−1
KE − MH	0	−1	0	1	0	0

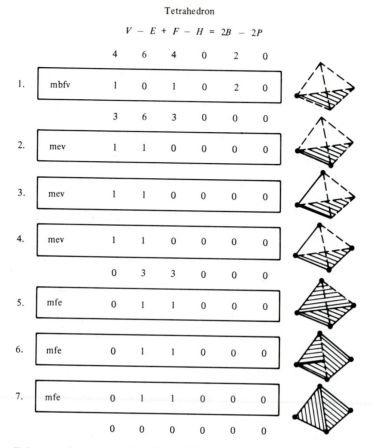

FIGURE 2.29 Euler operation construction. (From Ref. 17.)

2.29 [17] shows the procedure to build the tetrahedron. The order of operators is not unique for a solid. Variations are possible. Note that the MBFE creates an "acorn" which has no volume, one face and one edge, but no vertices. The MEV operator adds one edge and one vertex. There is a complementary set of "kill" operators for destroying or modifying an object. These operations can also be used to change an existing object rather than to construct a new solid.

2.5.4 Solid Modeler Storage Data Bases

After an object has been constructed, the modeler converts the input to a storage data structure, maintaining the geometry and topology of the part. Two major forms of storage exist: CSG and BREP. These are similar in concept to the CSG and BREP construction techniques mentioned above.

CSG stores the part as a tree, where the leaves are primitives and the interior nodes are boolean operators. Some modelers use CSG both to construct and to store the part, so no conversion is necessary between the model input and storage formats. The input goes directly into the CSG tree, the internal data base. Other modelers, on the other hand, allow different types of construction techniques and immediately convert each type to a BREP storage structure, maintaining linked lists of all vertices, edges, and faces which incorporate the geometry and topology.

The computation required to *evaluate* the faces, edges, and vertices from the user input is extensive. In return for this large number of calculations in the construction mode, the modeler gives a resultant data structure which is more detailed and more convenient for use by some of the analysis and output processors than the CSG storage method. The BREP stores the *actual part,* whereas the CSG stores the *instructions for how to make the part*.

A BREP data structure is represented in Figure 2.30*a*. In comparison, a CSG internal data structure is shown in Figure 2.30*b*. In addition to these single storage schemes, there are some modelers which store the object in multiple, redundant formats. These, of course, would require even larger storage space.

CSG STORAGE. A comparison of storage allocation necessary for each of these representations shows that the CSG tree is the smallest, containing the locations and definitions of the primitives and the boolean operators. Each node contains a boolean operation to unite, subtract, or intersect the solids in each of the subtrees. Each subtree can contain either another CSG tree or a leaf node which defines one of the primitives to be used. The leaf node contains an instance of a primitive in its default size and location. The leaf also contains a definition of scaling factors to make the primitive the correct size and proportions and a transformation to locate the object at the correct position and orientation needed to perform the boolean operation with the object on the other tree branch. Let these size and location transformations be designated as π_i. The tree, then, may look like the one in Figure 2.30*b*, with the result appearing as the finished part in Figure 2.31. The picture of the finished part is a result of evaluating the CSG tree for all faces, edges, and vertices produced by the boolean operations.

Usually the picture representation is calculated only for one viewpoint and is not stored for use by any other applications. A change in the viewer's eyepoint requires a new calculation of the faces, edges, and vertices as seen from this new location. This tree evaluation can take several minutes to several hours, depending on user-specified options such as picture resolution and face attributes, e.g., types of shading, shadows, transparency, and refraction. New developments in solid modeling and computer graphics are beginning to include hardware evaluators, VLSI geometry graphics engines whose sole purpose is to calculate surface–surface intersections and perform the hidden surface removal required to generate the pictures. It is assumed that within the very near future, solid modeling itself, whether CSG or BREP, will be performed entirely in hardware. Currently, there

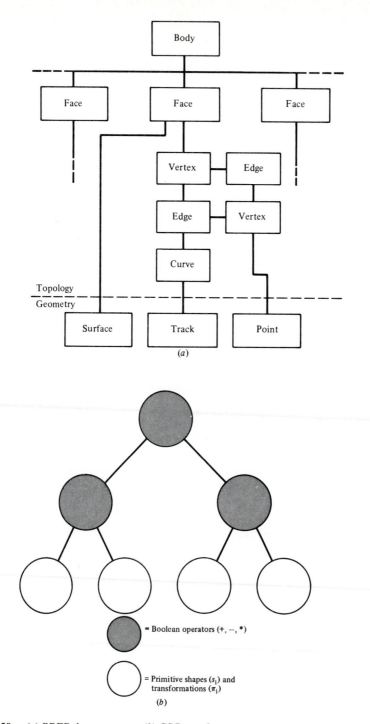

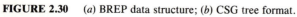

FIGURE 2.30 (*a*) BREP data structure; (*b*) CSG tree format.

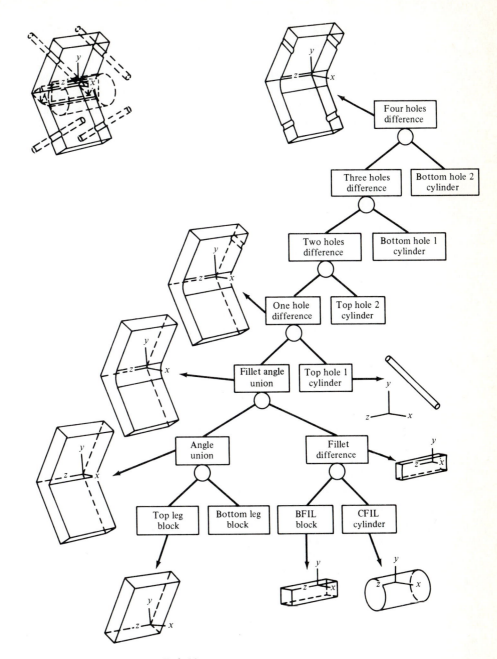

FIGURE 2.31   CSG object storage. (From Ref. 4.)

are solid modeling hardware modules that you can add on to a microcomputer to increase the model evaluation speed.

BREP MODEL STORAGE. BREP storage is larger than CSG and explicitly contains all vertices, edges, and faces. Also stored are surface normal directions, attributes of faces, dimensions, and the part topology. The actual storage size is highly dependent on the number of edges in the part.

One of the first proven storage structures for BREP was the winged-edge vertex format [2]. Other structures have been developed and are still being suggested as optimal. The suitability of a particular data structure can be measured by timing the modeler during object construction and object querying. The fastest response indicates the most efficient structure. Of course, storage space is also important, and the usual situation in developing an optimal storage method is to balance the speed against the modeler storage size.

An example structure is shown in Figure 2.30a. This is not necessarily the best framework for storing a solid, but it is a familiar form that is fairly easily understood. Note that the data structure is a linked list format with the links creating parent–child relationships. The overall framework is termed a graph because of the cyclic links in the loop definition. Topologic information is stored in the nodes in the upper half of the diagram and geometry in the lower-half nodes. Beginning at the top of the data structure, a BODY consists of a set of FACEs. Each FACE is bounded by one or more VERTEX and EDGE LOOPs.

Consider the object in Figure 2.32. Note that FACE F1 is bounded on the outside by a LOOP and on the inside by another LOOP. Each hole in a face is bounded by an additional LOOP. Each EDGE has one of two directions determined by the FACE in which the EDGE lies. To add consistency, a bounding LOOP has a direction; the direction of a LOOP is determined by tracing the LOOP such that the FACE is always on the tracer's left. Note that a FACE has only one side, the side exposed to air. The other side is inside the object and does not really exist. Because of this convention, each EDGE has two directions: one for each of its adjacent FACEs. In Figure 2.32, the FACEs F1 and F2 have EDGE E1 in common, but E1 actually has two directions, depending on which FACE is used. To store this directionality in the data structure, a CURVE node is added. The CURVE node contains the name of the EDGE and pointers to the two topological EDGEs with opposite directions. The EDGE nodes contain the directional information. EDGEs and FACEs are the most complex entities to follow in the data structure.

Geometry entities are the POINT, TRACK, and SURFACE. Each of these data nodes contains a geometric definition. In the case of the POINT, an (x, y, z) triplet is stored. For the TRACK, sufficient data is stored to define the line whether straight, circular, elliptical, cubic, or other-order curve. In the case of a straight line, two POINTS can be stored. A circle requires seven numbers: center (C_x, C_y, C_z), axis direction (A_x, A_y, A_z), and radius (R).

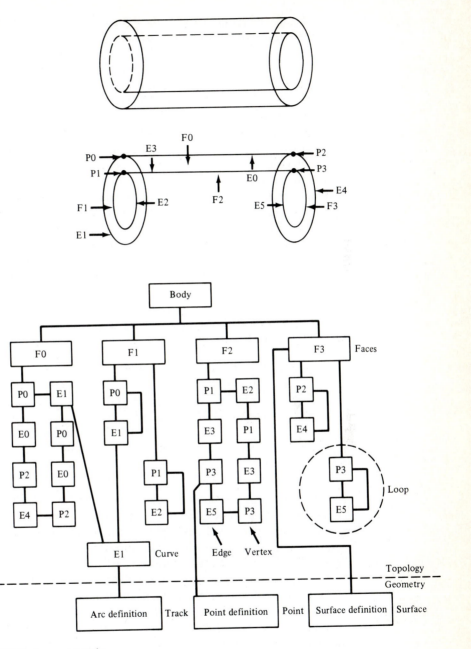

FIGURE 2.32   BREP storage example.

The SURFACE node contains the equation of the surface. A plane may be stored as four coefficients to satisfy the plane equation:

$$Ax + By + Cz + D = 0$$

Alternatively, it may be stored as a point on the plane (P_x, P_y, P_z) and a normal vector (N_x, N_y, N_z). A cylindrical surface requires a point on the axis (P_x, P_y, P_z), the axis direction (A_x, A_y, A_z), and a radius (R). Other, more complex SURFACEs and TRACKs, such as cubic curves and bicubic surfaces, can be stored as a set of control points or coefficients, as discussed in Section 2.4.3.

The importance of this data structure is revealed when a user asks a question about the solid. As an example, imagine that you, the manufacturing engineer, are calculating the NC tool path for machining FACE F3 and require the definition of all faces adjacent to F3. The user may generate the query to the data structure: What are all adjacent faces to F3? Expressed in the form of a software function invocation, this query may be ADJACENT(F3), which returns a list of faces. The real use of a solid modeler is just exactly this: the interrogation of a solid. In order to prepare for this question, the solid modeler may contain a function which, in pseudocode, may look like this:

```
      FUNCTION ADJACENT( FACE )
      DIMENSION ADJACENT_FACE(N)

      N = NUMBER_OF_EDGES(FACE)
      DO 100 I = 1,N
            EDGE = GET_EDGE(FACE,I)
100         ADJACENT_FACE(I) =
                  OTHER_FACE_WITH_SAME_EDGE(FACE,EDGE)
      RETURN(ADJACENT_FACE)
      END
```

Notice that the number of adjacent faces is equal to the number of edges bordering a face. It is clear that a subroutine must be written to determine the number of edges in the FACE, return a pointer to each EDGE one by one (GET_EDGE(FACE,I)), and trace through to find another face with the same edge (OTHER_FACE_WITH_SAME_EDGE(FACE,EDGE)). As an example of the algorithms used, consider the last subroutine mentioned. Given a FACE and EDGE, move through the data structure until another FACE is located which contains the same EDGE name (though opposite in direction). Only two faces are allowed per edge, and they may be the same face (think about a cylindrical face with a seam running its length). If one other face with the same edge can be located, we know that is the adjacent face.

Look at the BREP data structure (Figure 2.32) and start with an edge node. It is possible to trace to two faces from this node (the two adjacent faces). One, the beginning face, can be found by following the link to the vertex which leads up to the parent face. The other face can be found by tracing to the curve node and then to the other edge with the same name and then around the loop until the vertex is reached which leads to the face (the ADJACENT face). The subroutines mentioned above must perform exactly these operations. Other data structures exist for BREP solid models, and they are all linked list graph structures similar to this. They vary in the link connections and the types and number of nodes.

It is important to realize that any solid model useful in the integration of CAD and CAM must be able to answer questions similar to the preceding examples. More difficult queries include: "What is the volume of the solid?" "Where is an entrance to a hole?" [21] and "Is this part primarily a turned part or is it primarily prismatic?" [12]. Each of these queries requires traversal of the data structure. Therefore, the data base should be as small as possible to permit minimal storage, yet there should be many redundant links to make possible shortcuts through the data structure which in turn reduce execution time.

2.6 CHOOSING A SOLID MODELER FOR CAD/CAM INTEGRATION

In determining a satisfactory solid modeler for CAD/CAM integration, the two most important aspects appear to be the internal part representation and the modeling capabilities. Obviously, the solid modeler of choice should be able to model the gamut of parts an industry manufactures. The qualities of a good modeler include:

Flexibility: A solid modeler must address solids of all shapes. CSG modeling with primitive volumes is convenient, but inappropriate for complex objects such as airfoils and other sculptured surface objects. Multiple construction techniques increase the modeler's flexibility.

Robustness: The modeler should produce a consistent and proper solid. It should not allow the user by mistake to create nonrealizable objects.

Simplicity: The user interface is of utmost importance. People will not use a modeler that is unwieldy, difficult to use, or unpredictable. Pull-down menus and feature-based modeling have simplified today's solid modelers.

Performance: This is one of the drawbacks of today's modelers. There is hope for improvement in the future, though, with the incorporation of modeling procedures in the hardware.

Economy: Industry has viewed solid modeling as a large financial commitment. This hurdle will be reduced as software prices continue to come down and solid modeling migrates from mainframe computers to microcomputers and work stations.

It is not obvious that one type of object storage technique is superior for linking CAD with CAM. The choice depends on the information desired and the extent of calculations required.

2.7 FEATURE RECOGNITION: A CASE STUDY

Solid modelers have been used to generate automatically numerical control tool path coordinates, process plans, finite-element mesh geometry patterns, and, of course, realistic pictures of the part [1, 6, 8, 10, 11, 18, 20, 22]. Some of these applications require a series of software subroutines to identify features or patterns of geometry and topology. This case study will examine how to extract feature information from a solid model data base. A summary of solid modeler part representation schemes is given in Table 2.2.

The CSG tree stores a model in an implicit or "unevaluated" form; the edges and vertices of the final part must be calculated from the set operations on the primitives. An object's features, however, such as holes, slots, and so forth, may actually be present in the CSG tree. They may be represented explicitly, for example, subtracting a cylinder to create a hole. The definition of the hole can be derived directly from the definition of the cylinder (Figure 2.33a). It is possible, though, that the feature may not be present in the tree. The features may arise from evaluation of the CSG tree; for example, a slot may be formed by unioning two slightly separated blocks to a larger block.

The BREP graph contains an explicit, or "evaluated," model. The faces, edges, and vertices are represented explicitly, exactly as they are in the final part. In this case the features, though, are implicit, requiring *interpretation* of the geometry and topology. A hole may be present as a collection of faces and must be ascertained by interpretation of the part data (Figure 2.33b). This explicit/implicit difference has a great influence on suitability of a modeler in the recognition and extraction of feature information.

2.7.1 Feature-Based Design Using CSG Construction

The above comparison seems to suggest that manufacturing features may be more easily extracted from the CSG tree because it can describe them explicitly. For example, if the designer used only machining primitives, the objective could be created by beginning with a stock block of material and simulating the manufacturing steps during the design process by subtracting primitive volumes defined as slots, holes, or other routinely used features. The resulting CSG tree would contain explicit solid descriptions of all features in an unevaluated form.

TABLE 2.2
Comparison of modeler storage

Modeling technique	Model representation	Feature representation
CSG	Implicit	Explicit
BREP	Explicit	Implicit

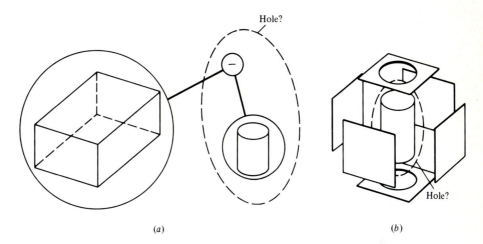

FIGURE 2.33   Interpreting a hole: (*a*) CSG hole representation; (*b*) BREP hole representation..

This approach has some obvious advantages over use of the BREP data structure, but there are also at least two major problems. The first difficulty is that it requires that the designer, in addition to optimizing the design, must also be involved with the manufacturing task by using only a limited selection of manufacturing primitives. Two problems arise from this approach. First, it is inconvenient for the designer to determine simultaneously a sequence of feature creation for all design iterations. In practice the designer is interested in the function of the design and will generate many intermediate designs before arriving at a satisfactory solution. Second, the use of machining volumes may be too restrictive. Many times the functional shapes used by a designer are not related directly to machining processes.

A possible procedure would be to allow the design to be completed for shape and functionality, and *then* determine the best way to make the part. This allows the designer and manufacturing planner to communicate in their own functional terms. This two-pronged design–manufacturing approach is closely related to traditional manufacturing planning, in which each person in the design and manufacturing stages interprets the same parts in his or her own way.

The second problem with interpreting the CSG tree to determine the process plan is the problem of nonunique trees. A feature can be constructed in multiple ways. A slot can result from subtraction of a block, or it may be an artifact of adding two blocks separated by a space (Figure 2.34). This example is a simple one, but there exist a myriad of possibilities more complex than this. To consider all possible primitive combinations for a certain feature type would be a monumental task.

An additional problem is related to tree complexity. To find or substantiate the presence of a particular feature, one approach would be to look for obvious primitive volume operations which would suggest that feature. Consider again the

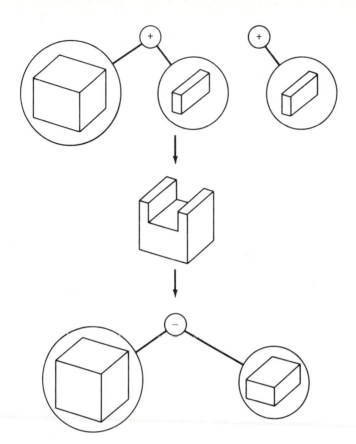

FIGURE 2.34   Nonunique feature construction.

case of subtracting a cylinder to create a round hole. In order to be sure that the hole does exist, the complete tree must be examined to see if any other primitive volumes overlap this cylinder and, if so, to determine if they affect the making of the hole. An example of this complication involves subtracting the cylinder, then filling the hole with another volume, and finally subtracting a volume of a different shape later in the CSG tree. A CSG tree can be a very large and complex structure, especially when considering such operations as chamfering and filleting. The process of proving the existence of a feature is essentially a partial evaluation of the tree to find surface–surface intersections. If this tree evaluation must be done for every suspected feature, the amount of calculating could surpass that required to generate the BREP data base. It is not clear that the advantage of having primitive feature elements as building blocks is worth the disadvantage of possibly large calculation times and the ambiguity of the primitive volume combinations.

As a final note on CSG modelers, a current trend among solid modeler developers is to change the user interface to allow the designer to create the part by using features. For example, one menu item may be "construct a hole." Invocation of this command may give the user tools to create a hole and then label the hole primitive so that it can later be used for automated process planning or NC path definition. This tagging of features during the design process shows much promise. Its greatest benefit is that it stores information in the part description that is usually lost in the modeling process: the designer's intent.

2.7.2 Using a BREP for Part Interpretation

Because of these concerns with the original CSG design combined with the BREP advantage of allowing designers to be free from process planning details, the BREP may be the structure of choice. The BREP data contains the evaluated part independent of the original building primitives. This format allows a fresh view of the part to perform manufacturing planning unrelated to the particular sequence used during the design stage.

As an example of automated feature recognition we will attempt to look for "holes" in the part [9]. A hole is defined as a set of faces which share a common axis and extend for 360° around the axis. A procedure for pattern recognition is to write a "rule-based" system with rules defining the patterns of interest and a set of facts, i.e., the part BREP model. A search routine can be written which examines the facts to satisfy the rules. A set of facts (faces) which does satisfy the rule for a hole is returned as a list of the faces in the hole. Here is how it works in practice. The rule for a hole can be written as follows:

```
A hole exists if:
        there exists an entrance to the hole and
        there is a face adjacent to this entrance which is a
            valid hole face and
        there is another face adjacent to this first hole
            face which is a valid hole face and
        there are more faces in the hole and
        the hole terminates with a hole bottom then
        this set of faces constitutes a hole
```

This rule can be written in pseudocode as follows:

```
hole(FACELIST) if
    entrance_face(FACE1) AND
    valid_hole_face(FACE1) AND
    adjacent(FACE1, FACE2) AND
    valid_hole_face(FACE2) AND
    more_hole(FACE2, FACELIST).
```

where the rule for "more_hole(FACE2,FACELIST)" looks for more faces, identifies the bottom of the hole, and returns this list of faces in the FACELIST. The details of these rules are not as important as the realization that the information exists in a solid model data base. You already know how to determine adjacent faces from a previous example. Finding a valid hole face is similar, except in this case the determination is made on the geometry of the face. The rule must check to see if the face wraps completely around a hole axis. This involves finding if the face is a cylinder, sphere, torus, cone, or symmetric bicubic patch, because these are the only surfaces which do wrap around an axis symmetrically. In addition, the rules must check to see if the surface wraps around to meet itself. If not, then the surface is not symmetric around the axis. The geometric information such as the surface type and location can be found in the surface node data structure in Figure 2.30*b*. The fact that the surface extends 360° around the axis can be found by examining the face loop to see if one of the faces adjacent to the hole face is the face itself. If so, then it must extend around the axis and meet itself.

Now, this is a somewhat simplified example. Surely you can think of situations where the above rule will fail. And we did not discuss all the special cases the rule must handle, such as holes which contain planes at the change of a diameter or even the definition of other features such as slots and pockets and turned parts, etc. In reality, feature recognition has many problems and a commercial system has not been developed. It is assumed that a good modeler would allow both feature-based design and automated recognition capabilities.

2.8 DATA TRANSFER STANDARDS

The final problem in making solid models available to many different applications consists of being able to pass model data between solid modelers themselves and from solid modelers to applications packages. Standards are being developed which dictate the format and content of a "neutral" solid model data base. Data transfer standards have three main objectives:

1. To enable the communication of data between two systems with no loss of information
2. To encode the data in as compact a manner as possible to minimize transmission costs
3. To avoid imposing unduly complex processing requirements on the sending and receiving systems to minimize processing development costs

There is disagreement on the correct way to handle this neutral data format, though. The two main standards being considered are the Initial Graphics Exchange Standard (IGES) [13], and the Product Definition Exchange Specification (PDES). PDES and IGES are being developed by the ANSI Y14 committee.

IGES was developed primarily for CAD data transfer and consists of a scheme to store and transmit geometric primitives such as points, lines, and surfaces. The PDES format is more comprehensive. It allows a richer set of primitives and includes the topology required of a solid model in addition to other product attributes such as material type and form features. IGES is in use world-wide now to transfer graphics data between CAD systems. PDES is still under development and may, in the future, be combined with the IGES standard.

It is hoped that these various standards will eventually come together under the auspices of the International Standards Organization (ISO).

2.9 CONCLUSION

The progression of this chapter has been from traditional engineering drawings through wire frames and surface models to the description and use of solid modeling schemes. CAD/CAM integration requires that a complete definition of a part be stored in a data base and, not only that, the model of the part must be robust enough to drive manufacturing applications routines such as the feature recognition example above.

This chapter has just skimmed the surface of the issue of solid modeling and model interrogation. The problem of tolerance specification and storage has not even been addressed. Modelers still cannot represent every part geometry which is manufacturable.

User interfaces are not ideal. The perfect user interface is one which assists a designer as a part is "molded" from a material. This "designer's assistant" gives freedom to the designer while at the same time acting as consultant to straighten edges, smooth surfaces, and guarantee that the object is in fact manufacturable.

As you progress through this book, keep in mind that most of the applications you will read about will involve the geometry of the part. Examples include group technology, part coding, automated robotic assembly, process planning, and computer-aided design applications. True integration of CAD and CAM comes from linking these processes together so that they can be performed with as little human intervention as possible. Geometric modeling is one of the threads.

EXERCISES

1. Draw the BREP solid model data structure for the cube in Figure 2.11 using the template shown in Figure 2.30a.
2. Describe an algorithm which, given an edge, can find an entrance to a hole. *Hint:* A hole entrance is a circular edge with a plane on one side and a cylinder on the other.
3. Given the two slopes and endpoints of a Hermite curve, sketch the curve. If the slope vector is increased in magnitude by a factor of 2, sketch the new curve to show the changes due to these new slopes. $p_0 = (0,0,0)$, $p_1 = (5,10,0)$, $p_0' = (1,1,0)$, and $p_1' = (-1,0,0)$.
4. Draw two possible CSG tree structures for the object shown in Figure 2.35.
5. For the polygonal data structure given in Figure 2.36 (edge-list structure) draw the two-dimensional polygons.
6. Show that Euler's formula works for the object shown in Figure 2.26c.
7. Plot four control points for a Bezier curve in a plane at (0,0), (1,2), (2,−3) and (5,0). Draw the Bezier approximation for these points by both sketching and plotting points.

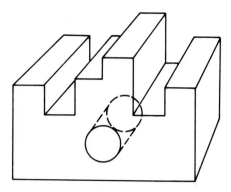

FIGURE 2.35

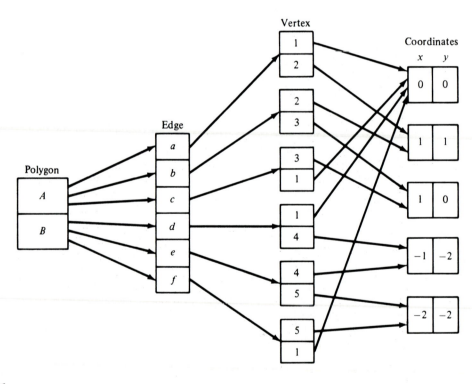

FIGURE 2.36

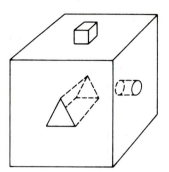

FIGURE 2.37

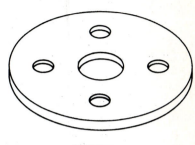

FIGURE 2.38

Sketch the convex polygon formed by these points. What happens if all four control points fall on a straight line, but not in order, for example, p_1, p_0, p_3, p_2?

8. Give the limitations of curves that can be represented by equations of first order; second order. Give an advantage of cubic curves.

9. Sketch the cubic B-spline curve formed by the following control points: (0,0), (2,2), (4,0), and (2,−2). Add the point (0,0) after (2,−2) and sketch the two segments. What type(s) of parametric continuity do they possess?

10. Discuss how to make a solid model of the object shown in Figure 2.37. There is more than one answer. Give two or three possibilities if you wish.

11. As an example of the use of solid modeling in CAM, explain what information is stored concerning the object shown in Figure 2.38. Tell how that information may be used to plan the manufacture of the part.

12. Write an "if . . . then" rule in English to recognize a square pocket. What are the conditions to find a general depression?

13. Hypothesize a list of information to be contained in a feature-based geometric model data base. What parameters of features may be important for design, analysis, and manufacture? For this example, choose the features: holes and slots. Subsequent chapters address the requirements of design, analysis, and manufacturing systems, but for this problem use your common sense and intuition.

REFERENCES AND SUGGESTED READING

1. Armstrong, G. T., G. C. Carey, and A. de Pennington: "Numerical Code Generation from Geometric Modeling System," in *Solid Modeling by Computers: Theory and Applications,* J. W. Boyse and M. Pickett, Eds., Plenum Press, New York, 1985.

2. Baumgart, B. G.: "Winged-Edge Polyhedron Representation," STAN-CS-320, Computer Science Department, Stanford University, Palo Alto, Calif., May 1974.

3. Bezier, P.: *The Mathematical Basis of the UNISURE CAD System,* Butterworths, London, 1986.

4. Boyse, J. W., and J. E. Gilchrist: "GMSolid: Interactive Modeling for Design and Analysis of Solids," *IEEE Computer Graphics and Applications,* March 1982, pp. 27–40.

5. Braid, I. C., R. C. Hillyard, and I. A. Stroud: "Stepwise Construction of Polyhedra in Geometric Modeling," CAD Group Document No. 100, Cambridge University, Cambridge, U.K., 1978.

6. Chang, T. C., R. A. Wysk, and R. P. Davis: "Interfacing CAD and CAM—A Study in Hole Design," *Computer and Industrial Engineering,* vol. 6, no. 2, pp. 91–102, 1982.

7. Farin, G.: *Curves and Surfaces for Computer-Aided Geometric Design,* 2d ed., Academic Press. San Diego, 1989.
8. Grayer, A. R.: "The Automatic Production of Machined Components Starting from a Stored Geometric Description," in *Advances in Computer-Aided Manufacture*, D. McPherson, Ed., Elsevier Scientific Publishers, New York, 1977, pp. 137–151.
9. Henderson, M. R.: "Extraction and Organization of Form Features," in *Software for Discrete Manufacturing*, J. P. Crestin and J. F. McWaters, Eds., Elsevier Science Publishers, North-Holland, 1986.
10. Henderson, M. R., and G. J. Chang, "FRAPP: Automated Feature Recognition and Process Planning from Solid Model Data," *Proceedings of ASME Computers in Engineering Conference*, San Francisco, Calif., vol. 1, August 1988, pp. 529–536.
11. Henderson, M. R., and S. Musti, "Automated Group Technology Part Coding from a Three Dimensional CAD Database," *ASME Transactions Engineering for Industry*, August 1988, pp. 278–287.
12. Hordeski, M. F.: *CAD/CAM Techniques*. Reston Press, Reston, Va., 1986.
13. Initial Graphics Exchange Specification (IGES) 3.0, National Bureau of Standards, Washington, D.C., 1986.
14. Jared, G. E.: "Shape Features in Geometric Modeling," in *Solid Modeling by Computers: Theory and Applications*, J. W. Boyse and M. Pickett, Eds., Plenum Press, New York, 1985.
15. Mantyla, M.: *An Introduction to Solid Modeling,* Computer Science Press, Rockville, In., 1988.
16. Miller, J., H. Steinberg, G. Jared, and G. Allen: "Tutorial Notes: Introduction to Solid Modeling," SIGGRAPH 1985, San Francisco, Calif.
17. Mortenson, M. E.: *Geometric Modeling*, John Wiley, New York, 1985.
18. Razdan, A., M. R. Henderson, P. Chavez, and P. E. Erickson, "Feature-Based Object Decomposition for Finite Element Meshing," *The Visual Computer*, no. 5, 1989, pp. 291–303.
19. Sata, Toshio: "Approaches to Highly Integrated Factory Automation," in *Software for Discrete Manufacturing*, J. P. Crestin and J. F. McWaters, Eds., Elsevier Science Publishers, New York, 1986.
20. Tapadia, R. K., and M. R. Henderson, "Using a Feature-Based Model for Automatic Determination of Assembly Handling Codes," *Computers & Graphics*, vol. 14, no. 2, May 1990.
21. Voelcker, H. A., and A. A. G. Requicha: "Addendum: Geometric Modeling Systems," Seminar on Geometric Modeling, SIGGRAPH 81, Dallas, Texas, August 3, 1981.
22. Woo, T. C., "Computer-Aided Recognition of Volumetric Designs," in *Advances in Computer-Aided Manufacture*, D. McPherson, Ed., Elsevier Science Publishers, New York, 1977, pp. 121–135.

CHAPTER
3

COMPUTER-AIDED DESIGN

Today, the use of CAD is in many areas the only way to obtain feasible solutions of . . . problems. . . . The problem complexity, the amount of data to be handled and the need for better user control over the problem-solving process call for system architectures like those that have been developed for CAD.

J. Encarnaçao and E. G. Schlechtendahl [2]

Good engineering design is necessary to guarantee that a part or mechanism functions correctly and lasts for a reasonably long time. Functional considerations during design involve, among other things, weight, strength, thermal properties, kinematics, and dynamics. The performance of a design can be evaluated by comparing its performance measurements with the required specifications. As important as satisfactory performance is, there are other areas of engineering design that are just as important. Besides functioning correctly, a part should be designed economically. This implies that the finished part should be designed as close to the specifications as possible. If the function of a supporting member of a structure requires the member to withstand 10,000 psi of compressive stress, then designing the member to withstand 30,000 psi is unnecessary and will probably be more expensive than the required design. Engineering design, therefore, should address functionality *and* economics. Functionality is determined by a part's

71

geometry, material properties, and environment. The economic factors include materials, processing costs, and marketing details.

As much as 70% of the production costs of a manufactured part are determined during the engineering design process. This means that by the time the part has left the designer's hands, the large majority of its production costs have been established. They are defined implicitly by the materials, dimensions, tolerances, surface finishes, and other parameters which determine processing costs. Therefore, only 30% of the part's cost is subject to money-saving efforts during the manufacturing planning stage. This 70/30 ratio emphasizes the importance of the design stage and should make it clear that computer assistance during the design can be a major help in assuring proper function and reasonable production costs.

Computer-aided design (CAD) is a term which means many things to many people. To some, it means computer-aided drafting or drawing. To others, it means computer-aided analysis. And to still others, it suggests totally automated design where the engineer need specify only the function of a part and the computer arrives at a satisfactory or even optimal design. CAD is all of the above, aiming mostly for the last description through techniques in artificial intelligence. CAD, however, still exists separately in each of the stages mentioned. Probably the most common and simplest CAD systems are limited to automated drafting capabilities. More advanced systems can perform analyses and even help guide the engineer to the equations necessary to determine the effectiveness of the design.

The thrust of this text is to demonstrate what components are required for computer-integrated manufacturing (CIM). CIM logically requires the integration of computer-aided design and computer-aided manufacturing. Most of the chapters in this text comment on the relevance of the chapter material to CAD/CAM integration. This chapter's emphasis is obvious.

3.1 KEY DEFINITIONS

Cathode-ray tubes (CRTs) are the display device of choice today. A CRT consists of a phosphor-coated screen and one or more electron guns to draw the screen image.

Computer graphics is the use of the computer to draw pictures using an input device to specify geometry and other attributes and an output device to display a picture. It allows engineers to communicate with the computer through geometry.

Computer-aided analysis allows the user to input the definition of a part and calculate the performance variables.

Computer-aided design (CAD) is the creation and optimization of the design itself using the computer as a productivity tool. Components of CAD include computer graphics, a user interface, analysis, and geometric modeling.

Computer-aided drafting is one component of CAD which allows the user to input engineering drawings on the computer screen and print them out on a plotter or other device.

Constraints are performance variables with limits. Constraints are used to specify when a design is feasible. If constraints are not met, the design is not feasible.

Criteria are performance variables used to measure the quality of a design. Criteria are usually defined in terms of degree—for example, lowest cost or smallest volume or lowest stress. Criteria are used to optimize a design.

Cursors are movable trackers on a computer screen which indicate the currently addressed screen position. The cursor is usually represented by an arrow or cross-hair.

Design can be a noun or a verb. In this chapter the verb "design" is the process of creating a specification for construction of a part. The process involves synthesis, analysis, and optimization. The noun "design" is the resulting specification and includes geometry, topology, tolerances, material, and other parameters necessary for manufacturing the part. The noun "design" is the source for all part information required to perform integrated CAD/CAM.

Design variables are the parameters in the design that describe the part. Design variables usually include geometric dimensions, material type, tolerances, and engineering notes.

Engineering work stations are self-contained computer graphics systems with a local CPU which can be networked to larger computers if necessary. The engineering work station is capable of performing engineering synthesis, analysis, and optimization operations locally. Work stations typically have more than 1 megabyte (1 mbyte) of random-access memory (RAM) and a high-resolution screen greater than 512×512 pixels.

Expert systems is a branch of artificial intelligence designed to emulate human expertise with software. Expert systems are in use in many arenas and are beginning to be seen in CAD systems.

Finite-element analysis (FEA) is a numerical technique in which the analysis of a complex part is subdivided into the analyses of small simple subdivisions of the part. The results are approximate because the object, instead of being a continuum, is now discontinuous.

Frame buffers store the raster image in memory locations for each pixel. The number of colors or shades of gray for each pixel is determined by the number of bits of information for each pixel in the frame buffer.

Interlacing is a technique for saving memory and time in displaying a raster image. Each refresh pass alternately displays the odd and then the even raster lines. In order to save memory, the odd and even lines may also contain the same information.

Kinematics/kinetics analysis is the measure of motion and forces of an object. This analysis is used to measure the performance of objects under load and/or in motion.

Mathematical models of an object or system predict the performance variable values based on certain input conditions. Mathematical models are used during analysis and optimization procedures.

Optimization occurs after synthesis and after a satisfactory design is created. The design is optimized by iteratively proposing a design and using calculated design criteria to propose a better design.

Parallel design process evaluates all aspects of the design simultaneously in each iteration. The design itself is sent to all analysis modules including manufacturability, inspectability, and engineering analysis modules; redesign decisions are based on all results at once.

Performance variables are parameters which define the operation of the part. Performance variables are used by the designer to measure whether the part will perform satisfactorily.

Pixels are picture elements in a digitally generated and displayed picture. A pixel is the smallest addressable dot on the display device.

Random-scan devices draw an image by refreshing one line or vector at a time; hence they are also called vector-scan or calligraphic devices. The image is subject to flicker if there are more lines in the scene than can be refreshed at the refresh rate.

Raster devices process pictures in parallel scan lines. The picture is created by determining parts of the scene on each scan line and painting the picture in scan-line order, usually from top to bottom. Raster devices are not subject to flicker because they always scan the complete display on each refresh, independent of the number of lines in the scene.

Refreshing is required of a computer screen to maintain the screen image. Phosphors, which glow to show the image, decay at a fast rate, requiring the screen to be redrawn or refreshed several times a second to prevent the image from fading.

Serial design is the traditional design method. The steps in design are performed in serial sequence. For example, first the geometry is specified, then the analysis is performed, and finally the manufacturability is evaluated.

Synthesis is the specification of values for the design variables. The engineer synthesizes a design and then evaluates its performance using analysis.

Transformations include translation, rotation, and scaling of objects mathematically using matrix algebra. Transformations are used to move objects around in a scene.

User interfaces are the means of communicating with the computer. For CAD applications, a graphical interface is usually preferred. User friendliness is a measure of the ease of use of a program and implies a good user interface.

3.2 THE DESIGN PROCESS

Before delving into CAD, it is important to understand the design process itself. Computer techniques are no more valuable than the theory on which they are based. After a short discussion of design in general, we will explore how the computer can help with each of the design stages.

Shigley [16] defines the design process as an iterative procedure consisting of six phases:

1. Recognition of need
2. Definition of problem
3. Synthesis
4. Analysis and optimization
5. Evaluation
6. Presentation

The first step, recognition of need, arises from an identified problem. Suppose, for example, that a new, more efficient jet engine is required because of a fuel shortage. Step 2, definition of the problem, is more specific. Defining the specifics of the jet engine involves knowing the thrust, horsepower, allowable weight, etc. Some of these specifications are *constraints;* i.e., they are requirements that *must* be met for the design to be feasible. The jet engine weight limit may be one constraint based on the available lift of the plane wings. A maximum weight, then, would be a constraint. Constraints are usually expressed as a maximum or minimum value or range of values. Other specifications, however, are termed *criteria*, i.e., measures of the goodness or quality of the design. Criteria are used to measure a design and compare designs against one another, as occurs during the optimization stage. One criterion of the jet engine may be gallons of fuel burned per hour or per mile of travel. Criteria have no maximum or minimum limits on values. A satisfactory design is one in which all the constraints are met. This best design, in addition to meeting the constraints, will have the best combination of criteria values. Selection of constraints and criteria is an integral part of step 2, definition of the problem.

Following definition of the problem, the next three steps (synthesis, analysis/optimization, and evaluation) are iterative. Synthesis and analysis are performed in conjunction with one another.

A design is synthesized, or created, to meet all constraints. The analysis determines (1) if the constraints have been met and (2) the values for the design criteria. If analysis shows that the constraints have not been met, or if the criteria can be improved, the design is modified and reanalyzed. Evaluation usually involves building a prototype for experimental testing. The actual model is tested and compared against the constraints for feasibility, and measurements are made of the criteria variables. If the constraints are not met, the design returns to the synthesis stage. If it is seen that improvements can be made in the design criteria values, the design may return to the synthesis stage depending on the economic advantage of a redesign. If the improvements will cost more than they will benefit, the redesign is not performed. This iteration may be performed many times before either a suitable or an optimal design is found.

Design is typically a serial process, in which the analyses are performed one after the other. Several possible types of analysis include stress levels, thermal performance, manufacturability, and esthetics. Serial design allows these analyses to be performed sequentially. The part may pass the stress-level test but fail the manufacturability requirements. A subsequent design modification to improve the manufacturability may cause the stress-level test to fail. The design may oscillate, satisfying one constraint but not the other, until a suitable design is found.

Parallel or concurrent design (Chapter 4) is another approach which specifies that all analyses be performed at once and any new design be based, not on the results of one analysis, but on the total analysis package. It is hypothesized that parallel design will reduce the design oscillation mentioned above. This supposition is especially true for functional versus manufacturability analyses. The prevalent design process today first examines and ensures conformance to functional performance constraints and then evaluates the part for manufacturability. In many cases, a functionally acceptable design will be found to be nonmanufacturable, and the designer must then redesign the part and reanalyze the functional characteristics before resubmitting to a manufacturing analysis. The parallel design concept includes manufacturability and, in fact, all pertinent analyses, to be performed on each design iteration simultaneously.

Several types of problems can arise during the design process. Synthesis requires determining the overall shape and system component specifications. Usually this takes the form of engineering drawings including geometry, tolerances, material types, and component part numbers. Chapter 2 discussed geometric modeling, so you know that this subject will enter into the computer-aided design scheme. Analysis may range from a simple esthetic evaluation to the solution of a set of complex differential equations which describe the performance of the part or assembly. The results of analyses can be expressed in tabular or graphical form. It is the job of the designer to evaluate the analysis data and to perform any redesign. Evaluation is also a type of analysis, but usually requires an expensive prototype and racks of instrumentation. The design process can be very costly and involve many people. Design cost may be lowered, however, by reducing the number of people and the prototyping process itself.

3.2.1 Applying the Computer to Engineering Design

The remainder of this chapter discusses ways in which the computer can help in the synthesis, analysis, evaluation, redesign, and presentation aspects of the overall design process.

We have discussed the way to design (a verb) above. The result of the design process is the design (a noun). The design (noun) of a part can be characterized as a data base. This data base contains the results of the synthesis stage—geometry, materials, etc.—which conform to the constraints and design criteria. This data base is necessary to manufacture the part. The data base can be partially represented as an engineering drawing. CAD can be described as the sequence of computer-aided steps to generate this design/manufacturing data base.

Many people identify the term "CAD" as standing for computer-aided drafting. Computer graphics is definitely an important part of CAD, but the major advantage of using the computer for design is the tremendous time saving during the analysis and redesign stages. Analysis, instead of being a time- and labor-consuming process, can be reduced to minutes on the computer and the result used to recall the design description and modify it according to the analysis results. Many iterations can be accomplished with the computer where previously they may not have been practical. This results in less expensive and better designs.

CAD modules can be divided into four categories [7]: (1) geometric modeling, (2) engineering analysis, (3) design review, and (4) evaluation and automated drafting/documentation. These four categories correspond to the last four modules in the serial design process shown previously.

Geometric modeling was discussed in Chapter 2. It implies the existence of a computer graphics screen and some interaction with the computer to generate the geometry and topology of the part. It also implies a data base and an appropriate data structure in which to store the part description.

Engineering analysis communicates with the data base to retrieve the part description and with the user to obtain the design constraints, boundary conditions, and other details of the analysis. Usually computer graphics is involved here. In addition, large central processing units (CPUs) are sometimes used in analysis. For example, finite-element analysis for stress levels sometimes requires parallel processors and supercomputers.

The design review and evaluation module allows the user to check the correctness, manufacturability, and processing details of the part. For example, this module may allow the user to overlay the part graphically with its finite-element stress results to inspect for high stresses. Other examples include kinematic animation and flow field display using colors for different temperatures. This module also implies the presence of computer graphics.

The drafting and documentation module contains some of both the oldest and newest technologies. Computer plotting of engineering drawings has been done for more than 20 years and is standard practice today. Documentation is also available in the form of desktop publishing, the merging of text and graphics for documents such as instruction manuals and reports. The output of the results can appear on high-quality devices such as laser printers. This module implies the presence of computer graphics and high-quality output devices.

The remainder of the chapter addresses the modules for analysis, evaluation, and documentation. Geometric modeling will be discussed only as it relates to these modules. Before proceeding with the modules, however, it is necessary to present sufficient details of CAD tools: hardware, application software, and computer graphics. These three components are present in every CAD module.

3.3   CAD HARDWARE

3.3.1   Computers

Computers can be divided into four types: micros, minis, mainframes, and super-computers. Microcomputers are sometimes called personal computers (PCs). They possess a dedicated central processor, a cathode-ray tube display (CRT), and a device (keyboard, mouse, etc.) to allow graphical input. The IBM PC and Apple Macintosh are examples of microcomputers.

Minicomputers (work stations) are more powerful than microcomputers and typically have more memory, a hard disk for storage, and a high-resolution color monitor for graphics. They are usually a single-user configuration, but have networking and multiuser capability. The Micro-VAX, SUN, and HP/Apollo computers are examples.

Mainframe computers are usually multiuser computers with even more memory and many hard disks for storage. The terminal type depends on the application and may include anything from alphanumeric terminals to sophisticated graphics displays.

Supercomputers are reserved for calculation-intensive applications such as weather forecasting, high-energy physics, and other similar uses. They may contain parallel computing capabilities and are usually networked to the user through a mainframe computer. They can have memory sizes in the hundreds of millions of bytes and speeds in excess of 1 billion floating-point operations per second (FLOPS). This compares with speeds of a microcomputer in the 1000- to 3000-FLOPS range.

Even given these four categories, however, it is not always easy to classify a given computer. The boundaries are fading with increases in on-board memory, CPU speed, and graphics capabilities.

CAD systems have been designed to run on each of the above types of computers. The AutoCAD system (Auto Desk, Inc., Sausalito, California) is PC-based and is the largest-selling CAD package in the world, with more than 50,000 units in place. Its primary function is automated drafting and documentation (see Figure 3.1).

A further category of CAD computers is the work station configuration. Work stations contain a local CPU, but are also networked to a larger computer which maintains large analysis programs and centralized design and manufacturing data bases. Work stations also share file server and output devices. Work stations can be networked in several ways, two of which are shown in Figure 3.2. The arrangement shown in Figure 3.2a shows a central host computer with each work station having access to the host CPU. Figure 3.2b shows a network in which all nodes, whether host, file server, work station, or output device, are shared and messages can proceed between any of the nodes. This latter configuration is typical of a local area network (LAN), a common way of connecting computers using network protocols such as Ethernet and MAP. Local area networks allow many different types of nodes to communicate, including machine tools, robots, and computer vision cameras as well as computers.

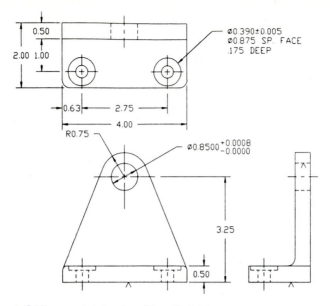

FIGURE 3.1 A CAD-generated drawing. (From Ref. 1.)

FIGURE 3.2 Networked work stations: (*a*) traditional; (*b*) LAN.

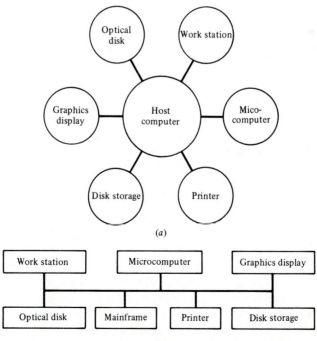

3.3.2 Input/Output Devices

The CAD work station communicates with the user through many types of input and output (I/O) devices. Each device can perform one or more functions. For example, a keyboard can be used to enter data, type text, or signal the computer to begin executing a command. The functionality of input devices can be divided into six logical categories: string, button, pick, locate, valuator, and stroke. These functions are defined below:

string   returns a string of characters. This is useful for typing text, of course, but also for entering commands. A keyboard is a type of string input device.

button   returns a 1 or a 0 (on or off). Buttons are useful to alert the computer to sample data or execute a command. A button on the mouse or even a keyboard key are examples of button devices.

pick   returns the identifier of an object on the screen. The object may be a line or point or a collection of lines and points. Picking is used to make an object active so that an operation can be performed on it. Picking can also be used to select menu items on the screen. Examples of picking devices are mice, joysticks, or even keyboards.

locate   returns an x–y screen coordinate. Locators are useful when identifying a point on the screen. For example, when moving a picked object to a new location, the locator will return the coordinates of the point to which the object is to be moved. Examples of locators include mice and tablets.

valuator   returns a value. A valuator is useful to enter values into a program. For example, a valuator may be used to dimension a part or to enter a force or density or other parameter. Examples are a dial, keyboard, and even a mouse if a graphical scale appears on the screen.

stroke   returns a string of positions. This logical type of input is used as shorthand in some CAD systems. To zoom in on an object, for example, the user may make a large, sweeping motion with a tablet stylus from the lower left to the upper right. The computer receives the string of positions and interprets the motion to mean zoom in. This type of input also allows the entry of values or letters by drawing them on the tablet and having the computer recognize the motion as a number or letter.

These logical input techniques are not necessarily equivalent to physical input devices such as a mouse, tablet, joystick, or keyboard. There is not a one-to-one correspondence between a logical function and a physical device. For example, a mouse can act as a pick, locator, or stroke device.

Now that we have looked at the logical functions that input devices can perform, let us examine some examples of particular types of physical devices used in input of graphical or geometric data into a CAD system. The devices discussed include the light pen, joystick, mouse, track ball, tablet, keyboard, and dial.

LIGHT PEN. A light pen is a device which allows interaction with a display from a computer screen. The light pen consists of a photodetector as diagrammed in Figure 3.3, which can be held in the hand and positioned on the screen. The photodetector senses the light coming from the screen and, through a timing circuit, can determine the x–y screen coordinates at which it is looking. The typical CRT display is refreshed by periodic sweeps of the electron beam in horizontal lines across the screen. If the light pen is placed on the screen and the electron beam moves by in its normal traverse of refreshing the screen, the photo-detector detects the lighting of the phosphor on the screen; through a flip-flop circuit, it signals the computer when the electron beam passes by the photodetec-tor. The computer can then calculate the x–y coordinates where the light pen is located. An advantage of the light pen is that it can be used as a drawing pen right on the screen. By holding the pen against the screen, one can place points and lines and perform drafting operations similar to drawing on paper. To place a point on a dark screen, the entire screen must be refreshed with white light (in a flash) in order for the light pen to sense its x–y location on the screen. Light pens are fading from use these days and are being replaced by tablets and mouse devices.

JOYSTICK. The joystick is a device which has been used for several years in video games. It allows the user to send a signal to the computer by moving the stick back and forth in two directions, either side to side, front to back, or at an angle, indicating x and y motions on the screen. The joystick usually contains two internal potentiometers, one for the x motion and one for the y motion (Figure 3.4).

FIGURE 3.3 Light pen operation.

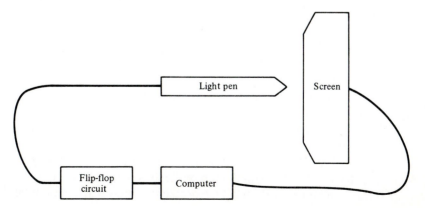

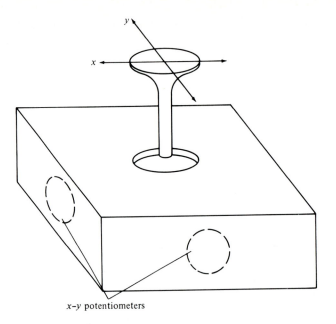

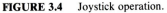
x–y potentiometers

FIGURE 3.4   Joystick operation.

There are two types of joysticks. One is a positional joystick, the other is called a velocity joystick. The positional joystick uses the feedback from the potentiometers to indicate an exact *x–y* coordinate.

The velocity joystick, on the other hand, uses the *x–y* outputs of the joystick to indicate *x* and *y* velocities for the cursor on the screen. By deflecting the joystick in the *x* or *y* direction, the computer is commanded to move the cursor in the direction the joystick has been moved. When the joystick is released, it springs back to the zero, or upright, position and the cursor stops. To reverse the cursor direction, the joystick must be pushed in the opposite direction. The cursor stops when the joystick is released and moves when the joystick is deflected. A velocity joystick is characterized by the springs which return it to an upright position when released. When the *positional* joystick is returned to the upright position, the cursor is usually returned to the center of the screen. As the joystick is deflected forward, backward, right, or left, the cursor moves to the indicated *x–y* coordinates.

The joystick can be adjusted through mechanical linkages and springs to give a "natural" feel. In fact, some joysticks used in airplanes for flight control are instrumented with strain gauges and involve no motion at all. If the joystick is pushed from the side, the strain gauge records the amount of force used and controls the plane accordingly.

Joysticks can also become three-degrees-of-freedom devices by the addition of a twistable knob on the top of the joystick. By deflecting the joystick in *x* and *y* positions, one can control the horizontal and vertical motion of the cursor; and by adjusting the knob on the top of the joystick, the z input can be changed. For additional input, a button can be added to the top of the joystick.

MOUSE. The mouse is an input device which allows two-dimensional input into the computer. It consists of a small hand-held box which can be slid on a table surface to indicate a change of $x–y$ coordinates. There are two mouse varieties: mechanical and optical. A mechanical mouse (Figure 3.5) contains two internal orthogonal potentiometers which are moved by sliding the mouse over a surface. The potentiometers are turned by wheels or a rubber ball which rolls on the surface.

FIGURE 3.5 Mechanical mouse: (*a*) top view; (*b*) transparent case. (Courtesy of Logitech, Fremont, California.)

(*a*)

(*b*)

An optical mouse is similar in function except that it contains a light source and a light sensor on its undersurface. By moving the mouse over a grid pattern, light is alternately reflected and blocked by the pattern on the grid lines. The light sensor can detect reflections for the side-to-side and fore-and-aft motions corresponding to x and y inputs. In addition, buttons can be added to a mouse device for further input.

The mouse has become very popular with microcomputers in recent years. It is inexpensive, small, and, for most people, easy to use. The sensitivity of the mouse can be set through hardware or software. If high sensitivity is needed for precise positioning of points on the screen, the ratio of mouse-to-cursor movement on the screen may be set to a number greater than 1, such that 1 in. of mouse motion can equal less than 1 in. of cursor motion. If gross and fast input-type motions are desired, the ratio can be set to less than 1. The speed of the cursor can also be related to the mouse sensitivity. Slow mouse motion may give precise control, while rapid mouse motion will give gross control of the cursor. The software to control this rate-sensitive motion must be able to measure how fast the user is moving the mouse itself. One method is to check the x and y coordinates of the mouse on subsequent loops through the program. Large x and y changes indicate rapid user motion, and the cursor velocity can be set accordingly.

The mouse can also be used as a string device. Consider the Apple Macintosh screen in Figure 3.6, in which the keyboard is shown as a window in the screen. By moving the cursor with the mouse and clicking the button when the cursor is positioned on a letter, the letter can be input into the computer program. Commands may be entered this way as strings of characters.

The mouse can also be used as a valuator input, as illustrated in Figure 3.7, where a calculator window is shown. The mouse can be moved around the window shown on the screen, and can be used to input either integer or floating-point numbers. In addition, the mouse can be used as a valuator input device using

FIGURE 3.6   Using a mouse as a string device.

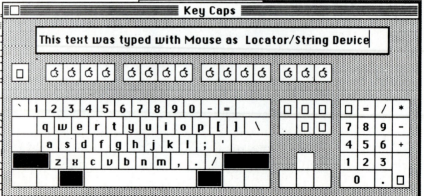

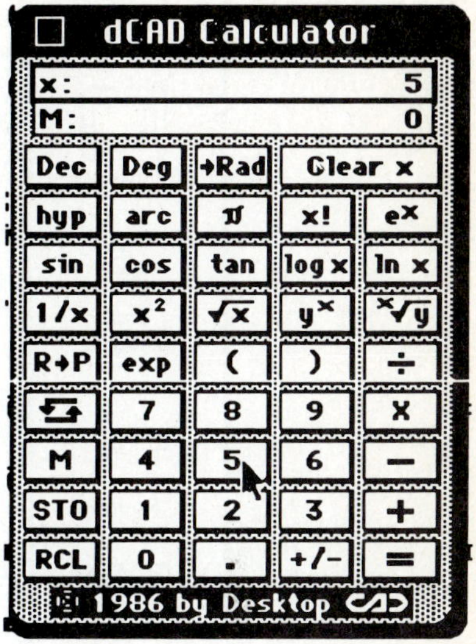

FIGURE 3.7   Using a mouse as a valuator.

active images or icons. One icon usable today is the thermometer gauge. The mouse can be used to move the cursor to grab the sliding thermometer scale and move it up and down to input a specific value into the program. The thermometer window device is shown among others in Figure 3.8.

TRACK BALL. A track ball is a device that is similar to a mouse. It consists of an "upside-down" mouse, in which the ball on the bottom of the mouse has been turned up to the top. By placing a hand on the ball and rolling it side to side or fore and aft, the potentiometers can be made to detect deflections and control the cursor accordingly. In addition, buttons can be added to the track ball. One advantage of the track ball is that it can be used by disabled people with reduced muscular control in their hands because it does not require large input motions.

TABLET. The most common and popular device for computer-aided design geometric input is the digitizer tablet. The tablet consists of a flat surface and a stylus. In moving the stylus along the surface, the sensors can detect the x and y coordinates in the surface. A button can be added to the tip of the stylus for on/off input. There are several types of tablets. Some use a magnetic field to determine x and y coordinates, while others use a strain wave, and still others use acoustical methods to determine the x–y position. An advantage of the acoustical tablet is that it can also be used in three-dimensional applications. Acoustical tablets consist of two linear microphones positioned at 90° to each other and intersecting at the origin of the tablet. In moving a sound-emitting stylus, the microphones detect the time required for the sound to travel from the stylus to each microphone and, therefore, determine the x and y coordinates. With the addition of a third, vertical microphone to the system, it can also detect the z coordinate of the input motion.

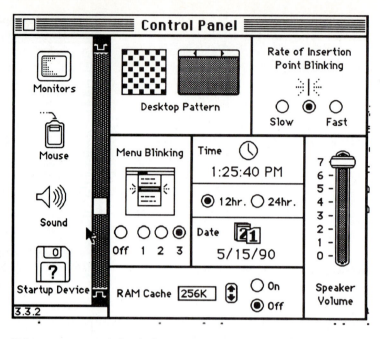

FIGURE 3.8 Using a mouse as a choice device.

Tablets have the advantage of having fairly high resolution, of the order of thousandths of an inch. They are also available in very large sizes for full-size engineering drawings. Variations of the stylus include a device called a *puck*, which consists of a small plastic cross-hair and one to four or more extra buttons. By moving the puck cross-hair around on an engineering drawing on the tablet surface, one can indicate very precisely the coordinates of points on a blueprint or other pattern being input into the computer.

KEYBOARD. The keyboard is the most common input device used for computers. It is, however, awkward for entering geometric data for computer-aided design. It can be used as a string device, a pick device, a valuator, a button, or a locator and, hence, is the most flexible of all the input devices. The keyboard is usually used to enter commands. Commands may be to read the file name for a previously stored part drawing, to input the x–y coordinates for a point, or to describe some other geometry which is to be picked or input into the computer.

DIALS. Dials are used as valuators and can exist in a bank or as a single dial incorporated into the keyboard. The dials can be programmed to input values into the graphics program. For example, one dial may be allotted to each of three types of rotations (x, y, and z) to control the position of an object in three-dimensional space on the screen. By rotating the dials, the values of x, y, or z are changed and the object moved to a new location on the screen. Other examples of dial input control are translations in space and zooming in and zooming out on objects.

Some dials have a user-programmable label. The label can indicate the function of the dial. As the functions of the dials change in the program—for example, from translating an x, y, or z to rotating around the x, y, and z axes—the labels can be changed interactively so the functions of the dials are always clear.

OTHER INPUT DEVICES. Other "on-screen" devices can also be used just as if they were physical devices controlled by the human operator. On-screen calculators, on-screen keyboards, and other active icons such as the thermometer scale and the dial scale can be used as devices to input data into a program.

Three-dimensional input devices exist, such as the previously mentioned three-dimensional tablet. Another three-dimensional input device senses changes in a magnetic field. A magnetic field emitter is set on a table and a sensor is moved through the field. The sensor can detect x, y, and z coordinates in three-dimensional space. In addition, it can also detect roll, pitch, and yaw, the angles around the three coordinate axes. By connecting the six outputs of this device—x, y, z, roll, pitch, and yaw—to the computer program, one can control the position and orientation of an object in space. A device which uses this type of sensor is the data glove (Figure 3.9). The data glove also contains fiberoptic sensors down the outside of each digit to determine the amount of finger flexion. The glove output can be input to a graphic hand simulator and used to manipulate objects in the computer's three-dimensional world.

It can be assumed that three-dimensional input devices such as this will be used for computer-aided design systems of the future. It is easier to input three-dimensional geometry initially than it is to input several views of two-dimensional geometry from the front, side, top, and other auxillary viewing planes.

SUMMARY. The advantages and disadvantages of various physical input devices are summarized in Table 3.1 [11]. This table indicates the advantages and disadvantages in terms of resolution, positioning speed, cost, and fatigue factor.

TABLE 3.1
Input device characteristics

Pointing device	Resolution	Positioning speed	Cost	Fatigue factor
Digitizing tablet	High	High	High	Low
Position joystick	Medium	High	Low	Low
Velocity joystick	High	Medium	Low	Low
Mouse	Medium	High	Low	Medium
Keyboard	Digital	Low	Low	Medium
Light pen	Low	High	Low	High

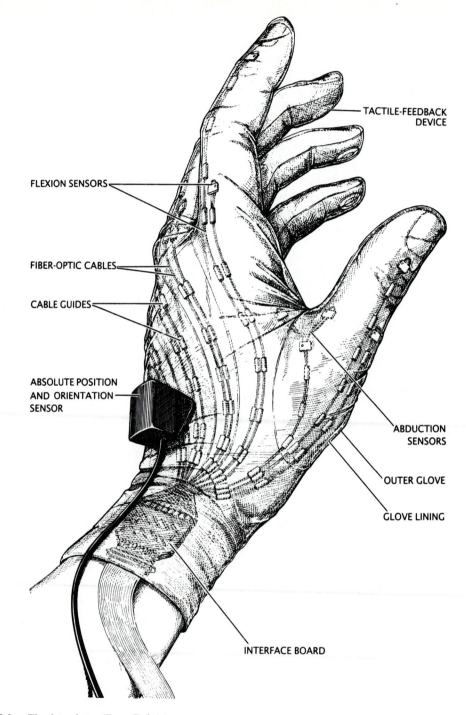

TACTILE-FEEDBACK
DEVICE

FLEXION SENSORS

FIBER-OPTIC CABLES

CABLE GUIDES

ABSOLUTE POSITION
AND ORIENTATION
SENSOR

ABDUCTION
SENSORS

OUTER GLOVE

GLOVE LINING

INTERFACE BOARD

FIGURE 3.9   The data glove. (From Ref. 3.)

3.3.3 Output Devices

Most computer graphics output devices can be thought of as analogous to drawing by hand with a pen or a pencil. This drawing may be done with an actual pen as in the case of the pen plotter, by some electrostatic device such as a laser printer or electrostatic printer, or by an electron beam as in the case of a cathode-ray tube.

The output devices to be discussed in this section include pen plotters, laser printers, electrostatic printers, storage tubes, and plasma screens.

PEN PLOTTERS. CAD results may be simply numbers, in which case a standard ASCII printer will be suitable, but most of the time CAD designs, in terms of geometry, topology, and other types of data, can be conveniently output to either a computer screen or a pen plotter–type device. Pen plotters are the simplest of output devices for computer-aided design.

Pen plotters have been available since the late 1950s, when they were called strip-chart recorders. A pen plotter consists of a device to hold paper and usually two orthogonal carriages which hold a pen. There are three inputs to the pen plotter: an x coordinate, a y coordinate, and a pen variable. The pen variable can specify the pen to be up, in the nondrawing position, or down, in contact with the paper in drawing position. It can also specify other commands, such as the pen color required. The most sophisticated pen plotters have a multiple-pen carousel which can be used to produce multicolor plots.

There are two varieties of plotters: flat bed and drum. Flat-bed plotters (Figure 3.10) use an orthogonal carriage arrangement so that the pen can be controlled using the two carriage motions simultaneously to move from point $x_1 y_1$ to point $x_2 y_2$. A simultaneous motion of the x and y carriages generates a line from

FIGURE 3.10   Flat-bed plotter. (Courtesy of Houston Instruments.)

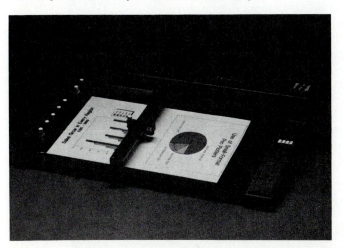

point x_1y_1 to point x_2y_2. One disadvantage of a flat-bed plotter is that the paper size is limited. The cost of a flat-bed plotter is approximately proportional to the size of the paper bed. Alternatively, drum plotters utilize a continuous roll of paper which rolls over the top of the drum. The orthogonal direction of the pen is controlled by a carriage which sits on top of the drum and moves back and forth. A line on a drum plotter requires simultaneous motion of both the carriage with the pen and the rotational motion of the drum. A drum plotter is shown in Figure 3.11. The use of a continuous roll of paper means that the only limit on plot length is the amount of paper available on the roll.

LASER PRINTERS AND ELECTROSTATIC PRINTERS. Electrostatic plotters use a system of electostatic tips or nibs which apply an electrostatic charge to the paper as it runs underneath the printing head. After the nibs contact the paper and leave an electrostatic charge, the paper moves on through a toner bath. The toner coats the paper wherever an electrostatic charge has been placed. The paper then runs between high-temperature, high-pressure rollers to fix the toner to the paper. The paper passes through the rollers only once; therefore, to draw a series of lines—for example, a box of side length L—the lines must first be processed from top to bottom so that the lines can be drawn in one pass as the paper is fed through the printer. This process of transforming random lines into horizontal scan lines is

FIGURE 3.11   Drum plotter. (Courtesy of Houston Instruments.)

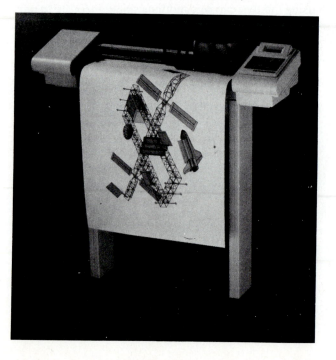

called *rasterization*. Rasterization is also used to display lines on a computer screen. In the case of a computer display, horizontal lines are painted by the electron beam as it scans line by line from the upper left corner to the lower right corner of the screen. In order to draw an angled line on the screen, it must first be determined which scan lines contain parts of the line. A similar process is used in electrostatic printers, so the image must be rasterized before it can be printed.

Laser printers are similar in concept to electrostatic printers except that instead of small nibs applying electrostatic charges to the paper, a laser writes the image onto a drum. The drum then picks up toner and transfers it to paper under high temperature and pressure. A laser printer is also a raster device, so a picture must be rasterized before it is drawn. Pen plotters are not raster devices: They are random-scan devices. Random scan means that the lines can be put on in any order or any sequence. To draw a line at 45°, the pen starts at one point and moves to the other end point of the line.

STORAGE TUBES. A relatively old technology (1960s) is the direct-view storage-tube display. Direct-view storage tubes, or DVSTs, were developed in the early 1960s by Tektronix, Inc. (Beaverton, Oregon) to draw graphical pictures from a computer without requiring the image to be stored in memory. DVSTs write an image which remains on the screen for an indefinite period of time, until the screen is erased. A schematic view of the operation of the DVST is shown in Figure 3.12. The storage grid is sensitive to high-energy electron beams coming from the writing gun. The writing gun writes the image to be drawn on the storage grid itself, behind the screen. Electrons from the flood gun, which are emitted continuously, go through the storage grid where the image has been written and move on to excite the screen phosphors illuminating the screen according to the storage

FIGURE 3.12   Direct-view storage tube.

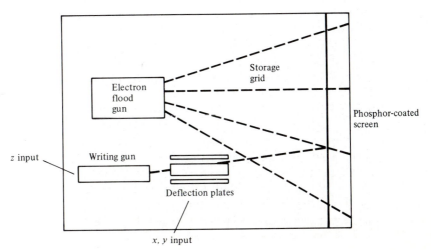

grid information. As long as the storage grid image remains, the screen image remains. To erase the DVST, a voltage must be applied to reset the storage grid. The DVST was a revolutionary device when it was developed, and it is still in widespread use. The main advantage is that the image does not have to be refreshed. In addition, the image can be drawn in random-scan order: It does not have to be rasterized. One disadvantage is that selective erase is not possible. If part of an image needs to be changed, the entire screen must be erased and the image redrawn.

PLASMA SCREENS. One of the newest display technologies is the plasma screen. This device consists of two plates of channeled, gas-filled glass. One plate has horizontal channels etched on it and the other has vertical channels. When the two plates of glass are put together, a small cavity is formed at every point where the two channels intersect. When a high voltage is applied to a horizontal and a vertical channel, the intersection point of the two grooves glows and a point appears on the screen. A line can be drawn from one point to another by sending voltages to the correct combination of horizontal and vertical grooves in the glass. Plasma screens have the advantage that they are flat; they require no electron tube. Consequently, these devices can be hung on the wall or placed on a desk, and require much less space than a standard CRT device. Plasma screens are not in widespread use at this time, but seem to hold some potential for devices which have space limitations and which require fairly high resolution.

3.3.4 Storing an Image

Of the output devices discussed, CRTs are the most commonly used in computer-aided design. The CRT allows users to interact with the screen through the input devices discussed earlier, and allows graphics to be drawn in either random-scan order for random-scan CRTs or in raster order for raster CRTs. CRTs can be monochrome or color, depending on the types of electron guns, the types of phosphors on the screen, and also the amount of information available to draw the image.

The storage of an image in a computer may require millions of bits of memory and depends on the resolution of the device and the number of colors or shades of monochrome intensity available for the image.

FRAME BUFFER. One method of storing an image is the use of a frame buffer [3]. The frame buffer contains n bits of information for each element of the picture. An element of the picture is called a *pixel*, and, for a CRT screen, corresponds to the smallest displayable dot on the screen. For the electrostatic printer the pixel usually corresponds to one nib or one electrostatic plotting point. The variable n determines how many colors or how many shades of gray are available for each pixel. One bit ($n = 1$) of information indicates that the beam or the pen is either on or off and indicates that each pixel will be black or white. Computer screens today which employ this technology are called *bit-mapped screens*. The monochrome screens of the Apple Macintosh, SUN Microsystems, and HP/Apollo computers are examples of bit-mapped screens. Shades of gray can be achieved by blending patterns of black and white pixels together on the screen. The number of colors possible on a screen is 2^n, where n is the number of bits per pixel. For example, 2 bits allows four levels of color, usually black, white, and two intermediate shades of gray. Four bits of information allows 16 different shades of gray, or shades of color, and "true color" displays allow as many as 24 bits of information giving approximately 16 million colors, or shades of gray. The existence of a frame buffer with 24 bits per pixel does not necessarily mean that the CRT hardware device is capable of achieving 16 million different shades of color. The hardware obviously becomes much more expensive when the number of colors is increased, mostly because of the increased resolution necessary for the electron guns or other devices which produce the image on the screen or on the paper. Common values for the number of bits per pixel on a microcomputer such as the IBM PC is three, with eight different shades of color: black, white, red, green, blue, cyan, magenta, and yellow. Some computers allow allocation of frame buffer memory between screen resolution (number of pixels) and number of bits per pixel. Software commands usually allow the user to increase resolution at the expense of number of colors and vice versa.

A frame buffer is made up of enough memory to store all n bits per pixel times the number of pixels on the screen or on the device. High-resolution computer graphics screens today can be as high as 2000×3000 pixels for a total of more than 6 million pixels on the screen. If the device allows eight colors ($n = 3$), then 18 million bits of storage are required for one picture on the screen. The phosphors on a CRT decay after being hit by an electron, causing the display to fade. The screen, therefore, must be refreshed. The human eye can detect a flicker from the phosphor decay if the screen is not refreshed faster than 30 times per second. The computer must process the 18 million bits of information 30 times per second for a total of 540 million bits per second. This translates roughly into 540 million baud, a measure of the transmission rate of data. To give an idea of how fast this data transfer is, computer terminals connected to a host computer transmit data between 300 and 19,200 baud. This incredibly high data transfer is for the ordinary *static* refresh of the device. To change the picture on the screen, potentially all of the bits in the frame buffer must be changed. A new image must be calculated very quickly, and all bits in the frame buffer must be updated to the new picture faster than the human eye can detect the picture changing. If the rate of update of the frame buffer is too slow, the user will see the picture change gradually from one part of the screen to the other as the pixels are updated.

DISPLAY PROCESSORS. Primarily because of this need to keep the screen refreshed and continually update the pixel memory in the frame buffer, standard CPUs were not able to do all of the functions required of the CAD system including interaction with the user, calculation of values, and updating the screen pixel values. The solution to this problem was the development of a device called a display processor unit or DPU.

A DPU is responsible only for refreshing the screen to keep the image visible and managing the input data to change the image on the screen. The CPU, then, is free to perform calculations and interact with the user. In addition to the DPU and CPU, a rasterizer may be present to transform the geometric information from the CPU into a raster image and output it to the frame buffer pixel memory as shown in Figure 3.13. Since the DPU is responsible for updating the screen, the cost of the device is somewhat dependent on the speed of the display processor unit. A low-resolution screen, with a small number of bits per pixel, can be inexpensive. But as the number of bits per pixel increases, and as the resolution of the device increases, the DPU must become more sophisticated and, of course, much faster to keep up with the large amount of data transfer information. A data- and time-saving technique used by some display processor units is interlacing. *Interlacing* is the alternate display of the odd and even lines in two different scan passes. In some cases, the same data is displayed on the even lines as is displayed on the odd lines. For the typical case of 480 scan lines, only 240 lines of information need to be stored for a 50% savings of memory space.

Display processors are becoming more complex. The SUN 3/163 work station contains a graphics buffer board and a graphics processor board. Both consist of VLSI chips to perform DPU-type functions. The graphics processor board contains floating-point transformations and painting processors for color images. The floating-point transformation takes input in the form of *x, y, z* translations and roll, pitch, and yaw orientations and outputs the view of the object from the correct eye point in the correct orientation. The graphics buffer board stores intermediate frames of the image seen on the screen. The painting processor in the graphics processor board takes polygons as input and performs hidden surface removal on these polygons. The advantage of VLSI chips is that standard graphics and arithmetic operations can be done much more quickly than in software. This results in processing that is a step closer to the goal of real-time animation and fast shaded-image generation of three-dimensional objects in space.

The trend in computer-aided design is away from two-dimensional drafting toward three-dimensional solid imaging. It is much easier for a manufacturing engineer to visualize an object in three dimensions with shaded surfaces than it is

FIGURE 3.13   Graphics data pathway.

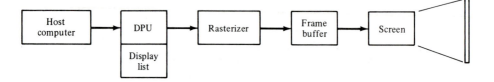

to visualize an object from a set of two-dimensional blueprints. A goal of VLSI chips and work stations is that a minimal amount of information must be sent to the chips to define the image on the screen or on the plot. For example, to display a cube, one alternative is to send a rasterized or pixel image of the object to the screen. The raster image must take into account that certain faces are visible on the cube and certain faces are not visible. It would be easier if, instead of sending the resulting pixel information to the DPU, concise descriptions of the cube faces were sent and the hardware processor could determine which faces are visible and calculate the resulting pixel values. Face polygon descriptions can be concisely transmitted in terms of vertex coordinates, whereas a complete hidden surface view of an object is sent in terms of pixels. Obviously, transmitting the small amount of information concerning the vertices of the polygon is preferable to sending the large amount of information for pixel values.

3.4 CAD GEOMETRY

The first part of this chapter discussed the design process and CAD hardware. But the primary source of information in a CAD system is a CAD data base. This CAD data base contains the description of the part in terms of geometry, tolerance information, material type—in sum, all the information necessary to analyze and manufacture the part. Typically, in the majority of today's CAD systems, the information representing the part is stored as a two-dimensional geometric representation. The user sits down at the terminal and draws lines and circles on a two-dimensional screen, duplicating what would be done on a conventional two-dimensional drafting board. The user can put notations involving tolerances and dimensions on the part. The user may also be able to draw in three dimensions. However, the primary content of a CAD data base in current CAD systems is two-dimensional wire frame geometry.

This section discusses ways to input part geometry and, more generally, a part description. We will begin with computer-aided drafting techniques, then expand to computer graphics in general, discussing two-dimensional object specification and transforms and then moving into three dimensions, where the part is described in terms of either a 3-D wire frame, a 3-D surface model, or a 3-D solid model as discussed in Chapter 2. As you read these sections, keep in mind that the end result of designing a part is the part description data base. This data base should be able to be used for *all* parts of the design and manufacturing process. Design consists of synthesis, analysis, and redesign. The information in the CAD data base must be suitable and complete enough to use in various types of analysis programs. The geometry must be sufficiently robust to transport to a finite-element package, or to a thermal analysis package, or to analyze flow around the object. In fact, one of the best potential uses of CAD data is in performing cost analysis. It must be possible from this CAD data to predict the machining processes and the functional performance of the part, and to forecast sales.

3.4.1 Computer-Aided Drafting

A common way of defining a solid part is by constructing several orthographic views from different viewpoints. Typical views include front, side, and top views of a part, and perhaps an oblique or auxiliary view for extra clarification of geometry. The orthographic technique has the advantage that the part can be scaled to its true size as long as the principal planes of the object are parallel to the paper. Figure 3.14 shows a part in three orthographic views plus a fourth auxiliary view (required because of the angular face). The only way this angular face can be scaled is by drawing a view on a plane parallel to that face.

Two-dimensional views can be drawn using two-dimensional drafting systems. Two-dimensional input devices, as mentioned before, can be used to draw points and lines. The reader is referred to Ref. 7 for a description of CAD input techniques. A complete description of a part, however, requires definition of the surfaces also. Occasionally, drafting systems will allow the user to input the geometry of the surfaces in addition to the edges and vertices. Surface definitions are required for numerical controlled machine tool path calculation. Figure 3.15 shows a computer-aided design system which allows the user to fit surfaces between three-dimensional edges and calculate the tool path with minimal input.

FIGURE 3.14   Orthogonal views of a part. (From Ref. 9.)

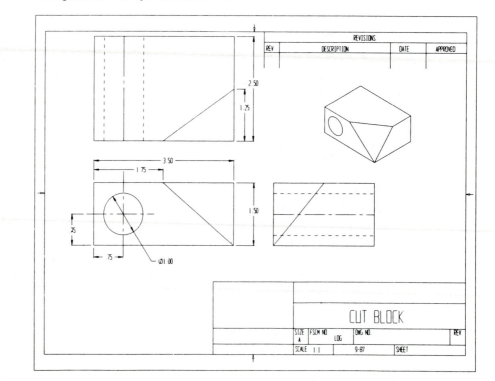

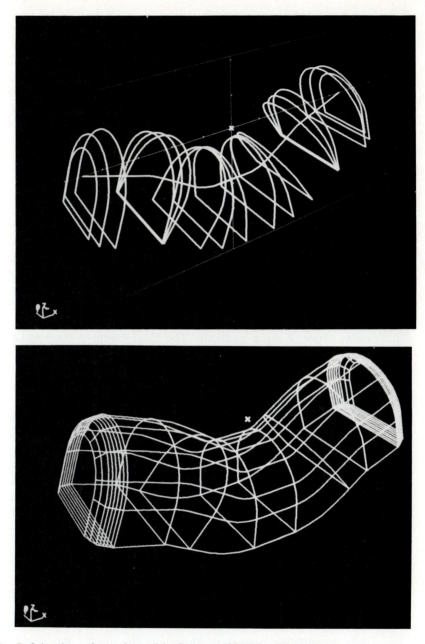

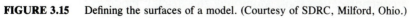

FIGURE 3.15   Defining the surfaces of a model. (Courtesy of SDRC, Milford, Ohio.)

3.5 COMPUTER GRAPHICS
AND THE PART MODEL

Computer graphics is the computer-to-user communication medium used in computer-aided design and drafting systems. An object's definition is primarily geometric; it is convenient to be able to display the geometry of the part on a computer screen or on a plot from any given viewing angle in a variety of sizes. This section explains techniques for 2-D and 3-D graphics and realistic image generation.

3.5.1   Interactive Graphics

Interactive computer graphics is used in five different areas: (1) to modify the display, i.e., to change the view, type of projection, or object attributes; (2) for data entry and data modification; (3) in command and monitoring; (4) for simulation; and (5) for design.

The display can be modified by interacting with the computer (1) to input new viewing parameters for an object; (2) to change the projection type from perspective to orthographic; and (3) to change attributes such as color, line type, line style, surface type, rotation angle, and other attributes pertinent to the definition of geometry. Data entry and data modification can be accomplished using many types of input methods. The most common form is keyboard entry. An example of data entry modification is the creation of a two-dimensional drawing by locating points on the screen and connecting them by lines. By creating many shapes and moving them around the screen with respect to each other, the data can be modified.

A typical graphics application involves electrical circuit design. Figure 3.16 shows an example of an electrical circuit which has been created by connecting points with lines. An alternative way of creating the circuit is to utilize menu items on the screen, representing such devices as resistors or capacitors. A mouse or a joystick can be used to move them into the circuit area and connect them to the rest of the circuit. It may be necessary at some point to modify the circuit and add a resistor or capacitor or to delete a circuit element. To do this, the user must employ a logical picking device to select an object on the screen. By choosing a menu command such as DELETE, the user can erase the picked element from the circuit.

It is important to realize that the graphical image on the screen is only a representation of the real circuit. For analysis purposes details of the connections and the circuit elements must be kept in a data structure. The data structure is used to perform analysis and redesign. The graphic picture, as important as it is, is only a communication tool. Modifying the display modifies the data structure. Changing the resistor on the screen changes the resistor in the data structure. This one-to-one correspondence between data structure and graphical representation must exist if the model is to be useful for further analysis.

Interaction can also be used to issue commands. Typically, commands are entered via the keyboard, but current interfaces allow the user to choose a menu

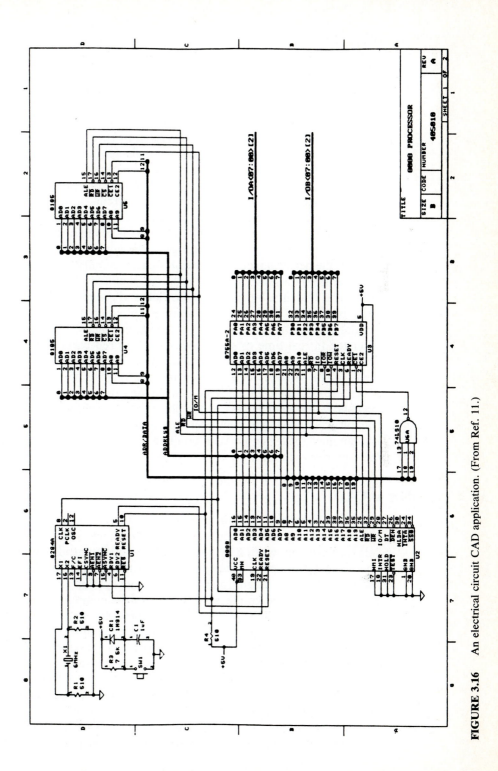

FIGURE 3.16   An electrical circuit CAD application. (From Ref. 11.)

item to initiate a command. In this case, the logical input device would have a picking functionality, and the physical device might be a mouse, a joystick, a tablet, or some other type of device which returns (x, y) coordinates to the program. Monitoring can be accomplished by displaying gauges and values on the screen to show the progress of the program. An example is a display showing the results of a finite-element stress analysis. Graphics may be used to indicate levels of stress on the part by giving high levels of stress a certain color, or by indicating deflections of the part by overlaying a deflected model over the original model geometry. In the case of the electrical engineering example above, monitoring might take place by showing a graphical voltage or current meter on the screen. The meter would fluctuate according to the circuit design parameters.

An example of a simulation interaction is the STEAMER program, written for simulation of steam power plants in naval submarines [9]. The control panel appearing on the screen shows gauges, thermometers, and other devices which indicate the performance of the steam system. Simulation occurs when new trainees operate the system and the program simulates a malfunction in the power plant. The students, by analyzing the dials and gauges on the screen, can come up with solutions to the problem and make the required corrections. Other types of simulations involve factory-floor processes, in which robots and machine tools are animated in real time to show how manufactured parts flow around the factory floor. A very common type of simulation, though a very expensive one, is aircraft flight. Flight trainees can operate the controls inside a simulated cockpit and watch the effects on graphical screens in the cabin. Some of the graphical screens may show dials and gauges, while others may show the simulated view from the cockpit window. Flight simulators are especially useful for training on take-offs and landings, because they can produce a graphical representation of airports on the computer screen.

3.5.2 Graphics in CAD

Every problem in computer graphics and, therefore, in computer-aided design, has three major parts: an application, hardware, and software. An application is a software program written by a user with a specific purpose in mind. One application would be computer-aided drafting. Another application may be finite-element analysis or circuit design. In any case, the application is specific to the particular use intended for the graphics system or for the computer-aided design system.

Hardware includes CRT terminals, pen plotters, CPUs, DPUs, and other hardware that has been discussed so far in this chapter.

The third category, software, includes the programs necessary to receive graphical data and put the required image on the screen. The image may be as simple as a 2-D wire frame, or as complex as a 3-D, real-time, animated, shaded image. The graphics software in the program receives geometric primitives such as lines, points, and surfaces and paints them on the screen or on the output device according to the object attributes. The result of this three-step process (i.e., application, hardware, and software) is graphical output. A block diagram of

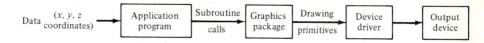

FIGURE 3.17   The layers of a CAD graphics program.

a computer-aided design program can be seen in Figure 3.17. As the diagram shows, data enters the application program. This data consists of numbers indicating geometry, data values, or other information pertinent to the application involved. Out of the application program come procedure calls to graphics routines which put the information on the output device. The graphics software usually consists of a graphics package which generates output primitives or ASCII sequences. The output primitives are in the form of drawing commands such as PLOT, MOVE, and LINE, which are understood by a display device. These output primitives may appear as decimal numbers, binary numbers, or even ASCII sequences of coded data. The output primitives are passed on to the display processor unit and are used to drive the display device.

3.5.3 Two-Dimensional Graphics

In order to draw 2-D images on the display, there must be at least two functions or subroutines available in the basic graphics package. The first is a MOVE(X,Y) subroutine, which moves the pen to an (x,y) location without drawing (with the pen up in the case of the plotter, or the beam off in the case of a CRT). The second command is the LINE(X,Y) subroutine, which moves the beam to an (x, y) location with the pen down, thus drawing a line. Alternative specific instances of these commands appear in various graphics packages such as the GKS package, in which the command may be "POLY_LINE(N,X,Y)," or the Calcomp package, as "PLOT(X,Y,N)." In the Calcomp package, the N indicates a pen code, either up or down, and in GKS the N indicates the number of points in a multiple-line-segment line.

Imagine the user wanting to draw a series of triangles rotated and shifted along an axis (Figure 3.18). There are several possible approaches. One is for the

FIGURE 3.18   Rotated and shifted objects.

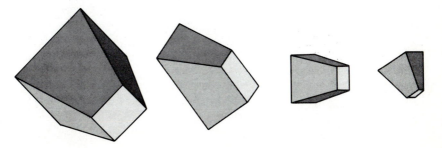

user to calculate the (x,y) coordinates of all the corner points of all the boxes in the row and use the MOVE and LINE commands to draw each box individually. This method may be acceptable for drawing a few boxes. However, if many thousands of boxes are to be drawn, each rotated a specific number of degrees and translated a specific number of units along the x or y axis, the amount of data transferred and calculated soon becomes very large.

An alternative to this individual box-drawing routine is to store the definition of one box defined with respect to the origin of the coordinate system and then use coordinate transformations to rotate, translate, or scale that box and place it in a new location, with a new size and a new orientation. This second alternative uses the concepts of separate object specification and object instancing. The "master" box is defined as an object. Copies of the box, in different sizes, orientation, and locations, are called *instances*.

Object specification is the definition of an object in parametric form. Figure 3.19 shows a picture of a box and the accompanying software code to draw the box using MOVE and LINE commands. Notice that the subroutine input is the parameter L, which is the length of one side of the box. This box is defined with its center at the origin. Therefore, all the MOVE and LINE commands can use relative coordinates to define the box. To draw a box of side length L in a certain location, the program is written first to move the device (either the electron beam or the pen) to an (x, y) location using the MOVE command and then to invoke the box subroutine with the parameter L of the correct dimension. This sequence of commands will draw a box of side length L centered at (x,y) on the screen. To draw a bigger box, a larger value of L would be passed to the subroutine.

FIGURE 3.19   A box-drawing program.

```
box(L , x , y ))
float L , x , y;

{       move(x,y);
        draw_line( -L/2, -L/2 );
        draw_line( L , 0 );
        draw_line( 0 , L );
        draw_line( -L, 0 );
        draw_line( 0 , -L );
        move( L/2 , L/2 );
}
```

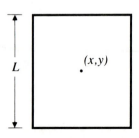

A sequence of boxes scaled and shifted in location can be drawn by using a sequence of MOVE subroutine calls, each followed by box subroutine calls of side length L.

Rotating the box is a bit more difficult, requiring the calculation of new vertex coordinates based on some rotation angle around the origin or around some arbitrary point. It is possible to congeal these three transformations (scaling, rotation, and translation) into a unified mathematical treatment using a homogeneous coordinate and matrix algebra. By applying homogeneous transformations to an x, y, z coordinate in space, the point can be translated, scaled, and rotated about an axis or the origin.

3.5.4 Two-Dimensional Transformations

In order to translate a two-dimensional coordinate (x, y) to a new position (x', y'), two equations can be written:

$$x' = x + \Delta x$$

$$y' = y + \Delta y$$

The Δx and Δy values are the amounts that the coordinate is to be translated in space. This equation can be written in matrix form as

$$\mathbf{P'} = \mathbf{P} + \Delta \mathbf{P}$$

where $\mathbf{P}$ is $\begin{bmatrix} x \\ y \end{bmatrix}$ and $\Delta \mathbf{P}$ is $\begin{bmatrix} \Delta x \\ \Delta y \end{bmatrix}$. By adding the vectors $\mathbf{P}$ and $\Delta \mathbf{P}$, a new vector $\mathbf{P'}$ is found which contains the translated values of the x and y coordinates.

Rotation is a bit more difficult. As shown in Figure 3.20, rotation of a point P to a point P' around the origin must take into account the initial angle around the

FIGURE 3.20 Rotating a point about the origin.

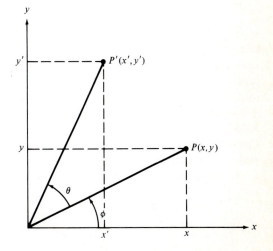

axis ϕ and the change in rotation angle θ. We can write an equation for the x coordinate as

$$x' = R \cos(\theta + \phi)$$

Using a trigonometric identity, the equation is expanded into:

$$x' = R \sin(\theta) \cos(\phi) + R \cos(\theta) \sin(\phi)$$

Recognizing that $x = R \cos(\phi)$ and $y = R \sin(\phi)$, the equation can be restated as

$$x' = x \cos(\theta) + y \sin(\theta)$$

A similar transformation for the y coordinate can be found. Two equations, one for x and one for y, result from rotating point P an angle ϕ around the z axis.

These two equations can be written in matrix form as

$$\mathbf{P'} = \mathbf{R} \times \mathbf{P}$$

where $\mathbf{R}$ is a rotation matrix defined below and $\mathbf{P}$ is the vector containing the original (x, y) coordinates.

SCALING. The scaling equations can be written as

$$x = S_x * x \quad \text{and} \quad y = S_y * y$$

where S_x and S_y are scaling factors in the x and y directions, respectively. For S values between 0 and 1, the point moves closer to the origin; and for S values greater than 1, the point moves farther away from the origin. Negative S values change the sign on the x or y coordinate and move the point to the opposite side of the coordinate axes, giving a mirror image. These scaling equations can be written in matrix form as

$$\mathbf{P'} = \mathbf{S} \times \mathbf{P}$$

where $\mathbf{P}$ is the vector containing the original x and y coordinates, and $\mathbf{S}$ is the scaling matrix as shown below.

HOMOGENEOUS COORDINATES. The above three operations (translation, rotation, and scaling) can be performed using matrix algebra. However, rotation and scaling are matrix multiplication operations, whereas translation is a matrix addition operation. For convenience, all the transformation operations can be made into matrix multiplication operations by introducing the concept of the *homogeneous coordinate*. The homogeneous coordinate is shown in the vector below as value w. w typically has the value of 1, but there are times when its value does not equal 1, and at those times particular care should be taken to retrieve the correct x and y values from this vector. $\mathbf{P}$ now becomes $[x \; y \; w]$.

The scaling matrix $\mathbf{S}$ then becomes

$$\begin{bmatrix} S_x & 0 & 0 \\ 0 & S_y & 0 \\ 0 & 0 & 1 \end{bmatrix}$$

and the rotation matrix **R** becomes

$$\begin{bmatrix} \cos\theta & -\sin\theta & 0 \\ \sin\theta & \cos\theta & 0 \\ 0 & 0 & 1 \end{bmatrix}$$

By adding the homogenous coordinate, we can now write the translation matrix as

$$\begin{bmatrix} 1 & 0 & \Delta x \\ 0 & 1 & \Delta y \\ 0 & 0 & 1 \end{bmatrix}$$

Now each of the transformation operations can be accomplished by multiplying a 1×3 coordinate vector containing the homogeneous coordinate times a 3×3 transformation matrix (translation, scaling, or rotation).

COMBINING TRANSFORMATIONS. Combinations of rotation, translation, or scaling operations can be combined by multiplying the required 3×3 matrices together to get one composite transformation matrix. An example may clarify this last statement.

Consider the box in Figure 3.21*a*. The user wants to move the box to a new position, rotate it around its center, and scale it by a factor of 2 about its center (Figure 3.21*b*). As mentioned before, each of the transformation operations is defined about the *origin*. Consequently, if the object is scaled by a factor of 2, its center point will move twice as far from the origin. In addition, a rotation of 45° around the origin would not only rotate the box clockwise or counterclockwise around its center, it would also rotate the center around the origin. The desired effect of rotating, scaling, and translating about the center can be accomplished by combining matrix operations into one composite 3×3 matrix. The technique is

FIGURE 3.21   Transformations applied to an object.

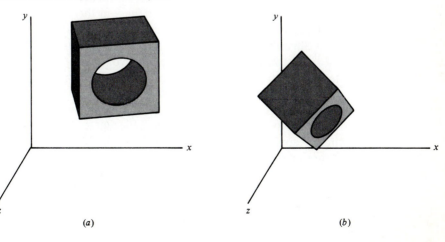

(a)   (b)

first to translate the box so that its center is at the origin. Then a scaling operation by a factor of 2 is performed to increase the size of the box, while leaving its center at the origin. Third, the box is rotated the required angle about the origin with its center remaining at the origin. Finally, the box is translated to the new location. This combination of operations can be expressed as

$$\mathbf{M} = \mathbf{T}_2 \, \mathbf{R} \, \mathbf{S} \, \mathbf{T}_1$$

where $\mathbf{T}_1$ is the translation to the origin and $\mathbf{T}_2$ is the translation from the origin to the new location. Note that the first matrix is placed farthest to the right to be applied first to the point $\mathbf{P}$. Matrix $\mathbf{M}$ is a 3×3 matrix, which, when multiplied by all vertices in the box, displays the box in the newly scaled, rotated, and translated position. This is the general technique for transforming an object about its center or about some desired point other than the origin.

Looking back at the original example of the rotated, scaled, and translated sequence of boxes described earlier, it is apparent that one way to perform this sequence is first to define a master box. The first instance is drawn in the desired location. The second box is drawn by sending the coordinates of the first box through the transformation matrix, which translates, scales, and rotates to give the second box position. Successive transformation matrices are calculated and each time a new instance of the box is drawn according to the scaling, rotation, and translation described by each successive $\mathbf{M}$ matrix. This concept is called *instancing*. A master object is defined in standardized coordinates. In this case the master object is a box of side length L defined around the origin. Each time this box is to appear on the screen, an instance of the box is copied. The master coordinates of that object are multiplied by a transformation matrix to generate a new object which has been transformed by a specified amount.

COORDINATE SYSTEMS. Once an object has been defined in terms of its vertex coordinates in two dimensions and the object has been transformed according to translation, rotation, and scaling matrices, then the job of the graphics package is to display the object on the output device in the correct location. Two types of coordinate systems are used to specify and draw an object. *World* coordinates are the coordinates used to define the object in its original master form. World coordinates can be in meters, feet, miles, micrometers—in fact, whatever coordinate system satisfies a particular application. If the designer is designing an airplane, the coordinates may be in terms of meters or millimeters. If the designer is working on a VSLI chip, the dimensions may be in micrometers. A good computer-aided design system allows the user to design in world coordinates suitable for a particular application. The software then translates to screen coordinates to draw a picture of the object on the screen.

The second set of coordinates are called *device* coordinates. Device coordinates may be in terms of inches on a pen plotter, electrostatic plotting points on an electrostatic plotter, or even pixels on a CRT screen. The *user* should not need to know the particular device coordinates being used. The software typically translates the world coordinates to appropriate device coordinates specified by the user or automatically adjusted to fit on the device. Any design should be able to be drawn or defined in terms of world coordinates appropriate for the application.

These two coordinate systems define two frames of reference. World coordinates can be used to define a *window* or frame through which the user looks to see the world. This window can be thought of as being similar to an empty picture frame which can be held up to view the world. By moving the frame around, different portions of the world can be viewed. The coordinates of the corner points of the window can be defined in world coordinates because the window exists in the world, not on the screen itself (Figure 3.22).

Device coordinates can be used to define a *viewport* or frame on the device in which the view will be drawn. The viewport appears on the screen or on the output device itself. The corner points of the viewport, which is usually rectangular, are defined in terms of the device or screen coordinates. That is, the corner points may be defined in terms of pixels on a CRT, or inches on a pen plotter.

The *window* dimensions and location determine the portion of the scene seen by the observer. The location and dimensions of the *viewport* determine the location on the screen where the picture is to be drawn. Recently, the term *windows* has been applied to certain programs which display information in different regions on the screen. For example, the Apple Macintosh uses the concept of "windows" to display pull-down menus and various applications programs or pictures on the screen as shown in Figure 3.6. Technically, these "windows" are really viewports on the screen. Windows, remember, are defined in the world, and viewports are defined on the screen.

FIGURE 3.22 The window in the 3-D world.

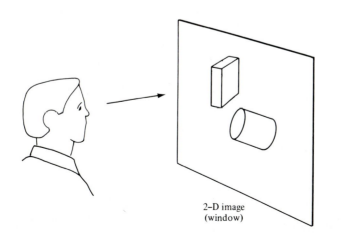

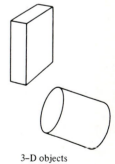

3–D objects
(world)

2–D image
(window)

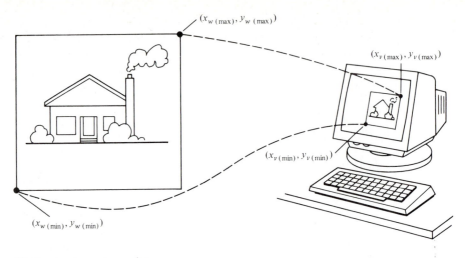

FIGURE 3.23   Window-to-viewport mapping.

Figure 3.23 shows an example of an object which appears in a window and has been plotted on an output device. A mathematical function can be derived which will transform world coordinate data to screen coordinates. As shown in Figure 3.23, to map a point (x_w, y_w) to a point (x_v, y_v) on the screen, one can take advantage of similar triangles, or create ratios of similar distances within the window and the viewport. The resulting mapping functions are shown below.

$$x_v = \frac{(x_{v(max)} - x_{v(min)})(x_w - x_{w(min)})}{(x_{w(max)} - x_{w(min)})} + x_{v(min)}$$

$$y_v = \frac{(y_{v(max)} - y_{v(min)})(y_w - y_{w(min)})}{(y_{w(max)} - y_{w(min)})} + y_{v(min)}$$

It is possible to write these equations in matrix format. Consequently, the scaling, rotation, and translation transformations, as well as the window-to-viewport mapping function, can be written as a sequence of matrix operations. This uniform treatment of coordinate information through matrix operations simplifies the process of taking input data and drawing it on the output device. Matrix multipliers to perform these operations can be incorporated into the hardware of the CPU or the DPU to increase the performance of graphics and CAD systems.

3.5.5   Three-Dimensional Graphics

In the case of three-dimensional graphics, a z coordinate is added to all points. Rotations must deal with angles around the x, y, *and* z axes instead of just the z axis. The following sections will discuss operations necessary to start with a three-dimensionally defined object, place it in space in the correct location, and draw it on a two-dimensional output device. The concepts involve three-dimensional transformations, perspective and orthographic projections, viewing transforms, and window-to-viewport mapping, all in three dimensions.

3.5.6 Three-Dimensional Transformations

The previous section discussed manipulation of objects in two dimensions. Similar concepts apply when manipulating three-dimensional objects in 3-D space. The translation matrix appears as follows:

$$\mathbf{T} = \begin{bmatrix} 1 & 0 & 0 & \Delta x \\ 0 & 1 & 0 & \Delta y \\ 0 & 0 & 1 & \Delta z \\ 0 & 0 & 0 & 1 \end{bmatrix}$$

Notice in this case that the translation matrix is 4×4 in size, including the homogeneous coordinate. Each point in space is defined by a 1×4 position vector, including x, y, z, and the homogeneous coordinate w. An object can be translated in three dimensions by placing the correct Δx, Δy, and Δz values in the matrix and multiplying the coordinates of each vertex of the part through the translation matrix.

The scaling matrix is also similar to the 2-D case. However, in this instance there are three scaling factors: S_x, S_y, and S_z. The reader should keep in mind that the scaling is defined about the origin. So, again, as in two dimensions, to scale an object about its center, it is necessary first to translate the object so that its center coincides with the origin and then perform the scaling operation before translating it back to the desired position.

Rotation in three dimensions requires the development of three separate rotation matrices, one each around the x, y, and z axes. These matrices are now 4×4 instead of 3×3 because of the addition of the z coordinate. The three rotation matrices are shown below:

$$\text{Rot}(x,\theta) = \begin{bmatrix} 1 & 0 & 0 & 0 \\ 0 & \cos\theta & -\sin\theta & 0 \\ 0 & \sin\theta & \cos\theta & 0 \\ 0 & 0 & 0 & 1 \end{bmatrix}$$

$$\text{Rot}(y,\theta) = \begin{bmatrix} \cos\theta & 0 & \sin\theta & 0 \\ 0 & 1 & 0 & 0 \\ -\sin\theta & 0 & \cos\theta & 0 \\ 0 & 0 & 0 & 1 \end{bmatrix}$$

$$\text{Rot}(z,\theta) = \begin{bmatrix} \cos\theta & -\sin\theta & 0 & 0 \\ \sin\theta & \cos\theta & 0 & 0 \\ 0 & 0 & 1 & 0 \\ 0 & 0 & 0 & 1 \end{bmatrix}$$

3.5.7   Composite Transformations
in Three-Dimensions

Many times in computer-aided design, the user desires to manipulate an object, moving it around in three dimensions. One particular application, especially in robotics, is to rotate an object around an arbitrary axis in space. In robotics, this axis would correspond to an axis of one of the robot arm joints. It may be, as shown in Figure 3.24, that the user can manipulate a robot arm graphically by turning an input dial on the computer console. The dial sends a value for the rotation angle to the applications program, which then calculates the new arm position the specified number of degrees around the desired axis. The actual operations taking place inside the applications program are usually transparent to the user. It is sometimes important, however, that the user understand the algorithm used for this type of manipulation.

Imagine rotating the robot arm A, shown in Figure 3.24, $\theta°$ about axis xx. Recall that the rotation matrices allow rotation only about the x, or y, or z axis

FIGURE 3.24   Robot animation: (*a*) robot configuration; (*b*) alignment of A with z axis.

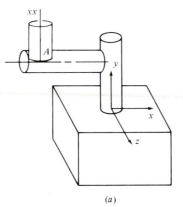

(*a*)

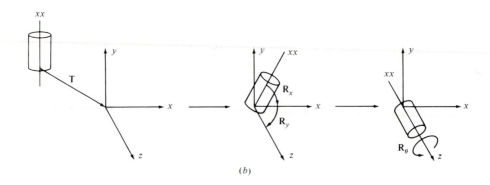

(*b*)

individually. Because of this limitation, a fairly easy way to think about the problem is to divide it into three steps. The first step is to transform the object so that its rotation axis coincides with either the x, y, or z axis. Step 2 is then to rotate the object around the coincident axis the required number of degrees. Step 3 is to move the axis and the object back to the place from which it came. These three steps can be combined into one 4 × 4 matrix.

Step 1, alignment of the object axis with one of the coordinate axes, involves three substeps: one translation and two rotations. First the object must be translated such that a point on axis xx lies on the origin of the fixed coordinate system. This is followed by a rotation around the x axis, and then a rotation around the y axis to align the object axis with the fixed z coordinate axis. This sequence of motions involves three matrices: the translation to the origin, the rotation around the x axis, and a rotation about the y axis. These three matrices can be combined into one matrix as shown below:

$$\mathbf{C_1} = \mathbf{R_y R_x T}$$

Step 2 rotates the object around the z axis a specified number of degrees (θ). This can be done using one 4 × 4 rotation matrix around the z axis.

Step 3 is the inverse of step 1. Three coordinate transformations are used to undo the first three transformations: an inverse rotation around the y axis, an inverse rotation around the x axis, and an inverse translation to put the coordinate axis back to the point from which it came. Notice that this process requires seven matrices:

$$\mathbf{C} = \mathbf{T^{-1} R_x^{-1} R_y^{-1} R(\theta) R_y R_x T}$$

These seven matrices can be concatenated into one 4 × 4 matrix. This 4 × 4 composite matrix ($\mathbf{C}$) is a function of the angle θ around the arbitrary axis in space. All other values are known from the geometry of the situation. For example, it is known how far to translate the axis back to the origin, and it also can be calculated how many degrees to rotate around the x and y axes.

Return for a moment to the situation where the user turns the dials that rotate the robot arm. The dial value is returned to the applications program and used to give a value to the angle θ in the 4 × 4 transformation composite matrix just derived. The arm appears to move when the vertices and lines defining the arm are multiplied through the composite matrix. By placing a loop structure in the program which repeatedly reads the dial and moves the robot arm, real-time animation can be achieved. As the user rotates the dial, the robot arm can move in real time *if* the computer processing capability is sufficiently fast.

The above example is not necessarily the most common application of the transformation matrices. However, it does serve to emphasize that matrices can be combined in many sequences in order to achieve a desired result. It is also important to realize that the result of these transformation matrices can be reduced to one 4 × 4 matrix—a major advantage in using matrix algebra for computer graphics and computer-aided design.

3.5.8   Projection

A difficulty in three-dimensional geometry is that eventually the 3-D coordinates must be mapped to a 2-D output device such as a computer screen or pen plotter. The question arises when going from three dimensions to two dimensions regarding which mathematical operations are necessary. The Greeks dealt with this same problem 4000 years ago, and solved it by a technique called *projection* [10]. *Projectors* emanate from the viewer's eye and extend into the scene. These projectors typically extend from the viewer's eye to each vertex in an object. If a canvas, or a transparent screen, is placed between the viewer's eye and an object, the projectors pierce the screen or canvas at intersection points. The viewer can then draw a picture of an image by connecting the intersection points of the projectors and the canvas in the same order that they are connected in the actual object. The projectors emanate from the eye point of the viewer. This is called *perspective projection*. The result of this technique is that objects that are farther away appear to be smaller, and objects that are closer appear to be larger. This effect gives the perception of depth to the scene. A disadvantage of perspective projection is that objects cannot be measured or scaled in the image. They are not either true size or proportional to one another as in the scene.

Engineering drawings typically use orthographic projection. An *orthographic projection* is one in which the projectors are parallel to one another. Again, if a canvas or a screen is put between the viewer and the object, these parallel projectors will pierce the screen and the intersection points can be connected in the order that the vertices are connected on the actual part. Because the projectors are parallel, all objects are in correct proportion to their actual size. An advantage of orthographic projection is that distances (but not necessarily object dimensions) can be scaled from the drawing or from the image. A disadvantage is that no depth perception is apparent. Objects are the same size regardless of their distance from the viewer or from the canvas.

Projections are similar to transformations in that they can also be handled as matrix operations. The computer graphics viewing pipeline in Figure 3.25 shows that an object can be transformed, projected, and mapped onto the viewport with a sequence of 4 × 4 matrix operations.

The one operation which is required to put 3-D data on a 2-D screen, but which cannot be performed using matrix algebra, is clipping. *Clipping* eliminates those portions of an object from the picture which lie outside the window or

FIGURE 3.25   The 3-D graphics pipeline.

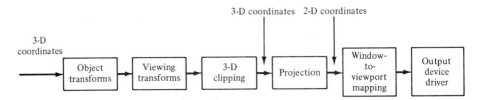

viewport. As might be expected, clipping involves calculating intersections between the lines of an object and the borders of the window or the viewport. Any lines which extend beyond the boundary are clipped off. Only those lines, or portions of lines, that lie within a window or a viewport are displayed.

Detailed discussion of projections and clipping can be found in any good computer graphics textbook.

3.5.9 Realistic Image Generation

One emphasis of today's CAD systems is the generation of realistic images of an object. Image generation involves not only three-dimensional transforms, projection, and clipping, but also image rendering. *Rendering* is the process of adding a lighting model and shading, reflection, transparency, refraction, and other surface and body attributes to a picture. The goal of image rendering is to generate a lifelike picture, as if a photograph had been taken of the object. The advantage of using image rendering is, of course, that the object exists only in the computer data structure. Image rendering bypasses the cost of prototyping objects and photographing them. *Realistic* image generation, however, is a very complex and computationally expensive process. And, although it bypasses prototyping costs, it adds costs of its own.

Consider an application where a part is to be displayed realistically on the screen. The geometry of the part may be represented as a set of planar polygons (see Chapter 2) or as a complete solid model with sculptured surfaces, e.g., Bezier or B-spline surfaces. Each of these representations implies certain algorithms for image rendering. In the case of planar polygons, the graphics system may be able to display them very rapidly by reading only the equation of the plane and a set of vertices which determines the polygon boundary. It is obvious, for example, that the object in Figure 3.26 is defined in terms of planar polygons. The polygon discontinuities are clearly visible. Many graphics work stations can display planar polygonal objects in rapid sequence by calculating the pixel colors using a hardware VLSI chip. Some currently available work stations, for example, can display such an object with 50,000 polygons in $\frac{1}{20}$ s. This makes real-time animation a practical reality. The object can be rotated, scaled and translated, and redisplayed with hidden surfaces removed at this rate. In addition, hardware algorithms are available to perform smooth shading, blending the polygon edges together, by interpolating the colors and intensities over the edges and surfaces.

It is not always possible to represent an object or scene using planar surfaces. Furthermore, the viewer may require certain special effects that are not available with the above techniques, such as generation of shadows from one or more light sources, reflection from shiny or metallic surfaces, refraction of light through transparent or transluscent objects, and removal of hidden surfaces which intersect. Various algorithms have been developed to handle special cases of the above effects, but, for the general case, only one method of rendering is used: ray tracing. *Ray tracing* simulates the path of light in the scene as it moves from the light source or sources and bounces around, off, and through objects until it

FIGURE 3.26   Rendering a realistic image. (Courtesy of Purdue University CADLAB, West Lafayette, Indiana.)

reaches the viewer's eye. In computer graphics, the viewer's eye, viewing window location and size, light sources, and objects are all defined in 3-D space using *x, y,* and *z* coordinates. Ray tracing fires a mathematical ray or line into the scene through every pixel (Figure 3.27). Intersections are calculated between the ray and all surfaces in the scene. The first surface struck by the ray is the surface the eye would see through the viewing window, and the pixel corresponding to that ray is given the correct value or color. For a screen of $n \times n$ pixels and a scene containing *k* surfaces, the maximum number of intersection calculations is n^2 times *k*. If the scene includes reflective or refractive objects, more calculations are required. For each reflective surface struck by a ray, the ray must be reflected according to the laws of optical reflection: The angle of incidence equals the angle of reflection. Assume that several mirror surfaces are present in the scene. As the ray is fired from a pixel into the scene, it intersects one of the reflective surfaces. The ray is reflected according to the law of reflection and may strike another reflective surface, and so on. Multiple reflective surfaces can drastically increase the number of calculations required for an image [5].

Refractive objects are treated according to Snell's law: The angle of refraction or bending of light when light moves from one medium to another is proportional to the ratio of the indices of refraction of the two media. When a ray strikes a surface that is transparent, the light is bent according to Snell's law.

Ray tracing can also be used to generate shadows. As a ray intersects a surface S_1, it can be determined whether that surface is in shadow or light by firing another ray toward the light source or sources. If the ray(s) strikes another surface

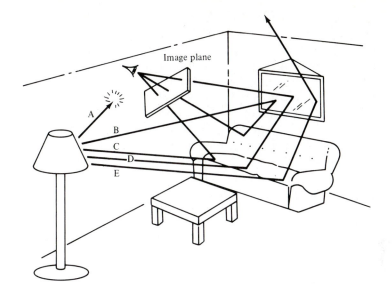

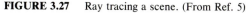

FIGURE 3.27   Ray tracing a scene. (From Ref. 5)

S_2 on its way to the light source(s), S_1 must be in shadow because S_2 is between S_1 and the light.

Each of these embellishments on ray tracing increases the computational effort. Let us look at a simple example of a scene with no shadows, reflections, or refractions. For a high-resolution screen or 2000 × 3000 pixels, and a typical mechanical part with 100 surfaces, the number of potential ray–surface intersection calculations required is 600 million. This is the number of calculations for a static picture! Animation requires that these calculations be performed for each animation frame. For standard video, this must be done 30 times per second, requiring more than 18 billion intersection calculations per second. Each intersection calculation requires several floating-point operations, so this translates into several times the 18 billion intersections per second. It is obvious that a powerful computer is required. Even a supercomputer, though, is capable of less than 10 billion FLOPS.

Ray tracing has the advantage, however, that each pixel calculation is independent of all other pixels. This characteristic makes it possible to use parallel processors to perform ray tracing. A machine with 64,000 parallel processors can perform animation in the above example if each processor is capable of only 300,000 intersections per second—a real possibility. Furthermore, other optimizations are possible, such as determining which surfaces are visible on certain areas of the screen. This bounding technique means that all surfaces need not be considered for all pixels. Surface bounding is especially valuable if a large portion of the screen is the background color and the object takes up only a small portion of the view.

The major point of this discussion is that real-time animation which generates realistic images is possible today only with specialized hardware. Flight simulators make use of parallel processing and efficient polygon data structures to allow real-time animation of scenes from the cockpit. These images are not ray-traced, however, but are generated using planar polygonal objects and shaded by interpolation of colors between vertices and edges. In the future, it is hoped that engineering work stations will also implement this technique to allow the engineer to view designs before the prototyping stage.

Realistic image generation can be used to save time and money in the first prototyping stages of product development. Usually, however, even though animating a realistic image of a part is useful, merely looking at an image of the object is not sufficient for an engineering evaluation. Analysis software must be used to perform functional performance calculations. For example, finite-element software can be used to determine the stress levels in a loaded part; kinematics and dynamics simulation software can calculate the trajectories and forces on an object or set of objects. In some cases, the animation itself must be generated by an analysis package.

The remaining sections of this chapter present some examples of how analysis software can aid the engineer is making design decisions. No attempt will be made to cover all types of computer-aided design analysis software. Each software package is highly application-dependent and is useful only for solving specific engineering problems.

3.6   ANALYSIS

This chapter so far has discussed the design process and the tools needed for computer-aided design, such as hardware, software, and computer graphics. The tools are important, but their value comes in helping the engineer perform analysis of the design model easily, consistently, and correctly. The major emphasis of computer-aided design is not the tools, but the generation of meaningful and accurate results. A computer graphics tool can help make results meaningful by showing pictures of objects or graphs and diagrams of results such as the material stress analysis results shown in Figure 3.28. If shown in color, high stress can be shown in red and low stress in blue, with a continuum of colors in between. The graphics makes it easy to pick out the stresses which exceed the maximum allowable stress in the material.

The analysis, however, is the cornerstone of CAD. The results generated by analysis modules are used by the engineer to determine the feasibility of a design or to optimize or redesign the object. Fundamental to analysis is the model developed for performance analysis of the design. The model is developed to accurately predict the performance based on the design variable values. Typically, sophisticated analysis programs are the performance model of the design. The user does not need to formulate equations which describe the performance or functionality. What the user must do instead is to model the object geometry (see Chapter 2) in an appropriate format.

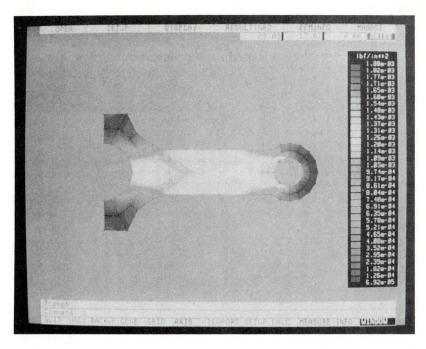

FIGURE 3.28   Color-coded analysis results shown here as shades of gray. (Courtesy of Aries Technology, Lowell, Massachusetts.)

Usually the engineer who is designing a commonly analyzed part or assembly can purchase analysis software to evaluate the design. It is only in the case of an unusual design or performance requirement, where off-the-shelf software is not available, that the engineer must formulate a unique model. In this case, the engineer will usually derive a mathematical model in the form of equations using the laws of physics and write a program to read the design description and calculate the performance parameters. This chapter will briefly discuss performance modeling, but the primary focus will be on the software which contains a model in the form of performance evaluation equations. These commonly available analysis modules read the design description and generate performance results in the form of numbers, graphs, charts, etc.

There are many types of analysis software. The most common are those which save the most time and money and perform calculations too cumbersome or difficult to do by hand. The interested reader can find many examples of computer-aided analysis software and techniques in Refs. 2, 7, and 11–15.

3.6.1 A Design Analysis Example

Analysis is used in the design process to prove the functionality of the part or system being created. Following input of the part definition, using computer-aided drafting or solid or surface modeling, an iterative loop is entered consisting of repeated analysis, design modification, and reanalysis. This cycle continues until either the constraints are met, indicating a satisfactory design, or the criteria are optimized, indicating an optimal design.

Let us look at a design example in order to understand the design process and how CADA (computer-aided design analysis) fits into the picture. Our example will be the design of a small, twin-hulled, ocean-going vessel with the following constraints:

1. The vessel must be no more than 20 ft long and 9 ft wide to allow on-the-road trailering.
2. The range of the vessel must be 3500 miles without refueling.
3. Sufficient rations must be stored for 2 weeks at sea for a maximum of four people.
4. The boat must survive mild sea turbulence without capsizing or fracturing.

The design criteria include:

1. Minimum cost
2. Minimum weight
3. Maximum fuel range

A typical design may have several constraints and criteria. The constraints, as mentioned before, are performance limits. Criteria, on the other hand, are measures of performance used to optimize the design. Occasionally, two criteria may conflict. For example, the minimum-weight criterion will dictate that a high-strength, low-density (and most probably high-priced) material be used for the pontoons. In direct conflict, the minimum-cost criterion says to use the least expensive material possible. Because of these potential conflicts, the criteria should be listed and used in order of decreasing importance. The primary criterion should be the driving force for optimization. Secondary criteria may be brought to bear when necessary. In the boat example, minimum cost is the overriding criterion. Therefore, the high-strength, expensive material would not be acceptable. The better of two equivalent-cost materials, however, could be chosen based on the secondary minimum-weight criterion.

The constraints and criteria suggest specific performance measures. Evaluating these measures is the primary function of the design analysis modules. We can look at the boat design variables in this context.

The size constraint is easy to measure, given the overall design dimensions. The dimensions may be stored in a design data base in the form of a solid model. The travel range is a function of the *volume* of the fuel tanks, the average *speed* of the vessel, and the *fuel consumption* of the engines. Each of these performance measures can be calculated from the design parameters. Recall that the design details exist in the design data base. The fuel tank volume is a function of geometry; the speed is a function of the pontoon drag, the engine horsepower, and the propeller efficiency; and the fuel consumption is defined by the type of engine chosen. The last constraint requires calculating the motion of the vessel in a "mild sea" by solving the differential equations of motion. The strength of the boat is found by determining the stress in the structural members. It is obvious that finding a feasible design requires many analysis modules.

Optimization of the design begins after a *feasible* design is found. It is then that functions for the cost, weight, and fuel consumption are used. The cost is a function of the type of materials and construction techniques used as well as the geometry of the components. Weight is a function of merely the materials and geometry. Fuel range depends on pontoon drag in the water (which is a function of geometry) and the engine characteristics.

It seems obvious that the vast majority of analysis input data can be found in the part or system geometry description. It is useful to take a look at how the geometry of an object or system, in this case the boat, can be stored.

Chapter 2 discussed various types of modeling techniques. Sculptured surfaces such as airfoils and liquid containers can be stored as surface models using one of the surface definition techniques defined in Chapter 2: Bezier or B-spline. The appropriateness of this representation depends on the types of analyses required.

The hulls of the pontoons, then, if defined as bicubic surfaces, would be defined as connected surface patches, each patch defined by 16 control points (Figure 3.29). The surface area of the pontoons can be calculated by integrating over the bicubic patch.

Let us assume that fuel is stored in the pontoons. We now must know the volume of the pontoons to find the volume of fuel that can be stored. Automatic calculation of solid volumes requires that the object be stored as a solid model, not merely as a collection of surfaces. Volume information includes the topology of

FIGURE 3.29 A pontoon hull.

Pontoon surface patches

Each patch defined by 16 control points

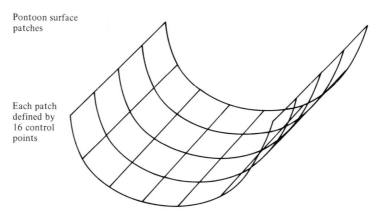

the object and contains data about the solidity, such as which side of the pontoon faces outward. An alternative to storing the exact surfaces is to store a polygonal approximation to the surfaces. A polygonal representation is useful for quick graphics display, but introduces errors in geometric calculations because it is, after all, an approximation.

Tolerances and dimensions are an integral part of the design data base. Tolerances are required to find the range of the craft because the smallest pontoon volume (that resulting from building the pontoon to the negative tolerance limits) should be used to calculate the minimum quantity of fuel available. Tolerances and dimensions are typically stored as drafting notes in an automated drafting system, but are not stored in a solid model. Using the example of the pontoon boat, we can examine a few types of analyses that may be performed to evaluate the design. A list of analyses is presented below:

1. Pontoon volume to calculate amount of fuel and buoyancy
2. Stress in the structural members to ensure no failure in moderate seas
3. Dynamics of the ship when excited by a moderate wave

This preliminary discussion of the boat design points out certain analysis require-ments. The three types of analyses mentioned above will be used as examples of the available modules for design evaluation. Of course, many more types of per-formance analyses are possible, including thermal heat transfer, fluid motion, and electrical circuit simulation.

PONTOON VOLUME CALCULATION. Because solid volume information is needed for this problem, it is advised to construct and store the pontoons, and, in fact, all of the components, as solids as discussed in Chapter 2. A solid model of the pontoon contains the information shown in Figure 2.30. The geometry in-cludes the definitions of surfaces, tracks or curves, and points. The topology contains the connectivity information for the geometric entities. Topology in-

FIGURE 3.30   A labeled cylinder.

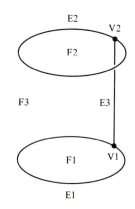

cludes the edge-vertex bounding loops of the faces and information on how these loops connect. As a simple example, assume that the pontoon is cylindrical. This is a poor shape for low drag, but it is an easy solid to model. More complex volumes would be stored similarly, but would contain much more data on the sculptured surfaces.

A cylinder consists of three faces (top, bottom, and a cylindrical side face), several edges, and vertices. We can assume that there are three edges and two vertices in this cylinder (Figure 3.30). The labels show the face, edge, and vertex names. A possible data structure for this object is shown in Figure 3.31. Notice that there are three edge-vertex loops, three faces, three surfaces, three edges, three tracks, three vertices, and two points. The surfaces, tracks, and point nodes represent the part geometry and consequently contain the three-dimensional defi-

FIGURE 3.31   The cylinder data structure.

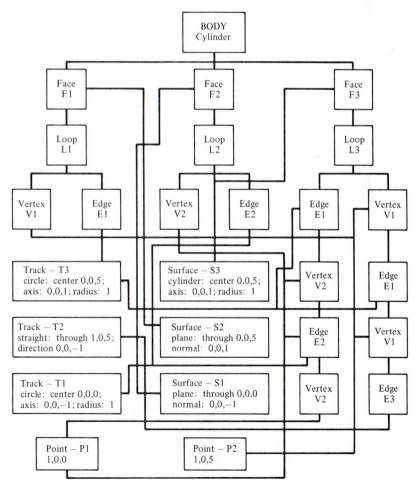

nitions of the entities. For the points, *x, y,* and *z* coordinates are stored. For tracks, the defining coordinates are stored. Note that a straight track can be defined by a point and vector and a circle is defined by a center point, axis direction, and radius.

Surfaces are defined according to the type. Planes are defined by a point and a normal vector; cylinders are defined by an axis direction, center point, and radius. These are merely representative ways chosen here to illustrate a possible data structure for a simple part; there are many alternative ways to define these entities. The best method depends on the applications required. Some definitions are better suited to some types of analyses than others. For example, defining a plane by three points, a common technique, is not as suitable for computer graphics as using a point and a normal direction. Polygon shading requires knowing the normal direction explicitly.

The topology is defined in terms of faces (or bounded surfaces), edges (or bounded tracks), and vertices (the entities required to join two edges at a point). A face may contain one or more boundary loops. Multiple loops indicate that a face has an interior hole such as an entrance to a cavity.

If it is known that this solid is a cylinder, the volume may be easily calculated by the formula:

$$V = \pi r^2 h$$

The radius *r* can be found from the surface data, and the value for the height *h* can be found by subtracting the *z* coordinates for the top and bottom surfaces or tracks.

It is not reasonable, however, to assume that the design system automatically knows that this object is a cylinder unless the user specifically tells the system. A computer knows only what you tell it. If the system cannot identify this object as a cylinder or, in the more general case, if this were not a cylindrical object, the volume would be calculated a different way. In general, the volume of an object can be found by evaluating the following expression:

$$\text{Volume} = \Sigma \left[\int (\text{area} \cdot dz) \right] \text{ for all slices}$$

where "area" is the area of one side of a constant-profile (or infinitessimally thin) slice, and *dz* is the differential height of each slice. This evaluation involves slicing the object into laminae of known approximate volume and then adding them together. The specific method used by a solid modeler itself to evaluate the volume depends on the modeler. In general, however, some form of the above expression is used. The accuracy of this approximation improves by taking more and thinner slices.

The mathematics of volumetric analysis are not important here. Instead, it is important that the reader understand how the geometry is stored and accessed to slice the object and calculate the volume. In order to mathematically slice the object into pieces, sectioning surfaces, usually planes, can be used. The positions of the sectioning planes can be calculated by accessing the geometry of the part. This example cylinder of radius 1, with the bottom face centered on the origin, occupies space along the y axis from $z = 0$ to $z = 5$. Limits or extents of the object can be found by fitting a box around the cylinder. The geometry of the edges, faces, and vertices is available in the data structure. Maximum and minimum x, y, and z values can be found for all entities. These maximum and minimum coordinates determine the bounding box. The bounding box defines the limits of integration or slicing. The bounding box around this cylinder extends from the minimum coordinates $(-1, -1, 0)$ to the maximum coordinates $(1, 1, 5)$.

Now that the limits are known, the orientation of the sectioning planes must be found. One method is always to section parallel to one of the axes. This method works well for objects oriented along one of the coordinate axes, but it is also fine for obliquely oriented objects or for objects with no particular orientation as long as a sufficient number of sections are taken.

The sectioning operation itself uses the model data structure to calculate the intersection edges and new faces after sectioning. It is assumed that the sections are sufficiently thin to be approximated as constant-profile extruded slabs. The volume calculation of a thin slab is

$$V = \text{area of slab face} \times \text{slab thickness}$$

The sectioning algorithm may proceed to find all intersections of all faces with the sectioning plane and find all edges resulting from the sectioning operation. These new edges create a polygon or polygons on the face of each sectioning plane. The area of the polygon can be calculated using a standard geometric polygon area formula. The areas of each sectioning operation are stored and multiplied by each section thickness to yield the laminar volume. Adding up all laminar volumes gives the object volume. The percent error can be estimated by knowing the number of slices taken. The more slices, the better the accuracy, but the slower the calculation will be. The volume for the object in Figure 3.32 was calculated using 99 slices and required approximately 120 s on a VAX 11/780 computer.

FINITE-ELEMENT ANALYSIS. The stress in the members of the pontoon can be found by employing a standard beam stress relationship if the members can be approximated as beams. Most irregular objects, like the pontoon, do not resemble beams and therefore must be treated differently. If a straightforward relationship does not exist for stress in the object, a technique called *finite-element analysis* (FEA) can be used. The theory is that an irregularly shaped object can be subdivided into smaller regular finite elements. Each element, then, can be treated by a standard stress formula, and the aggregate effect is the sum of the effects of all finite elements in the part. We will not get deeply into FEA in this treatment of CAD. We mention it merely because it is one tool (and a popular one) which can be used to analyze the strength of a part. For further information on the theory, the reader is referred to Ref. 12.

Upon entering a standard FEA program, the user is asked either to define the object geometry or to read in a file containing the geometry of the part. Some integrated systems, such as I-DEAS from SDRC, Inc. (Milford, Ohio), link mod-

Output of Mass Properties Module for Part Shown

Stepping pitch : 0.020202 number of steps: 99
Accuracy estimates:
C of G acc: 0.010101
          : 0.50505 percent (of tallest body height)
Volume acc: 0.040404
          : 0.50505 percent (of estimate for tallest body)

Volume: 81.8193
Center of gravity: −0.0,0.0,1.0
Inertia: 322.032 about: 1.0,−0.0
Inertia: 733.1 about: 0.0,1.0
Inertia: 1000.59 about: 0.0, 0.0,1.0
Moment of inertia: 731.128

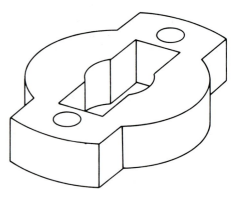

FIGURE 3.32   A mass properties calculation.

ules for object modeling and FEA analysis so that the user does not have to redefine the geometry specifically for the analysis.

In the case of the pontoon, first the user, or the software, if such software exists, must subdivide the pontoon into finite elements. This process is somewhat of an art and is best accomplished by an experienced FEA user. Certain restrictions apply to the number and types of elements chosen. For example, three-dimensional elements work best if they are either tetrahedral or hexahedral in shape. In addition, the most accurate elements have a unity aspect ratio.

Boundary conditions must be applied to the points of the pontoon which support other members of the boat and to the points where it is assumed that water will apply an external force. These forces are specified in terms of both node points on the elements and orientation vectors indicating the direction of the force. The FEA module uses the element definitions and boundary conditions to calculate the applied stresses at each node in the finite-element mesh and also to find the resulting object strain or deformation.

It is this finite-element strength data that the designer can use to determine if the design will perform satisfactorily. The stresses in the material are compared to the maximum allowable stresses dictated by the material used. If the stresses are too high, either the part must be made stronger by adding material, or the material must be changed to one with higher allowable stresses. Finite-element analysis is one of the most commonly used techniques in CAD analysis.

A color image can be generated by calculating the stresses at all finite-element nodes or vertices, assigning each stress to a color in the spectrum, and interpolating the stresses (and colors) along the polygon edges and across the polygon faces. Polygon edges can be smoothed using one of several computer graphics techniques including Phong and Gouraud shading algorithms.

Finite-element analysis is an example showing how computer graphics can be used as a tool in CAD. Another application in which computer graphics is used to show analysis results is kinematics and kinetics analyses.

KINEMATICS/KINETICS ANALYSIS. *Kinematics* is the study of motion. *Kinetics* is the study of why objects move. To put these two terms another way, kinematics deals with movement and kinetics deals with forces and moments causing the movement. These two disciplines become important when a designer wishes to prove the functionality of a moving assembly or analyze the motion of a projectile, rocket, automobile, or other object.

The mathematical analysis of the kinetics and kinematics of an object can be done by formulating and solving equations of motion. In a case of one-dimensional motion of a point object, Newton's second law of motion can be written as

$$F = \frac{m \cdot dv}{dt}$$

where F is the applied force, m is the object mass, and dv/dt is the rate of change of velocity with respect to time. Solving this equation with appropriate boundary conditions and assumptions yields an equation of motion such as

$$F(t_2 - t_1) = m(v_2 - v_1)$$

This equation can be used to solve for v_2, for example, if one knows when the motion begins and ends, t_2 and t_1, the initial velocity v_1, and the applied force F. This is just a simple example of a very simple problem. More complicated analyses, such as studies of objects which are not point masses, must be treated as rigid-body problems, in which case moments of inertia enter the analysis equations (See Figure 3.32.)

The results of motion analysis can be displayed as graphs of motion versus time, velocity versus applied force, or, in fact, any type of cause-and-effect relationship which applies to the problem at hand. Graphical output can show trends and maximum and minimum values at a glance. For example, consider the boat design discussed earlier. It may be desirable to find the maximum amount of pitching or rolling of the boat for certain roughnesses of the sea. The differential motion of the boat can be represented simply as

$$T = \frac{I \cdot d\phi}{dt}$$

where T is the applied moment rocking the boat caused by the wave force (as a function of time), I is the moment of inertia of the boat, and $d\phi/dt$ is the change of the roll angle of the boat with respect to time. Solving this equation requires knowledge of the force of the waves on the boat with respect to time, the moment arm or the distance from the central axis of the boat to the point of application of the wave force, and the moment of inertia.

The results of this analysis will be data giving the roll angle versus time. The CAD analysis routine which calculates this data can give the results in many ways. The obvious technique is to present a graph of ϕ versus time (Figure 3.33). From the graph it is possible to determine the amplitudes and frequencies of the boat oscillations. The designer may want to use this data to redesign the boat—perhaps changing the moment of inertia by redistributing the boat mass either closer to or farther away from the center of the boat. A recalculation will show how the boat response changes.

An alternative to a graph is an animation of the boat as it rolls. By using animation together with a vertical plumb line, the engineer can get a better feel for the boat action. The animation can show the amplitude and frequency of oscillation. In the case of boat building or other design projects in which the engineer may not have a maximum or minimum value for performance constraints, the animation may give a better ''feel'' for a good design than a pile of data or graphs.

One example of the use of animation in kinematics analysis involves the design, or rather the interpretation of a design, of a hacksaw mechanism. The drawing in Figure 3.34 shows a computer-generated image of the mechanism. The blueprints for the parts are complex, and it is difficult to gather from the drawings just exactly how the assembly is supposed to operate. By modeling the components as solid models, applying the motion and connection constraints, and sending the assembly description to an analysis package, it is possible to obtain a real-time animation sequence (Figure 3.35) which shows the expected motion in the

FIGURE 3.33   Roll angle versus time for the pontoon.

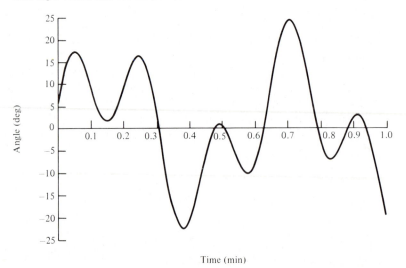

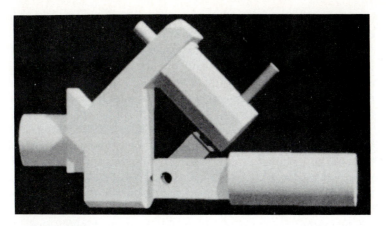

FIGURE 3.34   Hacksaw mechanism. (Courtesy of Purdue University CADLAB, West Lafayette, Indiana.)

FIGURE 3.35   Animated hacksaw mechanism. (Courtesy of Purdue University CADLAB, West Lafayette, Indiana.)

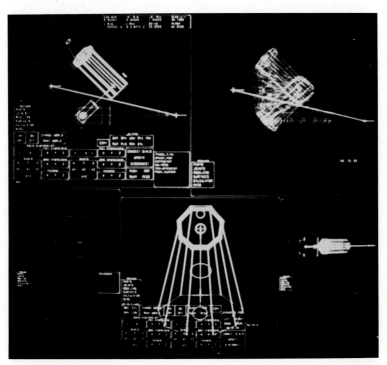

upper right quadrant as a time-lapse-blurred image: rotational input motion at one end and linear oscillatory motion at the output end.

Many times an animation sequence, while not as numerically illuminating as graphs and charts, gives insight into the subtle and not-so-subtle operation of mechanisms. The hacksaw mechanism, for example, may be checked for component interference or undesirable rotational output wobble by examining the animation frames individually.

The importance of geometry in kinematics/kinetics packages is obviously important. In the analysis of the rocking boat, the moment of inertia is important but can be calculated only if the solidity or mass distribution is known. This requires that the boat be modeled as a solid model with complete geometry and topology. A surface model of the boat stores only the collection of surfaces and has no knowledge of the solidity. The hacksaw example requires similar data to calculate the moments of inertia for accurate kinetic analysis. In addition, if interference calculations between assembly components are desired, it is mandatory that the objects be stored as solid models.

The exception to the solid model requirement is the finite-element analysis itself. A meshed object need not be modeled as a solid. FEA requires only the positions of the vertices. It should be mentioned, however, that newer CAD systems and FEA modules will cooperate to mesh the objects automatically in three dimensions. Automatic meshing *does* require a solid model. It must be known where the mass of the object is in order to divide it into finite elements.

3.7 INTEGRATED CAD

This chapter has discussed tools for CAD. The emphasis of the book, however, is integrated CAD/CAM. The integration aspect is accomplished when the design and analysis modules are united into one package. The design geometry entered by the designer can be used by the analysis programs and, as will be seen in later chapters, by the manufacturing, inspection, inventory, and other aspects of the product life cycle. The manifestation of integrating software is recognized when the user can sit down at a computer terminal, design a part, view it on the screen, analyze it for design requirements, and hit the proverbial *manufacturing* button on the keyboard to have the part automatically manufactured, inspected, and delivered.

Various stages of integration do exist, but none is as complete as the example just mentioned. Integration implies automation. Automation implies that the hardware and software can create a CAD data base and use it to analyze and manufacture the part totally automatically, with little or no human intervention. The question as to whether human intervention is advisable is still open. Certainly, human intelligence is valuable in developing a design in geometry and functionality, recognizing certain patterns and shortcomings of products, and formulating manufacturing plans. It has been suggested that artificial intelligence (AI) techniques can substitute for much of the human design/manufacturing input.

AI research has had success in the areas of pattern recognition, planning, natural language understanding, robotic control, fault diagnosis, and other applications. The most well-recognized area of AI is the development of expert systems. *Expert systems* simulate human expertise by storing a knowledge base of data about a specific domain. The architecture of an expert system is shown in Figure 3.36. The *knowledge base* contains knowledge about a specific problem domain and facts stating the problem. The *inference engine* is the program control which searches the facts about the problem to satisfy the knowledge base and solve the problem.

An example of an expert system in the design realm may be a program to specify the geometry for a support beam. The knowledge base would contain formulas and heuristics (or rules of thumb) concerning beam design. The problem facts would state the design parameters, such as the length of the beam span, the expected loads on the beam, and any other conditions which should be considered such as a corrosive environment. The inference engine would then use the formulas and heuristics to come up with a geometry (thickness, length, width) and a material type which satisfies the knowledge base.

The knowledge base may be in the form of rules, relationships, or other formats which allow knowledge to be stored. The source of the knowledge in an expert system comes from human experts. A "knowledge engineer" interviews an expert in the problem domain and formats the information into a form usable by the computer software.

Several expert system software packages called "shells" have been developed commercially. These shells contain the inference engine and methods for storing knowledge. The knowledge base itself, however, is empty, hence the name "shell." A shell can be filled with human expert knowledge and the inference engine invoked to attempt to solve a problem such as the beam design mentioned above.

The reason that expert systems are discussed here is that it is presumed that they will be used in future CAD/CAM integrated systems to perform tasks now done by humans, such as design optimization and planning the steps to manufacture a part. The knowledge base for each problem is specific. In the case of manufacturing planning, the knowledge base must contain information about available machine tools, raw material inventory, market forecasts of lot numbers, and other facts about the manufacturing environment. Of course, a major compo-

FIGURE 3.36   Expert system architecture.

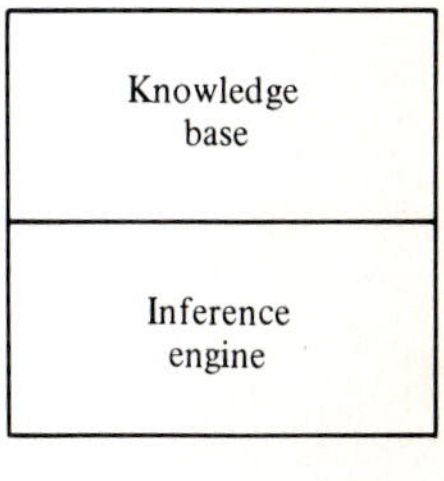

nent of the knowledge base is the geometric and topologic description of the part. Again, we return to the importance of the geometric data base, except that, in this case, the part description must incorporate dimensions and tolerances, surface finishes, and material type. The purpose of CAD is to supply this information to the engineering data base. There exist a myriad of modules in the path from design to shipping the product out the door. An integrated CAD/CAM system must address all modules. The importance of CAD is to prepare the product description for processing through the manufacturing modules and to do it as automatically as possible.

The qualities of a good integrated CAD system may be listed as follows:

1. Interactive, three-dimensional modeling
2. A choice of modeling types, either surface modeling or solid modeling
3. Ability to model assemblies and their connections
4. Capability to edit the geometry, including "undo" operations
5. Interfaces (closely) with analysis tools such as FEA, dynamics analysis, mass and section properties, and other modules
6. Storage of the part description in terms of geometry, topology, dimensions, tolerances, surface finish, and ability to store and attach engineering notes to geometric entities
7. Print out the part description in terms of standard engineering drawing format
8. Ability to design in terms of form features
9. Use of the latest hardware, including high-resolution color graphics output devices and multiple input devices such as dials, mouse, tablet, and joystick (a three-dimensional input device should be available if possible.)
10. Display of realistically rendered shaded images with hidden-surface removal
11. Command entry by various input techniques such as mouse, tablet, or keyboard
12. Automated finite-element mesh generator from the solid model
13. A convenient, window-oriented user interface
14. Networked work station capability to take advantage of programs on other computers
15. Integration with a desktop publishing module to allow engineering reports to be generated directly from the part data base, including text and graphics
16. Real-time animation display at high resolution
17. Capability to design using a multitude of geometric entities including sculptured surfaces
18. Availability to the designer of on-line engineering handbooks and formulas needed for analysis

This is a large wish list. No system on the market today contains all of the above components. Some systems are good from a graphics and solid model standpoint. Others excell in the report-generating or analysis areas. In an integrated system, an object can be designed as a solid model, edited, analyzed for strength performance, and part of the manufacturing plan generated from a single data base. Integration in the CAD arena is progressing. Linking of CAD with CAM is not progressing as rapidly. The remaining chapters discuss the requirements of a computer-aided manufacturing system. Taken together with the requirements in this and the previous chapter on geometric modeling, it can be seen that integrating CAD and CAM is a very large task.

EXERCISES

1. Proceed through the design process for a chair. Use the six steps listed in section 3.2. What are the constraints? criteria? Present a first design and use the criteria to make a better design. What types of analyses do you need?

2. List applications for the design in Exercise 1 that would be good modules in a CAD software package for chair design. Do the same for sailboat design. Use the four types of CAD modules described in Section 3.2.1.

3. If you have a CAD package available for documentaiton (typically a modeling or drafting package), use it to draw your chair design.

4. Investigate the types of computers used in your department and categorize them according to the types in Section 3.3.1. Ask your computer operator to categorize them also. You may notice a difference. The boundaries between categories are always moving and usually unclear.

5. Explain how a keyboard is a versatile input device by showing how it can be used as a string, button, pick, locate, valuator, and stroke device. How many uses has a mouse?

6. Output devices can be classified as stroke or raster devices. Categorize the output devices listed in the text and explain your reasoning. Which output device would be most appropriate when using CAD to evaluate the dynamic motion of an object? to show a full-size plot of the object?

7. Your job is to write a CAD package to interactively model the geometry of an object in three dimensions. Suggest some ways to input the geometric data. Now suggest some input techniques using the data glove.

8. An Apple Macintosh Plus screen is a monochrome display with 512 pixels horizontally and 342 pixels vertically. Draw your view of the screen frame buffer and calculate the size in bits and in bytes. How many bits would the frame buffer be if the scrcen allowed 16 colors? The actual Macintosh frame buffer has addressable pixels from $(-32768, -32768)$ to $(32768, 32768)$. How many total pixels is that? How can it have a frame buffer this large?

9. Make a list of functions you think would be appropriate for a DPU to handle.

10. Create the string of 2-D matrix transformations to take a square of side length 10 centered at $(100, -100)$ and make it three times as large and rotated at a $45°$ angle while leaving its center unmoved. Calculate the new square corner ponts.

11. Assume that you are an engineer designing a VLSI chip in micrometers and you want to view it on the computer screen. You set up a window from the point (0,0) to the point (100,200) in micrometers, and the viewport on the screen is defined between the points (50,100) and (100,150). Map the point on the VLSI chip located at (30,120) to a point in the viewport. What will happen to the shape of the VLSI chip when you map it into the viewport?

12. For realistic visualization, you wish to ray trace a scene of a newly designed automobile placed in a model showroom with spotlights overhead. Explain some problems with ray tracing this object, considering that the car shape is not planar and that it may be somewhat reflective. How are shadows generated?

13. List some other possible analyses that may be important in the boat design in Section 3.6.1. Suggest some other criteria.

14. Use a CAD system to model the pontoons of the boat and generate blueprints for construction. Calculate the volume of the pontoons if your CAD system has the capability to do it.

15. List designs that may have required a finite-element stress analysis. Sketch a cross-section of a part and attempt to construct a mesh of quadrilateals.

16. As expert engineers retire from the workforce or move to other jobs, it is sometimes important to capture their knowledge so that it can be used on subsequent designs. This is especially important for engineers with manufacturing skills. Explain briefly how an expert system could be used in this instance. How would the expertise be captured?

17. List all of the wish list items in Section 3.7 that deal with user-interface design, geometric part information, engineering analyses, computer hardware, and software.

18. Explain how a CAD system can interact with a manufacturing (CAM) system. How should files be transferred if they are separate software packages?

REFERENCES AND SUGGESTED READING

1. Bertoline, G.: *AutoCAD for Engineering Graphics*, Macmillan, New York, 1990.
2. Encarnaçao, J., and E. G. Schlechtendahl: *Computer-Aided Design: Fundamentals and System Architectures*, Springer-Verlag, New York, 1983.
3. Foley, J.: "Interfaces for Advanced Scientific Computing," *Scientific American*, vol. 257, October 1987, pp. 126–135.
4. Foley, J. D., and A. D. VanDam: *Fundamentals of Interactive Computer Graphics*, Addison-Wesley, San Francisco, 1982.
5. Glassner, A. S.: *An Introduction to Ray Tracing*, Academic Press, San Diego, 1989.
6. Goss, L.: *Fundamentals of CAD with CADKEY*, Macmillan, New York, 1990.
7. Groover, M. P., and E. Zimmers: *Computer-Aided Design and Manufacturing*, Prentice-Hall, Englewood Cliffs, N.J., 1983.
8. Hallquist, J.: Personal Communication, Lawrence Livermore National Laboratory, July 1987.
9. Hayes-Roth, F., D. A. Waterman, and D. B. Lenat, Eds.: *Building Expert Systems*, Addison-Wesley, San Francisco, 1983.
10. Hearn, D., and M. P. Baker, *Computer Graphics*, Prentice-Hall, Englewood Cliffs, N.J., 1986.
11. Hordeski, M. F.: *CAD/CAM Techniques*, Reston Publishers, Reston, Va., 1986.

12. Huebner, K. H., and E. A. Thornton: *The Finite Element Method for Engineers*, 2d ed., John Wiley, New York, 1982.

13. Pao, Y. C., *Elements of Computer-Aided Design and Manufacturing*, John Wiley, New York, 1984.

14. Ranky, P. G.: *Computer Integrated Manufacturing*, Prentice-Hall International, Englewood Cliffs, N.J., 1986.

15. Rembold, U., and R. Dillman, Eds.: *Computer-Aided Design and Manufacturing: Methods and Tools*, Springer-Verlag, New York, 1986.

16. Shigley, J. E., and L. D. Mitchell: *Mechanical Engineering Design,* McGraw-Hill, New York, 1983.

CHAPTER
4

CONCURRENT ENGINEERING

*Effective manufacturing systems form a true marriage
of the products and processes that comprise them.*

J. L. Nevins and D. E. Whitney [16]

Design may be defined as "all activities which transform a collection of inputs into a product satisfying a need" [7]. In a more global sense, design is the combination of processes, both economic (as in marketing) and technical (as in machining), that convert raw materials, energy, and purchased items into components for sale to other manufacturers or into end products for sale to the public [16]. Although one may think of the design process in traditional terms as a multitude of separate tasks, each with inputs which go together to generate a design, the design process is really just a *single process* made up of many related tasks including problem definition, idea generation, system modeling, component modeling, redesign, and documentation, as listed in Chapter 3. The design moves from conceptualization through functional analysis to production in a single thread of continuity.

We know that this process can be measured, managed, and improved. Measures of design include functional performance, production cost, production time, and maintainability. We manage the process by using documentation including engineering drawings, analysis results, and meetings among experts in manufacturing, assembly, quality control, finance, and marketing. Often, however, these experts do not communicate often enough, nor is the interaction efficient. It may be that the manufacturing experts are called in *after* the design is finalized. In fact, that is the typical and traditional approach to manufacturing planning. The design is "thrown over the wall" separating the design and manufacturing departments, and the manufacturing experts then have to figure out a way to make a part for which they had no input during the design stage. The same barrier exists between the designers and experts in assembly, field maintenance, quality control, marketing, and other subgroups in a manufacturing facility.

The traditional design process, as illustrated in Figure 4.1, is a serial process in which the design is passed through the various modules; should a design change be necessary, the design is returned to the top and the process is repeated. Note that the marketing experts give their needs to the designers, who determine product specifications, and, in turn, send their product design to the manufacturing experts, who specify the production system to make the design. The manufacturing experts get input from production system experts and make decisions on purchasing new equipment based on their return on the investment. Then, the production system (including fabrication and assembly together with quality control) is designed and a production cost is calculated. The input at the top, therefore, is a set of marketing needs (including a market value), and the output at the bottom is a production cost. If the cost is too high, then the process must be repeated by modifying the design at one of the stages. This iterative process has been carried out routinely in the United States for many years. The various domain experts are almost always located in physically separate departments, and communication among them is sometimes difficult. Modifications may be made in

FIGURE 4.1   The traditional view of design. (From Ref. 16.)

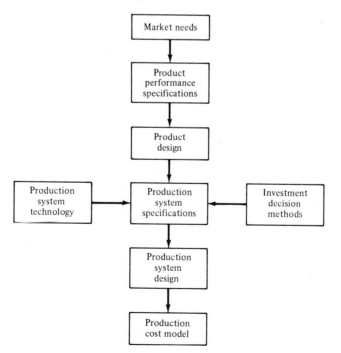

the product design or in the process design. There are usually many ways to manufacture a product, and the process actually selected depends on product function, dimensions and tolerances, available equipment, and costs.

Difficulties with the traditional design process can be characterized by the following factors [8].

- The design is driven by scheduled deliverable data items. There is pressure for drawings and specifications, which leads to a depth-first design search. Design alternatives are quickly eliminated in the interest of time, and one particular idea is pursued.

- The definition of design detail is costly in labor hours. Even with CAD/CAM tools, much manual effort is needed. It would be better to delay this task until the design is somewhat optimized.

- The design process is characterized by a rigid sequence of design decisions. The ultimate goal is usually lowest cost, when the goals should include optimal performance and ease of manufacture.

- Producibility and supportability issues are not considered until relatively late in the process, when a design change may be very costly.

- Production planning, support analysis, maintenance, and reliability are considered separately from the design process. The designers are left on their own to select the particular set of -ilities considered in the first design iteration.

- Design data is fragmented. Documentation includes CAD files, dimensioned component drawings, sketches, process drawings, 3-D solid models, etc. It is difficult, if not impossible, to maintain consistency at all times across these representations.

- Information is lost as the design progresses. The design intent may be lost by the time the documentation gets to the producibility experts. They then must rely on experience and luck in guessing what changes can be made to make the item producible and also functional. Ideally, the reasons for the design features would be included in the design documentation.

- Designers are usually not aware of cost information, so they cannot intelligently set cost reduction as a realistic goal. Companies do not release cost information routinely. There are no tools for estimating costs as there are for other design domains, and when the costs are calculated, it is often too late to make major design changes.

These are some of the many problems faced daily in the traditional design process. And these are some of the reasons why the United States is losing ground in the manufacturing of products to other countries such as Japan and Germany. These countries have realized the problems in current design practices and have modified the process to solve some of these problems.

Reasons given for this traditional serial approach to design include the following:

- Product development is traditionally its own organization and is physically and organizationally isolated.
- Process development and production operations are located together in a different organization labeled manufacturing.
- Most emphasis is on manufacturing operations, that is, shipping product out the door.

4.1 KEY DEFINITIONS

Following are some selected term definitions used in this chapter.

Assemblability   An evaluation of how easily and cheaply a product can be assembled.

Axiomatic design   The use of well-accepted truths (axioms) and corollaries in the concurrent engineering process.

CE (concurrent engineering)   Design of the entire life cycle of the product simultaneously using a product design team and automated engineering and production tools.

Computer-aided DFM   Use of computer tools to apply DFM.

Controllable factors   Those elements that can be controlled during the production process. Examples are dimensions and tolerances and material types.

Design science   The statement that design is a teachable science and not an art. Design science is used to design products by the use of design catalogs relating function to feature.

DFA (design for assembly)   A technique by which a product is designed for ease and economy of assembly.

DFM (design for manufacturability)   A technique by which a product is designed for ease and economy of manufacture.

DFM guidelines   The use of rules of thumb (heuristics) in the DFM process.

Domain expert   An expert in a particular domain of knowledge—for example, a domain expert in the area of process planning.

FMEA (failure-mode evaluation analysis)   Identification and prevention of various modes of product failure. The modes of failure are ranked from most to least impact on part function and then addressed one by one during a redesign process to reduce failures.

Functionality   An evaluation of the functional performance of a product. This includes meeting the functional specifications as determined by the product development team.

Group technology   Assignment of a code to a part which summarizes the pertinent part characteristics.

-ilities   A generic reference to the ease and economy of various stages in the life cycle of the part, e.g., producibility, maintainability, etc.

Inspectability   An evaluation of the ease and economy of a product to be inspected for dimensional and functional conformance to a set of specifications.

Liaison sequence   The establishment of relationships among components of an assembly in order to enumerate all possible assembly sequences for assembly analysis.

Manufacturability   An evaluation of whether a product can be manufactured. Good products should be manufacturable.

Orthogonal arrays   In the Taguchi method, a way of determining an experimental plan by separating the factors to allow experimental analysis of the cause-and-effect relationships of the input factors and performance.

PDT (product development team)   The core of CE; a team of individuals with many different life-cycle domains of expertise, who work together to design a product, the production processes, and, in fact, all aspects of the life cycle of the product.

Producibility   An evaluation of whether a product can be produced. This is also sometimes called *manufacturability*, but is more general and includes all facets of production, not just manufacturing.

Product life cycle   The time from product conception to final product disposal. The life cycle includes production, inspection, and all phases of a product's life.

SAPD (strategic approach to product design)   A concurrent engineering architecture emphasizing a thorough product analysis instead of the use of design rules.

Serviceability   An evaluation of the ease with which a product can be serviced in the field.

Signal-to-noise ratio   A measure of a system's resistance to being influenced by noise. The signal is the measure of performance and the noise is a measure of uncontrollable factors.

Taguchi method   A technique for designing robustness into a product design which establishes design parameters, system parameters, and tolerance parameters.

ULCE (unified life-cycle engineering)   A concurrent engineering architecture developed by the U.S. Air Force that emphasizes the integration of modules including design rules and metadesign knowledge about the designer's intent.

Uncontrollable factors   Those elements which are not controllable, e.g., noise, factors. Examples are the weather and the stock market.

Value engineering   A technique for measuring the quality of a product design as a ratio of performance to life-cycle cost.

4.2   DRIVING FORCES BEHIND CONCURRENT ENGINEERING

The driving forces for a new design approach include both economy and part functionality. Increased competition is coming from Europe and Japan in the form of lower-cost and better-quality products. Some progress has been made recently in design improvements in the United States. The coveted Milan Golden Compass

Award for Design was awarded in 1987 to an American company for its Automated Coagulation Laboratory, a medical analysis instrument. Its winning attributes were ease of use (geometrically coded vial shapes prevent life-threatening mixups) and good functionality while maintaining economical manufacture (the payback came after only 300 units were produced) [21]. This is just one of several recent design awards won by U.S. companies, who are now showing that they can improve products by improving the design process. Other examples are Ford Motor Company for its Taurus/Sable car design and IBM for its dot-matrix Pro-Printer. The Taurus was designed by a radical new approach in the auto industry: multidisciplinary design teams assigned to the product from the outset. These teams could examine all aspects of a subassembly or module before the final design was set. The IBM printer has 65% fewer parts and 90% reduced assembly time compared to its Japanese competitor. This is due to "design for assembly"; simplicity in style is matched by simplicity of assembly. These examples point to a fundamental flaw in the traditional design process. As Lew Veraldi, leader of Team Taurus, says: "Let me describe the traditional product development approach. What you have is each group of specialists operating in isolation of one another. . . . [B]y the time [the design] reaches manufacturing, there may be some practical problems inherent in the design that make manufacturing a nightmare. The people who actually build the vehicle haven't been consulted at all. And, marketing may well discover two or three reasons why the consumer doesn't like the product and it is too late to make any changes" [7].

The problem addressed above is the serial nature of design: One facet of the product is designed at a time. Functionality usually comes first, followed by manufacturing, then assembly, and last serviceability. This sequential approach must be modified if the design process is to be improved.

Other sources [8] have described design as addressing the complete life cycle of the product, from initial conceptualization through prototyping through production through marketing through maintenance to final disposal and scrap.

We must ask, if the design process is to be changed, at what points in the process are there likely candidates for modification? It has been shown that 70% of the production cost of a product is determined during the concept formulation stage (Figure 4.2). Since actual time and expense in product development during this stage is low, any changes at this point cost very little but can greatly affect production costs. Only about 20% of the production cost is affected by changes in later stages such as manufacturing planning of the part. After the conceptual design stage, changes are expensive because the documentation has begun and must be changed. Also, changes at this time in the *process* design do not greatly affect the overall production cost. This means that the conceptual design stage is by far the most profitable to address in order to reduce costs. The conceptual design stage, therefore, must be taken advantage of to create both low-cost product *and* process designs simultaneously.

Addressing all product aspects including cost during the design stage is a new approach to design that has been termed "design for manufacturing: getting it right the first time," or "concurrent design [or engineering] (CE)," or a "strategic

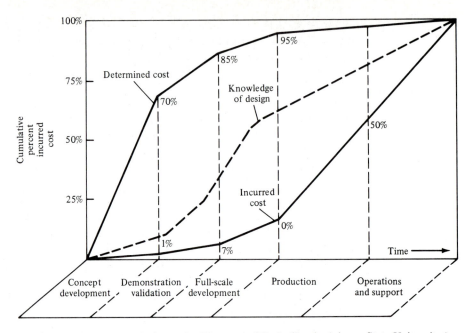

FIGURE 4.2 Costs incurred during the design cycle. (Courtesy of D. L. Shunk, Arizona State University.)

FIGURE 4.3 A better distribution of design-cycle costs.

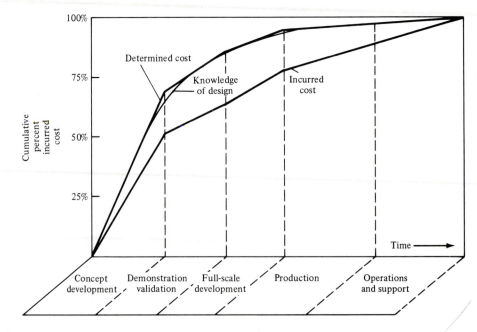

approach to product design (SAPD).'' What we will call *concurrent engineering* (CE) has evolved by thinking about all tasks as elements in an integrated design.

The difference between traditional design techniques and CE is that these tasks are performed, not by individual specialized groups, but by one multidisciplinary *team* of experts or a team of teams in which each domain expert has equal say in the design.

A more desirable distribution of life-cycle costs is shown in Figure 4.3 and is a result of concurrent engineering. Notice that the knowledge of the design from all standpoints occurs sooner in the process. The control of costs still lies in the first section of the design, but all factions of the engineering enterprise have the chance to affect the design before it is finalized.

4.3 THE MEANING OF CONCURRENT ENGINEERING

Concurrent engineering has as its purpose to detail the design while simultaneously developing production capability, field-support capability, and quality. It consists of a methodology using multidisciplined teams to carry out this concurrency: CE tools in the form of algorithms, techniques, and software, and the expertise and judgment of people who make up the complete design and production sequence. The essence of CE is the integration of product design and process planning into one common activity. Concurrent design helps improve the quality of early design decisions and has a tremendous impact on life-cycle cost of the product.

CE can be visualized as illustrated in Figure 4.4 [15]. In this figure, the designer, represented by the hub of the wheel, coordinates the comments and redesign suggestions from each of the domain experts around the circumference. Communication among the experts is indicated by the circumferential arrows. In this design procedure, a conceptual design is presented radially to the group of experts, at which time each can comment on the design relative to his or her own area. Assembly experts consider assemblability problems, process planning experts consider the process sequence, metal removal experts consider the available machine tools, new removal techniques, and the requirements of the design, and so on. The number of domain experts around the rim varies, but the typical domains are:

- Assembly
- Fabrication
- Inspection
- Field maintenance
- Marketing
- Domain-specific engineering functionality

These experts have the mission to conceptualize the product and optimize it until a consensus agreement is reached on the *functionality, producibility,* and *cost*

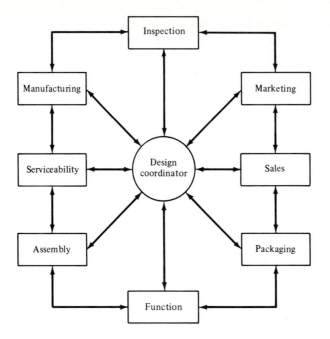

FIGURE 4.4 The concept of concurrent engineering. (From Ref. 15)

constraints. The design moves from the designer out to the experts, who discuss it and suggest design changes to satisfy these three constraints. The design is then passed back to the designer, who resolves conflicts in the suggested changes, modifies the design, and sends it out again for evaluation. Hopefully, the design will need fewer and fewer changes on each iteration until it finally arrives back at the designer with no new redesign suggestions. At this point, the design is considered to be feasible and, in all probability, somewhat optimized.

One can think of CE as accomplishing this purpose using five interrelated elements [16]:

1. Careful analysis and understanding of the fabrication and assembly processes. This allows the designers to predict the performance of the product and select production schemes from among alternative processes.
2. Strategic product design, conceived to support a specific strategy for making and selling the product. The product should be made to marketing specifications for market value, shelf life, and usability.
3. Rationalized manufacturing system design coordinated with product design.
4. Economic analysis of design and manufacturing alternatives to permit rational choices among design alternatives.
5. Product and system designs characterized by robustness. Robustness means resistance to unpredicted noise or errors in production and function. In other

words, the product function is as resistant as possible to variations in dimensions within the tolerance.

The goals of CE within these elements are:

- Avoiding component features that are unnecessarily expensive to produce—e.g., specification of surfaces smoother than necessary, wide variations in wall thickness of an injection-molded component, too-small fillet radii in a forged component, or internal apertures too close to the bend line of a sheet metal component.
- Minimizing material costs or making the optimum choice of materials and processes—e.g., can the component be cold-headed and finish-machined rather than machined from bar stock?

Another way to look at concurrent engineering is to examine the stages a product goes through during the design process. Figure 4.5 shows that product design and system design (including assemblies and production processes) and selling price are all worked out together. Notice that the selling price and production cost targets are known before the design details are developed. Also, note that the product and process are the wings of the concurrent design oval. Included in the oval are the domain experts listed in Figure 4.4. The oval output is a prediction of production cost. This is known precisely, because the product and process are arrived at simultaneously. If the production cost is excessive (and we know that from step 2), then redesign is done beginning with reevaluation of the selling price and production cost target. Notice that, after the details are produced, then the manufacturing system can be constructed. The essentials of the manufacturing system are an output of the CE oval because the processes are designed along with the product.

It is important to note that for this diagram and others in CE, the *design* is the complete product, including all components. The same scheme can be used at a lower level for each component of the total design. It is also important to realize that the diagram in Figure 4.5 represents discussions and compromises among people, all of whom have their own opinions, values, and attitudes. The people aspect of this process is a critical concern. The actual implementation of CE involves human relations and organizational and institutional arrangements which are now only dimly perceived. Many observers are convinced that these concerns outweigh the importance of technology in CE [20].

Several sources list steps for concurrent design [7, 9]. Here is a combined list of goals which are considered in the concurrent design process.

1. From the start, include all domains of expertise as active participants in the design effort.
2. Resist making irreversible decisions before they must be made.
3. Perform continuous optimization of product and process.

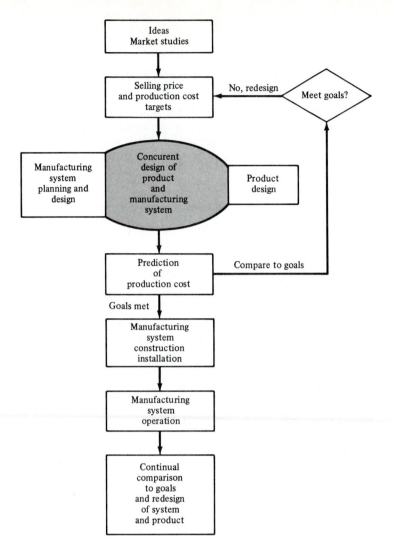

FIGURE 4.5 Product stages during concurrent design. (From Ref. 16.)

4. Identify product concepts that are inherently easy to manufacture.

5. Focus on component design for manufacturing and assembly.

6. Integrate the manufacturing process design and product design that best match needs and requirements.

7. Convert concept to manufacturable, salable, usable design by stating all constraints.

8. Anticipate fabrication and assembly methods and problems.

9. Reduce number of parts.

10. Increase interchangeability between models.

11. Define subassemblies to allow models to differ by the subassemblies.

12. Standardize fastener types and sizes; use low-cost, irreversible fasteners only where a skilled serviceperson would work.

13. Improve robustness of product and process.

14. Identify difficult process steps for which costs and process times cannot be predicted.

15. Use existing processes and facilities so that product yield is high.

16. Break down products and processes into self-contained modules and assembly lines.

17. Adjust tolerances to eliminate failures during assembly.

18. Identify testable areas.

19. Make assembly easier by minimizing setups and reorientations.

20. Design parts for feeding and insertion.

21. Determine character of product; what design and production methods are appropriate.

22. Subject the product to a product function analysis to ensure rational design.

23. Carry out design for producibility and usability study; can these two -ilities be improved without impairing function?

24. Design fabrication and assembly process.

25. Design assembly sequence.

26. Identify subassemblies.

27. Integrate quality control strategy with assembly.

28. Design each part so that tolerances are compatible with assembly method and fabrication costs are compatible with cost goals.

29. Design factory system to fully involve production workers in the production strategy, operate on minimum inventory, and integrate with vendor capabilities.

4.4 SCHEMES FOR CONCURRENT ENGINEERING

Many schemes have been devised to take care of all of the guidelines (and more) listed above. Each of these schemes has a common concern: measuring design quality. Quality can be expressed as in the guidelines, such as number of parts, ease of assembly, size of tolerances, functionality, etc. Measurement of the design is the key to CE. The design itself is considered to be complete when all domain experts sign off on it. That means that each measure of design completeness or sufficiency or quality has been satisfied. The following techniques will be covered in this chapter. They all evaluate different aspects of the design and measure its quality in different ways. Concurrent engineering is the application of a mixture of these techniques to evaluate the total life-cycle cost and quality.

1. Axiomatic design
2. Design for manufacturing guidelines
3. Design science
4. Design for assembly
5. The Taguchi method for robust design
6. Manufacturing process design rules
7. Computer-aided DFM
8. Group technology
9. Failure-mode and effects analysis
10. Value engineering

These will be defined and exemplified in the following sections and summarized at the end of the chapter. In addition, a sample system for injection molding will be shown.

4.4.1 Axiomatic Design

Optimizing a design involves consideration of the complete manufacturing, function, marketing, and maintenance of the product. Optimizing one domain such as manufacturing does not necessarily optimize the product. For example, increasing productivity of drilling holes may mean determining that the holes are not necessary for function and eliminating them. Optimization of the product involves maximizing the overall system productivity.

However, we cannot optimize the entire manufacturing cycle, because we do not have sufficient knowledge of all process details. That requires complete understanding of, for example, the physics of metal removal and tool wear. Because the knowledge base does not exist, we must rely on rules of thumb about what are generally considered to be correct machining practices. These rules are called *axioms*. It has been proposed that "there exists a small set of global principles, or axioms, which can be applied to decisions made throughout the synthesis of a manufacturing system. These axioms constitute guidelines or decision rules which lead to 'correct' decisions, i.e., those which maximize the productivity of the total manufacturing system, in all cases" [23].

Axioms have the fundamental properties that (1) they cannot be proven and (2) they are general truths, i.e., no counterexamples can be observed. Examples of design axioms for optimization are:

1. Minimize the number of functional requirements and constraints.
2. Satisfy the functional requirements from most important first to least important last.
3. Minimize information content.
4. Decouple the separate parts of a solution if functional requirements are coupled or interdependent.

5. Integrate functional requirements in a single part if they can be independently satisfied in the proposed solution.

6. Everything being equal, conserve materials.

7. There may be several optimum solutions.

These axioms are the basis of a complete library of axioms, some general such as those above and some very specific to a particular aspect of the design. The following are corollaries to the axioms, i.e., particular manifestations of the axioms.

1. Part count is not a measure of productivity.

2. Cost is not proportional to surface area.

3. Minimize the number and complexity of part surfaces.

4. If weaknesses can be avoided, separate parts.

5. Use standardized or interchangeable parts whenever possible.

One could imagine designing by applying each of the above axioms and corollaries as the design progresses. This is called *axiomatic design*. These axioms mimic the vast mental catalogs of detailed information kept in the mind of an experienced designer/process planner/assembly planner/marketing consultant, etc.

Use of design axioms in design is a two-step process. The first step is to identify functional requirements that are neither redundant nor inconsistent and arrange them in order from most to least significant. The second step is to proceed with the design, applying the axioms to each design decision. The design must not violate the axioms.

A difficulty with axiomatic design is that the axioms are neither straightforward nor easy to use. They are actually quite abstract and open to interpretation. Consequently, considerable practice as well as on-the-job experience are very helpful. In addition, the axioms offer no help in making decisions between trade-offs. Typically, design alternatives are not strictly good or bad. Compromises must be made. It is often not clear from the axioms how to make these compromises. Also, these guidelines lead designers to stress efficient processing over efficient assembly or other aspects of production. In following the axioms, the designer is encouraged to produce many simple components for an assembly instead of just a few multifunctional parts. This usually increases the cost of assembly, although it keeps down the cost of producing each part in the assembly. An example is the corollary that says "All bend lines in sheet metal should be in a single plane; avoid side holes and depressions." In reality, adding side depressions may allow the piece to be multifunctional, containing fastener holes or other functional features. This will increase the cost of making the sheet metal component, but reduce the number of parts needed, which will in turn reduce the assembly cost.

4.4.2 DFM Guidelines

Design for manufacturing (DFM) is the integration of product design and process planning into one common activity. The goal is to design a product that is easily and economically manufacturable. Consideration of manufacturability guidelines is an essential part of DFM. Examples of these DFM considerations include specification of reasonable-sized fillet radii in forged components, allowance of wide variations in the wall thicknesses of injection-molded components, and matching a hole diameter to available drill sizes.

DFM guidelines are statements of good design practice derived empirically from many years of design and manufacturing experience. They differ from design axioms in that the axioms are self-evident truths which are so obvious as to need no proof. Axioms are generated by examination of a complete process and generalizing principles. Corollaries are applications of the axioms. Alternatively, a guideline is a standard by which to make a judgment. Guidelines are generated by polling designers about specific rules used during design.

These guidelines are based on the fact that 40% of manufacturing cost is labor and non-material-related expenses, such as in-process transport, gauging, and machinery [16]. The guidelines attempt to reduce this portion of manufacturing cost by tapping into the experience of many designers and planners. The result is list after list of guidelines to be used during the design process. These guidelines typically do not address marketing or product function. Given that these two aspects of the design constraints can be satisfied, the DFM guidelines address the processing practices. Many publications have listed design guidelines for ease and economy of manufacture and assembly. The following list has been compiled from these guidelines [12, 22].

1. Design for a minimum number of parts.
2. Develop a modular design.
3. Minimize part variations.
4. Design parts to be multifunctional.
5. Design parts for multiuse.
6. Design parts for ease of fabrication.
7. Avoid separate fasteners.
8. Minimize assembly directions; design for top-down assembly.
9. Maximize compliance; design for ease of assembly.
10. Minimize handling; design for handling presentation.
11. Evaluate assembly methods.
12. Eliminate adjustments.
13. Avoid flexible components; they are difficult to handle.
14. Use parts of known capability.
15. Allow for maximum intolerance of parts.
16. Use known and proven vendors and suppliers.

17. Use parts at de-rated values with no marginal overstress.
18. Minimize subassemblies.
19. Use new technology only when necessary.
20. Emphasize standardization.
21. Use the simplest possible operations.
22. Use operations of known capability.
23. Minimize setups and interventions.
24. Undertake engineering changes in batches.

These design guidelines should be thought of as "optimal suggestions." They typically will result in a high-quality, low-cost, and manufacturable design. Occasionally compromises must be made, of course; in that case, if a guideline goes against a marketing or performance requirement, the next-best alternative should be selected [22].

An example of the use of DFM guidelines is the Nippondenso radiator design shown in Figure 4.6 [16]. Efficient manufacture was achieved by using modular design and interchangeability of components. The various radiator models use many common components, and changing models usually requires changing just one component. There are six parts in the assembly and three variations for each part. That gives a total of 3^6, or 729, different combinations or models. The assembly requires oven soldering of the base onto the heat exchanger. The use of jigless manufacturing allows the two components to be held together by a snap-fit during soldering. This saves the use of expensive fixtures to hold the parts in contact. In addition, to expedite part handling, all grip points on the various part variations are in the same place, so one robot gripper can be used to handle all variations of a part. Because of these considerations, the product is economical to produce and the production rate is a very fast 0.9 s per radiator. They key to this design was the use of DFM guidelines and, possibly more important, the design of the product and the factory simultaneously.

4.4.3 Design Science

An interesting alternative to guidelines and axioms is a concept called *design science*, which proposes that design can be taught by example. Creativity can be simulated by giving a designer a catalog of previously accepted designs. For example, if one is to design a product, a needs statement is developed first, followed by identification of generic functions such as force multiplication (levers), mass flow (pipes, fans, pumps), etc. The catalogs can then be used to find ways to implement these generic functions. The catalogs are arranged by morphology (shape), energy medium, or mechanical type. The resulting products are designed for function only. There is no consideration of manufacturability, and consequently this technique is not conducive to concurrent engineering. Design science has been most popular in Europe, especially Germany.

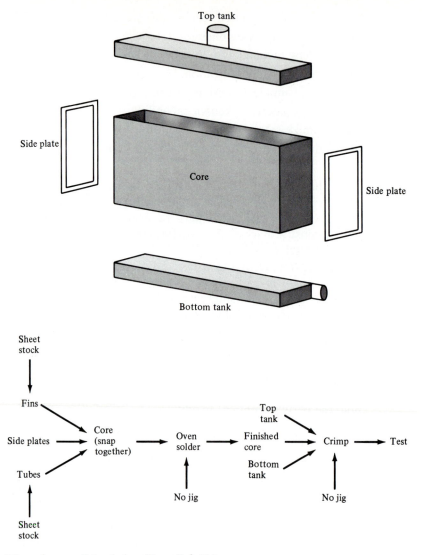

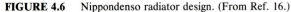

FIGURE 4.6 Nippondenso radiator design. (From Ref. 16.)

4.4.4 Design for Assembly

A subset of the total manufacturing process is assembly of parts. Boothroyd and Dewhurst [5] say that an important key to successful DFM is the process of design for assembly (DFA). It is estimated that a full 50% of manufacturing costs are tied up in the assembly process. Designers typically design for function and maybe even manufacturing, but rarely do they consider the assembly process. The most well-known method of DFA is a set of procedures developed by Boothroyd and

Dewhurst at the University of Massachusetts and later at the University of Rhode Island.

The two parts of DFA include, first, a catalog of generic part shapes and types classified by group technology methods according to ease of feeding by parts feeders and ease of assembly by manual or automatic means. Estimates are given for assembly times. Examples are parts that are assembled by pushing; pushing and twisting; pushing, twisting, and tilting; etc. A designer can estimate the time of assembly for all parts by consulting this catalog. Assembly times usually are directly proportional to assembly costs.

The second part of DFA is a source of rules, advice, or prompting questions concerning good DFA practice. The overall goals are those listed in Section 4.4.2 as DFM guidelines 1, 7, 8, 9, and 10. It is assumed that if these criteria are met, the product design will have achieved some savings in assembly costs.

The technique is manifested in a suite of software modules as described below. The user begins by knowing the design of each of the assembled components and an exploded view of the assembly. The first package assists the designer in determining if the product is a candidate for automatic assembly. The software queries the user as to the values of various parameters that are important in automated assembly. Example parameters include number of parts in the assembly, annual production per shift, capital expenditure, total number of parts for building different product styles, and annual cost of one assembly operator. Other parameters describe the costs of various assembly devices, such as robots and assembly times. The results are presented as a list of costs for using several assembly systems together with estimated costs of manual or automated assembly versus production volume.

Once it is determined which of the assembly methods will be suitable, the user enters a procedure to optimize the assembly itself. Part codes are given based on answers to questions about geometry, function, and foreseen problems, such as part tangling or nesting during feeding. The codes give the designer an idea of assembly costs and possible locations of bottlenecks in assembly or large costs. Finally, the designer is asked questions to determine the theoretical minimum number of parts in the assembly based on relative part motion, different required materials, and repeated assembly-disassembly of certain parts to allow assembly of other parts. The software will tell the user if parts should be eliminated or perhaps combined to be multifunctional in assembly. For example, can the fasteners (screws, pins, etc.) be attached to the housing and incorporated in a snap-fit? This would eliminate parts and will usually reduce assembly costs.

This last step is also concerned with individual part geometries and whether they contribute to ease of assembly. For example, a list of potential part features which create difficulty in simply feeding a part include nesting, tangling, fragility, etc. The system assigns penalty points for these parts in terms of increased manual assembly times, and this in turn increases assembly costs. Automated assembly requires that parts are suited to automatic handling. For parts to be oriented automatically, suitable features such as projections and grooves must be present for the gripper to use. The package realizes that small parts must also be assembled and sometimes cost much more to assemble than large, more expensive die castings. These small items can increase production costs by as much as 40% [3].

As an example, consider the latch mechanism in Figure 4.7, which shows an existing design (*a*) and a proposed design (*b*) for ease of assembly [11]. The new design eliminates 73% of the parts, reduces assembly time by 79%, reduces assembly cost by 79%, and reduces parts cost by 24% for an overall savings of 36% in total product cost. This overall savings was accomplished by removing fasteners (note the lack of screws and nuts in the improved design), redesigning components to be multifunctional, and changing part materials.

Another example, shown in Figure 4.8, is an automotive air conditioning module [16]. The old design includes many fasteners and brackets to mount the actuator, plus a separate shaft and door requiring assembly within the case to make a valve. The redesigned case achieved the goals of combining as many functions as possible into each component. This minimized the number of parts, a tenet of DFA. An especially interesting aspect of the redesign involves the case split line. Note that in design (*a*), holes must be drilled into the sides of the case. These same holes can be assembled into the part as shown in design (*b*). The case split line contains the holes, which are now stamped into each half of the case as semicircles. This allows the insertion of one-piece molded valves instead of the two-piece type. Note also that the actuator bracket is not molded into the case housing. All components were redesigned to snap into place instead of using the traditional nut-and-bolt fastening.

There are other DFA methods besides the Boothroyd-Dewhurst technique. The *assembly evaluation method,* developed by Hitachi, assigns penalty points for each assembly step [22]. A process begins with 100 points and is penalized for the number of separate motions required to achieve a mate together with the

FIGURE 4.7   Xerox latch mechanism design. (From Ref. 11.)

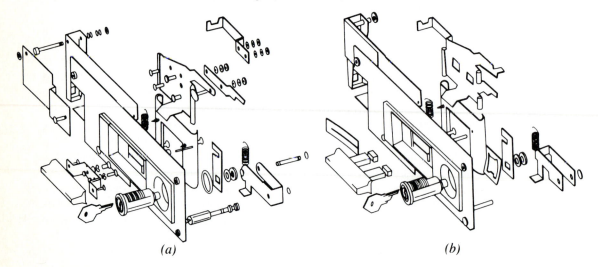

(a)   *(b)*

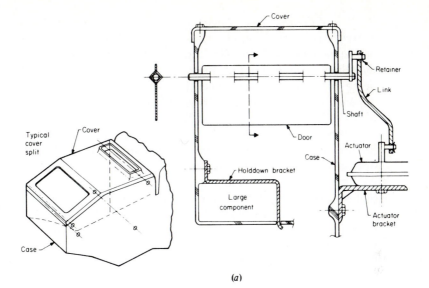

(a)

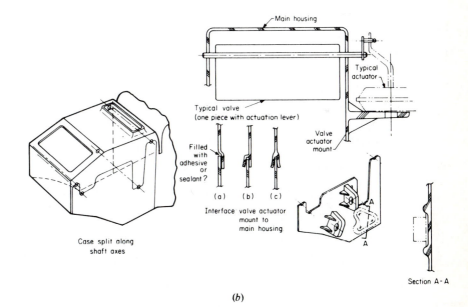

(b)

FIGURE 4.8 Automotive air conditioner assembly. (*a*) Old design. Note fasteners and brackets required to mount actuator plus separate shaft and door requiring assembly within the case to make a valve. Note location of split line between case and cover. (*b*) Redesign. Note new location of cover-case split line, permitting insertion of one-piece molded valves. Also note snap-on actuators and molded-on actuator bracket.

number of motion axes. A twist-and-turn operation would be penalized more than a straight push, for example. Mating parts with less than 90 points after deductions are subject to redesign. This method has been adopted by several companies, including General Electric, Ford, and Xerox. The method has the advantage of being educational in the sense that it sensitizes designers to the fact that cost depends on design. It also gives designers alternatives that are measurable in terms of relative assembly cost [16].

Another important part of DFA is the sequencing of the assembly steps themselves. Given a set of parts to assemble, the problem is to determine an optimal sequence. In this case, the word "optimal" means lowest cost and lowest cost means fewest assembly steps, fewest number of reorientations of the parts (called transformations), and the most foolproof sequence. We can think of a set of parts as having assembly liaisons, that is, relative motions and connections between two mating parts. As an example, consider the rear axle assembly shown in Figure 4.9 [16].

The assembly liaisons are shown on the bottom right of the figure and indicate how the various components are related. What is not shown is the relative precedence relationships among the components. For example, liaison 2 and ei-

FIGURE 4.9 Parts of rear axle and assembly liaisons. (From Ref. 16.)

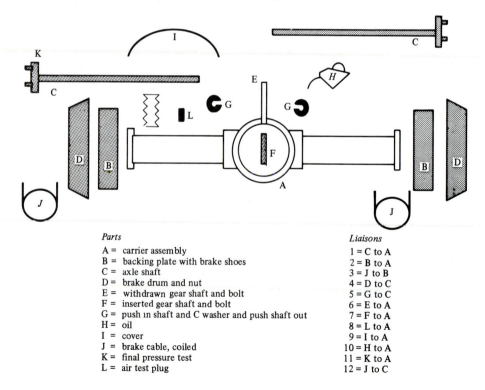

Parts	Liaisons
A = carrier assembly	1 = C to A
B = backing plate with brake shoes	2 = B to A
C = axle shaft	3 = J to B
D = brake drum and nut	4 = D to C
E = withdrawn gear shaft and bolt	5 = G to C
F = inserted gear shaft and bolt	6 = E to A
G = push in shaft and C washer and push shaft out	7 = F to A
H = oil	8 = L to A
I = cover	9 = I to A
J = brake cable, coiled	10 = H to A
K = final pressure test	11 = K to A
L = air test plug	12 = J to C

ther 5 or 6 should be done before liaison 8. That is, the brake shoe backing plates and the gear shaft must be assembled to the carrier assembly before the oil can be filled in the housing. Otherwise, the oil would leak out. Precedence constraints such as these can be written down and used to determine all possible and feasible assembly sequences allowed. A graph of the possible assembly processes can be seen in Figure 4.10. The 12-element matrix represents the 12 liaisons for the assembly, and the black filled squares indicate that a particular liaison has been accomplished. Note that all possible assembly sequences are shown as all possible vertical paths in the graph. There are, in fact, 330 different sequences possible. The problem is to come up with the best assembly sequence based on least cost. Reducing the number of sequences would make this analysis a bit less overwhelming. We can reduce the total number of paths in the graph by making some simplifying assumptions about the assembly process. For example:

1. The axles are to be fastened with C washers following axle-shaft insertion to avoid some unstable assembly states and reduce the risk of the assembly's coming apart.
2. Wrap and attach the brake cables as soon as possible to prevent damage to the cables.
3. Place the cover on immediately after filling the carrier with oil to prevent oil spillage.
4. Insert the plug immediately after the leak check to avoid leakage.

By applying these four common-sense restrictions on the process, the assembly graph becomes as shown in Figure 4.11. Note that now there are only 6 different assembly paths, and analyzing 6 sequences is certainly much more tractable than 330 sequences. Of course, other common-sense constraints could be applied besides the four listed above, and each would give a new set of possible sequences. This technique of establishing liaisons and using a graph to show possible assembly steps can help organize an assembly. The assembly evaluation method described above can then be used to evaluate each path for least cost.

An additional method for establishing an assembly sequence is to determine all possible disassembly sequences and then reverse them. This technique is sometimes easier to conceptualize than the assembly process because of its constrained nature. People can visualize disassembly more easily than assembly.

CASE STUDY: VW GOLF AND JETTA. The Volkswagen Golf was the first automobile from VW that underwent a form of concurrent engineering [16]. In fact, upper management allowed the introduction of the car to be delayed for a year while each part was examined from a processing standpoint. Usually, according to DFA standards one of the primary goals is to design with a minimal number of parts in an assembly. One result of CE on the Golf was the addition of a part in the fender to allow the front of the car to accept the engine and transmission installation by a robot in one quick motion. The engine installation was the bottleneck

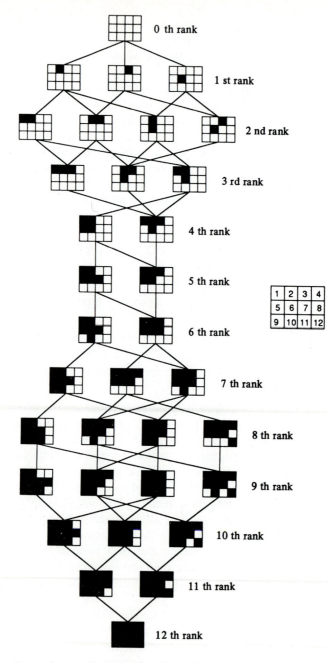

FIGURE 4.10   Rear axle assembly sequences. (From Ref. 16.)

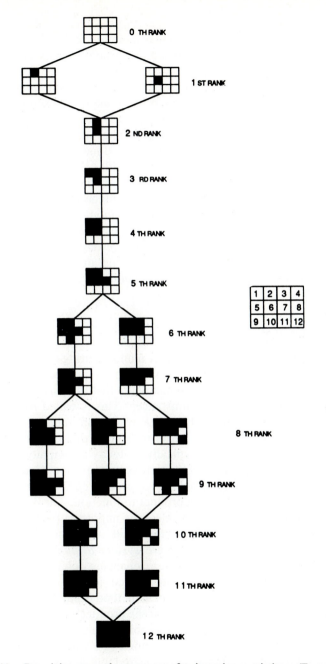

FIGURE 4.11   Remaining rear axle sequences after imposing restrictions. (From Ref. 16.)

operation in most prior car assembly operations, typically taking as long as 1 min. Addition of the bumper part and robotic installation reduced this time to 26 s. That meant the complete assembly line could be based on a 26-s cycle time instead of 1 min.

Another DFA principle is to reduce the cost of each item in the assembly. VW found that the use of a more expensive, cone-tipped screw, which cost 18% more than the standard flat-tipped screw, could speed up the line by going into holes where the two parts were not perfectly aligned. The use of the more expensive screws allowed more of the assembly to be done by robots instead of humans, saving time and money. As an interesting aside, these new screws became so popular in German auto factories that after 2 years the price was reduced to that of the older, flat-tipped screws because of the volume being ordered.

These two examples illustrate that even though procedures are established to save costs and optimize manufacture and assembly, sometimes those procedures must be reevaluated. Two rules of thumb were violated in the above example with the result being a better and lower-cost design. The moral here is that each individual case must be examined separately rather than depending blindly on the rules.

4.4.5   The Taguchi Method for Robust Design

So far we have discussed both design for manufacturing and design for assembly, but we have not addressed how to assure that the product is produced at a high level of quality. That is, we want to guarantee that the part performs its intended function no matter what the circumstances. This is called *robust design* and is addressed by the Taguchi method for optimization of products and processes. The goal is to determine a reasonable set of tolerances that reflect both part function and realistic manufacturability. Typically, in today's company, tolerances are set somewhat erratically and much of the time without reason. Usually the tolerances are set to be in the range that a machinist or process can hold, even if the tolerance is tighter than is necessary for the product to function. The Taguchi method uses statistical design of experiments to select the important parameters of the design and establish tolerances that, at least statistically, produce a good part. Additional components to the Taguchi method are *system design,* the development of a product concept, and *tolerance design,* changing tolerances and properties if parameter design does not produce the desired quality.

A primary component of the Taguchi method is *parameter design,* the determination of product parameters such that the product's functional characteristic is optimized and has minimal sensitivity to noise. Noise in this instance includes all uncontrollable design factors. Typically noise is a variable that is inherently unpredictable yet is large compared to the desired tolerances. An example of noise is the voltage in the outlet when you plug in an electrical appliance. That voltage can vary quite a bit, and the product should be able to be resistant to that fluctuation without becoming nonfunctional.

"The quality of a product is the (minimum) loss imparted by the product to the society from the time the product is shipped" [6]. This statement from Dr. Genichi Taguchi expresses a philosophy that products must minimize this loss to society to be of good quality. To do this, they must be functional even when exposed to noise. This loss function is expressed by Taguchi as a parabola, where the minimum is the target performance which deviates from ideal with deviations in the value of the quality characteristic. In the voltage example above, as the performance deviates from the nominal acceptable value, the loss to society (and the consumer) increases. The goal of parameter design is to minimize this loss by making the product functionally acceptable under variations in design parameters. The key is to identify the parameters which have the greatest effect on product performance and design the product to desensitize the function to variations in the parameter. To continue with the voltage example above, consider that a company wishes to market a television in both the United States (where line voltage is 105–120 V) and Europe (where line voltage is 210–230 V). The designer identifies a particular design parameter, in this case perhaps the value of a specific resistor in the voltage-regulation circuit, as being the primary component which is sensitive to input line voltage and causes malfunction if European voltage is applied. The Taguchi method uses statistical experiments to identify a value and tolerance for that resistor which makes the television less sensitive to input line voltage and, therefore, tolerant of either U.S. or European line voltage. The television works in either environment with the same circuit components. This design is now said to be robust with respect to line voltage.

The following discussion outlines the parameter design procedure by going through an example from a company having difficulty with a large kiln used to make tiles [6]. The job is to make the kiln functional and resistant to variations in operating conditions. The cause of unsatisfactory tiles being produced was determined to be uneven temperature inside the kiln. This uneven temperature was a fact of life in the kiln and had to be dealt with by desensitizing the tile quality to the temperature variation (the noise). Seven possible factors were identified as affecting tile quality: limestone content, fineness of additive, content of agalmatolite, type of agalmatolite, quantity of raw material, content of waste return, and content of feldspar. By conducting experiments in which these factors were varied and the quality of tile was evaluated, it was determined that the most significant factor was the content of limestone. The key to the analysis is to statistically separate the effects created by each of the individual factors by plotting the factors orthogonally. As a result of this analysis, changing the limestone content produced tiles for which the rate of unacceptable tiles produced went from 30% to less than 1%.

In another experiment more germane to manufacturing of parts, an experiment was conducted to determine a method to economically assemble an elastomeric connector to a nylon tube and deliver an optimal pull-apart performance in engine components [6]. The two goals were to minimize assembly effort and to maximize pull-off force. Four controllable factors and three noise factors were

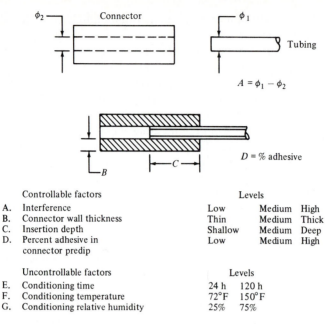

Controllable factors		Levels	
A. Interference	Low	Medium	High
B. Connector wall thickness	Thin	Medium	Thick
C. Insertion depth	Shallow	Medium	Deep
D. Percent adhesive in connector predip	Low	Medium	High

Uncontrollable factors		Levels
E. Conditioning time	24 h	120 h
F. Conditioning temperature	72°F	150°F
G. Conditioning relative humidity	25%	75%

FIGURE 4.12 Controllable and uncontrollable factors in the connector problem. (From Ref. 6.)

identified. These are shown in Figure 4.12. The time, temperature, and humidity were considered to be uncontrollable because they were difficult to control during the assembly process. Alternatively, the interference, wall thickness, insertion depth, and adhesive were easy to control. A series of experiments were performed varying both the controllable factors and the uncontrollable (noise) factors. One way to perform the experiments is to run tests with all possible combinations of all factors. If each of N factors has M possible values, then the number of tests to include all possible combinations is

$$\prod_{i=1}^{N} M_i$$

In the case here, where $N = 7$ and four of the factors have three possible values while the other three have two possible values, the total number of experiments is 648. This is probably an unmanageable number of experiments. The Taguchi approach uses a statistical experimental technique to give, in this case, 72 experiments followed by a statistical analysis to determine the effect of each factor on the pull-off strength.

Two experimental plan arrays were set up to determine what values of the factors should be used for each experiment. The arrays are shown in Figure 4.13. These two arrays were then combined to form a plan called an *orthogonal array*. The term orthogonal implies that this array can be used to determine the effect of each factor independently. Orthogonal in this case means independent. The lower

L9 ARRAY and Experimental Conditions

L9	A	B	C	D	Interference (a)	Wall Thickness (b)	Ins. Depth (c)	Percent Adhesive (d)
1	1	1	1	1	Low	Thin	Shallow	Low
2	1	2	2	2	Low	Medium	Medium	Medium
3	1	3	3	3	Low	Thick	Deep	High
4	2	1	2	3	Medium	Thin	Medium	High
5	2	2	3	1	Medium	Medium	Deep	Low
6	2	3	1	2	Medium	Thick	Shallow	Medium
7	3	1	3	2	High	Thin	Deep	Medium
8	3	2	1	3	High	Medium	Shallow	High
9	3	3	2	1	High	Thick	Medium	Low

L8 ARRAY (noise conditions) and measured data

	8	7	6	5	4	3	2	1	S/N Ratio (db)
E	2	2	2	2	1	1	1	1	
F	2	2	1	1	2	2	1	1	
E x F	1	1	2	2	2	2	1	1	
G	2	1	2	1	2	1	2	1	
E x G	1	2	1	2	2	1	2	1	
F x G	1	2	2	1	1	2	2	1	
Cond. Time (E)	120h	120h	120h	120h	24h	24h	24h	24h	
Cond. Temp. (F)	150F	150F	72F	72F	150F	150F	72F	72F	
Cond. R.H. (G)	75%	25%	75%	25%	75%	25%	75%	25%	
1	19.1	20.0	19.6	19.6	19.9	16.9	9.5	15.6	24.025
2	21.9	24.2	19.8	19.7	19.6	19.4	16.2	15.0	25.522
3	20.4	23.3	18.2	22.6	15.6	19.1	16.7	16.3	25.335
4	24.7	23.2	18.9	21.0	18.6	18.9	17.4	18.3	25.904
5	25.3	27.5	21.4	25.6	25.1	19.4	18.6	19.7	26.908
6	24.7	22.5	19.6	14.7	19.8	20.0	16.3	16.2	25.326
7	21.6	24.3	18.6	16.8	23.6	18.4	19.1	16.4	25.711
8	24.4	23.2	19.6	17.8	16.8	15.1	15.6	14.2	24.832
9	28.6	22.6	22.7	23.1	17.3	19.3	19.9	16.1	26.152

EXPERIMENTAL CONDITIONS

FIGURE 4.13 Experimental arrays to test pull-off force. (From Ref. 6.)

right portion of the table gives the results of the tests in terms of pull-off force for each combination of factors shown. For example, one can see that with low interference, thin wall thickness, shallow insertion depth, and low adhesive, and for time of 120 h, temperature of 150°F, and 75% humidity, the pull-off test gives a result of 19.1 lb. The controllable and noise factors are placed in different arrays so that the experiment can calculate the ratio of how the controllable factors affect the result when compared to the noise factors. This calculation is shown in the column to the far right (Fig. 4.13).

Use of this experimental method gives researchers a way to calculate the best combination of controllable variables to give maximum pull-off strength. The actual analysis to determine the effect of each of the factors on the pull-off strength is beyond the scope of this text. The reader is referred to Ref. 6 and a text on design of experiments. The results of the analysis, however, are worth discussing. Figure 4.14 shows the signal-to-noise ratios for each of the four controllable factors. The goal is to have a high signal-to-noise ratio, which means that the noise factors do not have as much control on the results. It can be seen that a good value for factor (a) is "medium" to resist noise; for factor (c), "deep" is a good value to resist noise factors. Factors (b) and (d) are equally sensitive to noise for all three of their levels; there is no optimum. In this case, choose the value that gives the lowest-cost design. Other types of analyses that can be performed are mea-

FIGURE 4.14 Signal-to-noise ratios for the controllable factors: (a) interference; (b) wall thickness; (c) insertion depth; (d) percent adhesive. (From Ref. 6.)

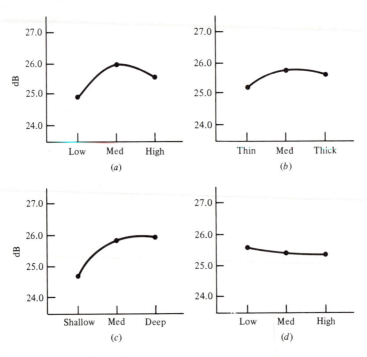

sures of interaction between two factors. For example, in this case it was determined that a medium value for interference was best (gave the greatest pull-off force) throughout the range of humidities (Figure 4.14).

The final step in this parameter design analysis is the confirmation of results. It may be that the optimal values of the factors which give the highest pull-off force were never tested together. That is because all combinations of values did not need to be used in this statistical analysis and, in fact, the optimal combination of values may never have been tested. Therefore, to check the results, experiments should be performed using the optimal values of the controllable factors together.

In summary, the Taguchi method of parameter design helps to produce a robust product which is minimally sensitive to noise factors and consequently should produce more consistent quality.

4.4.6   Manufacturing Process Design Rules

A key goal for manufacturers is to make designers aware of process-related constraints during the design process, before the design is irreversibly put down on paper [22]. Process design rules are typically process-specific guidelines or rules of thumb specialized for a particular industry. For example, the automotive industry would use process design rules for sheet metal forming to produce car bodies. Rules of thumb would include minimum radii of curvature, maximum angles of bend in a die, and the relationship between forming and rippling. The designer would consult these rules during the design process, before it is too late or expensive to incorporate DFM considerations.

4.4.7   Computer-Aided DFM

Design rules, DFA, Taguchi, and the other techniques can be implemented on the computer as a toolkit for designers. These usually are manufacturing method specific because of the great amount of knowledge needed to advise a designer on even one facet of manufacturing. An example would be a toolkit to aid the designer with parts to be made by the injection-molding process. Such a system has been proposed and developed. The effort [14] has produced a computer tool to guide a designer in design of injection-molded components by allowing sketching input to define the geometry, providing a user interface to answer queries about material type, gate and sprue placement, and information about specific features of the part.

The essence of this toolkit is a matching calculation between the desires of the product user (functionality), the process (manufacturability), and the design itself. A knowledge base contains information in the form of design rules together with a weighting and ranking scheme which gives an overall compatibility score. The various windows shown in Figures 4.15–4.18 allow the designer to define the product in sufficient detail for this score to be computed automatically.

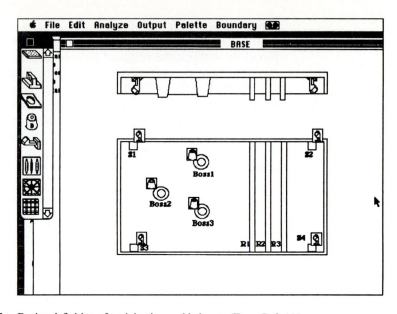

FIGURE 4.15   Design definition of an injection-molded part. (From Ref. 14.)

FIGURE 4.16   Assigning attributes and material properties to the molded part. (From Ref. 14.)

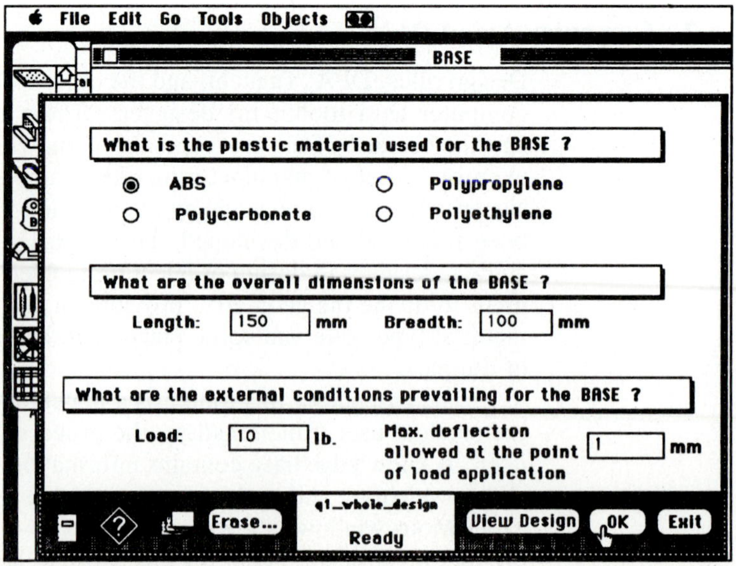

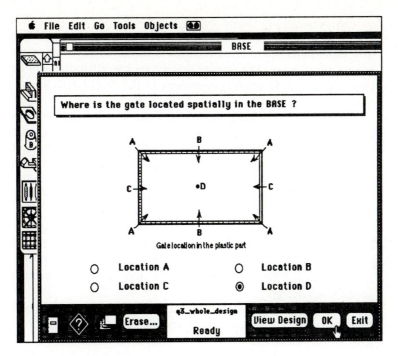

FIGURE 4.17 Defining the gates and sprues for injection. (From Ref. 14.)

FIGURE 4.18 Assigning attributes to the snaps for molding. (From Ref. 14.)

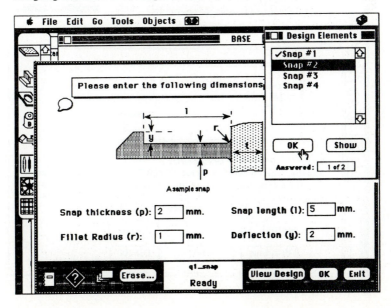

An example is shown below where the user has defined the geometry consisting of a flat part containing three ribs, three bosses, and four snap-fit connectors. Note that each of these features is identified by the designer so that attributes can be added later. The feature definitions and attributes will be used later by the knowledge base to make suggestions to the designer to avoid potential molding problems.

Figure 4.16 shows a window in which the designer is asked to specify information about material and overall dimensions. This information can be used to determine the volume and cost of each of the parts. All details that are not readily extractable from the sketch can be entered in a window such as this.

The window in Figure 4.17 allows the designer to define the locations of gates. Later the program will advise the designer of potentially better locations based on geometry and processing considerations.

Figure 4.18 shows a picture of a snap as designed and as it exists in a library of snap configurations. Note that the designer can choose from four different snap types and enter the exact definition of each snap in the proposed product. This information will be used to advise the designer on the acceptability of the proposed snap design.

After all information on geometry, material, and process details is input, the knowledge base calculates the compatibility and gives the reason if the score is low. In this case, the stress in the plastic base was too high (Figure 4.19).

FIGURE 4.19   Calculation of compatibility among design, user, and manufacturer. (From Ref. 14.)

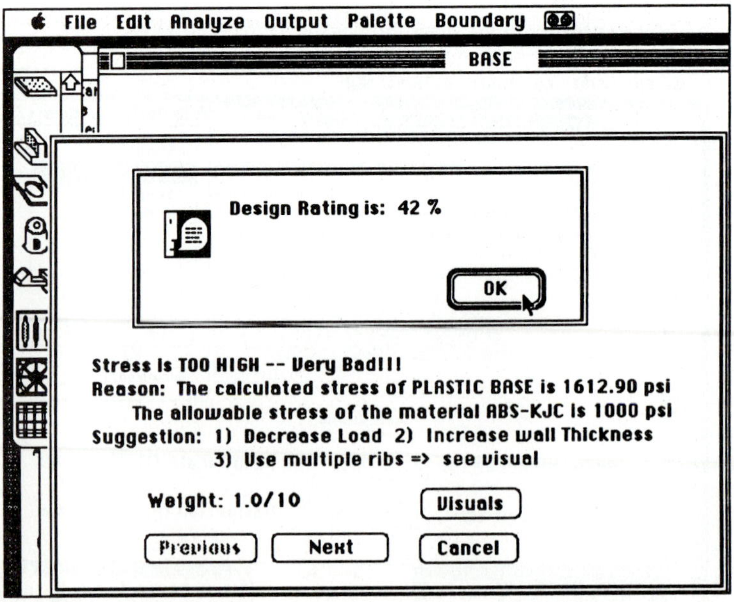

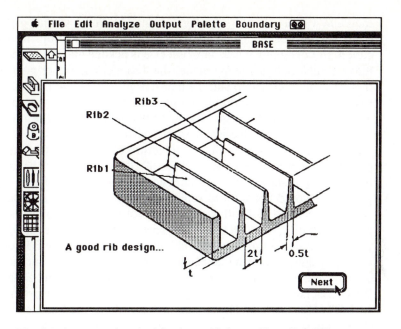

FIGURE 4.20 Visual design suggestions for injection-molded part. (From Ref. 14.)

The designer can request to see a visual (shown in Figure 4.20) with design suggestions to improve the score. The knowledge base takes care of presenting the appropriate visual to the designer. The visuals and input windows are prepared ahead of time for most situations likely to be encountered. Obviously, many rules of thumb must be extracted from professionals in the injection-molding industry. After all, design is somewhat of an art, and many guidelines are derived merely from experience. In that sense, this toolkit is very nice for inexperienced designers. It not only helps them to come up with a good design, it also teaches them the qualities of a good injection-molded part.

The Boothroyd-Dewhurst software for DFA also qualifies as a computer-aided DFM example.

4.4.8 Group Technology

Group technology (GT) concepts are covered in Chapter 5. The principle is that the characteristics of a part can be codified into a multidigit code. This code can then be used to classify parts into families for ease of design retrieval and variant process planning. GT is a DFM tool in the sense that it can be used to produce significant improvements in product quality and design efficiency. During the design process, the engineer can examine all parts currently made by a company to determine if a part with the desired characteristics already exists. If it does, then a new design is not necessary. It may be that a similar part exists, in which case a simple design change may bring it into specifications for the current application. For example, if a certain bolt size is needed and the GT code is used to find that a similar bolt is already made, the designer can either pull up the drawing of that bolt and modify it to suit the new design or redesign the part in which the bolt resides to accept the old bolt design, thereby removing the need for a new bolt design. This technique can obviously be used to limit the proliferation of new parts which are near duplicates of old parts.

GT can also help in the manufacture of parts by being used to extract the process plan of a similar part and modify it to be used on the new part. Whether this technique should be used depends on the quality of the old part's process plan. One would not want to replicate a poor plan for a new part.

An example of a software package for group technology is a research effort named CODER [13]. CODER is an expert system which reads the definition of the part to be manufactured and automatically determines the GT code. The resulting code may not agree with that calculated by a human user, but the software has the advantage that it is consistent. It will always find the same code for a given part. That is not necessarily true for humans. In fact, code inconsistency is one problem with GT. Human code determination is based on a person's own interpretation of the product.

4.4.9 Failure-Mode and Effects Analysis

The analysis of a part for correct function is a routine effort. In concurrent engineering, though, it is realized that there is a relationship between correct function and part geometry, assembly, and manufacture. Failure-mode and effects analysis (FMEA) provides the design team with an organized approach to evaluating the causes and effects of various modes of failure in a product. The goal is to increase overall part quality by anticipating failures and designing them out of the system.

The first step in FMEA is to list all possible types of failure [22]. These would include buckling, fracture, fatigue, excessive deformation, leakage, binding, etc. All of these modes of failure are then ranked according to their effect on the system. Failures that are catastrophic are ranked very high, and those that do not affect function are ranked low. Each failure is addressed one by one from most to least important. Design changes are then made to reduce the chance of failure. In some cases this means increasing the strength of members and in some cases it means simplifying the design. Simplification can result in a reduced number of parts or savings in material or manufacturing costs.

Design simplification begins with functional analysis: determining what the product should do rather than what it does now. Whitney and colleagues mention that they saved a company several million dollars per year by eliminating one part from a subassembly [24]. Originally five parts were used on a switch assembly to set the product's states from low to medium to high in three steps. Functional analysis showed that one of the parts always followed the actions of two other parts. The redundant part was eliminated. The three states are still attainable.

4.4.10 Value Engineering

Value engineering predates efforts at concurrent engineering. Developed in the 1950s and 1960s, *value engineering* is an evaluative technique which assigns a value to a product. The value is defined as the ratio of function or performance to cost. The idea is to maximize the value of the product by raising the functional worth or, more commonly, decreasing the cost of each function. The goal is to eliminate unnecessary features and functions by optimizing the value ratio. This value can be divided into two components: a use, or functional, value and an esteem value. The use value reflects how the product satisfies the user's needs, and the esteem value is a measure of the desirability of the product, a marketing and advertising concern. As a first step, the two values are investigated analytically by a team of experts based on a preliminary design. Then, with the values in hand, a creative attempt is made to maximize them. This second step depends on the innate abilities of the design team. There are really no guidelines except experience to direct the design optimization.

The primary component of value engineering is the functional statement. Each component of the product as well as the overall product itself is given a brief functional statement and then that function is given a numeric value based on a discussion by design team members. A cost for that function is also determined based on manufacturing and other costs. If a function of an assembly initially requires two parts but then a redesign reduces the number of parts to one, the value of the part will go up because production costs decrease.

The most difficult aspect of value engineering is estimating the values of the functions and costs. This is not a new problem. Other techniques in concurrent engineering must estimate costs to be able to evaluate design alternatives. Value engineering attempts to use the values in a systematic way.

4.5 SUMMARY OF CONCURRENT ENGINEERING TOOLS

Table 4.1 shows the particular activity addressed by each of the tools described in the paragraphs above [22]. Note that not all tools address functional issues. DFM guidelines, DFA, design rules, and the designer's toolkit do not directly address how well the part performs. These tools are obviously more process-related and should be used during the determination of process-oriented issues such as process planning and inspection. Other tools ignore processing issues. The fact that these tools are identified as being applicable to concurrent engineering does not mean that they are all comprehensive. In fact, a comprehensive CE system would make all of these tools available to the designer to use with discretion.

TABLE 4.1
DFM tools and their capabilities

DFM tools	Optimize concept	Simplify	Ensure process conformance	Optimize product function
Design axioms	●	●	●	●
DFM guidelines	●	●		
DFA method		●		
Taguchi method	●			●
Manufacturing process design rules		●	●	
Designer's toolkit			●	
Computer-aided DFM	●	●	●	●
Group technology	●	●	●	●
FMEA	●			●
Value analysis				●

Source: From Ref. 22.

Table 4.2 shows the strengths and weaknesses of each of these tools. They are ranked on the following features (with the topic numbers in the list below corresponding to the column numbers):

1. *Implementation cost and effort*
   Better   Merely a seminar or brief training required
   Worse   Extensive, companywide preparation required
   Average   Implementation efforts uncertain and may involve software, training, and organizational change

2. *Training and/or practice*
   Better   Little or no training required
   Worse   Extensive training or user experience required before any payback is achieved
   Average   Significant commitment to training required

3. *Design effort*
   Better   Little or no additional designer time or effort required for effective use
   Worse   Significant time must be allocated
   Average   A certain amount of commitment and perseverance may be required as well as additional design time

TABLE 4.2
DFM methodology comparison

DFM technique	1	2	3	4	5	6	7	8	9	10
Design axioms	●	◒	●	●	●	●	●	◒	○	●
DFM guidelines	●	●	●	●	○	●	●	○	○	●
Boothroyd-Dewhurst DFA	◒	◒	○	◒	◒	●	●	●	●	●
Taguchi method	◒	○	○	○	●	◒	◒	●	●	◒
Manufacturing process design rules	◒	◒	◒	●	◒	◒	◒	●	●	●
Designer's toolkit	◒	○	○	●	◒	○	◒	●	◒	◒
Computer-aided DFM	◒	◒	◒	●	◒	◒	◒	◒	●	◒
Group technology	○	◒	●	○	●	◒	◒	◒	◒	○
FMEA	◒	◒	○	○	●	◒	◒	◒	◒	◒
Value analysis	●	◒	○	○	●	○	●	◒	◒	●

Key: ● better; ◒ average; ○ worse.
Source: From Ref. 22.

4. *Management effort*

Better	Little or no management effort or expectation necessary
Worse	Significant management effort and commitment required
Average	Successful use requires management expectation that tool is used and gives good support

5. *Product planning team approach*

Better	Effective use requires good product planning and/or team approach
Worse	Method does not depend on nor encourage good product planning and/or team approach
Average	Can or may require good planning and/or team approach

6. *Rapidly effective*

Better	Likely to be rapidly effective for beneficial results
Worse	Benefits long in coming and long-term review required
Average	Can or may be rapidly effective depending on conditions

7. ***Stimulates creativity***

Better	Tends to require design innovation and creativity
Worse	Not likely to require or stimulate creativity
Average	Good potential for creativity stimulation

8. ***Systematic***

Better	Involves step-by-step systematic procedure to help ensure all relevant issues are considered
Worse	Little or no systematic procedure
Average	Systematic in some aspects

9. ***Quantitative***

Better	Primarily quantitative, generating one or more design ratings
Worse	Primarily subjective, with no design ratings
Average	Both qualitative and quantitative aspects, with some ratings generated

10. ***Teaches good practice***

Better	Teaches good DFM practice; reliance on method may diminish with use
Worse	Benefits always depend on formal use with no educational benefit
Average	Teaches good practice, but always must be formally applied

TABLE 4.3
DFM methods and their applications

DFM technique	Appropriate applications[a]
Design axioms	A, B, C, D, E
DFM guidelines	A
Boothroyd-Dewhurst DFA	A
Taguchi method	A, B, C, D, E, F
Manufacturing process design rules	C, E
Designer's toolkit	F
Computer-aided DFM	A, B, C, E, F
Group technology	A
FMEA	A, B, D
Value analysis	A, B, C, D, E, F

[a] Key: A, mechanical and electromechanical devices and assemblies; B, electronic devices and systems; C, manufacturing processes, other processes; D, software, instrumentation and control, systems integration; E, material transformation processes; F, specialized and/or unique manufacturing facilities such as flexible assembly systems.

Source: From Ref. 22.

Table 4.2 can be useful when determining which DFM strategy you should choose. Column 6 may be especially important if you are just beginning concurrent engineering and wish to get started quickly. The best choices then are design axioms, DFM guidelines, and DFA. If you are running a small company, then maybe columns 1 and 2 are the most important aspects, in which case you should choose design axioms, DFM guidelines, or value analysis.

Note that most of the methodologies are good in either being quick and easy to start or more formal and quantitative. For example, the Boothroyd-Dewhurst DFA is high in being quantitative and systematic, whereas the DFM guidelines, which are merely rules of thumb derived from experienced professionals, are more qualitative and less formal. An interesting exercise would be to propose various company scenarios and then use the table to choose reasonable approaches to automation and concurrent engineering. The methodologies can also be classified according to applications. That information is given in Table 4.3.

4.6 CONCLUSION

It should be obvious from our review of concurrent engineering methods that this new concept demands much change from the traditional enterprise. No more are the experts in various aspects of the product life cycle kept apart. No more is design a serial process, but a cooperative effort from the beginning which is optimized for function and process before production is begun. No more are all-important design decisions made by one designer who gives the result to frustrated process engineers who have no input in the design.

Now, with concurrent engineering, most design decisions are made early in the process by a design team comprised of experts in all facets of the product life cycle from marketing to maintenance. It may be that the most difficult part of the implementation of concurrent engineering is convincing upper management that a change must be made. Clausing suggests the following steps to enlist the support of upper management and incorporate concurrent engineering [7].

First, get top management commitment and involvement. Management must realize that this is an important change in the way the company will do business. All members of the organization must have respect for one another's jobs and take time to communicate with one another. Concurrent engineering requires a culture change where different functional groups mingle and cooperate instead of compete.

Second, establish a strategic vision. Know where you are going with concurrent engineering, and know its capabilities and limits. Announce to the company what the expectations of concurrent engineering are and what you hope to achieve and when. A schedule helps anchor the process and gives fellow workers an idea that each participant must be patient for a while until CE becomes a natural process.

Third, begin awareness projects. Start small pieces of the CE enterprise, such as interdepartmental meetings, to try out some of the techniques. DFA is a well-organized approach and may be a good way to start. Also, establish pilot projects. Take some products and run them through the DFM guidelines or the Taguchi method to evaluate their performance under CE. In addition, use simulation tools such as spreadsheets, CAE, and CAD/CAM software to make people aware of CE tools and their capabilities. Other software that may be tried includes project management scheduling packages and simulation languages to evaluate the performance of the current factory.

Last, and maybe most important, choose an internal champion. A key upper-management figure can do wonders to communicate your CE plan to the rest of upper management in a language that managers understand. The champion can also help keep the process of switching to CE going by calming company leaders when things may not appear to be moving as fast as they wish. Three types of people can become involved: those who make things happen, those who watch things happen, and those who wonder what happened. You want to choose the champion from the first group and eliminate those in the last group from the product development team.

Management may want to become overly involved in the conversion to CE by attending design review meetings. These meetings, which involve members from many company domain groups, sometimes become a place to impress other people, especially if upper management is present. The urge to impress should be contained. In fact, these review meetings should have clear objectives that keep you on schedule and on the subject at hand. Procedures for the review meetings should be developed. This is not a time for vindictive criticism. Nor is it a time for a member to show up at the end of the process, just as the design is about complete, and reveal his or her reasons why it will never work. It is not a time for revelations of closely held ideas. It is also not a time for executive mandates. All players on the design team must be treated with equal respect and value.

These guidelines are foreign in many companies, where workers are dependent on the dictates of upper management. CE is a group process.

Concurrent engineering has quite a bit of potential to reduce the time to market, increase quality, and address the product from concept to disposal. It has just begun to be implemented in the United States after many years of practice in Japan and Europe. The challenge in coming years will be to keep up with the many complementary tools of CE.

EXERCISES

1. Design axioms are listed in Section 4.4.1. Using the design of a kitchen food processor, apply several of the axioms to the product design and explain how the design would benefit.

2. Apply axiomatic design, using the two steps mentioned in Section 4.4.1, in designing a child's car seat that can be used in both very hot and very cold climates and that can protect the occupant in a 30 mph head-on collision with another car of equal mass. Propose your own design axioms in addition to those in text.

3. Apply the DFM guidelines to the design problem in Exercise 2. List those guidelines you used in your product design and specify how you satisfied them. Explain "design for compliance."

4. Which should be performed first in the design process (if they are not performed concurrently): DFM or DFA? Why?

5. Explain briefly how the Taguchi method of parameter design can help in the design of an automobile engine that must run on unleaded gasoline in the summer and, to cut winter pollution, a mixture of enthanol and gasoline in the winter.

6. Investigate CE tools by reading CAD/CAM journals and magazines to get an idea of the toolkit functions and their availability. The results of this investigation will get more involved as times goes on and CE becomes more supported.

7. Obtain a simple assembly—such as an automatic pencil, inexpensive camera, or child's toy—and disassemble it. Time yourself. Look for design modifications to ease the assembly process. Think about the assembly's being done by a robot, which does not have human dexterity. Are there fasteners you would change? Would you combine several pieces into one piece? If so, how would that affect the fabrication complexity and cost? Create drawings of your redesign and include justifications for each change.

REFERENCES AND SUGGESTED READING

1. Bancroft, C. E.: "Design for Manufacturability: Half Speed Ahead," *Manufacturing Engineering,* September 1988, pp. 67–70.

2. Billatos, S. B.: "Guidelines for Productivity and Manufacturability Strategy," *Manufacturing Review,* vol. 1, no. 3, October 1988, pp. 164–167.

3. Boothroyd, G., and P. Dewhurst: "Design for Assembly: Selecting the Right Method," *Machine Design,* November 10, 1983, pp. 94–98.

4. Boothroyd, G., and P. Dewhurst: "Design for Assembly: Manual Assembly," *Machine Design,* December 8, 1983, pp. 140–145.

5. Boothroyd, G., and P. Dewhurst: "Product Design for Manufacture and Assembly," *Manufacturing Engineering,* April 1988, pp. 42–46.

6. Byrne, D., and S. Taguchi: "The Taguchi Approach to Parameter Design," ASQC Quality Congress Transaction, Anaheim, Calif., 1986.

7. Clausing, D.: "Concurrent Engineering," Proceedings of ASME Winter Annual Meeting, San Francisco, Calif., December 1989.

8. Cralley, W., and E. Rogan: "Architecture and Integration Requirements for Unified Life Cycle Engineering (ULCE)." Paper presented at CAD/CIM Alert Conference on DFM, October 1987.

9. Crow, K.: "Ten Steps to Competitive Design for Manufacturability, CAD/CIM Alert," 2nd International Conference on Design for Manufacturability, Management Roundtable, Chestnut Hill, Mass., 1988.

10. Davidian, R., "Concept Analysis through Formal Design." Paper presented at CAD/CIM Alert Conference on DFM, October 1987.

11. Dewhurst, P., and G. Boothroyd: "Computer-Aided Design for Assembly," *Assembly Engineering,* February 1983, pp. 18–22.

12. Heidenreich, P.: "Designing for Manufacturability," *Quality Progress,* May 1988, pp. 41–44.

13. Henderson, M., and S. Musti: "The CODER System," ASME Winter Annual Meeting, San Francisco, Calif., December 1987.

14. Ishii, K., L. Hornberger, and M. Liou: "Compatibility-Based Design for Injection Molding," Concurrent Product and Process Design," *Proceedings of the ASME Winter Annual Meeting,* San Francisco, Calif., December 10–15, 1989, pp. 153–160.

15. McNeill, B. W., D. R. Bridenstine, E. D. Hirleman, and F. Davis: "Design Process Test Bed," *Proceedings of the ASME Winter Annual Meeting,* San Francisco, Calif., December 10–15, 1989, pp. 117–120.

16. Nevins, J. L., and D. E. Whitney: *Concurrent Design of Products & Processes,* McGraw-Hill, New York, 1989.

17. Norman, R.: "Concurrent Design Methodology: Making It Happen," International Technegroup, Inc., Milford, Ohio.
18. Rowe, G. W.: "An Intelligent Knowledge-Based System to Provide Design and Manufacturing Data for Forging," *Computer-Aided Engineering Journal,* February 1987, pp. 56–61.
19. Runciman, C., and K. Swift: "Expert System Guides CAD for Automatic Assembly," *Assembly Automation,* August 1985, pp. 147–150.
20. Savage, C.: *Fifth-Generation Management*, Digital Press, Bedford, Mass., 1990.
21. "Smart Design," *Business Week,* April 11, 1988, pp. 102–117.
22. Stoll, H. W.: "Design for Manufacture," *Manufacturing Engineering,* January 1988.
23. Suh, N., A. C. Bell, and D. C. Gossard: "On an Axiomatic Approach to Manufacturing and Manufacturing Systems," *Journal of Engineering for Industry (ASME),* vol. 100, May 1978, pp. 127–129.
24. Whitney, D. E., et al.: "The Strategic Approach to Product Design," Document CSDL-P-2742, The Charles Stark Draper Laboratory, Inc., Cambridge, Mass., December 1986.
25. Wilson, C. W., and M. E. Kennedy: "Some Essential Elements for Superior Product Development," Paper 89-WA/DE-7, *Proceedings of the ASME Winter Annual Meeting,* San Francisco, Calif., December 10–15, 1989.

CHAPTER
5

GROUP TECHNOLOGY

Order and simplification are the first steps toward the mastery of a subject.

Thomas Mann
Stories of Three Decades (1936)

The amount of data that a person's mind can readily work with at one time is relatively small. A computer can manipulate considerably more data than a human mind; however, even a computer has limits on the amount of data that can be manipulated at one time. For this and other reasons it is desirable to find ways to organize data so that only pertinent items need be retrieved and analyzed at a given time. To accomplish this, methods of structuring data have been devised. Some methods are very clever, such as data structures used in large computer data bases; others are relatively simple, such as listing words alphabetically in a dictionary. More elaborate systems have been devised, such as the taxonomy used in biology to classify all living things (see Figure 5.1). This latter example illustrates how thousands of items can be organized so that reasoning can be applied to a small group with some similar attributes. Another example is the coding and classification of books in a library catalog. Using this catalog, one can easily find all books written by an author, all books on a specific subject, or all books with a particular title.

The basis for group technology is analogous to these situations. A company may make thousands of different parts in an environment that is becoming more complex as lot sizes get smaller and the variety of parts increases. When they are examined closely, however, many parts are similar in some way. A design engi-

177

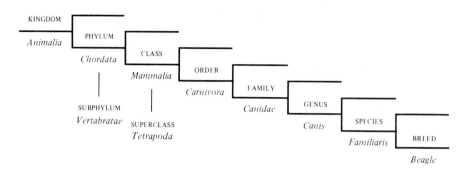

FIGURE 5.1 Biological classification.

neer faced with the task of designing a part would like to know if the same or a similar part had been designed before. Likewise, a manufacturing engineer faced with the task of determining how to manufacture a part would like to know if a similar process plan already exists. It follows that there may be economies to be realized from grouping parts into families with similar characteristics. The resulting data base would certainly be easier to manage; therefore, the manufacturing enterprise should be easier to manage. In 1969 V. B. Soloja defined *group technology* as "the realization that many problems are similar, and that by grouping similar problems, a single solution can be found to a set of problems thus saving time and effort." [25] This definition is very broad, but it is valid because group technology concepts have been applied to many environments.

This chapter will briefly recount the historical beginnings of formal applications of group technology in manufacturing. The basis for applying group technology is coding and classification of parts; consequently, this subject will be discussed in detail. Some cluster analysis techniques that can be used to arrange machines in a group layout will be described. In addition, economical models for production planning and tool analysis in a group technology environment will be explored. Also, benefits that can be achieved from a successful group technology implementation program will be discussed.

5.1 KEY DEFINITIONS

Attribute (polycode) code   Each part attribute is assigned to a fixed position in a code. The meaning of each character in the code is independent of any other character value.

Average linkage clustering algorithm   An algorithm for clustering things together based on the average similarity of all pairs of things being clustered. The similarity of each pair is measured by a similarity coefficient.

Bottleneck machine   In this chapter, a machine in a group (cell) that is required by a large number of parts in a different group.

Cell   In this chapter, a group of machines arranged to produce similar families of parts.

Classification   The process of categorizing parts into groups, sometimes called families, according to a set of rules or principles.

Cluster analysis   The process of sorting things into groups so the similarities are high among members of the same group and low among members of different groups.

Coding   The process of assigning symbols to a part to reflect attributes of the part.

Computer-aided process planning (CAPP)   An interactive computer system that automates some of the work involved in preparing a process plan.

Decision variables   In a mathematical model, values must be assigned to these variables. The objective is to select values that optimize the model's measure of performance, such as minimal cost. Initially the best values for these variables are unknown.

Dendrogram   A treelike graphical representation of cluster analysis results. The ordinate is in some similarity coefficient scale, and the abscissa has no special meaning.

Function layout   Layout of machines in a factory such that machines of a specific type are grouped together.

Group layout   Machines in a factory are arranged as cells.

Group technology   An engineering and manufacturing philosophy that groups parts together based on their similarities in order to achieve economies of scale in a small-scale environment normally associated with large-scale production.

Group tooling   Tooling designed such that a family or families of parts can be processed with one master fixture and possibly some auxiliary adapters to accommodate differences in some of the part attributes, such as number of holes and sizes of holes.

Hierarchical (monocode) code   The meaning of each character is dependent on the meaning of the previous character in the code.

Hybrid (mixed) code   A combination of an attribute and a hierarchical code. It combines the advantages of both code types.

Line layout   Machines in a factory are arranged in the sequence in which they are used. The work content at each location is balanced so that materials can flow through in a continuous manner.

Logic tree   A treelike graph that represents the logic used to make a decision. This differs from a decision tree in that the branches may contain logical expressions as well as calculations, data elements, codes, and keys to other data.

Machine–component chart   A matrix that denotes what machines a group of components (parts) visit.

Part family   A group of parts having some similar attributes.

Process plan   The detailed instructions for making a part. It includes such things as the operations, machines, tools, feeds and speeds, tolerances, dimensions, stock removal, time standards, and inspection procedures.

Production flow analysis A structured procedure for analyzing the sequence of operations that parts go through during manufacturing. Parts that go through common operations are grouped together as a family, and the associated machines are arranged as a cell.

Rotational part A part that can be made by rotating the workpiece. It is usually symmetrical along one axis, such as a gear.

Similarity coefficient In this chapter, a measure of how alike two machines are in terms of the number of parts visiting both machines and the number of parts visiting each machine.

Single-linkage clustering algorithm (SLCA) An algorithm for clustering together things that have a high similarity coefficient.

Threshold value A similarity coefficient value at which clustering is to stop. That is, no more clusters are to be formed if the largest remaining similarity coefficient value is below this value.

5.2 BACKGROUND

The small-lot manufacturing environment has been studied extensively since World War II. This environment is very difficult to manage well. From a casual observer's viewpoint, manufacturing activities seem to occur randomly. In the 1960s and early 1970s, some people thought that operations research (management science) techniques, such as queueing theory and mathematical programming, had the potential to improve management of an environment of this type. However, none of these techniques was practically successful. Until the late 1970s, no approach received widespread acceptance; in fact, very few new approaches were even proposed. In the late 1970s, however, a consensus emerged in the manufacturing countries that group technology provides a basis for better management of the small-lot manufacturing environment. Some people had advocated this philosophy since the early 1950s.

5.2.1 History of Group Technology

People have been informally using group technology concepts for centuries; one of the first recorded applications in manufacturing was done by F. W. Taylor, the scientific management pioneer [12]. In his attempts to improve productivity, he noted that there were similarities between some jobs, and he was able to categorize the similar attributes of the jobs. More recently, many companies have applied more formal group technology concepts, such as grouping machines into cells and establishing group tooling.

Still, no formal description of a group technology application in manufacturing appeared until 1959, when S. P. Mitrofanov, a Russian, published his book, *Scientific Principles of Group Technology* [18]. In the editor's forward to the

English translation of Mitrofanov's book, T. J. Grayson describes group technology as follows:

> . . . a method of manufacturing piece parts by classification of these parts into groups and subsequently applying to each group similar technological operations. The major result of this method of manufacture is to obtain economies which are normally associated with large scale production in the small scale situation and it is therefore of fundamental importance in the batch production and jobbing sections of industry.

If a company is going to produce several million units of a specific product, large expenditures can be justified for developing methods to manufacture the item for less cost, even if the reduction per unit is only a few cents. However, if only a small number are to be produced, it is not realistic to make such expenditures. The group technology philosophy advocates combining several similar parts so that the sum is large enough to justify expenditures to reduce design and manufacturing costs for the entire group of parts. In other words, any reduction in cost must be generic to the group to be cost-effective. Another premise is that group technology promotes standardization of parts, because design engineering strives to minimize new part designs by utilizing existing designs. These are some of the reasons that the acceptance of group technology as a manufacturing philosophy has spread throughout the manufacturing countries of the world.

In 1960, West Germany and Great Britain began serious studies into group technology techniques. Other European countries quickly followed. By 1963 the success of group technology applications in the Soviet Union prompted the government to promulgate a plan for increased implementation throughout Soviet industry. By 1970 the Japanese government had begun sponsoring group technology applications. In the United States, in contrast, group technology did not receive widespread acceptance until the latter part of the 1970s.

Current trends in manufacturing have set the stage for acceptance of group technology in the United States. These trends, as noted by Ham and Hitomi [11], include:

1. A rapid proliferation of numbers and varieties of products, resulting in smaller lot sizes
2. A growing demand for closer dimensional tolerances, resulting in a need for more economical means of working to higher accuracies
3. A growing need for working with increased varieties of materials, heightening the need for more economical means of manufacturing
4. An increasing proportion of cost of materials to total product cost due to increasing labor efficiency, thereby lowering acceptance scrap rates
5. Pressure from the above factors to increase communication across all manufacturing functions with a goal of minimizing production costs and maximizing production rates

It has been estimated that as much as 75% of all parts made in the United States will be produced on a small-lot basis, from one to a few thousand. Consequently, it is not unreasonable to predict that 50% to 75% of American manufacturing firms will be using some form of group technology by 1995.

There have been many recent advances in machine tool automation and metal removal technology, so modern manufacturing is often perceived as being productive and efficient. To the contrary, studies have determined that during fabrication of an average part in an average batch-type production facility, a part will be on a machine only 5% of the production through-put time [2]. This is exemplified by Figure 5.2. Of that 5%, less than 30% is spent in metal removal. The part is being moved or is waiting in a queue 95% of the time. These are additional reasons for batch-type manufacturing firms to consider some new concepts.

5.2.2 The Role of Group Technology in CAD/CAM Integration

The preceding discussion pointed out that competitive world market conditions are encouraging more and more batch-type manufacturing firms to consider adopting a group technology philosophy. Another major contributing factor to this acceptance is an increasing emphasis on the integration of CAD and CAM.

In this chapter it will become evident that group technology is an important element of CAD and CAM. An essential aspect of the integration of CAD and CAM is the integration of information used by engineering, manufacturing, and all the other departments in a firm. Group technology provides a means to structure and save information about parts, such as design and manufacturing attributes, processes, and manufacturing capabilities, that is amenable to computerization and analysis. It provides a common language for the users. Integration of many types of part-related information would be virtually impossible without group technology; consequently, group technology is an important element of CAD/CAM integration.

FIGURE 5.2   Distribution of work-in-process time during part fabrication.

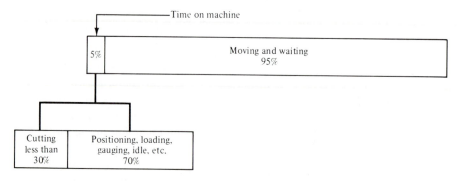

Another important aspect of the integration of CAD and CAM is automation. The next chapter, which discusses process planning, will explain how group technology is key to automating this function. Also, many manufacturing firms are automating their operations by arranging their machines into cells. The design of a cell is based on group technology. These observations reinforce the importance of group technology.

Section 5.9 of this chapter discusses many of the significant benefits that can be realized when a firm adopts a group technology philosophy. When this understanding is combined with the even greater benefits that can be gained from the integration of CAD and CAM, the attention given to group technology and its widespread acceptance is not surprising.

5.3 METHODS FOR DEVELOPING PART FAMILIES

Group technology is begun by grouping parts into families based on their attributes. Usually, these attributes are based on geometric and/or production process characteristics. Geometric classification of families is normally based on size and shape, while production process classification is based on the type, sequence, and number of operations. The type of operation is determined by such things as the method of processing, the method of holding the part, the tooling, and the conditions of processing. For example, Figures 5.3 and 5.4 show families of parts grouped by geometric shape and by production process. The identification of a family of parts that has similarities permits the economies of scale normally associated with mass production to be applied to small-lot, batch production. Therefore, successful grouping of related parts into families is a key to implementation of the group technology philosophy.

FIGURE 5.3   Parts grouped by geometric shape. (Reprinted with permission from Ref. 15.)

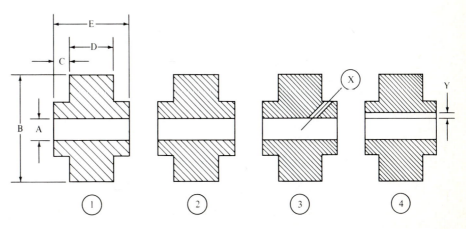

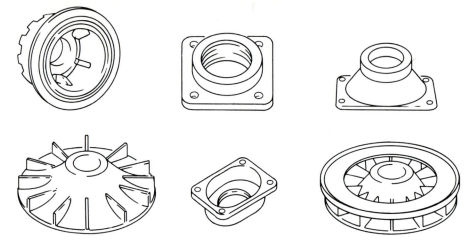

FIGURE 5.4   Parts grouped by manufacturing process. (Reprinted with permission from Ref. 15.)

There are at least three basic methods that can be used to form part families:

1. Manual visual search
2. Production flow analysis
3. Classification and coding

Each will be discussed briefly here, and the last method will be presented in more detail in Section 5.4. Manual visual search is not often used in formal group technology applications. It leads to very inconsistent results because seldom will two people group a set of parts into the same families. There are many reasons for this, such as the fact that each person will have a different knowledge of the processing capabilities of the factory, recognition of the significant part attributes may differ, and many different tools and machines can be used to perform a specific function although the costs of performing the work can be significantly different.

Production flow analysis (PFA) is a structured technique developed by Burbidge [2] for analyzing the sequence of operations (routings) that parts go through during fabrication. Parts that go through common operations are grouped into part families. Similarly, the machines used to perform these common operations may be grouped as a cell; consequently, this technique can be used in facility layout. Initially, a machine–component chart must be formed. This is an $M \times N$ matrix, where

M = number of machines

N = number of parts

$x = 1$ if part j has an operation on machine i; 0 otherwise.

If the machine–component chart is small, parts with similar operations might be grouped together by manually sorting the rows and columns. However, a more appealing method is to use a computer procedure to perform this work.

Figure 5.5 illustrates the use of PFA to form part families. For this technique to be successful, accurate and efficient routings must exist for each part. In many companies these routings do not exist. If routings exist, they are often inaccurate from lack of maintenance or they may be very inconsistent. The latter situation

FIGURE 5.5   (*a*) Component–machine chart; (*b*) example of production flow analysis.

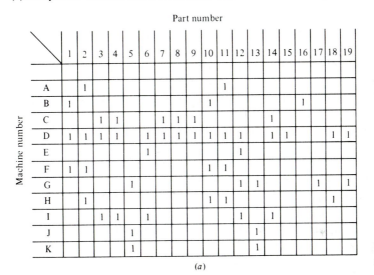

(*a*)

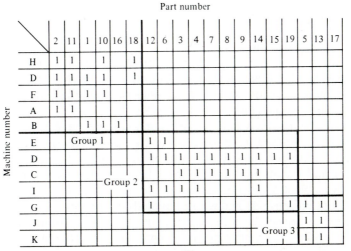

(*b*)

will occur if routings are established without using a coding and classification system. Also, using PFA involves judgment, because some parts may not appear to fit into a family when one or more unique operations are required. Furthermore, additional analysis is required to determine when a particular machine should be duplicated in another group. In Figure 5.5, for example, machine D is in groups 1 and 2. In this case, since machine D was visited by almost all of the parts, it was duplicated to keep the groups small. Otherwise, groups 1 and 2 might have been combined into one group having several parts with dissimilar routings. Likewise, you cannot determine how many machines of type D are required without evaluating demands and machine capacities. In addition, PFA does not consider part features and functional capabilities. Therefore, this technique should not be used to form part families for design engineering. One advantage of using production flow analysis compared to a coding and classification system is that part families can be formed with much less effort.

If the coding and classification technique is used, parts are examined and codes are assigned to each part based on the attributes of the parts. These codes can then be sorted so that parts with similar codes are grouped as a part family. Because these codes are assigned in a manner that does not require much judgment, the part families developed by this technique do not suffer from judgment inconsistencies. A disadvantage of using the coding and classification technique is that a large amount of time may be required to develop and tailor a code to meet the needs of a specific company. Afterwards, coding the parts will take an even larger amount of time. However, when properly applied, the results are much better than when other techniques are used. Consequently, coding and classification is the preferred approach and will be discussed in more detail than the other two approaches.

5.4   CLASSIFICATION AND CODING

Classification of parts is the process of categorizing parts into groups, sometimes called families, according to a set of rules or principles. The objectives are to group together similar parts and to differentiate among dissimilar parts. Coding of a part is the process of assigning symbols to the part. These symbols should have meanings that reflect the attributes of the part, thereby facilitating analysis (information processing). Although this does not sound very difficult, classification and coding are very complex problems.

Several classification and coding systems have been developed, and many people have tried to improve them. No system has yet received universal acceptance, however, because the information that is to be represented in the classification and coding system will vary from one company to another. This seems reasonable if one understands that the two greatest uses of group technology are for design retrieval and for group (cell) production, and that each company has some unique needs for these functions. Although all of these needs are not unique, enough are to prohibit the development of a universal system. Therefore, even though classification and coding systems can be purchased, a good rule of thumb is that 40% of a purchased system must be tailored to the specific needs of a particular company.

One reason that a design engineer classifies and codes parts is to reduce design effort by identifying similar parts that already exist. Some of the most significant attributes on which identification can be made are shape, material, and size. If the coding and classification system is to be used successfully in manufacturing, it must be capable of identifying some additional attributes, such as tolerances, machinability of materials, processes, and machine tool requirements. In many companies the design department does not exchange very much information with the manufacturing department. The analogy of "design engineering throwing the part design over the wall for manufacturing to make" is often used to describe the lack of communication between these departments. The classification and coding system selected by a company should meet the needs of both design engineering and manufacturing. A system that meets these combined needs will improve communication between departments and facilitate computer-integrated manufacturing.

Although well over 100 classification and coding systems have been developed for group technology applications, all of them can be grouped into three basic types:

1. Hierarchical, or monocode
2. Attribute, or polycode
3. Hybrid, or mixed

These systems will each be discussed in some detail.

5.4.1 Hierarchical Code

As we noted at the beginning of this chapter, people have been classifying and coding things for a long time. The hierarchical coding system supports this statement. It was originally developed for biological classification by Linnaeus in the 1700s. In this type of code, the meaning of each character is dependent on the meaning of the previous character; that is, each character amplifies the information of the previous character. Such a coding system can be depicted using a tree structure as shown in Figure 5.6b, which represents a simple scheme for coding the spur gear shown in Figure 5.6a. Using these figures, we can assign a code, "A11B2," to the spur gear.

A hierarchical code provides a large amount of information in a relatively small number of digits. This advantage will become more apparent when we look at an attribute coding system. Defining the meanings for each digit in a hierarchical system can be difficult, although application of the defined system is relatively simple. Starting at the main trunk of the tree, you need to answer a series of questions about the item being coded. Continuing in this manner, you work your way through the tree to a termination branch. By recording each choice as you

(a)

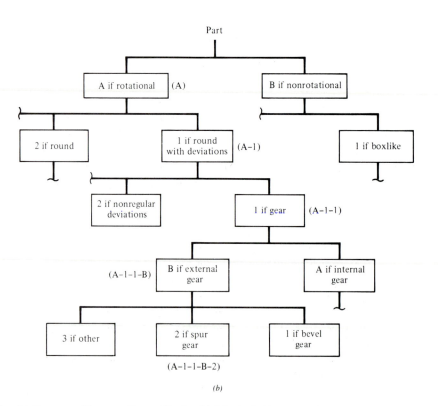

(b)

FIGURE 5.6   (a) Spur gear (Courtesy Garrett Engine Division); (b) hierarchical code for the spur gear.

answer each question, you will build the appropriate code number. However, determining the meaning of each digit in the code is complicated, because each preceding digit must first be decoded. For example, in the code developed in Figure 5.6*b* for the spur gear A11B2, a "1" in the second position means "round with deviations" because there is an "A" in the first position of the code. However, if there had been a "B" in the first position, a "1" in the second position would have meant "boxlike."

Design departments frequently use hierarchical coding systems for part retrieval because this type of system is very effective for capturing shape, material, and size information. Manufacturing departments, on the other hand, have different needs which are often based on process requirements. It is difficult to retrieve and analyze process-related information when it is in a hierarchical structure that will be equally useful to both the design and manufacturing organizations.

5.4.2 Attribute Code

An attribute code is also called a polycode, a chain code, a discrete code, or a fixed-digit code. The meaning of each character in an attribute code is independent of any other character; thus, each attribute of a part can be assigned a specific position in an attribute code. Figure 5.7 shows an example attribute code. Using Figure 5.7 to code the spur gear illustrated in Figure 5.6*a*, we would obtain the code "22213." Referring to Figure 5.7, we can see that a "3" in position 5 means that the part is a spur gear regardless of the values of the digits in any other positions.

If we had used this attribute code to code several parts and wanted to retrieve all spur gears, we would only need to identify all parts with a "3" in position 5 of the associated code. This becomes a simple task if a computer is used. Consequently, an attribute code system is popular with manufacturing organizations because it makes it easy to identify parts that have similar features that require similar processing. One disadvantage of an attribute code is that a position in the code must be reserved for each different part attribute; therefore, the resulting code may become very long.

5.4.3 Hybrid Code

In reality, most coding systems use a hybrid (mixed) code so that the advantages of each type of system can be utilized. The first digit, for example, might be used to denote the type of part, such as a gear. The next five positions might be reserved for a short attribute code that would describe the attributes of the gear. The next digit, position 7, might be used to designate another subgroup, such as material, followed by another attribute code that would describe the attributes. In this manner a hybrid code could be created that would be relatively more compact than a pure attribute code while retaining the ability to easily identify parts with specific characteristics.

Digit	Class of feature	Possible value of digits			
		1	2	3	4
1	External shape	Cylindrical without deviations	Cylindrical with deviations	Boxlike	• • •
2	Internal shape	None	Center hole	Brind center hole	• • •
3	Number of holes	0	1–2	3–5	• • •
4	Type of holes	Axial	Cross	Axial cross	• • •
5	Gear teeth	Worm	Internal spur	External spur	• • •
⋮	⋮	⋮	⋮	⋮	⋮

FIGURE 5.7   Attribute code example.

5.4.4   Selecting a Coding System

Because well over 100 coding systems have been developed, selecting the best one for a particular application can be a difficult and time-consuming task. The following are factors that should be kept in mind when selecting a system:

OBJECTIVE. The objective of installing the system will vary depending on who the user is—engineering, manufacturing, or both. Some typical engineering objectives are:

1. Provide an efficient retrieval system for similar parts
2. Provide part information in a standard form
3. Provide an efficient means to determine manufacturing capability and producibility

From the manufacturing viewpoint, some typical objectives are:

1. Provide information required to form part families
2. Provide for efficient retrieval of process plans
3. Provide an efficient means to form machine groups or cells for part families

ROBUSTNESS. The system selected should be capable of handling all parts now being sold or planned to be sold by the firm. This analysis will involve looking at planned group technology applications and the part attributes that might be needed. Table 5.1 provides some example applications and the associated attributes.

EXPANDABILITY. Because it is very difficult, if not impossible, to define everything that a coding system must be capable of handling during some indefinite future time period, ease of expanding the code is a very important characteristic.

DIFFERENTIATION. The amount of differentiation among codes developed by a system can vary a great deal. Taking an extreme case, after all parts manufactured by a company are coded, they might be classified as being in one family. On the other end of the spectrum, after all parts are coded, each part might represent a distinct family. In these two cases the coding system did not provide the appropriate amount of differentiation.

AUTOMATION. Most coding and classification systems in use today have been implemented using a computer. Therefore, when evaluating a potential system, sufficient time should be spent to determine how well the system has been automated. This evaluation should not be restricted to the coding and classification capabilities; the associated data base methodology and retrieval and analysis functions should also be considered.

EFFICIENCY. The code efficiency, the number of digits required to code a typical part, should be evaluated. If the number of available digits is too small, determine if this number can be increased.

TABLE 5.1
Applications versus part attributes

Applications	Part attributes										
	Shape	Form features	Treatments	Functions	Size envelopes	Tolerances	Surface finish	Material type/condition	Quantity	Next assembly	Raw material form
Design retrieval	X	X		X	X			X			
Generative process planning	X	X	X	X	X	X	X	X	X		X
Equipment selection	X	X	X		X	X	X	X	X		X
Tool design	X	X	X		X	X	X	X	X		
Time/cost estimation	X	X	X		X	X	X	X	X		X
Assembly planning	X	X			X	X	X	X	X	X	
Quality planning	X	X	X	X	X	X	X	X	X	X	X
Production scheduling	X	X	X		X	X	X	X	X	X	X
Parametric part programming	X	X	X		X	X	X	X			X

COST. Cost includes several facets: the initial cost of the system, the cost of modifying the system to meet the particular needs of a specific company, the cost of interfacing (integrating) the system to existing computer systems, and the cost of using the system.

SIMPLICITY. Ease of use is important. Many of the people who must use the system will not be very familiar with computer systems. Therefore ease of use is important for user acceptance, training considerations, and cost of use.

For a particular firm, there may be other considerations that could be added to this list. Even if none are added, the problem of selecting a coding and classification system may not have a trivial solution. Consequently, before a particular system is selected, a thorough evaluation should be performed of what is available. Because the problem is so complex, before a final decision is made it is recommended that some companies be visited that are using the system you are considering.

5.4.5 Developing Your Own Coding System

Many companies have opted to develop their own coding system. This is not something that can be done in a short period of time, even though it is not difficult to describe a series of steps to follow.

Initially a sample of parts should be selected. The coding system should accommodate purchased parts as well as those manufactured in-house; consequently, purchased parts should be included in this sample. The size of the sample will vary from a few hundred (25% of the active parts) out of a data base of 3000 parts to several thousand (1000 to 5000) if the active data base is large.

The next step is to assemble the drawings of the parts in the sample. Then these drawings can be sorted into families by manually examining each drawing and grouping together those that have similar features and therefore require similar processing. This procedure will enable you to identify which part features have a high frequency of occurrence. Once this step is completed it is relatively easy to identify the machine tool requirements for making these parts and link the groups of parts to existing machines in the factory. You may not be able to link some of the purchased parts to in-house capabilities. However, you should consider these parts, because sometime in the future you may purchase the equipment to make these parts. In addition, the coding system can be used to select vendors when the machining capabilities of the vendors are known.

Once the part features are identified, a hierarchy of these features should be established with the objective of minimizing the time required to code a part. As you code parts, you will quickly learn that a part can be coded in less time if certain attributes are identified before others. For example, if you identify the part as rotational, then all attributes that apply only to nonrotational parts can be ignored.

The next step is to test the coding system you have developed. This may be done by coding the sample of parts. When the coding is completed, you should analyze the results to ascertain how well the parts can be grouped into families. This process may involve several iterations before you are satisfied with the coding system. Because this process requires so much effort, many companies have purchased a coding system and then modified it to meet their special needs.

5.5   EXAMPLES OF CODING SYSTEMS

As was noted earlier, many coding systems have been developed. The codes that would be obtained by applying four of these coding systems to the bushing in Figure 5.8 are depicted in Figures 5.9 through 5.12. The DCLASS code is illustrated in Figure 5.9 [4]. This is an eight-digit hybrid code; the DCLASS system will be explained in detail later in this section. Figure 5.10 illustrates CODE, from Manufacturing Data Systems, Inc. [16]. This is an eight-digit hybrid code that can be expanded to 12 digits. Each position of the code can have one of 16 characters (0 through 9 and A through F). When CODE is expanded to 12 digits, the additional digits may be used for attributes such as heat-treat, hardness, finish, material, production cost, and time standards. The MICLASS code by TNO (The Netherlands Organization for Applied Scientific Research) is shown in Figure 5.11; the mandatory component of this system consists of a 12-digit hybrid code [26]. An additional optional 18 digits may be used to capture company-specific information. The MICLASS code will also be described in more detail later in this

FIGURE 5.8   Sample bushing to be coded.

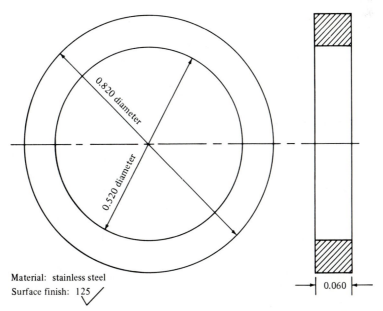

0.820 diameter

0.520 diameter

Material: stainless steel

Surface finish: 125

0.060

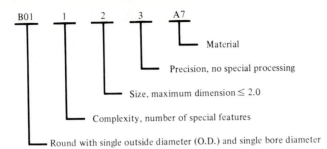

B01　1　2　3　A7

— Material

— Precision, no special processing

— Size, maximum dimension ≤ 2.0

— Complexity, number of special features

— Round with single outside diameter (O.D.) and single bore diameter

FIGURE 5.9　DCLASS eight-digit hybrid code for the bushing in Figure 5.8.

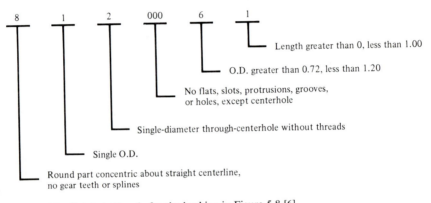

8　1　2　000　6　1

— Length greater than 0, less than 1.00

— O.D. greater than 0.72, less than 1.20

— No flats, slots, protrusions, grooves, or holes, except centerhole

— Single-diameter through-centerhole without threads

— Single O.D.

— Round part concentric about straight centerline, no gear teeth or splines

FIGURE 5.10　CODE eight-digit hybrid code for the bushing in Figure 5.8 [6].

section. Figure 5.12 illustrates the OPITZ code developed in Germany by H. Opitz [19]. It is a nine-digit hybrid code that can be extended by adding four more digits. The first nine digits represent design and manufacturing data. The extended four digits are referred to as the "secondary code" and are intended to represent manufacturing data. However, how these digits are used is up to the particular firm.

FIGURE 5.11
MICLASS 12-digit hybrid code for the bushing in Figure 5.8 [6].

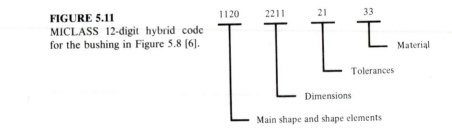

1120　2211　21　33

— Material

— Tolerances

— Dimensions

— Main shape and shape elements

Round part with single O.D. and I.D. without faces, threads, slots, grooves, splines, or additional holes, O.D. and length are within certain size ranges.

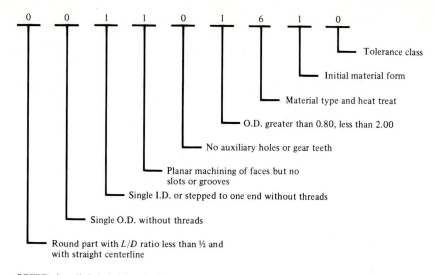

0 0 1 1 0 1 6 1 0

Tolerance class

Initial material form

Material type and heat treat

O.D. greater than 0.80, less than 2.00

No auxiliary holes or gear teeth

Planar machining of faces but no slots or grooves

Single I.D. or stepped to one end without threads

Single O.D. without threads

Round part with L/D ratio less than ½ and with straight centerline

FIGURE 5.12   OPITZ nine-digit hybrid code for the bushing in Figure 5.8 [6].

There are several reasons why so many coding systems have been developed. For example, the existing systems did not capture information that the developer thought was important, the manner in which a code was developed was inconvenient, or the code was difficult to use. Table 5.2 presents the information content obtained from each of the four coding systems used to code the bushing.

TABLE 5.2
Information content of group technology codes

Information content	DCLASS	CODE	MICLASS	OPITZ
End shape			X	X
Outside shape	X	X	X	X
Inside shape	X	X	X	X
Protrusions	X	X	X	X
Additional holes	X	X	X	X
Threads	X		X	X
Grooves or slots	X	X	X	X
Flats		X	X	
Gear teeth or splines	X	X	X	X
Splits, keyways, knurls, or swages	X			
O.D. range		X	X	X
I.D. range			X	X
Length range		X	X	X
Size ratios	X	X	X	X
Tolerance	X		X	X
Heat treat	X		X	X
Material form				X
Material type	X		X	X
Finish	X		X	

Note that for this very simple part the information content differs for each of the systems.

Table 5.2 also illustrates the different types of attributes that a firm may need to represent when a part is coded. As already noted, the required attributes may vary with the application. In general, as the number of applications increases, the number of required attributes generally increases. Consequently, the amount of information that a coding system can represent may be a very important consideration.

5.5.1 DCLASS Coding System

DCLASS (Design and Classification Information System) was developed at Brigham Young University. The part code portion of this system was developed because no commercial vendor was willing to provide such a system for educational and research purposes. Although its primary use to date has been in the university environment, many companies are using it for prototype development. Because of the availability of DCLASS to educational institutions, it will be described in some detail in this section.

Several premises were adopted and used as the basis for the development of the DCLASS code [4]:

1. A part may be best characterized by its basic shape, usually its most apparent attribute.
2. Each basic shape may have several features, such as holes, slots, threads, and grooves.
3. A part can be completely characterized by basic shape; features; size; precision; and material type, form, and condition.
4. Several short code segments can be linked to form a part classification code that is human-recognizable and adequate for human monitoring.
5. Each of these code segments can point to more detailed information.

After several years, an eight-digit hybrid part family code was developed.

The DCLASS part family code is comprised of eight digits partitioned into five code segments, as shown in Figure 5.13. The first segment, composed of three digits, is used to denote the basic shape. The form features code is entered in the

FIGURE 5.13 DCLASS part code segments [4].

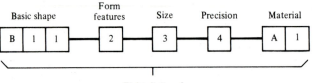

Eight-digit code

next segment; it is one digit in length. This code is used to specify the complexity of the part, which includes features (such as holes and slots), heat treatments, and special surface finishes. The complexity is determined by the number of special features. The one-digit-size code is the third segment of the part family code. From the value of this code, the user will know the overall size envelope of the coded part. The fourth segment denotes precision; it is one digit in length. The final two digits, which comprise the fifth segment of the part family code, are used to denote the material type.

The DCLASS code for the bushing in Figure 5.8 is "BO1 2 3 A7." The first three digits specify the basic shape. Normally we would code a part by answering questions posed by an interactive computer program. In this case, however, we will look at some figures to determine the appropriate values to assign to the code. First, look at Figure 5.14, which depicts a DCLASS part family classification chart. Note that this chart is structured as a logic tree. It could have been structured in some other manner, but logic trees have proven to be easy to work with. This is just one of many such charts in the DCLASS coding system. Using this

FIGURE 5.14   DCLASS logic tree. (Reprinted with permission from Ref. 4.)

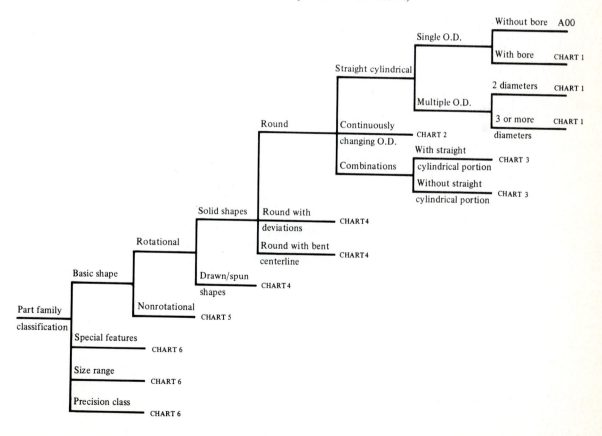

chart to code the bushing, we would take a path through the logic tree that utilizes the following branches: basic shape, rotational, solid shapes, round, straight cylindrical, single O.D. (outside diameter), and with bore. At this point we are referred to Chart 1, which appears in Figure 5.15. Looking at Chart 1, take the following branches: round concentric bore, and single bore diameter. The last branch terminates with the code value B01, which corresponds to the first three digits of the code for the bushing.

As previously noted, the second segment of the part family code describes the complexity of the special features of the part being coded, which in this case is a bushing. In the DCLASS system, special features include form features (holes, etc.), heat treatments, and surfacing finishing treatments (plating, painting, anodizing, etc.). Table 5.3 provides the various code values that can be used to code the complexity of the special features. For the bushing, we will specify a complexity code of ''1,'' as it has a hole and does not require heat treatment or surface finishing.

The third segment of the part family code refers to the size of the part. Table 5.4 is used to select the appropriate value. Using this table, a ''2'' would be placed in the code for the bushing.

The fourth segment denotes the precision of the part being coded. Precision in DCLASS represents a composite of tolerance and surface finish. Table 5.5 lists the five classes of precision used. Class 1 represents very tight tolerances and a precision-ground or lapped surface finish. At the other extreme, Class 5 represents loose tolerances and a rough-cast or flame-cut surface. A part with a precision code of ''1'' requires careful processing with careful inspection. The bushing under consideration requires no unusual tolerances or surface finish, so a code value of ''3'' will be specified.

The final segment of the part family code contains the material type. Referring back to Figure 5.8, we can see that the bushing is to be made out of stainless steel. The logic tree in Figure 5.16 is used to determine the appropriate code value for stainless steel. Looking at the logic tree in Figure 5.16, we would select the

TABLE 5.3
Complexity code for special features

Feature complexity code	No. of special features
1	1
2	2
3	3
4	5
5	8
6	13
7	21
8	34
9	>34

Source: Reprinted with permission from Ref. 4.

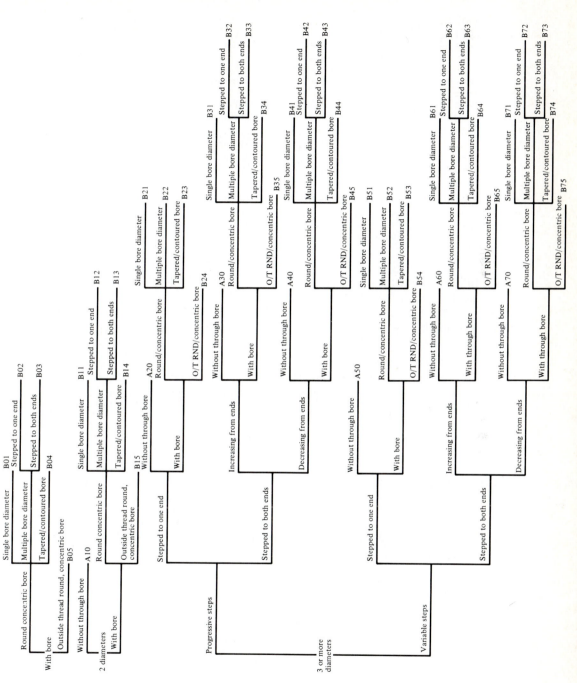

FIGURE 5.15 DCLASS Chart 1 logic tree. (Reprinted with permission from Ref. 4.)

TABLE 5.4
DCLASS size code

Size code	Maximum dimension		Description	Examples
	English (in).	Metric (mm)		
1	0.5	10	Subminiature	Capsules
2	2	50	Miniature	Paperclip box
3	4	100	Small	Large matchbox
4	10	250	Medium small	Shoebox
5	20	500	Medium	Breadbox
6	40	1,000	Medium large	Washing machine
7	100	2,500	Large	Pickup truck
8	400	10,000	Extra large	Moving van
9	1,000	25,000	Giant	Railroad boxcar

Source: Reprinted with permission from Ref. 4.

following branches: metals, ferrous metals, steels, high-alloy steels, and stainless steel. The appropriate code for the material type is "A7."

We have now completed the DCLASS part of family code for the bushing. This procedure was not difficult, and it would be easy to computerize. Dell Allen, at Brigham Young University, has developed a general-purpose computer system for processing classification and decision-making logic [5]. This system is available for several types of computers (micros, minis, and mainframes). In the next chapter it will be used to code some parts and to prepare some process plans. Most commercial coding systems are used similarly to the above DCLASS example, although some may be more difficult to use because more work is required to determine what the appropriate code should be.

5.5.2 MICLASS Coding System

MICLASS stands for Metal Institute Classification System; it was developed by the Netherlands Organization for Applied Scientific Research. MICLASS is one of the more popular commercial systems available in the United States. As noted

TABLE 5.5
DCLASS precision class code

Class code	Tolerance	Surface Finish
1	≤0.0005 in.	≤ 4 rms
2	0.0005–0.002 in.	4–32 rms
3	0.002–0.010 in.	32–125 rms
4	0.010–0.030 in.	125–500 rms
5	>0.030 in.	>500 rms

Source: Reprinted with permission from Ref. 4.

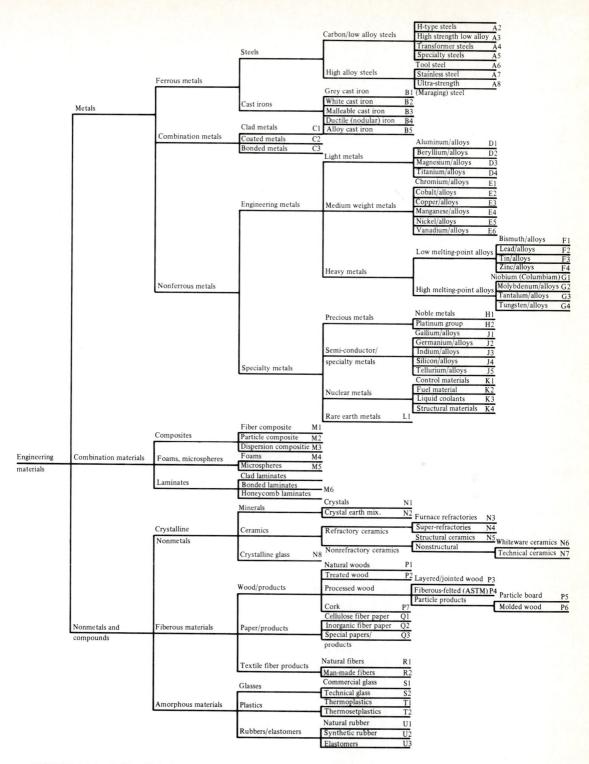

FIGURE 5.16 DCLASS logic tree for material code. (Reprinted with permission from Ref. 4.)

earlier, the MICLASS system consists of two major sections. The first section is a 12-digit code which is used to classify the engineering and manufacturing characteristics of a part: main shape, shape elements, position of the elements, main dimensions, ratio of the dimensions, an auxiliary dimension, tolerances, and material. This section is mandatory.

The first four digits deal with the form: main shape, shape elements, and the position of these elements (see Figure 5.17). The main shape of a part is the form of the final product as depicted in the drawing. It could be a rotational part, a boxlike part, a flat part, or some other nonrotational part. Shape elements are part features such as holes, slots, and grooves. The next four digits provide dimensional information: the main dimension, the ratio of the various dimensions, and the auxiliary dimension. The use of the auxiliary dimension varies with the main shape of the part; in general, it provides additional size information. Digits 9 and 10 contain tolerance information, and the final two digits provide a machinability index of the material.

The second section of a MICLASS code is optional. It can contain as many as 18 characters tailored to meet the specific needs of a company, such as vendors, lot sizes, costs, and producibility tips (similar to the expansion of CODE). The first 12 digits are universal in that the definitions used for the various digits should not change from company to company. The advantage is that a plant or division can read information contained in a part code received from another facility using the MICLASS code. A disadvantage arises when the code does not provide the necessary information in the first 12 digits; modifications may be difficult to make.

Manually coding several thousand parts using the 30-digit MICLASS code would be a very tedious and time-consuming job, and errors would probably be

FIGURE 5.17   MICLASS code structure [3].

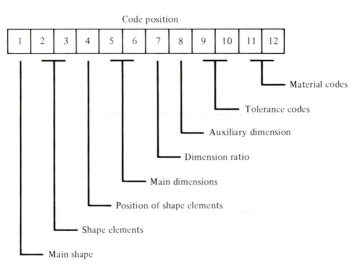

made. Consequently, the MICLASS system provides several interactive computer programs to assist the user. For example, one of the programs prompts the user with questions about a part to be coded. Figure 5.18 represents a part to be coded using the MICLASS interactive program, and Figure 5.19 lists the interactive dialog between the computer and the user. Note that the questions relate to characteristics of the part such as shape, size, materials, and tolerances. Responses are "yes" or "no" or dimensions; consequently, no in-depth computer knowledge is required. As the user answers questions, the program gathers the data required to code the part. The computer then assigns a code to the part and stores this information in the MICLASS data base. A code of "1271 3231 3144" is assigned to this part. Possible errors that could be made in coding a part are

FIGURE 5.18 Example part to be coded by MICLASS. (Reprinted with permission from Ref. 20.)

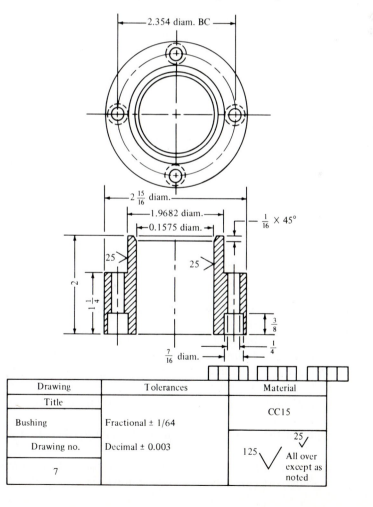

Drawing	Tolerances	Material
Title		
Bushing	Fractional ± 1/64	CC15
Drawing no.	Decimal ± 0.003	
7		125 / All over except as noted

```
VERSION-A-

3 MAIN DIMENSIONS (WHEN ROT. PART D.L AND O)?  2.9375  2  0
    DEVIATION OF ROTATION FORM?  NO
    CONCENTRIC SPIRAL GROOVES?  NO
TURNING ON OUTERCONTOUR (EXCEPT ENDFACES)?  YES
    SPECIAL GROOVES OR CONE(S) IN OUTERCONTOUR?  NO
    ALL MACH. DIAM. AND FACES VISIBLE FROM ONE END (EXC. ENDFACE + GROOVES)?   YES

INTERNAL TURNING?  YES
    INTERNAL SPECIAL GROOVES OR CONE(S)?  NO
    ALL INT. DIAM. + FACES VISIBLE FROM 1 END (EXC. GROOVES)?  YES

ALL DIAM. + FACES (EXC. ENDFACE) VISIBLE FROM ONE SIDE?  YES

ECC. HOLING AND/OR FACING AND/OR SLOTTING?  YES
    IN INNERFORM AND/OR FACES (INC. ENDFACES)?  YES
    IN OUTERFORM?  NO

ONLY KEYWAYING ETC.?  NO

MACHINED ONLY ONE SENSE?  YES
    ONLY HOLES ON A BOLTCIRCLE AT LEAST 3 HOLES?  YES

FORM-OR THREADING TOLERANCE?  NO

DIAM. ROUGHNESS LESS THAN 33 RU (MICRO-INCHES)?  YES
    SMALLEST POSITIONING TOL. FIELD?  .016
    SMALLEST LENGTH TOL. FIELD?  .0313

MATERIAL NAME?  CC15

CLASS.NR. = 1271  3231  3144
.........................

DRAWING NUMBER MAX 10 CHAR?  7

NOMENCLATURE MAX 15 CHAR?  BUSHING

CONTINUE [Y/N]?  N

PROGRAM STOP AT 4690
```

FIGURE 5.19   Interactive computer dialog for MICLASS coding example. (Reprinted with permission from Ref. 20.)

minimized by this procedure. Once stored in the data base, this information is available for analysis.

As stated earlier, the MICLASS system includes several programs to assist in the analysis of coding and classification information stored in the data base. These programs can be used for design classification, elimination of design duplication, manufacturing standardization, improving control and speed of material flow, optimization of machine tool purchase and use, and improving efficiency throughout design and manufacturing.

Some of the specific functions performed by the MICLASS programs are:

1. *Data management.* This provides the capability to manipulate data and files (including sorting), correct errors, and list data.

2. *Design retrieval.* Designs with the same or similar coded numbers can be retrieved. Note that similar parts as well as duplicate parts can be retrieved; the variation of the coded parts retrieved is controlled by the user.

3. *Production mix.* After hundreds or thousands of parts have been coded, analysis of the information contained in the data base can be very time-consuming. Programs are provided for assistance; for example, there are programs to produce graphs of the occurrence of a given classification code, or of a specific part of a code, or of combinations of a code.

One reason the MICLASS system has been so successful is the computer tools that are provided for the user.

5.6 FACILITY DESIGN USING GROUP TECHNOLOGY

Once parts have been grouped into families, determining how to arrange the machines in the factory can be a major problem. Facility layout is very important because, if the machines are poorly located, manufacturing costs can increase significantly.

There are three basic ways to arrange machines in a factory: by line, by function, and by group. Figure 5.20 illustrates each of these layouts. In a line layout, the machines and other workcenters are arranged in the sequence in which they are used. The work content at each location is balanced so that materials can flow through the line in a continuous manner. This type of layout is normally used in simple process industries, in continuous assembly, and for mass-produced components used in large quantities.

In a function-type layout, machines of a specific type are grouped together, as a lathe section, a grinder section, a drill section, etc. This layout can result in significant amounts of material handling, a large amount of work-in-process inventory, excessive setups, and long manufacturing lead times. All of these increase costs. In addition, function-type layouts are more difficult to manage because of the complexity of part routings. For instance, knowing the capacity of a manufacturing facility is fundamental to scheduling when parts are to be made. However, determining the capacity of a function-type layout involves considering all parts that require the use of each type of machine, which can involve a large amount of data.

In a group-type layout, machines are arranged as cells. Each cell is capable of performing manufacturing operations on one or more families of parts. Consequently, the capacity of a cell can be determined by considering only the families of parts that utilize that cell. As a result, this layout should be easier to manage. This is just one reason why a group layout may be more desirable; other reasons will be discussed in Section 5.9.

If a group-type layout is desired, it would be logical to define processes that correspond to one or more families of parts. Therefore, machines used to produce a family of parts might be grouped together in a cell. The procedure of forming

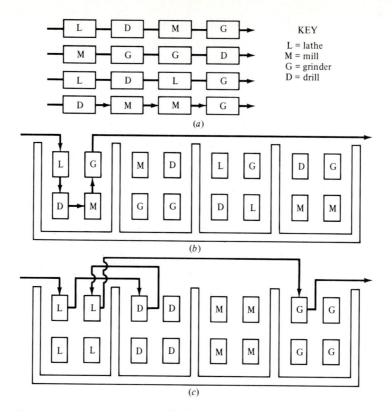

FIGURE 5.20   Three types of plant layouts: (*a*) line layout; (*b*) group layout; (*c*) functional layout.

cells is sometimes called machine–component grouping. As noted earlier in this chapter, production flow analysis is one method used to group parts (into families) and to locate machines in a factory. However, PFA can require considerable judgment. As a result, additional techniques have been proposed to aid in machine–component grouping. One of these is the single-linkage clustering algorithm (SLCA), which will be discussed in this section.

Production flow analysis (PFA) utilizes a machine–component chart. The objective is to group together parts that require a similar process. When PFA has been completed, parts having the same process will be grouped together as shown earlier in Figure 5.4. The 1's that denote the operations to be performed will be clustered together in the machine–component chart. This type of analysis provides an excellent opportunity to apply a clustering algorithm. McAuley [17] introduced a SLCA to the machine–component grouping problem. This algorithm utilizes a similarity coefficient method to group together parts that require a similar process. The SLCA was developed by Sneath [24] and is only one of many clustering algorithms that have been developed.

The machine–component chart in Figure 5.21 will be used to describe the SLCA utilized by McAuley to form part groups. A similarity coefficient is calcu-

Component

		1	2	3	4	5	6	7	8	9	10
	A	1	1				1	1	1		
	B			1	1	1				1	1
Machine	C			1	1	1					1
	D			1	1		1			1	
	E	1	1					1	1		

FIGURE 5.21 Machine–component chart.

lated for each pair of machines to determine how "alike" the two machines are in terms of the number of parts which visit both machines and the number of parts that visit each machine. A 2×2 table is a convenient way to show the different alternatives, as illustrated in Figure 5.22. For instance, the a in Figure 5.22 denotes that a part visits both machines, and the b denotes that the part visits machine i but not machine j. McAuley defined the similarity coefficient between two machines as

$$S_{ij} = \frac{a}{a + b + c} \tag{5.1}$$

where
S = similarity coefficient between machines i and j
a = number of parts common to both machines
b, c = number of parts that visit one or the other of machines i and j, but not both

FIGURE 5.22 A 2×2 table.

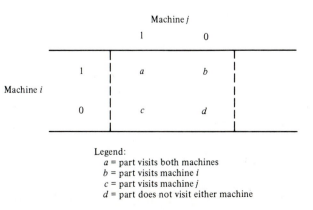

Machine j

		1	0
	1	a	b
Machine i	0	c	d

Legend:
a = part visits both machines
b = part visits machine i
c = part visits machine j
d = part does not visit either machine

The similarity coefficient between machines B and C in Figure 5.21 is $\frac{4}{5}$, or 0.8.
The SLCA consists of the following steps:

1. A pairwise similarity coefficient is calculated for each machine. These coefficients could be displayed in a similarity matrix like the one in Figure 5.23. Since the matrix is symmetric, only the lower triangular portion is needed.
2. The similarity matrix is scanned to locate the largest similarity coefficient. This designates the two machines that will form the initial cluster.
3. The similarity matrix is then scanned to locate the largest remaining coefficient. The associated machines are grouped together.
4. Steps 2 and 3 are repeated until all the machines are clustered together into one group.

The algorithm is terminated when all the machines are clustered into one group or until the remaining similarity coefficients are below some specified level. This level is sometimes called a threshold. The threshold level can be used to control the number of clusters formed.

The results of applying the SLCA to the machine–component chart in Figure 5.21 can be seen in Table 5.6. A dendrogram, as shown in Figure 5.24, provides a more descriptive means of showing the results. The abscissa of a dendrogram has no special meaning; in this example it denotes machines. The similarity coefficient scale, usually having a range of 0 to 1.0, is represented on the ordinate. The dendrogram in Figure 5.24 depicts the same results as Table 5.6. In Figure 5.24, each branch at the lowest level represents one machine. Moving toward the top of the dendrogram, the branches merge into new branches representing clusters of machines. The value of the similarity coefficient at which this occurs is denoted on the left scale of the dendrogram. Looking at Figure 5.24, we can see that at a similarity value of 0.8, machines B and C are grouped together and machines A and E are likewise grouped together. At a value of 0.5, machine D is clustered with machines B and C. The machines are clustered together into one group at a similarity coefficient value of 0.12.

FIGURE 5.23
Similarity matrix.

Machine

Machines		A	B	C	D	E
	A	–				
	B	0	–			
	C	0	0.8	–		
	D	0.12	0.50	0.33	–	
	E	0.80	0	0	0	–

TABLE 5.6
Results of SLCA

Cluster	Machines	Similarity coefficient
1	BC	0.8
2	AE	0.8
3	BCD	0.5
4	BCD	0.33
5	BCDAE	0.12

Looking at Table 5.6 or Figure 5.24, we can see that the SLCA will group machines together into four different sets of clusters. The algorithm, however, does not denote which of these is the ideal way to group the machines. This is one of the drawbacks of using this type of clustering algorithm, because more information is needed to determine when to stop forming groups. Some of the items that must be examined are: the number of intergroup/intragroup movements, machine utilization, planning and control factors, and managerial considerations. In addition, no insight is provided into when a "bottleneck" machine should be duplicated. A bottleneck machine is a machine that is required by a large number of parts from a different group. Another weakness of this algorithm is that two or more clusters may be joined together because two of their members are similar even though the majority of their members may have relatively low similarity coefficients. This is called *chaining*.

An algorithm of this type does provide several advantages. One is that the algorithm provides a structured way to group machines together, which is very significant, especially if many machines are involved. Also, the algorithm is easy to computerize. Furthermore, to overcome some of the disadvantages noted above, several enhancements have been proposed.

One enhancement is to define a different similarity coefficient. Considerable work has been done in this area. The initial coefficient presented was defined as

$$S_{ij} = \frac{a}{a + b + c}$$

FIGURE 5.24
Example dendrogram.

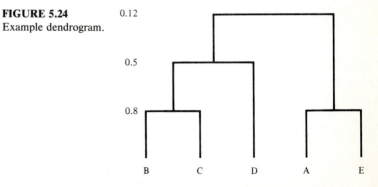

where equal weights are given to a, b, and c. We could, however, assign more weight to the number of operations that the machines have in common. Dice, as reported in Anderberg [1], for instance, developed a similarity coefficient defined as

$$S_{ij} = \frac{2a}{2a + b + c} \tag{5.2}$$

By varying the weighting factor, this coefficient can take on many forms. See Ref. 1 for a good discussion on similarity coefficients.

It was mentioned above that one problem with the SLCA is a potential for chaining. This can be reduced by using an average-linkage clustering algorithm (ALCA). In this type of algorithm, the similarity coefficient is based on the average similarity of all pairs involved. There are many ways in which an average similarity coefficient can be computed. One of these is the unweighted pair-group method developed by Sokal and Michner. This algorithm is explained in Ref. 24. They defined the similarity coefficient between two groups i and j as

$$S_{ij} = \frac{s_{ij}}{N_i * N_j} \tag{5.3}$$

where s_{ij} = the sum of pairwise similarity coefficients between all members of the two groups

N_i, N_j = the number of entities (machines) in groups i and j, respectively

The term S gives a measure of the average similarity for links between the two groups, i and j. The number of pairwise similarities between groups i and j is the product of N_i and N_j. For instance, if group i consists of machines A and B, and group j consists of machines C, D, and E, then S is computed using the expression

$$s_{AB,CDE} = \frac{s_{A,C} + s_{A,D} + s_{A,E} + s_{B,C} + s_{B,D} + s_{B,E}}{N_{AB} * N_{CDE}}$$

$$= \frac{s_{A,C} + s_{A,D} + s_{A,E} + s_{B,C} + s_{B,D} + s_{B,E}}{2 \times 3}$$

The machine–component chart in Figure 5.21 will be used to illustrate an ALCA that utilizes a similarity coefficient defined by Eq. (5.3).

The ALCA is composed of the following steps (the first two steps are the same as for the SLCA):

1. A pairwise similarity coefficient is calculated for all the machines in Figure 5.21. Figure 5.23 shows the results of these calculations.
2. The similarity matrix is scanned to locate the largest similarity coefficient. This designates the two machines that will form the initial cluster.
3. The average similarity coefficients between the new cell and the remaining cells (machines) are calculated. The similarity matrix is then revised using these new values and the new cell.
4. The revised similarity matrix is then scanned to locate the largest coefficient. The associated machines are grouped together.

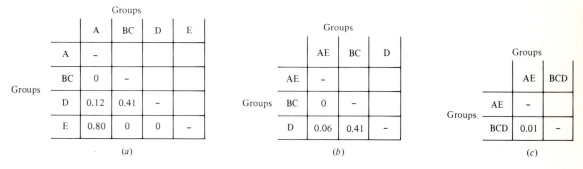

FIGURE 5.25   Revised similarity matrix interations: (a) interaction 1; (b) interaction 2; (c) interaction 3.

5. Steps 3 and 4 are repeated until all the machines are clustered together into one group.

After performing steps 1 and 2, we see that machines B and C are the first machines to be grouped together. The similarity matrix is revised by calculating the similarity coefficients between cell BC and the remaining machines: A, D, and E:

$$S_{BC,A} = \frac{s_{A,B} + s_{A,C}}{N_{BC} * N_A} = \frac{0 + 0}{2 \times 1} = 0 \tag{5.4}$$

$$S_{BC,D} = \frac{s_{D,B} + s_{D,C}}{N_{BC} * N_D} = \frac{0.5 + 0.33}{2 \times 1} = .415 \tag{5.5}$$

$$S_{BC,E} = \frac{s_{E,B} + s_{E,C}}{N_{BC} * N_E} = \frac{0 + 0}{2 \times 1} = 0 \tag{5.6}$$

The revised similarity matrix is shown in Figure 5.25a.

Repeating steps 2 and 3, we see that machines A and E are grouped together. The new similarity matrix is depicted in Figure 5.25b.

During the third iteration, groups D and BC are clustered to form cell BCD. The revised similarity matrix is shown in Figure 5.25c. During the last iteration groups AE and BCD are clustered to form cell AEBCD.

Seifoddini [22] has incorporated in a model the ALCA described above and the intracell/intercell cost of materials handling to determine the best way to group machines into cells. His model identifies bottleneck machines and duplicates them when it is economically wise.

5.7   CELL EXAMPLE

In this section we will describe an automated prefinish gear machining cell being developed at a prominent aerospace firm for machining spur gears, like the gear shown earlier in Figure 5.6a. This cell is illustrated in Figure 5.26. In this cell are two fully automatic machining centers for turning the faces and the outside and

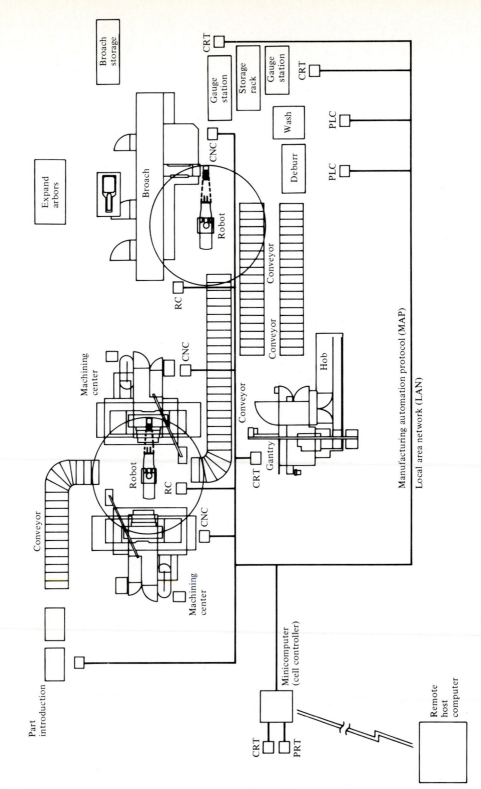

FIGURE 5.26 Spur gear cell. (Courtesy Garrett Engine Division).

inside diameters, a broach for machining internal splines, a hob for gear cutting, and inspection stations. In addition, robots and conveyors are used to fully automate the material handling. At present, parts are processed in a functional shop layout with many different routings, and the parts are moved manually from machine to machine on carts. In contrast, all parts processed in this cell will have similar routings within the cell, and all material handling will be fully automated, including tool changing.

Figure 5.27 contains an outline of the process plan for the parts traveling through this cell. Material in the form of rough blanks will be manually loaded on the conveyor in front of the first machining center by a cell operator. At that time

FIGURE 5.27   Prefinish process plan for spur gear.

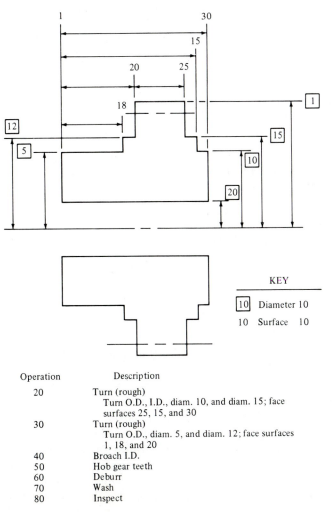

Operation	Description
20	Turn (rough) Turn O.D., I.D., diam. 10, and diam. 15; face surfaces 25, 15, and 30
30	Turn (rough) Turn O.D., diam. 5, and diam. 12; face surfaces 1, 18, and 20
40	Broach I.D.
50	Hob gear teeth
60	Deburr
70	Wash
80	Inspect

the operator will also enter part-related information (part number, date, work order number, etc.) into a computer terminal. The cell control computer will then take over control of processing until the last operation is completed in the cell. It should be noted that not all operations required to make a spur gear will be performed in this cell, because, after the gear teeth are cut, the gears will be moved to a heat-treat facility and then to some finish-grinding operations. The heat-treatment operation requires a special facility.

Group technology was used to classify and code parts that have a potential for being processed in this cell. Approximately 250 different parts were selected and grouped into seven families. The cell did not have the capacity to handle the demand for all of these parts. Consequently, this list of parts was reduced to the parts with the highest demands, approximately 50 parts. At that time manufacturing engineers spent a lot of time studying the operations performed on each part in each family with the objective of developing group tooling and eliminating operations by utilizing the flexibility of the new equipment. These analyses resulted in a 70% reduction in the number of different perishable tools required and eliminated an average of seven detailed operations for each part. Combining these results with the improved efficiency of the new equipment, and reductions in setup times, material handling, and work in process, it is estimated that the production time for this group of parts will be reduced by 80%. The annual savings have been predicted to be in the millions of dollars. But these savings will not be free, because the cost of developing the cell will also be in the millions of dollars.

5.8   ECONOMIC MODELING IN A GROUP TECHNOLOGY ENVIRONMENT

Several types of analytical models have been developed to improve the understanding of what benefits group technology provides in the manufacturing environment. Two types of models will be discussed in this section to illustrate what has been accomplished. The first model assumes that the reader is familiar with mathematical programming concepts. If this is not a valid assumption, reading the material will provide some insight into cost modeling and optimization.

5.8.1   Production Planning Cost Model

The production planning cost model assumes that the objective is to develop a manufacturing production plan that will minimize production costs over a planning horizon of T time periods. Normally a time period will represent a month. Manufacturing costs can be divided into three types: direct labor costs, direct material costs, and burden (overhead) costs.

Direct labor costs consist of all types of labor costs for production workers who are engaged directly in manufacturing operations to convert raw materials into finished products. For example, it is usually possible to charge the time of a machine operator directly to each job on which he or she works. However, it is much more difficult—practically impossible, in fact—to charge toolroom attendents' time directly to each job handled in the shop. Consequently, the toolroom attendent's time is usually classified as an indirect cost and is allocated to all parts made in some time period, such as a month. Other names for this type of cost are burden or overhead cost. The same type of explanation can be applied to direct material costs and indirect material costs. The cost of the metal from which a gear is fabricated is an example of a direct material cost. The costs of the coolant, grease, and oil used by the machine tools are examples of indirect costs.

The model described in this section can be used to minimize the production costs of a group technology cell while satisfying product demand. The decision variables are the production quantities, overtime, inventories, and setups for parts and part families for each period over the planning horizon. A decision must be made regarding what value should be assigned to each decision variable. The best values for the decision variables are those that minimize the value of the objective function, Eq. (5.7):

Minimize:

$$Z = \sum_t^T \left[L \left(\sum_i V_{it} P_i + W_{it} S_i \right) + \sum_j \left(F_{jt} G_j + o O_t \right) \right. \quad \text{(labor)}$$

$$+ \sum_i V_{it} \left(M_i + B_i \right) \quad \text{(material and burden)}$$

$$\left. + \sum_i h_i I_{it} \right] \quad \text{(inventory holding)} \quad (5.7)$$

subject to:

$$V_{it} + I_{it-1} - I_{it} = d_{it} \qquad \forall i,t \quad (5.8)$$

$$\sum_i V_{it} P_i + \sum_i W_{it} S_i + \sum_j F_{jt} G_j - O_t + U_t = R_t \quad \forall i,t \quad (5.9)$$

$$\sum_{i \varepsilon j} W_{it} \le CF_{jt} \qquad \forall t,j \quad (5.10)$$

$$V_{it} \le CW_{it} \qquad \forall i,t \quad (5.11)$$

$$F_{jt}, W_{it} \ge 0 \qquad \forall i,j,t \quad (5.12)$$

$$F_{jt}, W_{it} \le 1 \qquad \forall i,j,t \quad (5.13)$$

$$F_{jt}, W_{it} = 1 \text{ or } 0 \qquad \forall i,j,t \quad (5.14)$$

$$O_t \le E \qquad \forall t \quad (5.15)$$

$$U_t \le A \qquad \forall t \quad (5.16)$$

$$V_{it}, U_t, O_t, I_{it} \ge 0 \qquad \forall i,t \quad (5.17)$$

where V_{it} = production quantity of part i in period t
I_{it} = inventory of part i at the end of period t
U_t = undertime associated with the plan in period t
O_t = overtime associated with the plan in period t
W_{it} = number of setups for part i in period t
F_{jt} = number of setups for family j in period t
t = period in the planning horizon
T = planning horizon
L = labor rate
P_i = unit processing time for part i
S_i = standard setup time for part i
G_j = standard setup time for part family j
o = percent increase in the labor rate for overtime
M_i = standard material cost per unit for part i
B_i = burden cost per unit of part i
h_i = holding cost per unit for part i
R_t = regular hours scheduled for period t
d_{it} = demand for part i in period t
C = an arbitrarily large constant
E = maximum overtime permitted
A = maximum undertime permitted

The objective function measures for a given production plan the three types of costs noted above. Direct labor is computed by applying a single labor rate to the production time. The production time includes run and setup times, and considers possible differences between regular and overtime pay rates. Differences in setup times between parts in a family and between families are also permitted. If group tooling has been developed, there may be virtually no setup time required between parts in the same family; however, the setup time between families may be significant. This situation can be accommodated. Material and overhead costs are assumed to be linearly related to production volume. Inventory holding costs are applied on a per-unit per-period basis.

Looking at the constraints, we see that constraint (5.8) ensures that the demands are satisfied for all parts in all periods in the planning horizon. Constraint (5.9) limits the capacity of the cell and computes the amount of overtime and undertime utilized in a production plan. The next two constraints, (5.10) and (5.11), ensure that setup time is incurred for a part and for a part family when appropriate. Note that these setups are limited to one per period and must be integer values by constraints (5.12), (5.13), and (5.14). Constraints (5.15) and (5.16) limit the amount of overtime and undertime.

Examining this model in some detail, we can see that some simplifying assumptions were made. For example, it represents a cell as a single resource with limited capacity and considers the machine and setup times as aggregate values. Consequently, if there is more than one machine in the cell, machine loading and balancing must be considered when the cell is designed and when parts are assigned to be processed in this cell. As a result of including part and family setups as decision variables, this model is a mixed-integer linear programming problem. Although some simplifying assumptions were made to keep the model simple, solving the model is very difficult. It belongs to a class of problems termed NP-hard, which means that in general no method exists for solving problems of a practical size.

Graves [7] used this model to study the effects of group technology on production planning; it is similar to a production planning model proposed by Manne in 1958 and later refined by others [13]. Faced with an NP-hard problem, Graves examined what some others had done to develop a more tractable formulation. He found that Hax and Meal [14] examined a similar formulation for planning production and end products. They simplified the required computations by aggregating some of the data. Parts were aggregated into families based on similarities in setup requirements. These families were aggregated into types based on similar seasonal demand patterns and production rates. With these aggregations, the modified model could be used to solve practical problems. However, the solution obtained was nonoptimal.

The fact that the solutions were nonoptimal does not mean that they are worthless, because a good solution is usually acceptable to management in a practical environment. Graves also observed that there is a similarity in the aggregates of "families" and "types" in this model with parts and families in group technology. This led to an approach for aggregating some of the variables in the above production planning model. Following is the resulting model:

Minimize:

$$Z = \sum_{t}^{T} [\, L\,(\,V_tP) + oO_t + V_t\,(M + B\,) + HI_t] \qquad (5.18)$$

subject to:

$$V_t + I_{t-1} - I_t\ = D_t \qquad \forall_t \qquad (5.19)$$
$$V_tP - O_t + U_t = R_t \qquad \forall_t \qquad (5.20)$$
$$O_t \leq E \qquad \forall_t \qquad (5.21)$$
$$U_t \leq A \qquad \forall_t \qquad (5.22)$$
$$V_t,\ U_t,\ O_t,\ I_t \geq 0 \qquad \forall_t \qquad (5.23)$$

where t = period number
   T = planning horizon
   L = labor rate
   V_t = aggregate production volume
   P = aggregate processing time per unit
   o = overtime rate
   O_t = overtime hours
   M = aggregate material cost per unit
   B = aggregate burden cost per unit
   H = aggregate holding cost per unit

I_t = aggregate inventory
D_t = aggregate demand
R_t = regular hours available
E = maximum overtime per period
A = maximum undertime per period

Graves provides some procedures for performing the aggregation computations. Basically, weighted averages based on the planned demands are calculated for the material, holding, and burden costs. The aggregate processing time is a weighted average; however, it also includes an estimate of setup time requirements for the entire planning horizon. The production quantities and inventories are simple summations.

This formulation is identical to the fixed workforce-linear cost model used with end products [14]. Graves used this model to analyze the cost structure of a group technology cell, to perform sensitivity analysis of the model parameters, and to analyze modifications to the family and machine compositions. An interesting observation has to do with the different approaches used to arrive at the same model formulation. Originally the model was developed without consideration of group technology concepts. Hax and Meal, in the process of making their model more computationally tractable, aggregated the parts into families and types based on setup requirements, production rates, and seasonal demand patterns. Graves demonstrated that group technology provides a logical process for doing this. Although group technology was initially developed to assist in engineering design and manufacturing engineering, following the path taken to develop the above model gives some insight into how group technology aids in simplifying the management of the manufacturing environment.

5.8.2 Group Tooling Economic Analysis

Group tooling can provide benefits in a manufacturing environment. However, one should not assume that group tooling will result in lower costs than conventional tooling. This section, which is based on material presented in Ref. 24, will explain how to determine which is the least-cost strategy in a particular case.

The cost of conventional tools required to produce a batch of parts may be expressed as

$$C_{ct} = \sum_{i=1}^{P} C_i \tag{5.24}$$

where p = number of fixtures used
C_i = cost of a fixture
C_{ct} = total tooling cost using the conventional method

In a similar manner the cost of the group tooling required to produce a family of parts is

$$C_{gt} = C_g + \sum_{a=1}^{q} C_a \tag{5.25}$$

TABLE 5.7
Conventional/group tooling data

Items	Conventional tooling	Group tooling
Cost of fixture	$500	$480
Cost of each adapter	—	80
Number of pieces/family to be produced	100	100

where q = number of the adapters used
C_g = cost of group fixture
C_a = cost of group adapter
C_{gt} = total tooling cost using group tooling

These results can be used to calculate the unit tooling costs:

$$C_{uc} = \frac{C_{ct}}{N} \tag{5.26}$$

$$C_{ug} = \frac{C_{gt}}{N} \tag{5.27}$$

where N = number of parts produced
C_{uc} = unit tooling cost for conventional method
C_{ug} = unit tooling cost for group tooling method

The following example will provide some additional insight into how the cost of these two methods can differ.

Example. A family of 100 parts are to be produced. Table 5.7 contains the costs of the individual tools. Table 5.8 compares the tooling costs and the unit tooling costs

TABLE 5.8
Tooling cost companion

Number of different parts in part family	C_{tc}	C_{uc}	C_{tg}	C_{ug}
1	$ 500	$5.00	$ 560	$5.60
2	1000	5.00	640	3.20
3	1500	5.00	720	2.40
4	2000	5.00	800	2.00
5	2500	5.00	880	1.76
6	3000	5.00	960	1.60
7	3500	5.00	1040	1.49
8	4000	5.00	1120	1.40
9	4500	5.00	1200	1.33
10	5000	5.00	1280	1.28

for both methods as the number of parts in a family varies. The following illustrates how the computations were made.

1. When $p = 1$, $q = 1$,

$$C_{ct} = C_1 = 500$$

$$C_{uc} = \frac{C_{ct}}{N} = \frac{500}{100} = 5.00$$

$$C_{gt} = C_g + C_1 = 480 + 80 = 560$$

$$C_{gt} = \frac{C_{ug}}{N} = \frac{560}{100} = 5.60$$

2. When $p = 2$, $q = 2$,

$$C_{ct} = pC_1 = (2)(500) = 1000$$

$$C_{uc} = \frac{C_{ct}}{2N} = \frac{1000}{2 \times 100} = 5.00$$

$$C_{gt} = C_g + qC_1 = 480 + (2 \times 80) = 640$$

$$C_{ug} = \frac{C_{gt}}{2N} = \frac{640}{2 \times 100} = 3.20$$

FIGURE 5.28 Tooling costs.

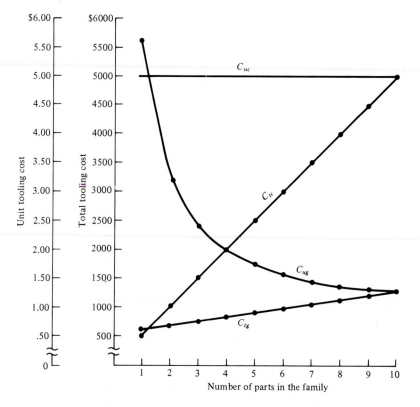

Figure 5.28 is a plot of the total tooling costs of these two methods in relation to the number of different parts in the family. Figure 5.28 also plots unit tooling costs instead of total costs. The rate of increase in total cost for conventional tooling is much greater than for group tooling. Also, as the number of parts in a family increases, the unit tooling cost for group tooling become more economical.

5.9 ECONOMICS OF GROUP TECHNOLOGY

As more and more companies successfully implement group technology programs, significant saving are being reported. The following savings are typical:

1. 50% in new parts design
2. 10% in number of drawings
3. 60% in industrial engineering time
4. 20% in plant floor-space requirements
5. 40% in raw material stocks
6. 60% in in-process inventories
7. 70% in setup times
8. 70% in through-put time

In addition, a successful group technology program provides many benefits that are difficult to quantify, such as simplification of the manufacturing environment, improvement of the work environment, better quality, and improved product designs. A brief explanation of why these benefits might be expected will now be given.

5.9.1 Benefits in Design

Whenever a new part is required, the design engineer is faced with the problem of finding a similar part that has been designed before and modifying the design, or designing a new part. If a similar part exists, it is probably filed by part number with a short descriptive title. Since most companies have many thousands of drawings on file, it is virtually impossible to retrieve the part drawing unless the engineer can remember the part number or descriptive title. So the engineer usually designs a new part, which increases design costs and complicates the manufacturing environment by introducing yet another part number into the system. A new design will result in several thousand dollars in manufacturing preparation costs. Table 5.9 lists some activities that are affected by the introduction of a new part.

In addition, manufacturing management becomes more difficult with the introduction of each new part. For example, the complexity of sequencing parts through machines increases, machine tools must be acquired and delivered, and long and frequent setups may reduce machine utilization and result in large

TABLE 5.9
Some activities affected by a new design

Engineering activities	Manufacturing activities
Design	Equipment and facilities planning
Analysis	Process planning
Drawing	Tool design
Testing	Tools and gauges
Auditing	Time standards
Documentation	Production planning and control
	Scheduling
	Quality control
	Accounting
	Cost estimating

through-put times. This situation occurs quite often in many batch-type manufacturing organizations. Figure 5.29 depicts 2100 randomly selected drawings taken from all components moving into production at one company during a 10-week period [22]. Note that after 6 or 7 weeks the number of new shapes has nearly leveled off. Although the curve will never become level, it illustrates that a classification and coding system can save a great deal of design work by facilitating the retrieval of designs of identical or similar parts. This will reduce the number of new parts that need to be designed, so that a "new" design may only require changes to an existing design. The same type of savings can be achieved in manufacturing by reducing the number of process plans that must be prepared for new part designs. In addition, this type of environment encourages standardization of design features, simplification of the design process, and improvement of cost-estimating systems.

FIGURE 5.29   New component shapes.

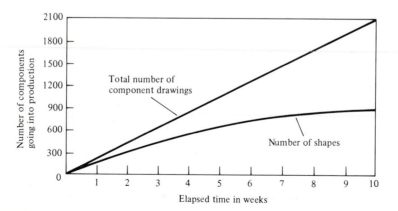

The integration of design engineering and manufacturing will produce significant benefits for a manufacturing firm. The associated benefits may dwarf benefits from anything else a firm may do. This integration should begin with design, and group technology is a very important element in most successful integration efforts. Group technology is concerned with the structuring of information so that it can be readily analyzed and transferred. Since a large portion of the integration of a firm involves the integration of information, group technology is a principal component of an integration program. This subject will be discussed in more depth in Chapter 12.

5.9.2 Benefits in Manufacturing

Large cost savings can be achieved by the application of group technology to production planning and control, manufacturing processes, tool design, and facility design. Consequently, many manufacturing organizations are implementing some type of group technology program.

Component variety complicates batch-type manufacturing to the point that many companies become job shops. In this type of manufacturing environment, there are usually many similar components used in different products. As noted above, implementation of a classification and coding system can facilitate a reduction in the variety of parts. In addition, if group scheduling is utilized, the number of components that must be considered is significantly reduced.

Most manufacturing firms have the capability to fabricate a relatively wide variety of parts in small lots. In order to stay competitive, most firms must be able to meet special customer requirements by providing customized features on their products. This must be done within lead times that are getting shorter and at a cost that provides enough profit for the firm to remain viable. As a result, many firms are attempting to automate as many functions as possible. Computer-aided process planning (CAPP) represents this reasoning. This subject will be discussed in depth in the next chapter; however, it should be noted that most CAPP systems are based on group technology concepts. Consequently, because group technology facilitates the development of CAPP, some of the benefits associated with CAPP should be credited to group technology.

Group technology facilitates a group layout of the factory. A group of machines is formed to perform all the operations required by one or more families of parts. The amount of time needed for each operation for each part can be determined; this is the basis for calculating the capacity required for each machine in the group. Often these machines are arranged as cells, which can reduce material handling costs by reducing travel and facilitating increased automation. Figures 5.30a and 5.30b illustrate part flow before and after group technology concepts have been utilized to streamline the production process. Cells also simplify scheduling by reducing the number of different types of parts that utilize a given group of machines. Since setup times are usually very short between different parts in a family, a group layout can also result in dramatic reductions in setup times. As setup times approach zero, batch sizes of parts can be reduced to a number

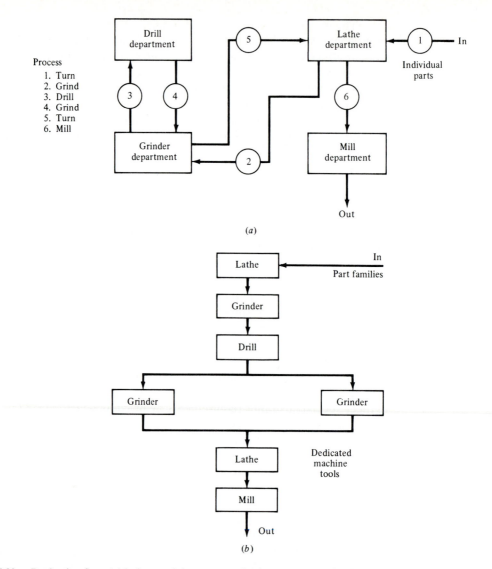

Process
1. Turn
2. Grind
3. Drill
4. Grind
5. Turn
6. Mill

(a)

(b)

FIGURE 5.30 Production flow: (a) before applying group technology concepts; (b) after applying group technology concepts.

approaching 1. This provides a manufacturing company more flexibility in meeting changing schedules. In addition, reduced setup times result in shorter through-put times, which lead to shorter manufacturing lead times and smaller inventories. All of these factors can provide some very important benefits in the competitive manufacturing environment.

Group tooling represents another possible source of cost reduction. Once cells and associated families have been established, it is normally possible to design group tools, such as group fixtures, which can be used with most if not all parts in a family. A set of tools for machining a part family can be grouped together and identified with a number. These tools could be preset on an automatic machine tool or kited in a tool crib. In some cases, whole turrets of tools could be preset and then interchanged with the current turret on the machine tool. Also, it should be noted that a numerical control (NC) program can be considered as a tool. An NC program is a type of program used to control an NC machine. A machine in this category does not need an operator to control each operation manually. The operator loads the part and possibly the appropriate tools, if the machine does not have an automatic tool handler. Then the operator pushes a start button to initiate the NC program, which controls the specific operations performed, including indexing the part. Most machines being installed in modern manufacturing firms are of the NC type. This subject will be discussed in detail in Chapter 9. Some companies assign tool numbers to NC programs. Group technology can reduce NC programming time, because once an NC program has been developed for one part in a family, it is usually a minor task to modify that program to make another part in the same family. Part family programming increases the productivity of NC operations by saving programming time, labor, and tape prove-out time. Therefore, group tooling can reduce tooling costs, NC programming costs, tooling inventory costs, and setup times. In addition, lead times are reduced, and, since fewer tools are required, management of tooling services is simplified.

Group technology simplifies production planning and control. The complexity of the problem has been reduced from a large portion of the shop to smaller groups of machines. Within a cell the problem is reduced to scheduling a relatively small number of parts through the machines in that cell. If setup times have been reduced using group tooling, the lot size may approach 1, and parts can be processed through the cell without consideration of lot size.

Costs of materials and purchased parts can also be reduced. Many manufacturing firms purchase a large number of their parts. Selecting the best vendor to supply a particular part is not an easy task. Use of group technology can assist in planning and controlling these purchases. By grouping purchased parts into families and identifying what vendors should be considered for a particular family, significant savings can be achieved and management of this function can be simplified.

5.9.3 Benefits to Management

It has been noted that group technology simplifies the environment of the manufacturing firm, which is a very significant benefit to management. One result of this simplification is a reduction of paper work. It can also improve the work environment. With a group layout, workers may be assigned to a machine cell in which a relatively small number of parts are completely processed. In this environment, the supervisor has in-depth knowledge of the work performed and better control; consequently, he or she can be more effective. From the worker's viewpoint, working in a machine cell means that the worker must know how a number of tasks are performed, and all of the tasks being completed can be seen by individual workers. This realization leads to better performance, higher morale, and better work quality.

Some additional factors of a social nature provide significant benefits to management. A group technology environment fosters greater job satisfaction because it leads to more worker involvement in decision making, personalized work relationships, and variety in tasks.

5.9.4.   Group Technology Advantages/ Disadvantages Summarized

Although group technology concepts are gaining widespread acceptance, one should understand the advantages and disadvantages when undertaking an application. Some of these are summarized below.

ADVANTAGES
1. A good coding and classification system provides design engineering with a system that facilitates:
   a. Efficient retrieval of similar parts
   b. Development of a data base containing effective product design data
   c. Standardization of designs
   d. Avoidance of design duplication
   e. Forming of part families
   f. Use of producibility tips
   g. Incorporation of engineering design changes into the engineering and manufacturing systems
2. A good coding and classification system provides manufacturing with a system that facilitates:
   a. Development of a computer-aided process planning system
   b. Retrieval of process plans for part families
   c. Development of standard routings for part families
   d. Development of machining cells
3. Standard routings facilitate the development of tooling groups, NC (numerical control) program groups, and standard setups for part families.
4. Production planning and control can be simplified.
5. Because production planning can be simplified, it can be more comprehensive.
6. Production scheduling can be simplified.
7. Machining cells can reduce in-process inventory, resulting in shorter queues and shorter manufacturing through-put times.
8. Improved machine utilization yields shorter setup times and better scheduling.
9. Part family data facilitates improving plant layout, which in turn can reduce materials handling costs.

10. Purchasing can be more effective. It is easier to choose the proper vendor because the many different parts and materials have been grouped into families, which reduces the complexity of the problem.
11. Management can be more effective because the environment has been simplified.

All of the above should lead to lower costs, and many will facilitate improved quality.

DISADVANTAGES
1. Installing a coding and classification system requires a large amount of time and effort; it is expensive.
2. If communication between design engineering and manufacturing is poor, as is often the case, difficulties may be encountered in installing a coding and classification system. It may not be very successful.
3. There are no accepted group technology standards. Consequently, there is no common implementation approach, and implementation is often difficult.
4. Groupings of machines may lead to poor utilization of some machines in the group. This is difficult for management to accept even though overall costs are reduced.
5. Large costs may be incurred in rearranging the plant into machine cells or groups.
6. Group technology concepts require changing how people work; therefore, employee resistance may be encountered.
7. Without strong support from top management, implementation of group technology will be difficult.

The benefits that can be realized from implementing group technology concepts in a company are very great. However, implementation should not be initiated without considering the problems that might be encountered, because they can cause the effort to fail.

5.10 SUMMARY

Within the last decade, American industry has realized that most parts are produced in small batches and that this manufacturing environment will extend into the foreseeable future. Any changes in this environment are expected to be toward even more production dedicated to small-batch production. Within this same time period international competition has increased significantly, forcing manufacturing firms to look for new ways to reduce costs. In this search, industry has rediscovered a philosophy that had been used informally for centuries. The group technology philosophy was formalized by Mitrofanov in the Soviet Union in the 1950s. However, widespread acceptance did not occur until industry realized that it needed some new ways of doing business.

In the 1980s CAD and CAM began to receive notable emphasis because of technology advances in reducing the cost of a computer and the promise that computers can reduce manufacturing costs. In this chapter, we have learned that group technology is a key element of CAD and CAM. Group technology has also been called the "glue" in the integration of CAD and CAM, because it provides a formal way for capturing information about parts. Once this information has been captured, it can be stored in a data base and retrieved for analysis. So it is not by chance that group technology achieved widespread acceptance in the 1980s.

Coding and classification is the best way to apply group technology. Although several systems are available, no single system has found universal acceptance because of the unique aspects of each manufacturing firm. One decision that each firm must make is whether to purchase a system or develop one. Success or failure can occur either way. However, the difficulties of implementing a coding and classification system should not be minimized. In addition, the costs and time required for implementation are not insignificant.

The application that utilizes group technology the most is process planning. This is the topic of the next chapter.

EXERCISES

1. Describe two examples of informal uses of group technology that have not been discussed in this chapter.
2. Given the data in Figure 5.31, apply the production flow analysis technique to determine what machines should be grouped together as one or more cells.

FIGURE 5.31   Machine–component chart for Exercises 2 and 4.

Part number

Machine number	1	2	3	4	5	6	7	8
1	1					1		
2		1	1		1			1
3			1		1			1
4				1				
5	1					1		
6		1	1		1			
7				1		1		

FIGURE 5.32
Part for Exercise 3. (Courtesy Garrett Engine Division).

3. Use the DCLASS coding system to develop codes for the parts shown in Figures 5.6a and 5.32. You may need to refer to Figure 5.33. Material is stainless steel.

4. Given the data in Figure 5.31, use the average-linkage clustering algorithm to determine what machines should be grouped together. What threshold value would you recommend using?

5. Develop and document an attribute coding scheme for coding pencils.

6. Develop and document a hierarchical coding scheme for coding pencils.

7. Develop and document a hybrid coding scheme for coding pencils.

8. Develop a plan for validating the coding scheme that you developed in Exercise 5, 6, or 7.

9. A family of 150 parts are to be produced. A conventional fixture costs $650 and a fixture for group tooling costs $1000. The adaptors for the group tooling cost $80. Compute the tooling costs and the unit tooling costs for conventional and group tooling for 1, 5, and 10 parts in a part family.

10. Develop a plan for installing group technology in a metal fabrication company that makes parts for automobiles. Assume that the company has 5000 active part numbers.

11. Implementing group technology concepts within a company can be difficult and expensive. Write a justification for implementing group technology in the company described in Exercise 10. This justification must be approved by the company president, and it should not be more than two pages in length because the president's time is limited. Remember that the president will want to know how group technology will affect the company's "bottom line" (profits).

12. Why do you think that poor communications exist in many companies between design engineering and manufacturing engineering?

13. What will the typical manufacturing company in the United States look like in the year 2000?

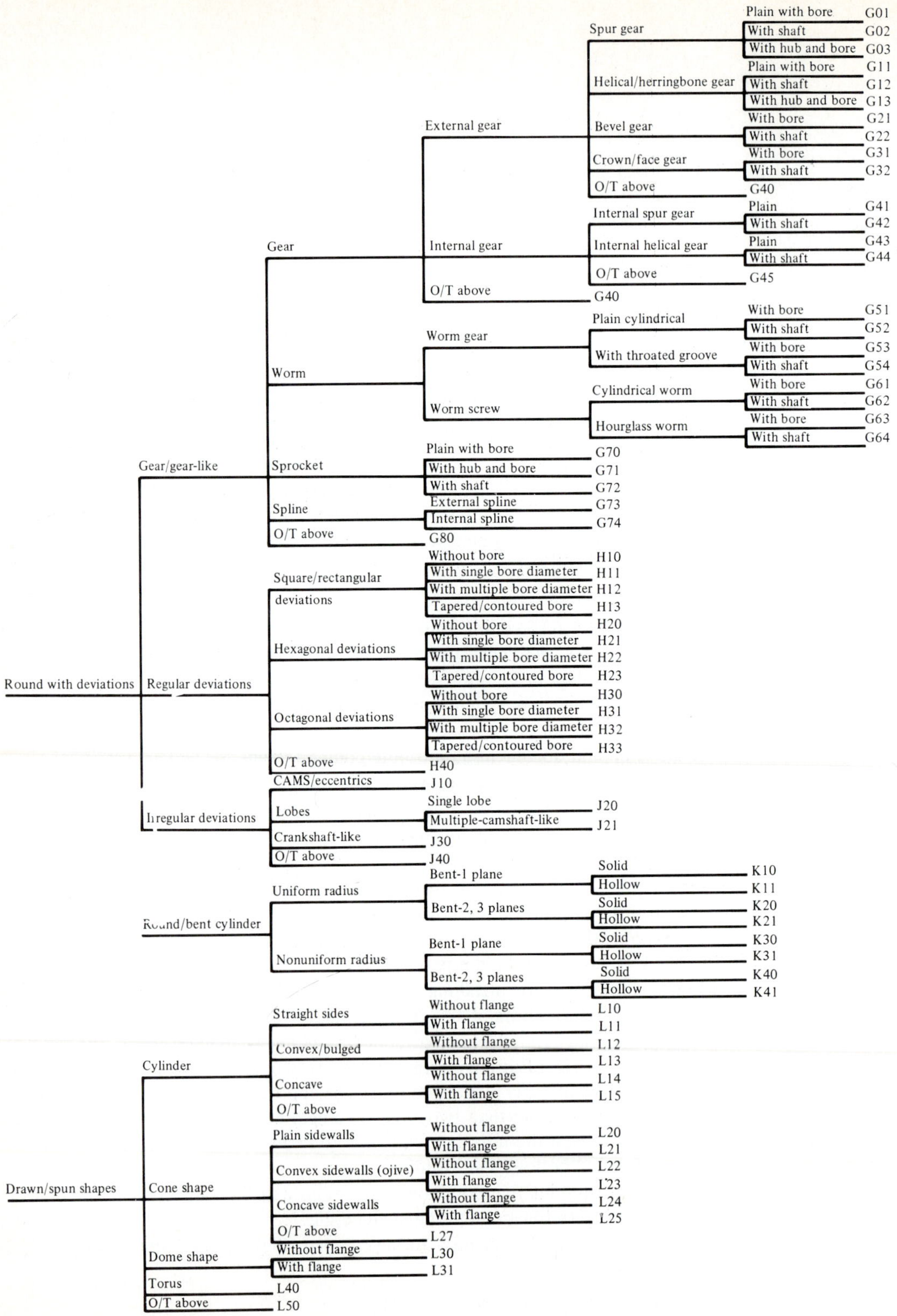

Spur gear — Plain with bore — G01
With shaft — G02
With hub and bore — G03

Helical/herringbone gear — Plain with bore — G11
With shaft — G12
With hub and bore — G13

Bevel gear — With bore — G21
With shaft — G22

Crown/face gear — With bore — G31
With shaft — G32

O/T above — G40

External gear

Internal gear — Internal spur gear — Plain — G41
With shaft — G42
Internal helical gear — Plain — G43
With shaft — G44
O/T above — G45

O/T above — G40

Gear

Worm gear — Plain cylindrical — With bore — G51
With shaft — G52
With throated groove — With bore — G53
With shaft — G54

Worm screw — Cylindrical worm — With bore — G61
With shaft — G62
Hourglass worm — With bore — G63
With shaft — G64

Worm

Sprocket — Plain with bore — G70
With hub and bore — G71
With shaft — G72

Spline — External spline — G73
Internal spline — G74

O/T above — G80

Gear/gear-like

Square/rectangular deviations — Without bore — H10
With single bore diameter — H11
With multiple bore diameter — H12
Tapered/contoured bore — H13

Hexagonal deviations — Without bore — H20
With single bore diameter — H21
With multiple bore diameter — H22
Tapered/contoured bore — H23

Octagonal deviations — Without bore — H30
With single bore diameter — H31
With multiple bore diameter — H32
Tapered/contoured bore — H33

O/T above — H40

Regular deviations

CAMS/eccentrics — J10
Lobes — Single lobe — J20
Multiple-camshaft-like — J21
Crankshaft-like — J30
O/T above — J40

Irregular deviations

Round with deviations

Uniform radius — Bent-1 plane — Solid — K10
Hollow — K11
Bent-2, 3 planes — Solid — K20
Hollow — K21

Nonuniform radius — Bent-1 plane — Solid — K30
Hollow — K31
Bent-2, 3 planes — Solid — K40
Hollow — K41

Round/bent cylinder

Straight sides — Without flange — L10
With flange — L11
Convex/bulged — Without flange — L12
With flange — L13
Concave — Without flange — L14
With flange — L15
O/T above

Cylinder

Plain sidewalls — Without flange — L20
With flange — L21
Convex sidewalls (ojive) — Without flange — L22
With flange — L23
Concave sidewalls — Without flange — L24
With flange — L25
O/T above — L27

Cone shape

Dome shape — Without flange — L30
With flange — L31

Torus — L40
O/T above — L50

Drawn/spun shapes

FIGURE 5.33 DCLASS Chart 4 logic tree. (Reprinted with permission from Ref. 4).

14. Explain why group technology is important to achieving the integration of CAD and CAM.

15. Compare the DCLASS part family code and the MICLASS code.

REFERENCES AND SUGGESTED READINGS

1. Anderberg, M. R.: *Cluster Analysis for Applications*, Academic Press, New York, 1973.
2. Burbidge, J. L.: *The Introduction to Group Technology*, John Wiley, New York, 1975.
3. Chang, T., and R. A. Wysk: *An Introduction to Automated Process Planning Systems*, Prentice-Hall, Englewood Cliffs, N.J., 1985.
4. Computer Aided Manufacturing Laboratory: *Part Family Classification and Coding*, Monograph No. 3, Brigham Young University, Provo, Utah, 1979.
5. Computer Aided Manufacturing Laboratory: *DCLASS Information Processing Systems Technical Manual*, Brigham Young University, Provo, Utah, 1975.
6. Eckert, R. L.: "Coding and Classification Systems," *American Machinist*, December 1975.
7. Graves, G. R.: "Considerations for Group Technology Manufacturing in Production Planning," Ph.D. Diss., Oklahoma State University, Stillwater, Okla., 1985.
8. Groover, M. P., and E. W. Zimmers: *CAD/CAM: Computer-Aided Design and Manufacturing*, Prentice-Hall, Englewood Cliffs, N.J., 1984.
9. Ham, I.: "Group Technology—The Natural Extension of Automated Planning," in *Proceedings of CAM-I's Executive Seminar—Coding, Classification and Group Technology for Automated Planning*, San Diego, Calif., 1976, pp. 192–218.
10. Ham, I.: "Introduction to Group Technology," Technical Report, Society of Manufacturing Engineers, Dearborn, Mich., 1978.
11. Ham, I., and K. Hitomi: "Group Technology Applications for Machine Loading under Multi-Resource Constraints," *North American Manufacturing Research Conference Proceedings, 9th*, 1981, pp. 515–518.
12. Hathaway, H. K.: "The Measurement System of Classification: As Used in the Taylor System of Management," *Industrial Management*, September 1930, pp. 173–183.
13. Hax, A., and D. Canden: *Production and Inventory Management*, Prentice-Hall, Englewood Cliffs, N.J., 1984, pp. 69–124, 393–464.
14. Hax, A., and H. Meal: "Hierarchial Integration of Production Planning and Scheduling," in *Studies in the Management Sciences, Vol. I, Logistics*, North Holland, New York, 1975.
15. Hyer, N. L., Ed.: *Group Technology at Work*, Society of Manufacturing Engineers, Dearborn, Mich., 1984.
16. Manufacturing Data Systems, Inc.: *CODE: The Parts Classification Data Retrieval System for Computer-Aided Manufacturing*, Product Information PI-30-6000-0, Manufacturing Data Systems, Inc., Ann Arbor, Mich., 1977.
17. McAuley, J.: "Machine Grouping for Efficient Production," *The Production Engineer*, vol. 52, February 1972, pp. 53–57.
18. Mitrofanov, S. P.: *Scientific Principles of Group Technology*, English Translation, National Library for Science and Technology, Washington, D.C., 1966.
19. Opitz, H.: *A Classification System to Describe Workpieces*, Pergamon Press, Oxford, England, 1970.
20. Organization for Industrial Research, *MICLASS, MIGROUP, MIPLAN, MIGRAPHICS* (marketing brochure), Waltham, Mass.
21. Schaffer, G. H.: "Implementing CIM," *American Machinist*, August 1981, pp. 151–174.
22. Seifoddini, H. S.: "Cost Based Machine-Component Grouping Model: In Group Technology," Ph.D. Diss., Oklahoma State University, Stillwater, Okla., 1984.
23. Singh, C. K., and P. Kumar: "An Economic Analysis of Group Technology," *Journal of the Institution of Engineers* (India), September 1978, pp. 64–69.

24. Sneath, P. H., and R. R. Sokal: *Numerical Taxonomy*, W. H. Freeman and Company, San Francisco, 1973.
25. Soloja, V. B., and S. M. Ursoevic: "Optimization of Group Technology Lines by Methods Developed in the Institute for Machine Tools and Tooling (IAMA) in Belgrad," *Group Technology—Proceedings of the International Seminar*, September 1969, pp. 157–176.
26. TNO (Netherlands Organization for Applied Scientific Research), *An Introduction to MICLASS*, Organization for Industrial Research, Inc., Waltham, Mass.

CHAPTER
6

PROCESS PLANNING

Civilization advances by extending the number of important operations which we can perform without thinking about them.

Alfred North Whitehead
An Introduction to Mathematics (1911)

Process planning consists of preparing a set of instructions that describe how to fabricate a part or build an assembly which will satisfy engineering design specifications. The resulting set of instructions may include any or all of the following: operation sequence, machines, tools, materials, tolerances, notes, cutting parameters, processes (such as how to heat-treat), jigs, fixtures, methods, time standards, setup details, inspection criteria, gauges, and graphical representations of the part in various stages of completion. It is obvious that process planning can be a very complex and time-consuming job requiring a large amount of data. In addition, several people may participate in developing a process plan, because no one person may have the broad expertise required. This is further complicated by the fact that the plan is a critical element in making the part correctly and economically.

Process planning received little attention until the latter 1970s. Informally, process planning has been performed for hundreds of years—ever since someone first developed instructions to make something. However, the industrial revolution fostered a need to formalize process planning in the manufacturing environment. Initially, manufactured parts had few components and were made by a small number of people. In this setting, formal process plans were not required. As the number of parts and complexities increased, a need for formal process plans was recognized. Nevertheless, until the 1970s the importance of process planning was understood primarily by those closely involved in making the individual parts. Consequently, little was done to automate this process.

Today's manufacturing environment has become very competitive and complex. This complexity is a function of more intricate parts and factors, such as machining technologies that permit making a part several different ways, small lot sizes that neither support long setup times nor provide frequent learning reinforcement for the machine operator, increased government regulation that requires documentation of process plans, many types of materials that may require special tools and/or processes, and less skilled machinists. These factors combined with increased emphasis on reducing manufacturing costs have affirmed the significance of process planning, and corroborated that substantial savings can be achieved by automating the preparation of process plans. Consequently, this function has been receiving widespread attention.

The benefits that have been reported from successful applications of automating some of the process planning functions are impressive and have one of the shortest payback periods of all CAD/CAM technologies. Some typical benefits [23, 44] include:

1. 50% increase in process planner productivity
2. 40% increase in capacity of existing equipment
3. 25% reduction in setup costs
4. 12% reduction in tooling
5. 10% reduction in scrap and rework
6. 10% reduction in shop labor
7. 6% reduction in work in process
8. 4% reduction in material

Some of these benefits may not appear to be related to the automation of process planning. However, consider what can happen if the process planner's productivity is significantly improved:

1. More time can be spent on methods improvements and cost-reduction activities.
2. Routings can be consistently optimized.
3. Manufacturing instructions can be provided in greater detail.
4. Preproduction lead times can be reduced.
5. Responsiveness to engineering changes can be increased.

Benefits such as these are not going unnoticed.

This chapter presents some of the methodologies used to automate process planning. An example will be used to illustrate some of the more popular systems used. Finally, future directions of process planning will be discussed.

6.1 KEY DEFINITIONS

Computer Managed Process Planning (CMPP)   A generative process planning system developed by United Technologies and the U.S. Army.

Decision table   A tabular method for expressing the actions that should follow if certain conditions exist. The table is composed of four major components: condition stubs, condition entries, action stubs, and action entries.

Design and Classification Information Retrieval (DCLASS)   A general-purpose computer system for processing classification and decision-making logic.

Decision tree   A graphical method for expressing decision-making logic (the actions that should follow if certain conditions exist). The graph is composed of a root, branches, and nodes.

Generative process planning   A process planning system, including a data base and decision logic, that will automatically generate a process plan from graphical and textual engineering specifications of the part.

Geometric Modeling Applications Interface Program (GMAP)   A project funded by the U.S. Air Force to expand upon the results obtained in the Product Definition Data Interface (PDDI) project. PDDI is extended to cover two new part types (turbine disks and blades), and applications in inspection, analysis, and product support.

Hierarchical tree   A method for expressing decision information and logic as a graphical tree.

Initial Graphics Exchange Specification (IGES)   A standard sequential file format for interchanging product definition data (wire frame edge-vertex geometry, annotations, and structure) among CAD systems.

Operation code (op code)   An alphanumeric code used to represent a series of operations performed at one machine or one work station.

Operation plan (op plan)   A detailed description of operations represented by an op code. All operations performed at a machine or work station may not be described, only those deemed necessary to concisely and clearly describe what tasks are to be performed.

Part family matrix   A matrix used to represent the group technology codes for parts in a family. The columns of the matrix represent character positions in the group technology code, and the rows represent the number of characters that can be assigned to any one position in the code.

Process planning   The preparation of a set of instructions that describe how to fabricate a part or build an assembly which will satisfy engineering design specifications.

Product Data Exchange Specification (PDES)   A project initiated by the IGES organization to develop specifications that will facilitate transferring a complete product model with sufficient information as to be interpretable by advanced CAD/CAM applications, such as process planning. The product model includes data relative to the entire life cycle of a product, encompassing design to field support.

Product Definition Data Interface (PDDI)   A project funded by the U.S. Air Force to define and demonstrate a prototype system that replaces the engineering drawing as the interface between design and all manufacturing functions, including process planning, numerical control programming, quality assurance, tool design, and production planning and control. This concept is based on a neutral file exchange format that can be used to translate data among dissimilar CAD and CAM systems.

Standard for Transfer and Exchange of Product Model Data (STEP)   A worldwide standard being defined by the International Standards Organization (ISO) to capture the information comprising a computerized product model in a neutral format that can be used throughout the life cycle of the product.

Standard process plan   A sequence of op codes representing all operations included in a process plan for any part in a particular part family.

Tolerance chart   A procedure for calculating the dimensions and associated tolerances that must be maintained for each cut so that the finished part will satisfy blueprint specifications.

Variant process planning   Computer-aided process planning involving a library of standard process plans; interactive editing programs; and storage, retrieval, and documentation capabilities. The plan for a new part is created by retrieving and modifying the standard process plan for a given part family.

6.2   THE ROLE OF PROCESS PLANNING IN CAD/CAM INTEGRATION

With the emergence of CAD/CAM integration as a predominant thrust in discrete parts industries, the communication between design engineering and manufacturing engineering has become a very important consideration. Most firms are using CAD techniques extensively to design their products; similarly, they are using some CAM techniques, such as computerized numerical control (CNC) machines, to manufacture products. However, in many of these firms there is very little communication between design and manufacturing. As was noted in the previous chapter, the norm is well described by "engineering designs the product and then throws the drawings over the wall for manufacturing to make the product."

Process planning emerges as a key factor in CAD/CAM integration because it is the link between CAD and CAM. After engineering designs are communicated to manufacturing, either on paper or electronic media, the process planning function converts the designs into instructions used to make the specified part. CIM cannot occur until this process is automated; consequently, automated process planning is the link between CAD and CAM.

As discussed in Chapter 5, group technology is an important element in CAD/CAM integration, because it provides a basis and a methodology for engineering and manufacturing communications. Automated process planning provides a means to facilitate this communication and remove the "wall" between engineering and manufacturing. The significance of group technology to this integration will be reinforced in this chapter with the realization that most CAPP systems utilize group technology concepts. As a result, it is easy to understand why process planning and group technology have received a great deal of attention as CAD and CAM technologies are implemented and integrated.

6.3 APPROACHES TO PROCESS PLANNING

Traditionally, process planning has been performed manually by most companies. Figure 6.1 depicts the information flow as a process plan is developed and portions are sent to other departments within a company. As companies seek to automate this function, two approaches are considered: variant and generative. Each of these approaches will be discussed in this section.

FIGURE 6.1   Process plan information flow.

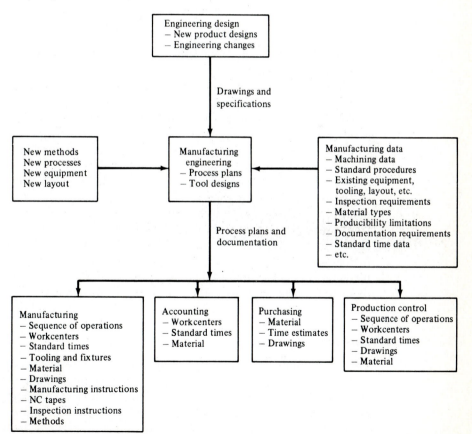

6.3.1 Manual Approach

Under the manual approach, a skilled individual, often a former machinist, examines a part drawing to develop the necessary instructions for the process plan. This requires knowledge of the manufacturing capabilities of the factory: machine and process (such as heat-treat) capabilities, tooling, materials, standard practices, and associated costs. Very little of this information is documented; often this information exists only in the minds of the process planners. When a process plan is being prepared, if the planner has a good memory, a process plan for a similar part might be retrieved and modified. In a more organized company, some "workbooks" might be used to store and provide limited retrieval capabilities. This approach relies almost entirely on the knowledge of the individual planner. Consequently, process plans developed for the same part by different planners will usually differ unless the part is simple to make.

The same planner may develop a different process plan for the same part if there is a long time lag between the analyses for that part, because the planner's experience may change during the time interval and/or shop conditions may change significantly. For instance, a critical machine might have been under repair and because the part was needed as soon as possible, a different machine might have been specified, even though the cost to make the part might be considerably greater. In this case, when the critical machine is repaired, the process plan probably will not be modified because of the manual effort involved.

Manual preparation involves subjective judgments that reflect the personal preferences and experiences of the planner; consequently, plans prepared by different planners for similar parts can vary significantly. Furthermore, as much as 40% of the task involves the preparation of documentation for the plan. As a result, this approach is very labor-intensive, time-consuming, and tedious. For example, it is not uncommon for a process plan for one part in the aerospace industry to contain more than 100 pages.

Despite these disadvantages, the manual approach is generally preferred by small firms that have few process plans to prepare. However, as the volume of plans to be prepared increases, a point is reached where some type of computerized system should be considered to assist in this task. The exact point will depend on the cost of the system and the benefits that can be realized. These factors are rapidly changing as the cost of computing decreases and process planning system capabilities increase with changes in technology. The capabilities of some of these systems will be discussed in this chapter.

6.3.2 Variant Approach

The variant approach is one of two approaches, the other being generative, used to develop a CAPP system. A variant system is much simpler than a generative approach, but it can require more human interaction. It is important to realize that all CAPP systems in use outside research environments require someone to input the specifications (shape, features, dimensions, tolerances, instructions, etc.) of the part being planned. One of the difficult obstacles to overcome in fully automating process planning is the development of computer software capable of reading part specifications directly from a CAD data file. This topic will be addressed later, when future directions in automated process planning are discussed.

Variant CAPP systems are designed to utilize the fast storage and retrieval capabilities of a computer, and to provide an interactive environment between the planner and the computer. This type of system is developed so that a planner with limited computer knowledge can effectively prepare a process plan. Thus, the planner is prompted for the necessary data, and the inputs are edited. If errors are detected, the user is prompted to correct the erroneous entry.

A variant system requires a data base containing standard process plans. A standard process plan for a family consists of all instructions (such as operations, tools, and notes) that would be included in a process plan for any part in that family. Initially, group technology concepts are usually applied to form families of parts. Then a standard process plan is created for each of these families and stored in the computer.

The process planning task for a new part starts with coding and classifying the part into a part family using group technology. The standard process plan for this family can then be retrieved from the computer. Since this plan contains instructions for all parts in this family, some editing may be required for a specific part. Therefore, variant systems provide editing programs to facilitate this procedure. Often, very little editing is required, because the new plan is a variation (variant) of the standard process plan. As a result, considerable time is saved in preparation of the plan, and a significantly greater consistency in plans is obtained. After editing, the plan can be stored and/or printed out.

If the part being planned cannot be classified into an existing family of parts, the planner can develop a new standard process plan using interactive computer programs. In essence, a variant CAPP system attempts to create a work environment for the planner that utilizes the computer to "remember" (store and retrieve) similar process plans. Interactive programs are provided so that any required editing can be performed expediently. Also, programs are provided to create any documentation required.

One of the first variant systems developed was called CAM-I Automated Process Planning [8]. Initial development was completed in 1976 under the sponsorship of CAM-I (Computer-Aided Manufacturing-International) to demonstrate the feasibility of CAPP systems. It was so successful that several companies modified this system to meet their specific needs and are still using it. Table 6.1 provides a summary of some of the variant CAPP systems that are available commercially. Many more variant systems have been developed at academic institutions for research purposes.

6.3.3 Generative Approach

A generative CAPP system will automatically generate a process plan from engineering specifications (graphical and textual) of the finished part. Many times

TABLE 6.1
Some commercial variant process planning systems

System name	Company	Part data input form	Decision logic	Planning functions	Ref.
CUTPLAN	Metcut	Code	Standard plans and decision tree[a]	Process sequence Materials Machines Tools Fixtures Feeds and speeds	56
CAPP	CAM-I	Code	Standard plans	Process sequence	8
COMCAPP V	MDSI	Code	Standard plans	Process sequence	
DCLASS	CAM Lab. BYU	Interactive part description	Decision tree[a]	Process sequence Materials Machines Tools	7
INTELLICAPP	CimTelligence	Interactive part description	Decision tree[a]	Process sequence Materials Machines Tools	23
MAYCAPP	Maynard	Interactive part description	Decision tree[a]	Process sequence Materials Machines Tools Standard times Speeds and feeds	

[a] Some limited generative capabilities.

when engineering specifications are considered, only the graphical drawings come to mind; however, quite often extensive textual information is also required, such as material type, special processing details, and special inspection instructions. Considering the many details contained in some process plans and the complexity of some parts, it is understandable that a truly universal generative process planning system has not been developed. However, systems of this type have been developed for special classes of parts with limited types of geometric features.

The first step in generating a process plan for a new part using a generative system is to input the engineering specifications into the system. Ideally, these specifications would be read directly from a CAD system. For this to occur, the CAPP system must have the ability to recognize the features of a part, such as a hole, slot, gear tooth, and chamfer, as stored in raw data form in the computer. Although this is being done in the laboratory for simple parts, it is pragmatically beyond the state of the art. Consequently, the simpler approach of coding the physical features of the part is used. The coding scheme utilized must define all geometric features and associated details such as locations, tolerances, and sizes. Additionally, it must describe the part in its rough state, because the system must be able to determine what material must be removed to obtain a finished part. In a metal fabrication environment the initial form of a part may vary considerably; for example, it may be a solid block of metal or a "near-net-shape" casting. The near-net-shape casting may require very little machining.

The second major component of a generative system is a set of programs that can transform the coded data and accompanying textual information into a detailed process plan. In general, it may be impossible to develop such a program; however, we will consider some of the things that a generative system must do.

TABLE 6.2
Some generative process planning systems[a]

Author	Name of system	Part data input	Decision logic	Planning functions	Part shape	Commercial or experimental	Ref.
Wysk	APPAS	G	Decision tree	1, 4	H	E	55
Evershiem, Fuchs, and Zons	AUTAP	D	Decision table	1, 4	R	C	22
BYU	BYUPLAN	G	Decision tree	1, 4	B	E	5
Chang	CADAM	D	Decision table	1, 4	H	E	11
Sack	CMPP	D	Decision model	1, 2, 3, 4	R	C	44
Kung	GAPPS	D	Decision model	1, 4	R	E	30
Kung	FREXPP	D	AI	1, 4	N	E	29
Descotte and Latombe	GARI	D	AI	1, 4	N	E	19
Tulkoff	GENPLAN	G	Decision model	1, 2, 3, 4	B	C	48
Darbyshire and Davies	EXCAP	D	AI	1, 4	R	E	18
Logan	LOCAM	D	Decision model	1, 2, 3, 4	B	C	36
MGS	OPTA-PLAN	D	Decision model	1, 2, 3, 4	R	C	40
Phillips	PROPLAN	D	AI	1, 4	R	E	42
Choi	STOPP	D	Decision model	1, 4	H	E	16
Chang	TIPPS	D	AI	1, 4	N	E	12
Sack	XPS-1	D	Decision model	1, 4	R	E	43

[a] Key:

G = group technology	B = rotational/nonrotational
1 = sequence of operations	D = part description approach
2 = tolerance control operations	3 = reference surface selection
H = hole	4 = process plan output
N = nonrotational	C = commercial
R = rotational	E = experimental

Source: From Ref. 33.

The best sequence of operations for transforming a part must be specified. For each operation, directions must be specified for such things as machines, tooling, fixtures, gauges, clamping, feeds and speeds, setup and cycle times, and inspection. A large data base and some very complex algorithms are required for a computer to be able to accomplish this. The development of a universal generative system appears to be overwhelming. As a result, the approach taken to date has been to simplify the problem statement by developing generative systems for special classes of parts—for example, cylindrical parts such as gears. These parts are normally symmetrical in one dimension and have a relatively limited set of part features.

Li [33] conducted a survey of existing generative process planning systems to determine the input requirements and the capabilities of these systems. The results are summarized in Table 6.2. Only one system (FREXPP, an experimental system) links with a CAD system (PADL-1, an experimental solid modeling system). Although Li lists several commercial generative systems, none has found widespread acceptance. All require an individual to look at a graphical representation of the part being planned and then code the features of the part in some manner for input to the generative process planning system. In this chapter we will look at the CMPP (Computer Managed Process Planning) system to gain a better understanding of a generative system. The fact that none of these systems has found widespread acceptance suggests that each has severe limitations. Some of the major limitations are that extensive data bases and decision logic that are unique to each firm must be developed and maintained.

6.4 EXAMPLE PROCESS PLANNING SYSTEMS

In this section we will examine some representative process planning systems, which will provide an awareness of what is available. It will also provide a basis for selecting a process planning system, or a good starting point for developing one.

6.4.1 CAM-I Automated Process Planning (CAPP)

As noted earlier, CAM-I sponsored development of a prototype variant system; the initial development was completed in 1976. Revisions have been made to enhance the system and to adapt it to other computer systems (versions exist for IBM, DEC, HP, and possibly other computers by now). This system was a prototype sponsored by several different firms (members of CAM-I) from diverse industries. Because of the extreme variety of process planning requirements in industries such as electronics, consumer goods, and aerospace, the design specifications had to be limited. The fabrication of machined detail parts (as opposed to assemblies) was selected. The resulting system was successful in creating an awareness of how a computer system could aid the process planner. This is attested by the fact that CAM-I has distributed several hundred copies, and, in addition, a large number of companies are using process planning systems whose origins can be traced to CAM-I's system.

This system can be used to generate a sequential list of individual manufacturing operations and associated information needed to produce a part. The associated information for each operation may include such things as machine department, machine code and location, time standards, work instructions, work elements, and work element parameters such as feeds and speeds. Figure 6.2 provides a graphical description of this CAPP system.

A process planner operates the system via an interactive computer terminal. One design premise was that parts would be coded and classified into families external to the CAPP system using a group technology code with as many as 36 alphanumeric characters. Within the system one standard process plan, a sequence of operation codes (op codes) for each family, is maintained in a system file (standard sequence file). The detailed description of each operation code is stored in the operation plan file. The standard plans and operation plans must be developed for each installation because they are a function of the machines, procedures, and expertise of a particular company.

Each part family is defined and stored as a matrix. One such matrix is depicted in Figure 6.3. Columns of the matrix represent character positions within the group technology code; rows represent the number of characters that can be assigned to any one position with the code. In the example in Figure 6.3, the group technology code has five positions; each position can have a value of 0 through 9.

FIGURE 6.2 Flow diagram of CAM-I's CAPP system.

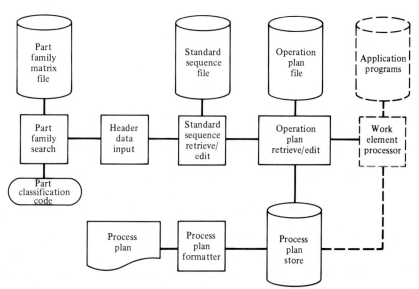

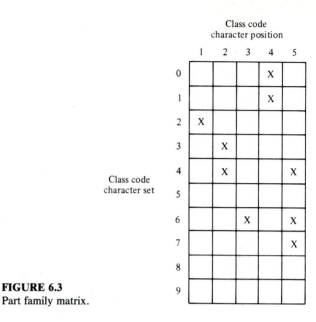

FIGURE 6.3
Part family matrix.

Looking at Figure 6.3, we can see that the following group technology codes represent a family:

24607

23614

24616

Figure 6.3 is called a part family matrix. This data is stored in the part family matrix file of the CAM-I system.

Creation of a process plan using this system can be broadly described as consisting of four steps. Initially, the part being planned would be classified as belonging to a group technology part family. The second step involves inputting the header information used to identify a process plan; it consists of data elements such as part number, material, planner, part name, and revision. In the third step, the operation sequence required to manufacture the part is specified; this sequence is sometimes called a routing. The operation sequence is created by inputting and/or modifying a standard plan operation sequence consisting of mnemonic abbreviations (op codes) of available machines. In the last step, the text describing the work performed at each operation is created and/or edited. Once this is completed for each operation, the plan is stored, after which it can be printed and edited as desired.

A typical process planning session using the CAM-I system would start with the planner coding the part with a group technology code. Then the system would be initiated. The main menu is shown in Figure 6.4. Abbreviations for the operable commands for this menu are displayed across the top. The operable commands

```
                        CAPP SYSTEM MAIN MENU
                FS   FP   DP   HD   OS   OP   PP   LO
         ──────────────────────────────────────────────────────

         PART FAMILY SEARCH              FS

         FORMAT PLAN                     FP

         DELETE PLAN                     DP

         CREATE NEW PLAN                 HD

         RETRIEVE HEADER                 HD/(PARTNO, PLAN TYPE, STATUS)

         RETRIEVE OF CODE SEQ            OS/(PARTNO, PLAN TYPE, STATUS)

         RETRIEVE OP PLAN                OP/(PARTNO, PLAN TYPE, STATUS)

         PROCESS PLAN REVIEW             PP/(PARTNO, PLAN TYPE, STATUS)

         LOG OFF                         LO
         ──────────────────────────────────────────────────────
         FS
```

FIGURE 6.4   CAM-I CAPP main menu.

will vary depending on the menu being displayed on the CRT. More than 30 commands are provided. All of the menus have the same basic format. The top line on the CRT display contains a list of two-letter mnemonics representing the commands operable for that particular menu. The bottom lines on the screen are reserved for user input (note that the command mnemonic FS has been input in Figure 6.4). The middle portion of the screen is reserved for the following types of information: tutorial information (as in Figure 6.4), retrieved process planning data (such as a standard plan), the results of data editing performed by the user, and data fields that are to be filled in with part-specific data.

Some of the mnemonic command codes appearing on the main menu need further explanation. The code FS denotes a part family search. When this option is specified, another menu is displayed with options that facilitate searching for a specific part family. The second code, FP, indicates that a copy of a specific process plan is to be printed out. The next code, DP, is used to delete a process plan from the system. The CREATE NEW PLAN command, HD, signifies that a new part is being planned and that header information must be input. This command invokes the header data input/edit menu which displays the fields for which data are to input. The exact data input will vary with each company; therefore, the types of header data needed are designated when the system is installed at a company. Figure 6.5 depicts some of the header data that might be required; the system requires only the part number. The remainder of the data is used to identify the part in greater detail for the planner and to provide instructional information on the process plan. The next command in the menu in Figure 6.4, RETRIEVE HEADER, has the same mnemonic as the previous command followed by some arguments which denote the part number for which the header data is to be retrieved. After the header data is retrieved it can also be edited. The mnemonic OS invokes the op code sequence display/edit menu and displays the op codes for the process plan specified. The next command, OP, invokes the operation plan edit menu and displays the standard operation sequence. This menu and some associated submenus are used for editing and/or creating standard operations and associated information, such as machine code and work element. The mnemonic PP, process plan review, permits the user to review and edit each operation in a process plan one operation at a time. The initial information displayed is the header data; after that the process plan review/edit menu is displayed. How these commands and associated menus might be used is best illustrated by an example.

In this example, the planner has already analyzed the new part and determined that it should have a group technology code of 20456. The next step involves querying the data base to see if a part having the same or a similar code has been previously planned. If so, that plan might be retrieved and reviewed. If only slight modifications are required, that plan could be the basis for the process plan for the new part. In that case, a new header would be created and the modified

FIGURE 6.5   Example of CAM-I CAPP header data elements.

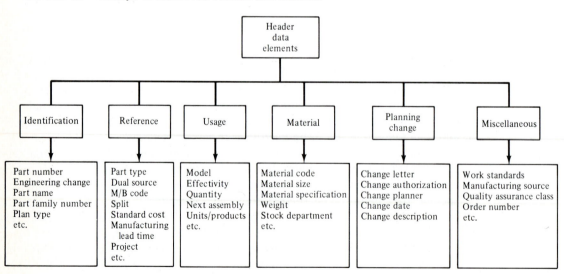

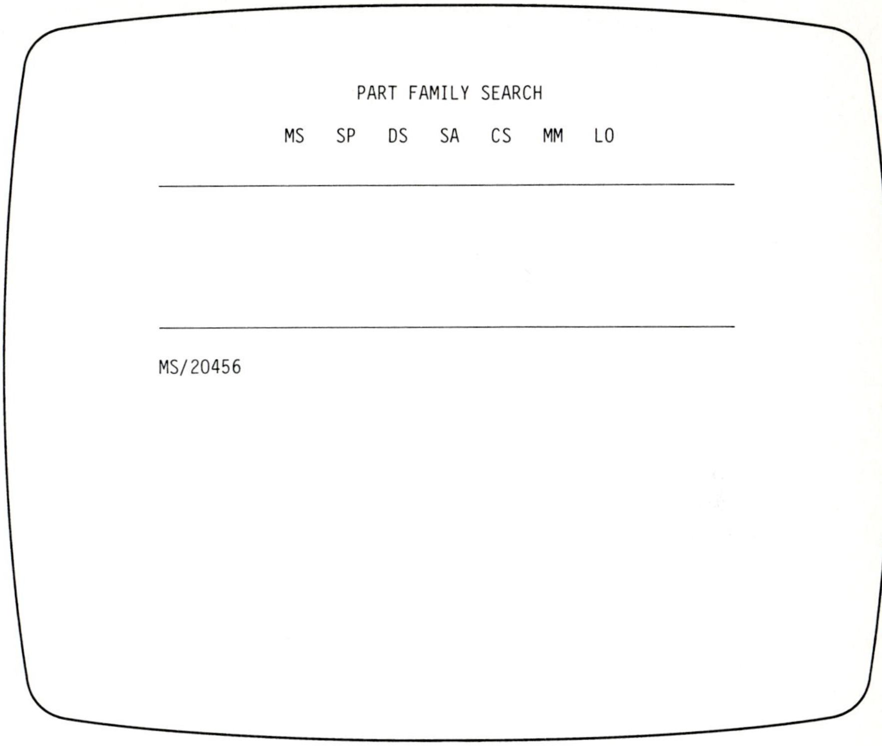

```
                          PART FAMILY SEARCH
                 MS   SP   DS   SA   CS   MM   LO
    _____

    _____

    MS/20456
```

FIGURE 6.6   Part family search menu.

plan would be stored. Figure 6.6 shows the command, MS/20456, to perform a part family matrix search to determine if such a situation exists. Figure 6.7 illustrates the possible responses. In this case an existing part family was found. If a similar code had not been found, then the planner would create a new standard process plan and any new operation plans that might be required.

When a part family is retrieved, as has been done here, the system temporarily stores the information about the part family. Then it displays the header input data screen for identification of the part being planned (see Figure 6.8). One should remember that the header data elements are defined when the system is installed at a particular firm. The input command, FI, in Figure 6.8 specifies that the cursor is to be positioned so that the respective header data elements can be entered.

Figure 6.9 shows the screen after the header data and the command OS (RETRIEVE OPCODE SEQUENCE) have been input. CAPP then displays the standard sequence of operations codes for part family 15 and permits this plan to be edited using commands such as DL/XX--IS/YY, OPCODE, which specifies

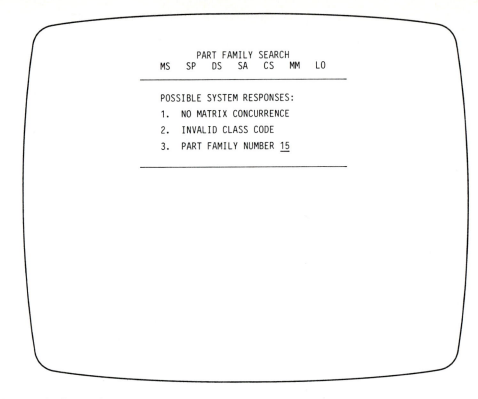

```
                    PART FAMILY SEARCH
            MS    SP    DS    SA    CS    MM    LO
            ─────────────────────────────────────────

            POSSIBLE SYSTEM RESPONSES:
            1.   NO MATRIX CONCURRENCE
            2.   INVALID CLASS CODE
            3.   PART FAMILY NUMBER 15

            ─────────────────────────────────────────
```

FIGURE 6.7   Part family search response.

that op code XX is to be deleted and op code YY is to be changed. If several changes are made and the op codes are no longer in a desired increment, the system will renumber them in the desired increment using the command RN/ *increment* (where *increment* is an integer number).

After the planner is satisfied with the op code sequence for the part being planned, the specific task performed for each op code can be reviewed. The system provides the data structure shown in Figure 6.10 for each operation plan. The command OP causes the operation plan edit screen to be displayed. Figure 6.11 depicts operation 30, CNCLATH, being edited. The process plan edit menu shown in Figure 6.12 can be used to edit the detailed process plan.

By now you should have an understanding of how a variant process planning system works. Essentially, this type of system is designed to use the retrieval and storage capabilities of a computer and the cognitive powers of a human being. The interface is facilitated through interactive computer programs. In some systems the editing capabilities resemble a word processor. Most companies implement variant systems because they are relatively simple and the payback is significant.

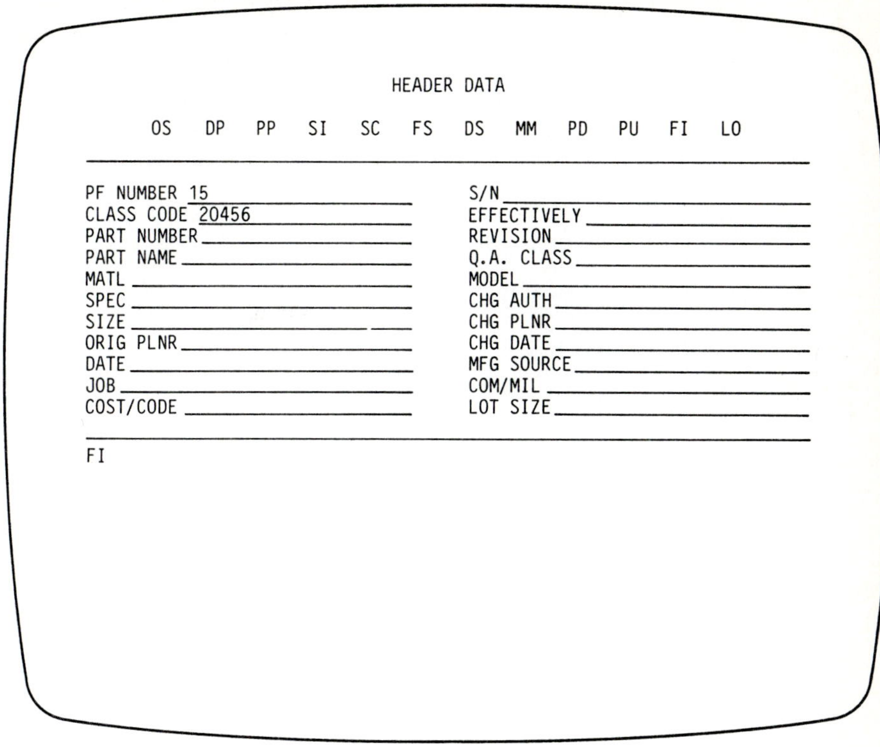

```
                          HEADER DATA

        OS   DP   PP   SI   SC   FS   DS   MM   PD   PU   FI   LO


PF NUMBER 15 _____      S/N _____
CLASS CODE 20456 _____   EFFECTIVELY _____
PART NUMBER _____   REVISION _____
PART NAME _____   Q.A. CLASS _____
MATL _____   MODEL _____
SPEC _____   CHG AUTH _____
SIZE _____ ____   CHG PLNR _____
ORIG PLNR _____   CHG DATE _____
DATE _____   MFG SOURCE _____
JOB _____   COM/MIL _____
COST/CODE _____   LOT SIZE _____


FI
```

FIGURE 6.8   Header menu.

6.4.2 DCLASS

The DCLASS part family code was introduced in Chapter 5. In this section, the term DCLASS, by itself, relates to a "general-purpose computer system for processing classification and decision-making logic." This description needs some clarification; however, a few related terms should first be explained.

A *decision table* is a tabular method for expressing the actions that should follow if certain conditions exist [4]. This technique permits the representation of many logical if-then expressions in a compact, structured manner. Decision tables have been used for many years as an aid to developing and documenting computer programs that involve several different actions depending on the values of some program variables.

Figure 6.13 illustrates the four major components of a decision table. The upper half of a decision table contains the decision conditions, which are expressed as condition stubs and entries. The condition stubs are the criteria that the decision maker wants to apply in the decision process. The condition entries are the responses for each condition stub.

```
                          OPCODE SEQUENCE

        IS   IN   DL   RV   RT   RN   OP   DS   FS   MM   SI   SC   LO
       ───────────────────────────────────────────────────────────────
           PART NUMBER  357789-1              PART NAME   SPUR GEAR
           PF NUMBER   15                     CLASS CODE   20456
       OPCODE SEQ
       10  RAWMTL
       20  SAWBAR
       30  CNCLATH
       40  INSP
       50  HTRT
       60  SAND
       70  HOB
       80  NCLATH
       90  DEBUR
       100 CNCGRIND
       110 SLURRY
       120 INSP
       130 LASERMK
       140 PKG
```

FIGURE 6.9   RETRIEVE OPCODE SEQUENCE menu.

FIGURE 6.10   CAM-I's CAPP op plan data elements.

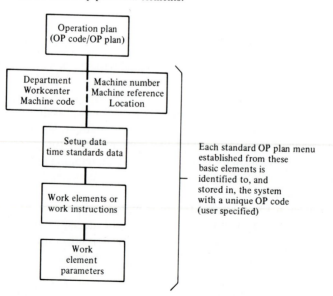

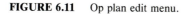

```
                          OP PLAN EDIT
        EX   SU   WE   DL   WX   WP   MM   FS   DS   OS   SI   SC   FI   LO
       _____

          PART NUMBER  357789-1            PART NAME  SPUR GEAR
          PF NUMBER  15                     CLASS CODE  20456

    10 RAWMTL        OP 30 CNCLATH              MACHINE SETUPS  2
    20 SAWBAR        DEPT 96-30                 WORKCENTER G3JL
    30 CNCLATH       MACH REF CNC J&L           MACH CODE 0123
    40 INSP          LOCATION 30/AA19
    50 HTRT
    60 SAND
    70 HOB           SETUP 1                    WORK ELEMENTS
    80 NCLATH        FIXTURE NO_____         10 TURN EDGE
    90 DEBUR         INSTRUCTIONS _____        20 TURN INSIDE DIA
    100 CNCGRIND       _____
    110 SLURRY       CUTTER _____
    120 INSP         TOOL/CODE _____
    130 LASERMK
    140 PKG
```

FIGURE 6.11   Op plan edit menu.

The bottom half of a decision table contains the actions that are to be taken when the specified conditions are satisfied. The components are action stubs and action entries. The stubs denote all actions that can be taken, and the entries designate the specific actions to be taken for specific conditions.

Each column of entries is called a rule and is numbered so that it is easily identified. The number of possible rules in a decision table is related to the number of condition stubs. In general, there are 2^N possible rules, where N represents the number of condition stubs.

As an example of a decision table, we will investigate what logic is required to determine which of two process plans was created first. Assume that the date of creation is one of the data elements recorded on each process plan. Let D1 and D2 represent the creation dates for process plans 1 and 2, respectively. Figure 6.14 shows how this logic might be displayed in a decision table format.

A limited-entry type of decision table has been described. There are at least three other types (extended, mixed, and open-ended). An extended-entry decision table lists the possible conditions in the condition stubs; the condition entries, however, contain values of the condition being evaluated. For instance, one condition for determining if a process plan should be developed for a new part might

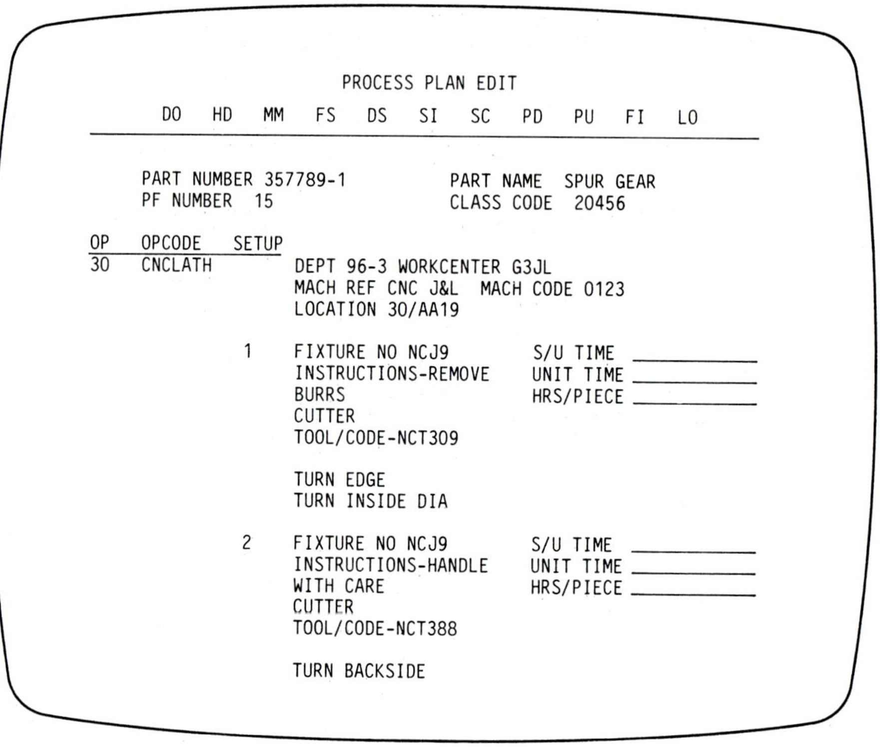

```
                          PROCESS PLAN EDIT
           DO   HD   MM   FS   DS   SI   SC   PD   PU   FI   LO
         ─────────────────────────────────────────────────────────

           PART NUMBER 357789-1          PART NAME   SPUR GEAR
           PF NUMBER   15                CLASS CODE   20456

         OP    OPCODE    SETUP
         30    CNCLATH            DEPT 96-3 WORKCENTER G3JL
                                  MACH REF CNC J&L  MACH CODE 0123
                                  LOCATION 30/AA19

                        1    FIXTURE NO NCJ9        S/U TIME  _____
                             INSTRUCTIONS-REMOVE    UNIT TIME _____
                             BURRS                  HRS/PIECE _____
                             CUTTER
                             TOOL/CODE-NCT309

                             TURN EDGE
                             TURN INSIDE DIA

                        2    FIXTURE NO NCJ9        S/U TIME  _____
                             INSTRUCTIONS-HANDLE    UNIT TIME _____
                             WITH CARE              HRS/PIECE _____
                             CUTTER
                             TOOL/CODE-NCT388

                             TURN BACKSIDE
```

FIGURE 6.12   Process plan edit menu.

be whether the part is to be purchased or made. The entry associated with the condition of ''Make/Buy?'' might be ''M'' or ''B.''

A mixed-entry decision table is a combination of limited- and extended-entry tables. Any condition stub row must be either all limited entry or all extended entry. However, the rule columns are a mixture of limited and extended entries.

The open-ended decision table allows access to other decision tables. For example, the action ''Error'' in Figure 6.14 might be replaced by ''Go to error table,'' which might control the printing of error codes in a computer program. Open-ended tables allow a complex decision involving many conditions to be broken down into a series of smaller decisions.

Decision table methodology might be used to develop process plans. Figure 6.15 illustrates such an application. In this example, open-ended decision tables were utilized. Computer-aided process planning systems have been developed based on decision table methodology; in practice, however, such a system is relatively difficult to maintain. Other approaches have proven to have more benefits.

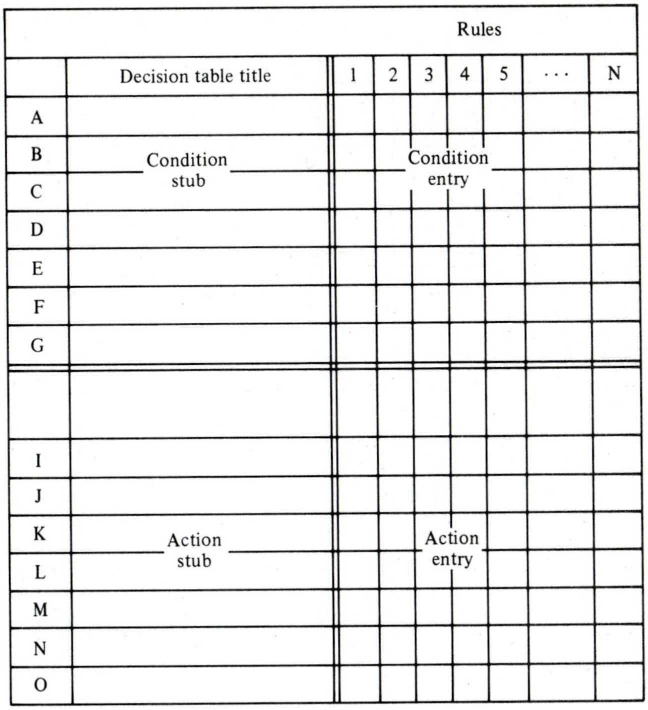

	Decision table title	Rules						
		1	2	3	4	5	· · ·	N
A								
B	Condition stub			Condition entry				
C								
D								
E								
F								
G								
I								
J								
K	Action stub			Action entry				
L								
M								
N								
O								

FIGURE 6.13   Components of a decision table.

FIGURE 6.14   Example decision table.

	Earliest date	Rules			
		1	2	3	4
A	$D1 < D2$	Y	N	N	Y
B	$D1 = D2$	N	N	Y	Y
C	D1 earliest	1			
D	D2 earliest		2		
E	Same date			3	
F	Error				4

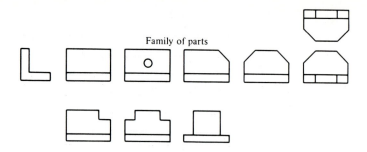

Family of parts

	Table 1	1	2	3	4
A	Sheet metal	Y	N	N	N
B	Cylindrical	N	Y	N	N
C	Prismatic	N	N	Y	N
D	Other	N	N	N	Y
E	Go to Table 100	1			
F	Go to Table 20		1		
G	Go to Table 30			1	
H	Go to Table 40				1

	Table 100	1	2	3	4	5	6
A	Hole	Y	Y	Y	N	N	N
B	Angle cuts	N	Y	N	Y	N	N
C	Notches	N	N	Y	N	Y	N
D	No features	N	N	N	N	N	Y
E	Go to Table 110	1	1	1			
F	Shear		2		1		
G	Notch			2		1	
H	Go to Table 140	2	3	3	2	2	1
I	Re-enter table	X	X	X	X	X	X
J	Store and end	3	4	4	3	3	2

	Table 110	1	2
A	Thickness ≤ 0.5	Y	N
B	Punch	1	
C	Drill		1
D	Return to Table 100	2	2

	Table 140	1	2
A	Thickness ≤ 0.5	Y	N
B	Bend 1	1	
C	Bend 2		1
D	Return to Table 100	2	2

FIGURE 6.15   Decision tables for part family process plans.

A decision table can be converted into a structure known as a *decision tree*. Such a conversion is shown in Figure 6.16. Each decision tree, composed of a root, branches, and nodes, is a type of graph. The starting point is called a root; each tree should have only one root. Emanating from the root and the nodes are branches. The branches may represent a value or an expression; however, a

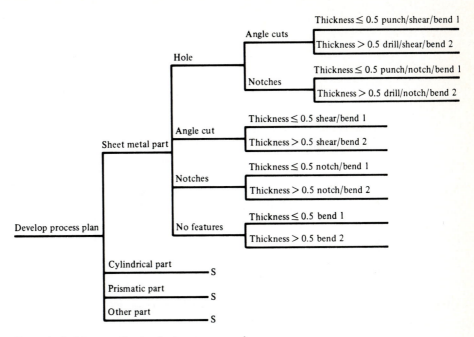

FIGURE 6.16 Example decision tree for developing a process plan.

branch can have one of two logical values, true or false. Relative to tables, trees have several benefits:

1. They are easy to maintain.
2. Selected branches may be extended while other branches may be restricted.
3. They are easy to understand and visualize.

Similar to decision tables, several types of trees have been developed. We will discuss some of them.

An E-tree has mutually exclusive paths, as shown in Figure 6.17. Although an E-tree may have many branches at each node, only one path is selected at each

FIGURE 6.17
E-tree example.

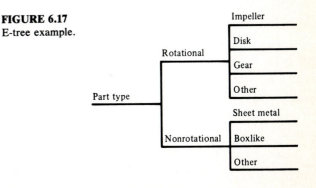

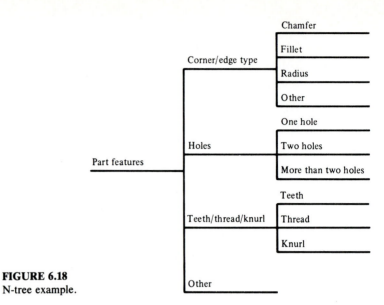

FIGURE 6.18
N-tree example.

FIGURE 6.19 DCLASS part coding and process planning example application. (From DCLASS training material. Reprinted with permission.)

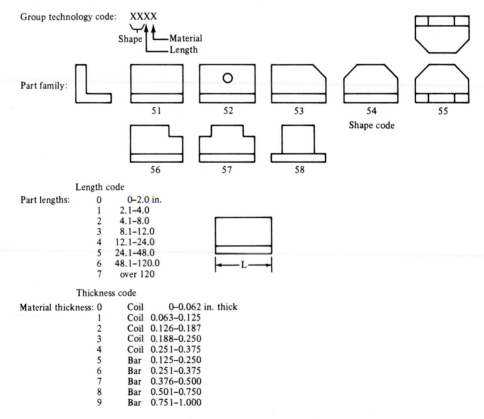

node until a terminal point is reached. This type of tree might be used to code part material, assuming that the part is made from one type of material.

Another type of tree, the N-tree, has nonmutually exclusive paths (see Figure 6.18). From any one node, any number of paths may be selected concurrently. A combination tree results from combining an E-tree and an N-tree.

From the discussion of decision tables and trees, it is easy to see how logic can be captured by both of these techniques. However, for either technique to be effective, computer software must be developed to make it easy to define and maintain this type of logic. The DCLASS system is a general-purpose tree processor developed by Allen and Millett at Brigham Young University to facilitate the use of decision trees. See Ref. 7 for a detailed explanation of DCLASS. Well over 100 companies are using this system, which runs on most of the major computer systems and the IBM PC. It is written in Fortran and can be easily interfaced with other application programs.

An example will be used to describe this system. We will assume that a process plan needs to be prepared for a new part and that all of the necessary decision logic has already been entered into the DCLASS-based system. Classifying and coding of the part is the initial step. We will use the four-digit code shown in Figure 6.19. The first two digits of this code correspond to the different part shapes shown in this same figure. The part designated by shape 52 is the one we will be coding; at this point, however, we will assume that we do not know the specific shape number. All of the parts depicted in Figure 6.19 belong to the same family. The length and material of a specific part can vary; the permissible variations and associated codes are shown in this same figure. The standard op codes and op plans are listed in Table 6.3.

FIGURE 6.20
DCLASS interactive session to code and plan example part.

```
ENTER PART ID NO.
>> 30980-2

ENTER DESCRIPTION
>> SHEET METAL PART

TYPE OF PART?
1-SHEET METAL
2-CYLINDRICAL
3-PRISMATIC
4-OTHER
>> 1

PART FEATURES?
1-HOLE
2-ANGLE CUT
3-VERTICAL NOTCHES
4-NO FEATURES
>> 1

PART LENGTH (IN.)
>> 13.5

RAW MATERIAL TYPE?
1-COIL STOCK
2-BAR STOCK
>> 1

THICKNESS OF COIL STOCK (IN.)
>> .125
```

TABLE 6.3
Op codes and op plans for DCLASS example

Op codes	Op plans
Barcut	Department 513 Machine no. C317 Cut barstock to _____ length (1-in. thickness cap.)
Coilcut	Department 513 Machine no. C334 Cut coilstock to _____ length (0.375-in. thickness cap.)
Bend 1	Department 604 Machine no. B116 or B123 Bend to _____ degrees at _____ in. from long edge ($\frac{1}{2}$-in. thickness; 120-in. length cap.)
Bend 2	Department 604 Machine no. B145 Bend to _____ degrees at _____ in. from long edge ($\frac{1}{2}$-in. to 1-in. thickness cap.; 48-in. length cap.)
Punch 1	Department 604 Machine no. P336 Punch _____ holes of _____ dia. ($\frac{1}{2}$-in. diam.; $\frac{1}{2}$-in. thickness cap.)
Drill 1	Department 604 Machine no. D185 Drill _____ holes of _____ diam. (1-in. diam.; 2-in. thickness cap.)
Shear 1	Department 513 Machine no. S461 Shear _____ corners on _____ angle at _____ in. from edge (1-in. thickness cap.)
Notch 1	Department 604 Machine no. P350 Notch _____ corners _____ in. × _____ in. ($\frac{1}{2}$-in. thickness cap.)

Figure 6.20 lists the interactive session that a planner might have as this part is planned. Underlines denote data that are input from the keyboard. The decision trees that contain the process planning logic to be used are shown in Figure 6.21. We have assumed that this logic was previously entered into the computer.

The session begins by prompting the planner for a part number and description. Next, the part type is identified as "Sheet metal." Note that a "1" appears in the Figure 6.21 node before the part type branches. This designates the type of branch selection control that is specified at this node. In this case the "1" indicates that only one branch is to be selected.

DCLASS provides three major types of branch selection controls: (1) manual, (2) automatic, and (3) override. Manual branch selection control allows the operator to select branches manually. Using automatic control, DCLASS performs the branch selection based on the specified decision tree logic. Override control allows the operator to override previously specified decision logic. Table

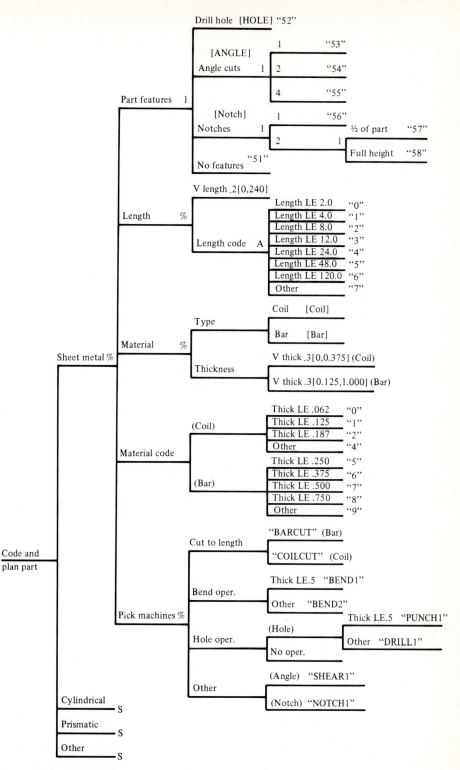

FIGURE 6.21   Example DCLASS decision tree.

6.4 contains a list of the branch selection command capabilities provided in DCLASS.

After selecting the sheet metal type of part, the planner is then prompted to enter the part features. Note the branch control symbol, "%," at the sheet metal node. This denotes that all branches emanating from this node are to be selected in the order in which they appear, starting with the top branch, "Part features." The next branch, "Length," will not be traversed until the path emanating from the "Part features" branch is completely traversed. We will proceed to do that.

After selecting the shape branch, the planner selects the "Drill hole" branch, which specifies the code "52." Note the output flag "[HOLE]." DCLASS provides the capability of setting flags that can be used in decision making. Consequently, if the flag "Hole" is encountered again as an input flag, "(HOLE)," the branch containing this flag will be selected automatically. The brackets, [ ], denote an output flag, and the parentheses, ( ), denote an input flag in the decision tree in Figure 6.21.

TABLE 6.4
Some DCLASS branch selection commands

Command type	Variations	Symbol	Description
Manual	Single branch	1	Allows user to control the solution of a single branch manually.
	Multiple	None	User may manually select several branches in any sequence.
Automatic	All branches	%	Selection occurs with operator input. All branches are selected at a node from top to bottom. When top path is traversed, next branch below in decision tree is selected.
	Comparison check	A	Selection is based on condition between a constant and a variable. Comparisons can be made for the following conditions: LT, LE, EQ, GE, GT, or NE.
	Descending check	D	Selection is based on comparison of variable with constants. Branches must be arranged so that constants are in descending order. Branch is selected when variable is GE to the constant.
Override	Manual	None	Overrides automatic selection command, displays the menu with the branches that would have been selected, and allows the operator to continue tree traversal by pressing ⟨CR⟩ or to change the selection by entering a command.
	Debug	None	Similar to manual but used during debugging of decision tree.
	Backup	B	Backs the operator up to the previous backup command or, if none exists, 10 node levels.

The next branch we encounter is "Length." Emanating from the node at the end of this branch are two branches; the top branch contains the description "V LENGTH.2[0,240]." The "V" denotes that a variable is being defined. The remainder of the description designates that the variable, LENGTH, will maintain two digits of accuracy after the decimal point, and that the permissible range of values for this variable is 0 to 240, inclusive. We are prompted to enter the length of the part in inches.

Proceeding on, we traverse the "Length code" branch. Emanating from this branch is a node with the automate branch control symbol "A." This designates that one branch will be selected based on a specified condition, which, in this case, is the value of the variable LENGTH. Since we input a value of 13.5 in. for the length, the fifth branch is selected and the code "4" is set, which completes this path.

Next, we traverse the "Material" branch. We are asked to specify the raw material type; "Coil stock" is selected and the flag "[COIL]" is set.

Continuing on, we traverse the "Thickness" branch, after which we are prompted to enter the thickness of the coil. This latter branch was selected automatically by DCLASS because the flag "(COIL)" had previously been set.

In a similar manner, the path starting with the "Material code" branch is traversed, and the code "1" is selected since we entered .125 for the thickness of the coil.

The next path negotiated starts with the "Pick machines" branch. Figure 6.21 denotes the branches that are traversed from this point. Note that no more input from the planner is required, because DCLASS will automatically select the correct machines based on the flag and variable values that have been specified.

Based on the decision logic depicted in Figure 6.21, we have completed this portion of the process plan for the new part. The results can be printed out and stored in the system data base. The printed output might look like that in Figure 6.22.

DCLASS is described as a system that facilitates the processing of decision trees. The term *logic trees* would be more appropriate, because the DCLASS system provides some unique capabilities, such as three different types of branch-

FIGURE 6.22
Example DCLASS output.

```
ID # 30980-2

DESCRIPTION:  SHEET METAL PART

***CODES***
   5241

***KEYS***
SM              HOLE          COIL

***VARIABLES***
LENGTH = 13.5          THICK = .125

***PROCESS PLAN***
COILCUT
BEND1
```

ing control. In the case of process planning, what is captured in the logic trees is the knowledge of the process planner. Consequently, some people refer to this type of application of DCLASS as a type of "expert system." From the above example, you can begin to appreciate the capabilities of DCLASS and understand why it has been applied to many types of applications that can be represented in a DCLASS logic tree.

6.4.3 Computer Managed Process Planning (CMPP)

The Computer Managed Process Planning (CMPP) System, developed by the United Technologies Research Center with some funding from the U.S. Army, is

FIGURE 6.23 Example CMPP interactive display. (Reprinted with permission from Ref. 44.)

```
    PART  ARMY-TEST-1              INTERACTIVE PART INPUT          071382  1343
                                FINISHED DIMENSIONING SESSION


    ABBR DESCRIPTION                                    VALUE       FLD SIZE
         SURFACE NAME                                   D009           4
         SURFACE TYPE                                   DIAMETER       3
     MS= MATERIAL SIDE (L,R,A, OR B)                    BELOW          1
     RD= DIR FROM REF SURFACE (+=ABOVE,-=BELOW)         +              1
     DV= DIMENSION VALUE                                .625           7
     DT= DIMENSION TOL                                  .004           5
     RS= REFERENCE SURFACE                              DIA            4
     BC= BLEND CONFIG(B=BREAK, F=FILLET)                B              1
     BV= BLEND VALUE                                    .0125          6
     BT= BLEND TOL                                      .025           4
     SF= SURFACE FINISH                                 125            3
    SN1= SURFACE NOTE1                                                 2
    SN2= SURFACE NOTE2                                                 2
    SN3= SURFACE NOTE3                                                 2
      R  REVIEW--REPAINT SCREEN WITH UPDATED INFO                      0
      Q  EXIT TO SELECT ANOTHER SURFACE                                0
    SELECT THE DATA ABBREV. TO BE UPDATED FOLLOWED BY THE NEW DATA
    DT=.005,R
```

a generative system capable of automatically making process decisions [37]. However, an extensive interactive capability is provided which permits the user to examine and modify a process plan as it is developed. Figure 6.23 shows one of the interactive displays provided. This system was developed specifically for machined cylindrical parts (such as gears) that are characterized by complex manufacturing processes and tight tolerances. As a result, this system can be used to plan the fabrication of cylindrical parts involving processes such as:

1. Turning, grinding, and honing
2. Broaching, milling, electrical discharge machining, and drilling used to produce flats, slots, holes, gear teeth, and other noncylindrical features on a cylindrical part
3. Plating, heat treating, and other nonmachine-oriented operations performed on parts without consideration of cylindrical restrictions.

Other process capabilities can be added to the system as long as each process is performed on a cylindrical part.

CMPP is manufacturer independent; that is, process capabilities and associated planning logic for a specific firm are loaded into the system data base at implementation time. The CMPP system is depicted functionally in Figure 6.24. The data base that must be developed contains specifics about available machines, tools, stock removal allowances for cutting materials, type of cuts that can be

FIGURE 6.24   CMPP system overview. (Reprinted with permision from Ref. 44.)

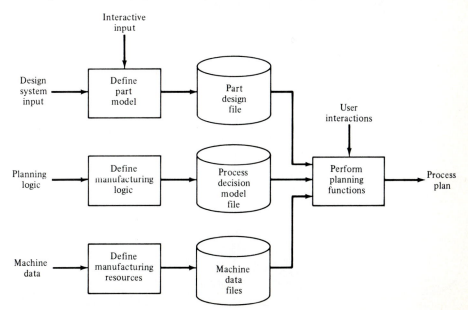

made by a particular machine, tolerance information with regard to cutting conditions, machinability information, and process decision models.

This latter type of information, process decision model (PDM), is a technique used to define local manufacturing rules. Essentially, each PDM is comprised of the logical rules that stipulate how a family of parts is to be produced. To enhance the user/computer interaction required to define these rules, a computer process planning language (COPPL) was developed. The language was designed to utilize process planning terminology and to be simple to use. Consequently, the language resembles a restricted form of English rather than a computer programming language, and it can easily be read by manufacturing people. Because the vocabulary used may differ from one company to another, capability is provided to expand the vocabulary. Figure 6.25 illustrates a portion of a PDM for generat-

FIGURE 6.25 Process decision model (PDM) illustrated. (Reprinted with permission from Ref. 44.)

```
 0010 DRAW MATERIAL AT MC0100 (BENCH) $
N
N            THE NEXT RULE ESTABLISHES 'NORMAL' AS THE ORIENTATION IN
N            WHICH THE LONGEST OD IS 'EXPOSED' FOR MACHINING.
N
 0020 ORIENT PART FOR LONGEST OD EXPOSURE $
N
N            RULES 0030-0080 CUT OD'S AND THE THRU BORE.  IF THE
N            BORE'S L:D RATIO IS SMALL ENOUGH, IT CAN BE CUT ALONG
N            WITH OD'S ON A LATHE. IF NOT, IT MUST BE GUN DRILLED.
N
 0030 DO 0040 IF THRU BORE L:D RATIO IS .LT G, ELSE 60 $
N
 0040 TURN ON MC0200 (LATHE) IN NORMAL IF
            SURFACE IS LARGEST OD (OR)
            FEATURE IS AN OD STEP,
             FEATURE IS EXPOSED,
             STEP HEIGHT  IS .GT .050 (OR)
            SURFACE IS AN END,
             SURFACE IS EXPOSED (OR)
            SURFACE IS THRU BORE $
N
 0050 DO 0080 $
N
 0060 TURN ON MC0200 (LATHE) IN NORMAL IF
            SURFACE IS LARGEST OD (OR)
            FEATURE IS AN OD STEP,
             FEATURE IS EXPOSED,
             STEP HEIGHT IS .GT .050 (OR)
            SURFACE IS AN END,
             SURFACE IS EXPOSED $
N
 0070 DRILL THE THRU BORE ON MC0300 (GUN DRILL) IN NORMAL
N
 0080 TURN ON MC0200 (LATHE) IN REVERSE IF
            FEATURE IS AN OD STEP,
             FEATURE IS EXPOSED,
             STEP HEIGHT IS .GT .050 (OR)
            SURFACE IS AN END,
             SURFACE IS EXPOSED $
```

ing a summary of operations. The lines in this figure ending with an "N" denote commentary to improve readability, and the underlined words denote vocabulary terms. Note that each rule is numbered and may generate an operation, define part orientation, or transfer control to another operation. Using the interactive capabilities of CMPP, the planner can review the sequence of operations generated; Figure 6.26 is an example of one of these displays. Figure 6.27 also illustrates some rules from a PDM that specify how some reference surfaces are to be selected, and Figure 6.28 shows the results of applying this logic.

After the manufacturing logic and the resources have been defined for a particular firm, the planner may begin creating process plans. CMPP requires part information equivalent to the information content of a blueprint. This includes final part shape, dimensions, tolerances, surface finish, geometric form conditions, material specifications, and surface treatments and/or coatings. Since CMPP is a generative system, the type of raw material, geometry, and dimensions must be specified. Defining this much information for a part with many features is not a small task. The ideal system would read the design data from a CAD system and require virtually no user input; however, this is currently beyond the state of

FIGURE 6.26 CMPP example interactive display for modifying operation 20 details. (Reprinted with permission from Ref. 21.)

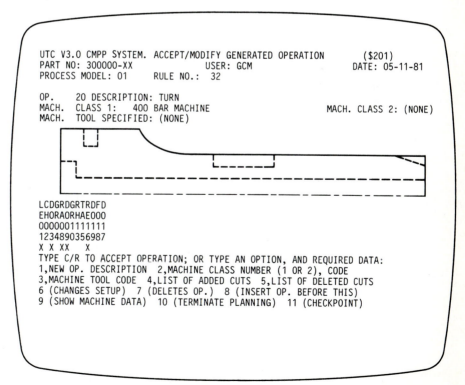

```
UTC V3.0 CMPP SYSTEM. ACCEPT/MODIFY GENERATED OPERATION        ($201)
PART NO: 300000-XX              USER: GCM            DATE: 05-11-81
PROCESS MODEL: 01    RULE NO.:  32

OP.   20 DESCRIPTION: TURN
MACH. CLASS 1:   400 BAR MACHINE              MACH. CLASS 2: (NONE)
MACH. TOOL SPECIFIED: (NONE)


LCDGRDGRTRDFD
EHORAORHAEOOO
0000001111111
1234890356987
X X XX    X
TYPE C/R TO ACCEPT OPERATION; OR TYPE AN OPTION, AND REQUIRED DATA:
1,NEW OP. DESCRIPTION  2,MACHINE CLASS NUMBER (1 OR 2), CODE
3,MACHINE TOOL CODE  4,LIST OF ADDED CUTS  5,LIST OF DELETED CUTS
6 (CHANGES SETUP)  7 (DELETES OP.)  8 (INSERT OP. BEFORE THIS)
9 (SHOW MACHINE DATA)  10 (TERMINATE PLANNING)  11 (CHECKPOINT)
```

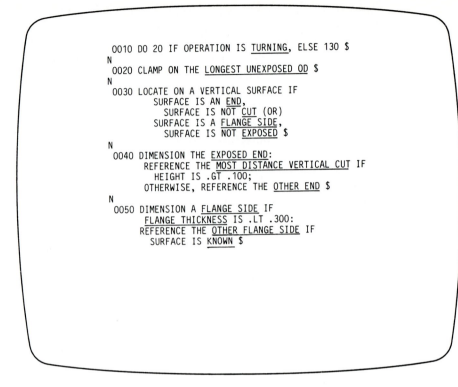

```
     0010 DO 20 IF OPERATION IS TURNING, ELSE 130 $
   N
     0020 CLAMP ON THE LONGEST UNEXPOSED OD $
   N
     0030 LOCATE ON A VERTICAL SURFACE IF
               SURFACE IS AN END,
                  SURFACE IS NOT CUT (OR)
               SURFACE IS A FLANGE SIDE,
                  SURFACE IS NOT EXPOSED $
   N
     0040 DIMENSION THE EXPOSED END:
             REFERENCE THE MOST DISTANCE VERTICAL CUT IF
               HEIGHT IS .GT .100;
             OTHERWISE, REFERENCE THE OTHER END $
   N
     0050 DIMENSION A FLANGE SIDE IF
             FLANGE THICKNESS IS .LT .300:
             REFERENCE THE OTHER FLANGE SIDE IF
               SURFACE IS KNOWN $
```

FIGURE 6.27   Process decision model (PDM) for reference surface selection. (Reprinted with permission from Ref. 21.)

FIGURE 6.28   CMPP results from applying reference surface selection process decision model. (Reprinted with permission from Ref. 21.)

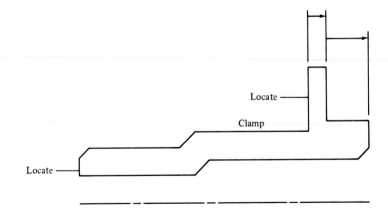

technology. We will discuss the desired capability later in this chapter under the topic of future trends in process planning.

Consider the problem of describing the shape of a part to a computer system. In some ways, it is analogous to describing a part to a person over a telephone using only x and y coordinates and instructions to connect these coordinates with lines. In addition, one is permitted to use a limited vocabulary of feature descriptors, such as a hole, in the conversation. Utilizing Figure 6.29, we will briefly describe how this might be done using the CMPP system.

Remember that CMPP is applicable only to rotational parts. Therefore, in order to conserve time in developing a drawing, Figure 6.29 shows only one-half of the part relative to its axis of rotation. Nothing more needs to be shown, since the part is concentric about its axis of rotation. This illustrates, in a graphical manner, why process plans for rotational parts are usually easier to develop than plans for prismatic parts.

Identification numbers are used to identify each surface and feature in Figure 6.29. Each number is assigned in clockwise order beginning with the left intersection of the centerline. In this figure, surface "2" denotes the outside diameter of the part. The number "3" denotes a feature (a recess), and the surfaces making up that feature are "4," "5," another feature "6" (composed of the surfaces "7," "8," and "9"), and "10." The number "11" represents a feature (a radial hole). As each surface/feature number is input to the computer, other descriptive data (as many as 26 elements) must be included, such as feature type, dimensions, tolerances, and finish. Table 6.5 lists features the CMPP system is capable of

FIGURE 6.29   CMPP feature specification methodology.

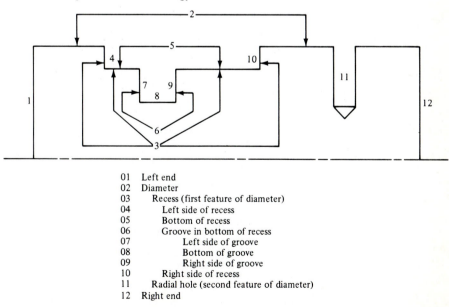

01	Left end
02	Diameter
03	Recess (first feature of diameter)
04	Left side of recess
05	Bottom of recess
06	Groove in bottom of recess
07	Left side of groove
08	Bottom of groove
09	Right side of groove
10	Right side of recess
11	Radial hole (second feature of diameter)
12	Right end

TABLE 6.5
CMPP feature capabilities

Diameter	Radial hole
Face (not an end)	Bolt hole
Left or right end	Axial slot, round bottom
Taper	Axial slot, square bottom
Chamfer	Lug
Centerdrill (for turning center)	Flat
Drill point	Window
Countersink (for turning center)	Thread
Countersink (not for turning center)	Cross slot, round bottom
Groove	Cross slot, square bottom
Relief	Tab
Recess	Scallop
Half fin	

handling. Raw material geometry is described to the system in a similar manner. Therefore, not only is data entry tedious, but many system definitions, such as the difference between a recess and a slot, and procedural rules, such as how to number the surfaces and features, must be understood. If the computer could read this part feature data directly from a CAD file, a great deal of time could be saved and many potential mistakes could be avoided.

At this point, assuming that the part descriptive information has been entered, the process decision and analysis functions can be performed. The CMPP system performs the four process planning functions shown in Figure 6.30. The

FIGURE 6.30 CMPP process planning functions. (Reprinted with permission from Ref. 21.)

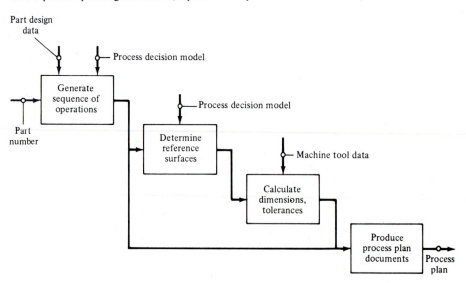

first function generates a sequence of operations in a summary format. The part number input by the planner is used by the system to retrieve the relevant part design data, which includes a part family code that is used to retrieve the appropriate process design model. The sequence of operations generated includes for each operation an operation number and description, type of machine, orientation of the workpiece on the machine, and surfaces cut or otherwise processed (such as heat-treated).

The second function involves the selection of dimensioning reference surfaces for each cut in each operation. In addition, clamping and locating surfaces may be selected. Process decision models are used to perform the logic required to accomplish this function.

In the next planning function, analysis is performed to determine machining dimensions, tolerances, and stock removals for each surface cut in each operation. Tolerance adjustments are made within machining capabilities and design limitation so that all cuts are made in a cost-effective manner. This analysis is done using a tolerance charting procedure (a variation of the manual procedure described by Wade [50]) which ensures that:

1. Blueprint dimensions and tolerances can be achieved by the sequence of operations that has been generated.
2. The best possible tolerance is assigned to each cut, considering the machine capabilities and costs.

Tolerance charting will be explained in Section 6.5.

The fourth planning function is the generation of process plan documentation. Three documents are produced: a printed summary of operations, a printed tolerance analysis, and dimensioned workpiece sketches for each machining operation. The dimensioned workpiece sketches are produced automatically (some editing may be required), which saves significant time because, in environments where parts are made to tight tolerances, sketches are often produced for each metal-cutting operation as part of the process plan to aid the machinists. Figure 6.31 shows a sketch produced by CMPP.

The CMPP system has received limited use in the industrial environment; it is used mostly in the aerospace engine industry, where many rotational parts with close tolerances are made. Some experience with CMPP has been reported in the literature. For instance, Sikorsky, a major manufacturer of helicopters, reported that this system is used for 10% of all cylindrical parts, resulting in a 75% reduction in labor devoted to preparing process plans [51].

Although the system is in use, some aspects need further development, such as the ability to perform operation detailing (determining the best machine, cut sequences, feeds, speeds, etc.). This involves some complex logic that might be best solved using artificial intelligence. As further enhancements have been pursued, CMPP has been installed in research environments. For instance, at Arizona State University, research has been performed on integrating CAD and CAPP systems [33]. A feature-recognition algorithm is used to extract pertinent

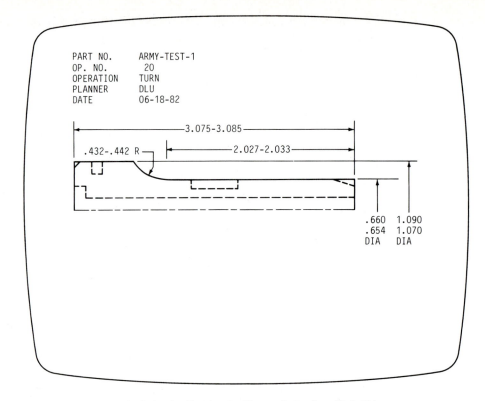

PART NO.   ARMY-TEST-1
OP. NO.   20
OPERATION   TURN
PLANNER   DLU
DATE   06-18-82

FIGURE 6.31   Example CMPP-generated sketch. (Reprinted with permission from Ref. 44.)

part data from the CAD data file (Figure 6.32). The extracted data is then transformed to an IGES (Initial Graphics Exchange Specification) standard data format, which is the predominant method currently used to exchange part definition data between different CAD systems. Next a postprocessor converts the part definition data from the standard format into the formats required by CMPP. Thus, even though the use of the CMPP system is limited, the system is significant because of the extent to which the process planning function is automated, because it represents one of the most successful attempts at developing a generative

FIGURE 6.32   Integration of CAD and CAM using CMPP. (From results of Li's research [33].)

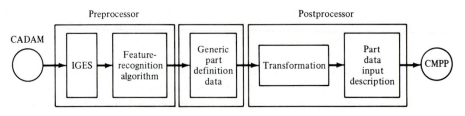

system, and because it is being used as a basis for further research into automated process planning.

6.5 TOLERANCE CHARTS

We will explain the tolerance charting procedure using an example taken from some of the CMPP descriptive material [37]. However, we will describe the procedure as if we were doing it manually instead of automatically.

Figure 6.33 shows a form that can be used in the tolerance charting procedure. The operations of a potential process plan are listed on the left side of the form. The vertical lines in the center extending up to the part drawn at the top of the figure represent part surfaces. Horizontal arrows drawn in the central portion of the chart represent cuts; each is associated with an operation listed on the left side of the chart. An arrowhead denotes a surface to be cut, and a circle at the other end of the arrow denotes the reference surface. At the bottom of the chart, below line 21, are the part print dimensions, represented by lines with circles at both ends, and to the right of these lines are the blueprint dimension values with their respective tolerances. We will assume that the turning operations, AUTO BAR and TURN LA, limits are 0.011–0.003, and the grinding operations, O.D. GRD, SURF GRD, and CRUSH GRD, limits are 0.005–0.001. With this information, we can proceed to use the tolerance charting technique to evaluate the feasibility of using this process plan to make the part to the blueprint dimensions. You might note that only lateral dimensions are considered in this procedure. Since a diametral dimension is measured from one side to the other of a cylindrical surface or hole, there is no locating reference surface. Consequently, the tolerance on the final cut is the value on the part blueprint unless the cutting tool was not capable of holding the desired tolerance.

The tolerance charting procedure is usually initiated from the final part dimensions; this is how we will begin. As we proceed, keep in mind that some analysis has been performed before this procedure is used. In other words, we know that the operations specified will fabricate the desired part; however, we do not know if we can maintain the desired tolerances. Blueprint dimensions 24–27 correspond to cuts 16–19; therefore, the print dimensions can be entered in the "DIM" column under "LATERAL DIMS." Moving up to cut 13, we see that no blueprint dimension corresponds directly to this cut; however, we can calculate this dimension by summing dimensions 29 and 30 to get 1.375. Cut 13 is a surface finish operation performed by a grinder, which means that, from time and cost considerations, it would be desirable to remove very little material at this operation. Using this logic, we will assign a stock removal value of 0.010 in the "DIM" column under "STK. REM." With this stock removal value we can calculate the dimension after cut 6 is performed; it is 1.385. We can now look at cut 6, which is performed on a lathe. Relative to a grinder, a lathe can remove larger amounts of material in one cut, but it cannot hold as small a tolerance. A reasonable value to assign for stock removal for cut 6 is 0.035. This is the last value needed to determine the length of our raw stock after cut 4 is performed. Using similar logic,

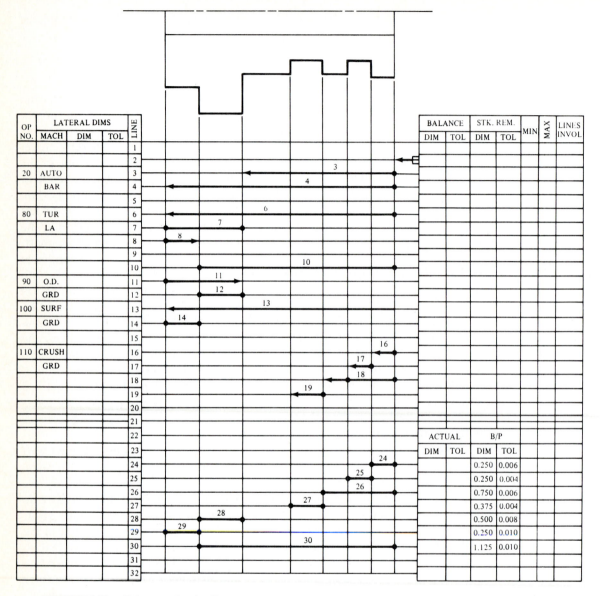

FIGURE 6.33 Tolerance charting form.

the dimensions can be obtained for the remaining cuts. These are shown in Figure 6.34. Up to this point we have not tried to calculate the tolerances that must be held at each cut. We can now do that.

Tolerances on cuts 3 and 4 do not affect the blueprint tolerances; therefore, a value of 0.011 will be chosen to minimize machining costs. Since cuts 16–19 correspond to blueprint dimensions, the blueprint tolerances will be used. The remaining tolerances are not as simple to assign, because they are the result of more than one cut. For example, the tolerance on print dimension 30 is the outcome of cuts 6 and 8. We can represent this type of affect with a "balance line." A balance line represents a dimension calculated by adding or subtracting

FIGURE 6.34 Tolerance charting example with dimensions calculated.

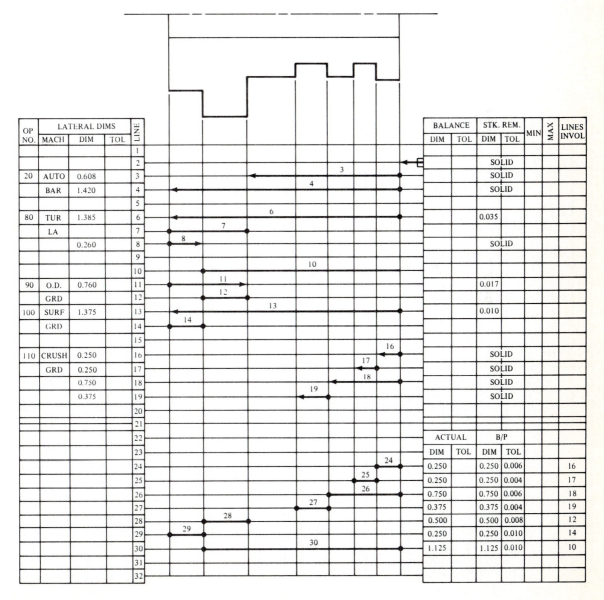

two other cut or balance lines above it on the tolerance chart. Lines 7, 10, 12, and 14 in Figure 6.35, with circles at both ends, represents balance lines. The lines involved in calculating the dimensional values of a balance line are denoted under the column "LINES INVOL" on the chart. We can now compute the dimensional value for each of the balance lines.

FIGURE 6.35 Tolerance charting example showing cut tolerances. (Reprinted with permission from Ref. 44.)

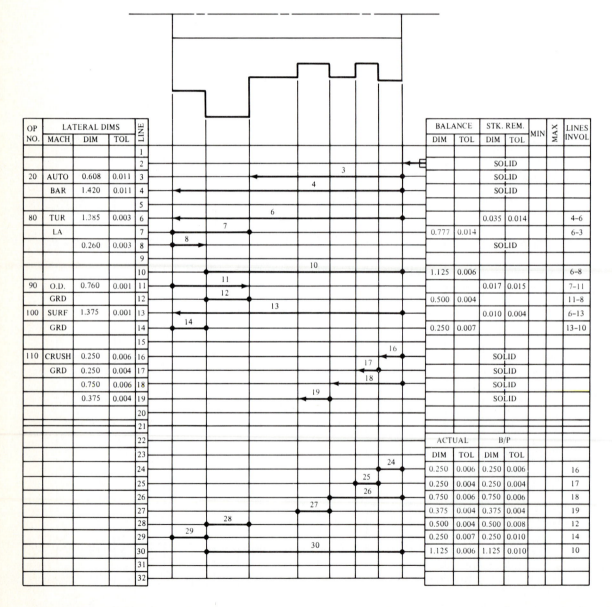

In computing the tolerances that must be held for each of the balance lines, we will first list what is known about each of the blueprint dimensions being considered:

Dimension number	Balance line	Associated cuts	Blueprint tolerance
30	10	6 + 8	0.010
29	14	13 + 6 + 8	0.010
28	12	11 + 8	0.008

The best solution denotes as large a tolerance as possible for each cut. However, if some cuts must have smaller tolerances than others, it is best to assign the smaller ones to the grind operations because of machine capabilities.

We will start our evaluation by using the largest values of machine limits, 0.011 and 0.005, respectively, for the turning and grinding operations.

$$30: \quad 0.011 + 0.011 \qquad\qquad = 0.022 \quad \text{(fail)}$$
$$29: \quad 0.005 + 0.011 + 0.011 = 0.027 \quad \text{(fail)}$$
$$28: \quad 0.005 + 0.011 \qquad\qquad = 0.016 \quad \text{(fail)}$$

Not a very encouraging start. We will next decrease each of the turning and grinding tolerances by 0.004 and 0.002, respectively.

$$30: \quad 0.007 + 0.007 \qquad\qquad = 0.014 \quad \text{(fail)}$$
$$29: \quad 0.003 + 0.007 + 0.007 = 0.017 \quad \text{(fail)}$$
$$28: \quad 0.003 + 0.007 \qquad\qquad = 0.010 \quad \text{(fail)}$$

Repeating this process should produce favorable results.

$$30: \quad 0.003 + 0.003 \qquad\qquad = 0.006 \quad \text{(pass)}$$
$$29: \quad 0.001 + 0.003 + 0.003 = 0.007 \quad \text{(pass)}$$
$$28: \quad 0.001 + 0.003 \qquad\qquad = 0.004 \quad \text{(pass)}$$

This demonstrates that a valid set of tolerances can be found. Not all of the allowable tolerances have been used; consequently, a succeeding step might involve loosening these results. We will stop at this point and enter our results in the tolerance chart in Figure 6.35. Although the part used in this example was relatively simple to make, the computation of the allowable tolerances for all of the cuts involves a substantial amount of detail. Consequently, a procedure that organizes these details is very useful.

6.6 CRITERIA FOR SELECTING A CAPP SYSTEM

Selecting the best CAPP system, like most software-selection decisions, is not an elementary process. The initial step in the selection process should be to establish the objectives that are to be achieved by implementing a CAPP system. Although this will vary from one company to another, the following are representative of the objectives that might be established:

1. Capture the manufacturing engineering expertise and make it available to all process planners.
2. Reduce the amount of manual effort and time required to prepare a process plan.
3. Make it easier to introduce new manufacturing technologies.
4. Provide the capability to make mass changes to existing process plans.
5. Develop optimum and consistent process plans for part families.
6. Provide a data base of manufacturing knowledge that can be readily accessed.
7. Facilitate the integration of CAD and CAM.
8. Reduce the skill required to develop a process plan.
9. Reduce the number of errors made in preparing process plans.
10. Achieve the many cost reductions (such as tooling, shop labor, and scrap costs) that are directly and indirectly realized from implementing a process planning system.
11. Provide the ability to retrieve and store standard text and graphics.

This list of objectives can become long, since the potential benefits can be realized by many functions within a firm.

Once objectives have been established, criteria should be developed for selecting the process planning system. The criteria presented below should provide a basis for structuring a selection procedure.

1. Classification and coding capabilities
    1.1 Classification and coding methodology
    1.2 Ability to alter part attributes
    1.3 Process plan retrieval capabilities using keys such as part attributes and codes
2. Graphics capabilities
    2.1 CAD
    2.2 CAD data base management
3. Work instruction creation and maintenance
    3.1 Word processing
    3.2 Text and graphics storage (combined or separate) and editing
    3.3 Global update
    3.4 On-line help
    3.5 Forms generation
    3.6 Automatic revision history
4. Process planning capabilities
    4.1 Variant, generative, or hybrid
    4.2 Part family identification

4.3 Capture knowledge of process planners

4.4 Selection of tools, machines, and sequence of operations

5. Time standards capabilities

5.1 Calculation of standards

5.2 Access to other vendors' standard time modules

6. Machining parameters

6.1 Selection of speeds and feeds

6.2 Determination of stock removal

6.3 Tolerance charting

7. Total system considerations

7.1 Hardware required

7.2 Data base management capabilities

7.3 Interfaces to other software systems, such as MRP and shop-floor control

7.4 System cost

7.5 Ease of use

7.6 Future enhancements

8. Vendor qualifications

8.1 Years in business

8.2 Number of employees to support software

8.3 Experience of employees

This list is not exhaustive, and may need to be modified to meet specific concerns of a particular firm. However, it has proven to be an excellent starting point for developing CAPP system selection criteria.

6.7 RESEARCH IN CAPP

From previous discussions, you will be aware that the integration of CAD and CAM through a generative process planning system seems to be a logical approach that promises large benefits. However, we know that this is not an easy task. In this section we will examine technological developments underway that may have an important impact on this integration. Also, we will discuss some related research being performed.

6.7.1 Product Definition Data Standard

There are several good CAD systems in widespread use; however, as was noted in Chapter 2, there is no standard for representing the design data in a file. Consequently, transferring data between dissimilar CAD systems can be difficult at best. To solve this problem, several different approaches have been developed. Of these, IGES (Initial Graphics Exchange Specification) has received the most industrial support to date.

As noted previously, IGES is a neutral, vendor-independent sequential file format for interchanging product definition data among CAD systems [6]. Since 1980 the National Bureau of Standards has coordinated work performed by sev-

TABLE 6.6
IGES Version 2.0 geometry entities

Circular arc	Parametric spline surface
Composite curve	Plane
Conic arc	Point
Copious data	Rational B-spline curve
Finite element	Rational B-spline surface
Flash	Ruled surface
Line	Simple closed area
Linear path	Surface of revolution
Node	Tabulated cylinder
Parametric spline curve	Transformation matrix

eral committees composed of vendors and users to develop IGES. The IGES format is in the public domain, and consists of geometry, structure, and annotation entities. The geometry entities define the geometry of the product being described (see Table 6.6). Annotation entities correspond to the information used by a draftperson to describe how the product is to be manufactured (see Table 6.7). Structure entities are used to communicate the data base structure from one CAD system to another (see Table 6.8). The part is represented as a wire frame edge-vertex data model and is intended for human interpretation at the receiving end. IGES was developed to facilitate the exchange of graphics data between systems being used in the 1970s and early 1980s. Consequently, 2-D and 3-D wire frame models and some generative-type surfaces were emphasized. As a result, most entities are generic to geometric representation, such as line, arc, composite curve, and associativity. Therefore, if the data is to be used by a process planning system, some human interpretation will be required to transform these entities into features, such as holes and slots.

Although IGES is a valuable tool that is being used more and more for transferring data between dissimilar CAD systems, it is not a perfect specification. Getting it to work, depending on the CAD systems involved in the data transfer, may require careful planning and testing. There are many CAD systems in use today, and, since there is no industry standard data format, there are almost as many data formats as there are different CAD manufacturers. Consequently,

TABLE 6.7
IGES Version 2.0 annotation entities

Angular dimension	Linear dimension
Centerline	Ordinate dimension
Diameter dimension	Point dimension
Flag note	Radius dimension
General label	Section
General note	Witness line
Leader	

TABLE 6.8
IGES Version 2.0 structure entities

Associativity	Property
Circular array	Rectangular array
Drawing	Subfigure definition and instance
Line font definition	Text font definition
Macro definition and instance	View

special software for converting from each system to another must be incorporated into the IGES specifications. This is complicated by the fact the formats may change as the systems are being enhanced. Boeing and Pratt & Whitney (two firms that exchange CAD data with many vendors) suggest some general policies that should be adopted if IGES is to be used as an exchange specification [17]:

1. Reduce the number of systems exchanging data. When possible, use the same type of CAD system. If this is not possible, encourage suppliers to standardize on at most three types of CAD systems. Preferably, all these systems should also be used by the prime company.
2. Develop standards for CAD data. Adopt written standards for symbols, text fonts, dimension types, and mathematical surfaces. Establish conventions for which layers will hold specific data types, how symbol or detail libraries will be organized and labeled, and the use of color.
3. Work out standards for intersystem exchange. Document what data types can be sent and received, and the organization of the data.
4. Write test procedures and develop test data for all data conversion software.

Although they were developed to facilitate the use of IGES, these policies will apply to any data exchange specification.

For CIM to become a reality, it has been recognized that a more complete product data exchange methodology is needed. In 1984 the Initial Graphics Exchange Specification (IGES) Organization initiated a long-term project to develop the Product Data Exchange Specification (PDES) [28]. Product data includes data that might be used in the life cycle of a product; besides engineering design, this includes such activities as manufacturing, quality assurance, testing, and field support. One objective of this project is to be able to transfer a complete product model with sufficient information content as to be interpretable without human assistance by advanced CAD/CAM applications such as process planning. In addition to geometric data, including solid models, PDES will support a wide range of nongeometric data such as part features, tolerance specifications, surface finishes, and material specifications. Another objective of this project is to recommend this specification to the International Standards Organization (ISO) as a worldwide standard for the exchange of product data, because one objective of the ISO Technical Committee TC 184 (Industrial Automation Systems) is to develop a standard, which has been named Standard for Transfer and Exchange of

Product Model Data (STEP), for capturing the information comprising a computerized product model in a neutral format that can be used throughout the life cycle of the product. The United States hopes that PDES and STEP will be identical.

The IGES Organization plans to maintain a conversion path between IGES and PDES. Consequently, the long-range intent is to ensure that software and data investments in systems that utilize the IGES formats will not be lost as PDES becomes an accepted standard.

Some specifications for PDES will be based on knowledge gained from similar, more narrowly focused efforts sponsored by several different organizations. One such effort is the XBF-2 format (Experimental Boundary File—Version 2) developed by CAM-I as a standard for transmitting 3-D solid models. Another is the ESP (Experimental Solids Proposal) developed by IGES as a means for transmitting 3-D solid models between CAD systems. Extensive work has been funded by the CIM program of the U.S. Air Force to develop PDDI (Product Definition Data Interface) [38]. The objective of the PDDI effort is to define and demonstrate a prototype system that replaces the engineering drawing as the interface between design and all manufacturing functions, including process planning, NC programming, quality assurance, tool design, and production planning and control. The PDDI concept is based on a neutral exchange format that can be used to translate data among dissimilar CAD/CAE/CAM systems. This system has been developed by aerospace firms using four part types typically found in an airframe. GMAP (Geometric Modeling Applications Interface Program) [20, 41] is a similar effort but larger in scope, as it extends PDDI to cover two new part types, disks and turbine blades, and applications in inspection, analysis, and product support. This work is being performed by jet engine aerospace firms with funding from the U.S. Air Force.

The PDDI system includes entities for boundary representation and topology. The topology entities permit a part to be decomposed into faces, edges, and vertices which describe the connectivity of the part. In addition, PDDI identifies part features entities such as holes, flanges, threads, webs, pockets, and chambers. Within PDDI the feature entities are related to specific topology and geometry entities so that information for a particular feature can be identified explicitly. This is necessary to support factory automation where features of a part are denoted by such things as holes and flanges. Significant strides were also made toward developing a complete part model in PDDI. For example, tolerance information is maintained in a form that is directly interpretable by a computer rather than in a computerized text form intended primarily for interpretation by a human being. Consequently, this information must be associated with those entities in the model affected by the tolerances. Other nonshape entities addressed include such things as specifications, materials, notes, and effectivity.

For CAD and CAM to be integrated, PDES or something like it must be available. Otherwise, a human must interpret the part information contained in the CAD system and then input the required information into the CAPP system and other CAM systems.

6.7.2 Part Feature Recognition

We are just beginning to understand what is involved in integrating CAD and CAM systems. Preparing specifications for systems like PDES will be a significant aid to this integration. The subject of feature recognition represents some of the research that has greatly assisted this understanding. The objective of this work has been to bridge the gap between CAD data bases and automated process planning systems by automatically distinguishing the features of a part from the geometric and topological data stored in the CAD system. Once the features are classified, the automated planning system could develop the required process plan to make the part, thereby eliminating the need for a human to translate the CAD data into something the process planning system can understand. Although this research has not yet produced a practical process planning system, the work has greatly enhanced our understanding of what is required to develop an automated planning system. Table 6.9 summarizes feature-recognition research. We will discuss some of these systems.

TABLE 6.9
Some feature-recognition research

Author	Part feature-recognition system	Recognizable features	Ref.
CAM-I	Form feature taxonomy	—	10
Choi	CAD/CAM-compatible, tool-oriented process planning system	Holes, slots, pockets	16
Henderson and Anderson	Extraction of feature information from 3-D CAD data	Holes, slots, pockets	24
Jalubowski	Syntactic characterization of machine part shapes	Rotational part family	25
Kakazu and Okino	Pattern-recognition approaches to GT code generation	Rotational GT code	26
Kakino et al.	A method of parts description for computer-aided production planning	Grooves, steps, flanges	27
Kung	Feature-recognition and expert process planning system	Holes, slots, planes, and pockets	29
Kyprianou	Shape features in geometric modeling	Rotational part family	31
Lee	Integration of solid modeling and data base management for CAD/CAM	—	32
Liu	Generative process planning using syntatic pattern recognition	—	35
Srinrasan, Liu, and Fu	Extraction of manufacturing details from geometric models	Rotational part family	46
Woo	Computer-aided recognition of volumetric designs	—	54

Source: From Ref. 33.

Solid modeling systems provide the most robust part descriptions; therefore, these appear to be a logical starting point for developing a feature-recognition system. Henderson and Anderson [24] developed a system called FEATURES to perform automatic feature recognition using data from a solid modeling system. FEATURES simulates the human interpretation of part features. The system consists of a feature recognizer, extractor, and organizer. At present this system performs well for objects containing swept features, which are volumes created by sweeping a 2-D lamina through space. This approach is encouraging, because conceptually it can be applied to more complex (nonswept) parts.

The FEATURES system uses the boundary representation (BREP) of a part, which denotes the faces, edges, and vertices (FEV). Thus features such as holes must be derived from more primitive data representing FEV contained in a BREP graph. Consequently, a hole may be present as a collection of faces that must be recognized from the part data.

This observation led to the formalized procedure of defining a feature as a production rule. For example, if $P_1, P_2, P_3, \ldots, P_n$, then A; where $P_1 \ldots P_n$ are conditions which, if satisfied, imply that A is the specified feature. This procedure corresponds to one form of knowledge representation used in expert systems. Henderson and Anderson provide the following example rule (another version of this rule was presented in Chapter 2) for a simple cylindrical hole (see Figure 6.36):

> If a hole entrance exits
> and the face adjacent to the entrance is cylindrical,
> and the face is convex,
> and the next adjacent face is a plane,
> and this plane is adjacent only to the cylinder,
> then the entrance face, cylindrical face, and plane
> comprise a cylindrical hole.

It would be necessary to develop other rules to define terms such as entrance, face, convex, and cylindrical. In this research the ROMULUS solid modeler was

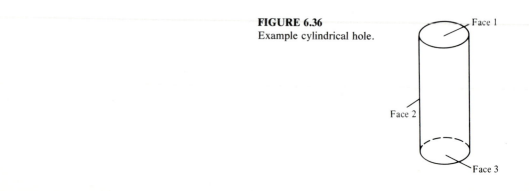

FIGURE 6.36
Example cylindrical hole.

used to create the BREP data. The feature rules were formalized and applied using the PROLOG language.

The part appearing in Figure 6.37 illustrates an application of the FEA-TURES system. In this example, a compound cavity (a cavity composed of multiple features) is recognized, and then, using boolean subtraction, each feature is extracted from the compound cavity. The result is a set of separate cavities which form the original compound cavity. The separated features are then linked in a feature graph according to their adjacency. Features with entrance faces common to the stock material are linked directly to the stock material node. Other features are linked to their adjacent features as determined during the feature-extraction process. The structure of the graph shows feature definition, feature adjaceny, and, to some extent, feature accessibility. This feature graph could then be passed to an automated process planning system to determine what operations are required to make the part.

Kung [29] performed some comprehensive research in feature recognition and demonstrated the feasibility of automatically generating a process plan using a

FIGURE 6.37
Example application of the FEA-TURES system.

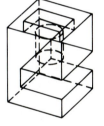

Original part   Cavity volume

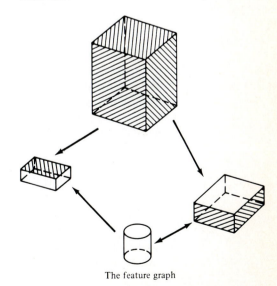

The feature graph

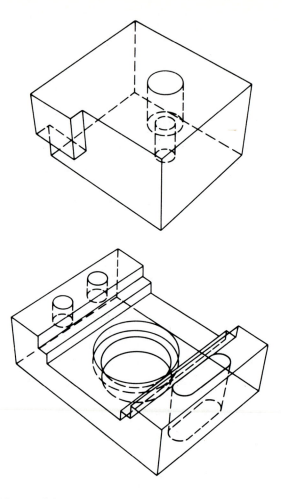

FIGURE 6.38
Parts used by Kung [29] to verify
feature-recognition algorithms.

BREP data file created by the PADL-1 solid modeling system. The parts depicted
in Figure 6.38 were used to verify this research. Kung's work paralleled Henderson and Anderson's in that Kung constructed a set of logical rules for identifying
form features from a BREP data file. However, Kung used Fortran rather than
PROLOG to develop the necessary computer procedures. Logical procedures
were developed to recognize the form features listed in Figure 6.39. As the features are identified, the associated information is stored in a feature file, which is
very similar to what Henderson and Anderson's FEATURES system does. Next,
Kung used an expert system and the data in the feature file to prepare a process
plan automatically. Some operator assistance is required for textual material such
as material type and surface finish.

Although the research in feature recognition to date has been restricted to
relatively simple parts, the successes offer encouragement. Also, the results provide a basis for understanding how we should proceed in the future.

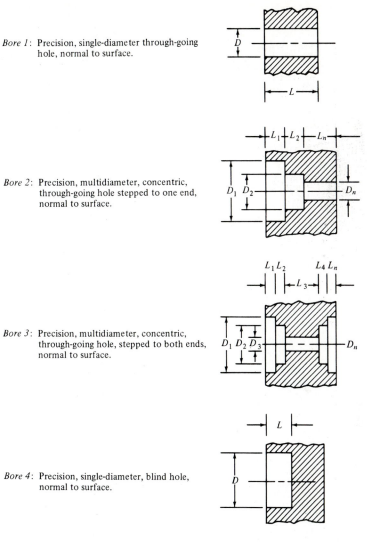

Bore 1: Precision, single-diameter through-going hole, normal to surface.

Bore 2: Precision, multidiameter, concentric, through-going hole stepped to one end, normal to surface.

Bore 3: Precision, multidiameter, concentric, through-going hole, stepped to both ends, normal to surface.

Bore 4: Precision, single-diameter, blind hole, normal to surface.

Bore 5: Precision, multidiameter, concentric, blind hole, stepped to one end, normal to surface.

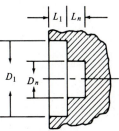

FIGURE 6.39   Features recognized by Kung's algorithms [29].

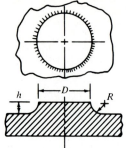

Boss: A short, cylindrical projection from the surface or a fabricated workpiece. Used to provide a seat for another mating part or reinforcement around a hole. The outer surface of the boss is usually machined.

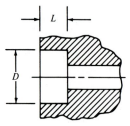

Counterbore: Cylindrical enlargement on the end of a hole.

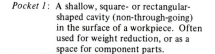

Pocket 1: A shallow, square- or rectangular-shaped cavity (non-through-going) in the surface of a workpiece. Often used for weight reduction, or as a space for component parts.

R_2 = bottom fillet radius

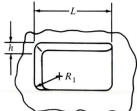

Pocket 2: A shallow, circular-shaped cavity (non-through-going) in the surface of a workpiece.

R_2 = bottom fillet radius

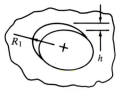

Slot 1: A long, narrow through-joining opening. May have square or radiused ends.

FIGURE 6.39   (*continued*)

Slot 2: A long, narrow channel with a T-shaped cross section.

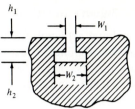

Hole 2: A through opening of square shape, normal to the surface of the workpiece.

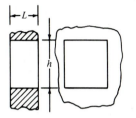

Hole 1: A through, round, single-diameter opening, perpendicular to the surface of the workpiece.

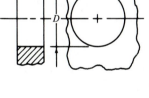

Notch: A shallow angular cut or rounded recess in the surface or edge of a workpiece. Often used to provide clearance or locative surfaces.

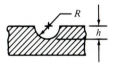

Pad: A slight projection, rectangular in shape, on the workpiece surface. Often found around fastening holes to provide a seat for other components. The pad surface may be formed by casting, molding, welding, etc. It is usually machined or ground to provide a precision surface.

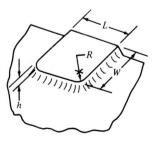

FIGURE 6.39   (*continued*)

6.7.3   Artificial Intelligence in Process Planning

Applying artificial intelligence (AI) techniques to process planning promises to provide significant results, especially for developing generative systems where a new process plan is to be created by synthesizing process information (procedural knowledge) and decision logic (declarative knowledge). This is a relatively new area of research, and the problem of developing a process plan for all but the simplest part is not trivial. Consequently, most of the AI-based process planning systems developed so far, such as TOM [3], GARI [19], and Hi-Mapp [3], are still in the experimental stages .

TOM (Technostructure of Machining) is a knowledge-based system developed at the University of Tokyo. Only hole-making processes are considered, and only one hole at a time can be planned. Therefore, features in conjunction with one another are not considered. TOM performs a backtrack search procedure with the capability of using the machining time to evaluate alternate sequences of actions. Although TOM has limited capabilities, it is significant because it represents one of the first applications of AI technology to process planning.

The GARI system represents a more ambitious effort, because the part features are not limited to holes. A set of production rules and processes with weights is used to develop a process plan. The part is described to GARI using a special language. Plan generation proceeds through successive refinements, where each iteration produces new assertions that apply to the set of potential processes (cuts). The assertions that have been produced up to that iteration represent a set of plans and imply a partial order on the set of potential cuts. If two pieces of advice imply contradictory results, the advice with the lower weight is rejected, as are all assertions based on that advice. Successive refinements are terminated when the list of active rules by the current set of assertions has been exhausted. The GARI system has provided considerable perspective into how a generative process planning system might be developed using AI. Production rules have been developed for only a limited number of part features; consequently, it is not used outside a research environment.

Hi-Mapp, a more recent AI application, utilizes the experience gained from earlier AI work. A set of system words is used to describe the part form features. A hole is described by specifying the type of hole, the face from which it starts, the face it opens into, and the diameter. Figure 6.40 illustrates how a countersunk hole, labeled as H1, might be described. The Hi-Mapp word for countersunk hole is FTYPE, the diameter is 20, the countersink diameter is 30, it starts from face FZP, it opens into face P1, it has a tolerance of 11, and the countersink depth from face FZP is 6.5. A direction of positive or negative must be specified for each form feature. Feature direction is important in determining the relationships of form features. Kung, in his feature-recognition research, used much of the same geometric descriptive data as is used in Hi-Mapp, which implies that Kung's feature-recognition algorithms might be combined with Hi-Mapp's plan-synthesizing capabilities. The experimental version of Hi-Mapp contains 45 production rules in seven major categories: process selection, cut selection, machine selection, rest-

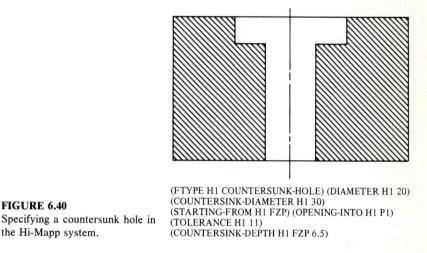

(FTYPE H1 COUNTERSUNK-HOLE) (DIAMETER H1 20)
(COUNTERSINK-DIAMETER H1 30)
(STARTING-FROM H1 FZP) (OPENING-INTO H1 P1)
(TOLERANCE H1 11)
(COUNTERSINK-DEPTH H1 FZP 6.5)

FIGURE 6.40
Specifying a countersunk hole in the Hi-Mapp system.

ing face selection, tool selection, feed selection, and miscellaneous action. The control strategy permits a plan to be developed in a hierarchical manner; that is, for efficiency, certain tasks should be planned before others. In addition, actions can be done simultaneously (e.g., moving a part to the desired location while it is being rotated). Starting from an unordered set of goals, Hi-Mapp uses a backward search strategy. The inference engine is based on a revised form of DEVISER [3], which was originally developed to generate command sequences for spacecraft. In addition to the revised DEVISER, the system uses the following files: a problem file, containing information about the part and a statement of the desired goals; machines, tools, and materials files, including characteristics and availability specifics; a production file, containing the production rules (knowledge base); and the domain functions file, containing the LISP functions used by the production file. Hi-Mapp is a significant step toward the development of a practical generative process planning system. Successes such as this are providing encouragement for others to perform research in this area; consequently, future accomplishments should come more rapidly and with greater frequency.

6.8 SUMMARY

In the early 1970s the function of process planning received very little attention. Today, however, as the manufacturing environment has become more complex and competition has become more intense, process planning has been accepted as critical to the success of many companies. In addition, it has been acknowledged to be the link between CAD and CAM and, as a result, process planning is recognized as a vital element in CIM.

Many firms are in the process of implementing CAPP systems. Two approaches are being utilized: variant and generative. The variant approach is relatively simple, and the payback is significant; as a result, the variant approach has been more popular. Most successful commercial systems have been of the variant type. However, some generative systems have been developed for specialized types of parts, primarily rotational parts that are symmetrical in one axis. Also, many research projects have been initiated with the objective of developing a generative system.

If the process planning function is to be completely automated, techniques must be developed to recognize part features from a CAD data file. Some research has been performed on this subject. Closely related to this subject is product data modeling. IGES represents some of the initial work on this subject. Although IGES has not been the success that was initially envisioned, it is the most popular ''neutral'' format used for transferring data between CAD systems. However, a more encompassing format is needed for product modeling—one that will remain valid from design to field support. PDES specifications are being developed with that objective in mind. The end objective is an international standard for product modeling.

Although process planning has only recently received widespread attention, a large amount of work is now being done on this topic and related subjects. In the next decade we should see some significant developments that will facilitate the integration of CAD and CAM.

EXERCISES

1. When would you advocate using each of the following approaches to process planning: manual, variant, and generative?
2. What are the major components in each of the following types of process planning systems: variant and generative?
3. Describe three ways of inputting part specifications into a process planning system.
4. Describe what CAM-I's CAPP system would do for you if you were a manufacturing engineer responsible for developing new process plans.
5. What is the difference between an op code and an op plan?
6. Briefly describe the DCLASS system, assuming that your audience knows nothing about the system.
7. Explain why the CMPP system requires a large amount of time to implement; include a description of the types of data you think would be in the system data base.
8. Explain when you would use the tolerance charting procedure. Illustrate your explanation with a drawing of a part, surfaces to be cut, and reference surfaces.
9. Why will a CAPP system facilitate the integration of CAD and CAM?
10. Eleven objectives were listed in this chapter for implementing a CAPP system. Develop three more objectives for this list.
11. Eight major criteria were provided in this chapter as a basis for developing a structured approach to selecting a CAPP system. Using this list, develop a quantitative methodology for selecting a CAPP system.
12. Briefly explain the following: IGES, PDES, PDDI, and GMAP. Also, explain how they differ.
13. Some research is being performed on developing a CAD system that permits the designer to define the part in terms of its form features. From both the design engi-

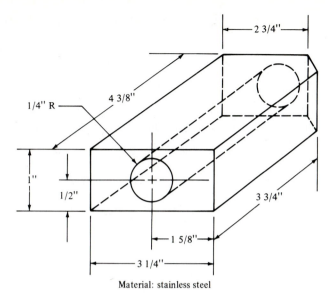

2 3/4"

1/4" R

4 3/8"

1"

1/2"

3 3/4"

1 5/8"

3 1/4"

Material: stainless steel

FIGURE 6.41   Part for Exercise 18.

neer's and the manufacturing engineer's viewpoints, explain the pros and cons of a system of this type.

14. Develop a decision table for the following problem statement. Tomorrow is a holiday, and you want to decide what recreational activity to do. If the temperature is above 50°F and the probability of rain is less than 0.3, you will play golf. If the temperature is above 50°F and the probability of rain is greater than or equal to 0.3, you will go to a movie. If the temperature is 50°F or less, you will read a novel.

15. Develop a decision tree for the problem statement in Exercise 14.

16. Develop a decision table for deciding what to wear to class today.

17. Develop a DCLASS logic tree for deciding what to wear to class today.

18. Develop a DCLASS logic tree for determining how the part in Figure 6.41 is to be made.

19. Using the data in Figures 6.19 and Table 6.3, develop a part family matrix for the family of parts with a shape code of 52 and made from coil material.

20. Considerable research is being performed to develop generative process planning systems. Explain why you think this approach will or will not be successful.

6.9 REFERENCES AND SUGGESTED READINGS

1. Allen, D. K.: "Generative Process Planning Using the DCLASS Information System," Monograph No. 14, Computer-Aided Manufacturing Laboratory, Provo, Utah, February 1979.

2. Allen, D. K., and P. R. Smith: "Computer-Aided Process Planning," Computer-Aided Manufacturing Laboratory, Provo, Utah, 1980.

3. Berenji, H. R., and B. Khoshnevis: "Use of Artificial Intelligence in Automated Process Planning," *Computers in Mechanical Engineering*, September 1986.

4. Burch, J. G., F. R. Strater, and G. Grudnitski: *Information Systems Theory and Practice*, John Wiley, New York, 1983.

5. *BYUPLAN Generative Process Planning System User Manual*, Computer-Aided Manufacturing Laboratory, Brigham Young University, Provo, Utah, 1985.

6. CAD/CIM Alert, *The Manager's Guide to CAD/CAM Standards for Integration*, Management Roundtable, Inc., Chestnut Hill, Mass., 1986.

7. CAM Software Research Laboratory: *DCLASS Technical Manual*, Brigham Young University, Grove, Utah, 1985.

8. CAM-I: *CAPP 2.1 User's Manual*, #PS-76-PPP-03, Computer-Aided Manufacturing-International Inc., Arlington, Tex., 1976.

9. CAM-I: "Training Material for CAPP, CAM-I Automated Process Planning," Computer-Aided Manufacturing-International, Arlington, Tex., 1978.

10. CAM-I: *Form Features Taxonomy,* #R-86-PPP-01, Computer-Aided Manufacturing-International, Inc., Arlington, Tex., 1987.

11. Chang, T. C.: "Interfacing CAD/Automated Process Planning," *AIIE Transactions*, vol. 13, September 1981, pp. 223–233.

12. Chang, T. C.: "TIPPS—A Totally Integrated Process Planning System," unpublished Ph.D. dissertation, Department of Industrial Engineering and Operations Research, Virginia Polytechnic Institute, Blacksburg, Va., November 1982.

13. Chang, T. C., and R. A. Wysk: "Integrating CAD and CAM through Automated Process Planning," *International Journal of Production Research*, December 1983.

14. Chang, T. C., and R. A. Wysk: *An Introduction to Automated Process Planning Systems*, Prentice-Hall, Englewood Cliffs, N.J., 1985.

15. Chevalier, P. W.: "Group Technology: The Connecting Link to Integration of CAD and CAM," in *AUTOFACT 5 Conference Proceedings*, Detroit, Mich., 1983, pp. 1.26–1.29.

16. Choi, B. K.: "CAD/CAM Compatible Tool-Oriented Process Planning System," unpublished Ph.D. dissertation, School of Industrial Engineering, Purdue University, West Lafayette, Ind., December 1982.

17. Computer-Aided Design Report, "Exchanging Data with Suppliers," CAD/CAM Publishing, Inc., San Diego, Calif., February 1987.

18. Darbyshire, I., and B. J. Davies: "EXCAP: An Expert Systems Approach to Recursive Process Planning," in *Proceedings of 16th CIRP International Seminar on Manufacturing Systems*, Tokyo, 1984, pp. 73–82.

19. Descotte, Y., and J. C. Latombe: "Making Compromises among Antagonist Constraints in a Planner," National Polytechnic Institute of Grenoble, France, January 1984.

20. Disa, R., and L. Phillips: Geometric Modeling Applications Interface Program, Interim Technical Report, August 1985 to October 1985, #FR19038-01, United Technologies, East Hartford, Conn., February 1986.

21. Dunn, M. S., "Computerized Production Process Planning for Machined Cylindrical Parts," CASA/SME Process Planning Seminar, Orlando, Fla., January 1986.

22. Eversheim, W., H. Fuchs, and K. H. Zons: "Automatic Process Planning with Regard to Production by Application of the System AUTAP for Control Problems," in *Computer Graphics in Manufacturing Systems*, 12th CIRP International Seminar on Manufacturing Systems, Begrad, 1980.

23. Granville, Charles S.: "The Impact of Applying Artificial Intelligence with Group Technology and Computer-Aided Process Planning System," Autofact Conference, Detroit, Mich., 1986.

24. Henderson, M. R., and D. C. Anderson: "Computer Recognition and Extraction of Form Features: A CAD/CAM Link," *Computers in Industry*, vol. 4, no. 5, 1984, pp. 315–325.

25. Jalubowski, R.: "Syntactic Characterization of Machined Part Shapes," *Cybernetics and Systems*, vol. 13, 1982, pp. 1–24.

26. Kakazu, Y., and N. Okino: "Pattern Recognition Approaches to GT Code Generation on GSG," in *Proceedings of 16th CIRP International Seminar on Manufacturing Systems*, Tokyo, 1984, pp. 10–18.

27. Kakino, Y., F. Oba, T. Meriwaki, and K. Iwata: "A New Method of Parts Description for

Computer-Aided Production Planning," in *Advances in Computer-Aided Manufacturing*, D. McPherson, Ed., North-Holland, Amsterdam, 1977, pp. 197–213.

28. Kelly, J. C.: "The Product Data Exchange Standard (PDES)," Federal Computer Conference, September 1985.

29. Kung, H. K.: "An Investigation into the Development of Process Plans from Solid Geometric Modeling Representation," unpublished Ph.D. dissertation, Oklahoma State University, Stillwater, Okla., 1984.

30. Kung, J. S.: "Integrated CAD and CAM—A Study of Machined Cylindrical Parts Design," unpublished MS Engineering Report, Department of Industrial and Management Systems Engineering, Arizona State University, Tempe, Ariz., August 1984.

31. Kyprianou, L. K.: "Shape Classification in Computer-Aided Design," unpublished Ph.D. dissertation, University of Cambridge, Cambridge, England, July 1980.

32. Lee, Y. C.: "Integration of Solid Modeling and Database Management for CAD/CAM," unpublished Ph.D. dissertation, School of Electrical Engineering, Purdue University, West Lafayette, Ind., August 1984.

33. Li, R. K.: "A Conceptual Framework for the Integration of Computer-Aided Design and Computer-Aided Process Planning," unpublished Ph.D. dissertation, Arizona State University, Tempe, Ariz., 1986.

34. Lin, L.: "A Classification and Coding Scheme and Computer Aided Process Planning System for Rotational and Gear Parts Using DCLASS," unpublished Master's Report, Arizona State University, Tempe, Ariz., 1986.

35. Liu, D.: "Utilization of Artificial Intelligence in Manufacturing," in *AUTOFACT 6 Conference Proceedings*, Anaheim, Calif., October 1984, pp. 2.60–2.78.

36. Logan, F. A.: "Process Planning—The Vital Link Between Design and Production," *CAPP Computer-Aided Process Planning*, SME, Dearborn, Mich., 1985.

37. Mann, W. S., M. S. Dunn, and S. J. Pflederer: "Computerized Production Process Planning (Final Report, Contract DAAK4076-C-1104), UTRC Report R77-942625, August 1977.

38. McDonnell Aircraft Company: "PDDI Overview," Materials Laboratory, Wright-Patterson Air Force Base, September 1985.

39. Nilson, E. N.: "Integrating CAD and CAM-Future Directions," CASA/SME Advanced Computer Techniques for Design and Manufacturing Conference, April 1977.

40. *Opta-Plan Generative Process Planning*, Product Bulletin, Manufacturing Generative Systems, Milwaukee, Wis., 1985.

41. Phillips, L., P. Kieffer, K. Perlotto, J. Wright, and J. Rirok: "Geometric Modeling Applications Interface Program," Interim Technical Report, August 1986 to October 1986, #FR19038-05, United Technologies, East Hartford, Conn., February 1987.

42. Phillips, R. H.: "A Knowledge-Based Approach to Generative Process Planning," *AUTOFACT 7 Conference Proceedings*, Chicago, October 1985.

43. Sack, C. F.: *CAM-I's Experimental Planning System, XPS-1*, Preliminary Publication, Computer-Aided Manufacturing-International, Arlington, Tex., 1983.

44. Sack, C. F.: "Computer Managed Process Planning—A Bridge between CAD and CAM," CASA/SME Autofact Conference, November 1982.

45. Schwartz, S. J., and M. T. Shreve: "GT and CAPP—Productivity Tools for Hybrids," *The International Journal for Hybrid Microelectronics*, vol. 5, no. 2, November 1982.

46. Srinrasan, R., C. R. Liu, and K. S. Fu.: "Extraction of Manufacturing Details from Geometric Models," *Computers & Industrial Engineering*, vol. 9, 1985, pp. 125–133.

47. Steudel, H. J.: "Computer-Aided Process Planning: Past, Present, and Future," *International Journal of Production Research*, vol. 22, no. 2, 1984.

48. Tulkoff, J.: "Process Planning in the Computer Age," *Machine and Tool Blue Book*, November 1981.

49. Tulkoff, J., and R. Subrin, Eds.: *CAPP Computer-Aided Process Planning*, SME, Dearborn, Mich., 1985.

50. Wade, O. R.: *Tolerance Control in Design and Manufacturing*, Industrial Press, New York, 1967.

51. Waldman, H.: "Process Planning at Sikorsky," *CAD/CAM Technology*, Summer 1983.

52. Wolfe, P. M.: "Computer-Aided Process Planning: A Link between CAD and CAM," *Industrial Engineering*, August 1985.

53. Wolfe, P. M., and H. K. Kung: "Automating Process Planning Using Artificial Intelligence," *1984 Annual International Industrial Engineering Proceedings*, Chicago, 1984.

54. Woo, T. C.: "Feature Extraction by Volume Decomposition," *Proceedings Conference on CAD/CAM in Mechanical Engineering*, MIT, Cambridge, Mass., March 1982.

55. Wysk, R. A.: "Automated Process Planning and Selection—APPS," unpublished Ph.D. dissertation, School of Industrial Engineering, Purdue University, West Lafayette, Ind., May 1977.

56. Zdeblick, W. J.: "Computer-Aided Process Planning and Operation Planning," *Commline*, September–October 1984.

CHAPTER
7

INTEGRATIVE MANUFACTURING PLANNING AND CONTROL

Consumption is the sole end and purpose of production; and the interest of the producer ought to be attended to only as far as it may be necessary for promoting that of the consumer.

Adam Smith, *Wealth of Nations* (1776)

Each year the typical manufacturing environment becomes more complex as types of parts, materials, machines, tools, and job skills proliferate. Governmental regulations and worldwide competition have confounded the situation. In the last few years, the awareness of the importance of manufacturing competence has increased significantly [4, 22, 29]. In fact, the survival of many manufacturing firms may be at stake. Table 7.1 illustrates well what can happen if manufacturing is not emphasized. America leads the world in inventing new products; however, many of these new products are or will be manufactured by other countries. The inability of U.S. manufacturers to compete cannot be blamed on lower-cost labor, because more than one-half of the U.S. trade deficit comes from foreign industries that pay higher wages [1].

Dramatic changes have occurred in manufacturing during the twentieth century. From the beginning of this century through World War II, large investments were made in mass production facilities which resulted in division of labor and mechanization. These facilities concentrated on removing inefficient labor and

TABLE 7.1
Technology invented in the United States but now made elsewhere

U.S.-invented technology	1987 market size (millions $)	U.S. producers' share of domestic market (%)			
		1970	1975	1980	1987
Phonographs	$ 630	90	40	30	1
Color TVs	14,050	90	80	60	10
Audiotape recorders	500	40	10	10	0
Videotape recorders	2,895	10	10	1	1
Machine tool centers	485	99	97	79	35
Telephones	2,000	99	95	88	25
Semiconductors	19,100	89	71	65	64
Computers	53,500	N.A.[a]	97	96	74

[a] N.A. = not available.

Source: Data from Council on Competitiveness, Commerce Department, reprinted from Ref. 1.

skill from the manufacturing processes. Frederick Taylor and Henry Ford are well-known pioneers of this era. Taylor eliminated waste from work. Ford applied mechanization to simplify repetitive tasks, eliminating the worker where possible. As a consequence, mass production became dominant in major segments of industry. In many cases, survival of a firm depended on successful adoption of such approaches. Another result was a reduction in the intelligence and skill level of the worker from a previous era characterized by craft-based skills and individual judgments. These developments contributed immensely to the competitive advantage of America until about 1960.

Two important developments in the 1960s resulted in some people looking at manufacturing from a new perspective. Practical computers became a reality, and internationalism changed the competitive environment. People such as John Diebold started looking at the manufacturing process instead of the worker; processes could be automated using the computer. Others, such as W. Edwards Deming, begin emphasizing the importance of quality; large productivity improvements could be achieved by making the product right the first time. At first management, particularly in America, did not pay much attention to these developments. Computer usage increased, but, in general, these applications were in accounting and other business applications rather than in areas related to manufacturing. Current levels of quality seemed adequate, because people bought what was produced.

In the 1970s, however, the microprocessor was developed and true international competition became a fact. The Japanese took the lead in demonstrating the importance of quality and in automating manufacturing processes. Computer applications in manufacturing mushroomed. Table 7.2 lists some of these applications. In addition, the number of new products proliferated and product life cycles became relatively short. As a result, management around the world began to realize the importance of manufacturing as a competitive weapon. Manufacturing competitiveness had reached a global scale.

With the beginning of the 1980s came a growing awareness that a firm can no longer rely just on a product containing the latest technology. The firm must also manufacture a quality product in a timely manner at a low cost. Also, the manufacturing capabilities must be flexible enough to handle proliferation of new products with short life cycles. This perspective led to consideration of manufacturability during product design, increased manufacturing flexibility, and greater worker skills. In addition, since quality and improved productivity are of paramount importance, the philosophy of continuous improvement must be reflected in all functions of a firm. Although manufacturing capabilities can be a competitive

TABLE 7.2
Example computer-assisted manufacturing applications

Accounting
Bill of materials
Capacity requirements planning
Data base management
Design
Drafting
Digitizing and/or scanning
Drawing or engineering information retrieval
Finite-element analysis
Cell control
Geometric modeling
Group technology
Inspection/quality control
Inventory/purchasing control
Job cost estimation
Machine setup
Machining utilization data base
Management reporting
Materials requirement planning
Manufacturing resource planning
NC programming
NC postprocessor generation
Nesting, sheet or plate
Order entry, scheduling, tracking
Plant or department layout
Process planning
Production scheduling
Proposal development
Shop-floor control
Simulation analysis
Tool management
Work measurement and standards
Work-order tracking
Word processing

weapon, the many functions of a firm must be integrated. Thus, a much broader concept of manufacturing is required.

Integration is difficult within a manufacturing firm because several organizations have conflicting objectives. Sales would like to meet any demand on request, which suggests that a large inventory would be desirable. Finance would like to minimize investments and capital requirements, meaning that the less inventory the better. Manufacturing would like to meet all schedules, suggesting that a large inventory with long lead times would be desirable. Management must balance these and many other conflicting objectives in a way that meets the objectives of the firm as a whole.

In this chapter we will discuss the primary functional departments within a typical manufacturing organization (see Figure 7.1), and present some methodologies for planning and controlling the operations of these departments. The industrial and manufacturing engineering departments have the responsibility of working with design engineering to ensure that the product designed can be manufactured in a cost-effective manner. Once a design is approved to be manufactured, these departments are responsible for developing the process plans, developing the numerical control programs, designing and acquiring necessary tooling and fixtures, and monitoring the manufacturing process to work out any problems with the process that might develop. Other functions include designing the facilities, justifying and acquiring the appropriate equipment including materials handling devices, and designing and acquiring management information systems and shop-floor automation. The exact functional responsibilities of a typical industrial and manufacturing department may vary; however, those noted here are representative of functions that must be performed.

Production planning and control deals with developing plans for producing product, implementing these plans, monitoring progress toward achieving these plans, and developing and implementing corrective actions when the original plans need modification. Consequently, this function controls production and provides a mechanism for the manufacturing organization to support the overall objectives of the firm.

Manufacturing operations is directly responsible for manufacturing the product. This includes scheduling and manning the manufacturing equipment, capacity planning, shop-floor control, maintenance, cost management, and quality management.

FIGURE 7.1   Primary departments in manufacturing.

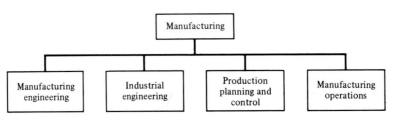

7.1   KEY DEFINITIONS

Aggregate production plan   A plan where production requirements per time period, usually months, are expressed in terms of some aggregate unit, such as direct labor hours or dollars.

Bill of materials (BOM)   A treelike structure denoting the detailed parts and the quantities of each that comprise an end product. A BOM is structured to represent how a part is made. Material or parts represented in the structure at a lower level go into the fabrication or assembly at the next higher level. The end product is assigned the level number 0, which is the top of the treelike structure. Subsequent levels are assigned higher integer levels starting with 1.

Business plan   A plan that states the value in dollars to be shipped, the expected value of inventory and production, and the value of the order backlog for monthly intervals in the planning horizon. Other expenditures not directly connected to manufacturing, such as research and development, may also be included in this plan.

Capacity planning   The process of evaluating the feasibility of a manufacturing plan considering all of the important production constraints, such as labor and equipment.

Cellular manufacturing   The organization of manufacturing machines and people into groups responsible for producing a family of parts.

Just-in-time (JIT) philosophy   A management philosophy with the goal of eliminating anything that does not add value to the product.

Linear regression   A mathematical technique for fitting a line through a set of data points to minimize the sum of the squared deviations of points not residing on that line.

Manufacturing resources planning (MRP II)   A closed-loop MPR system that, at a minimum, includes detailed capacity analysis. Some MRP II systems include the business plan in this closed-loop system.

Master production schedule   A plan that specifies how many of each type of end product must be produced in each period of the planning horizon.

Material requirements planning (MRP)   The process of comparing the master production schedule with on-hand and in-process inventories and orders to determine how many of each item in the bill of materials must be manufactured. The order quantity and latest order release date are also determined.

Order release   The process of releasing production orders to production and purchase orders to vendors.

Part effectivities   A designation of when (date or part number) a specific part design (usually designated by drawing number and revision number) is to be used.

Planning horizon   The time into the future that the aggregate plan and business plan cover. This horizon is composed of time intervals; each interval usually represents one month.

Rough-cut capacity planning   The process of using some broad guidelines, such as direct labor requirements, to ascertain capacity requirements and limitations.

Shop-floor control   The process of monitoring activity on the production floor and taking corrective action when appropriate. The level of control can vary considerably, depending on the type of production processes involved and the management philosophies of the particular firm.

Statistical process control (SPC)   A technique that involves statistically establishing control limits on some measured characteristic of a process and the subsequent monitoring of the process using these control limits.

Total quality management (TQM)   The management philosophy of continually striving to improve all tasks.

7.2   THE ROLE OF INTEGRATIVE MANUFACTURING IN CAD/CAM INTEGRATION

Industrial and manufacturing engineering provide the bridge between design engineering and manufacturing operations in terms of ensuring that new designs and design changes can be produced in a cost-effective manner (see Figure 7.2). Production control develops the plans for producing the specific items; these plans include quantities of each part per time period. Previous manufacturing eras were marked by a lack of understanding of the importance of design engineering working closely with manufacturing. As a result, getting a new product into production was an interactive process. The product would be designed and then given to manufacturing to make. Manufacturing engineering, after reviewing the new design, would suggest to design engineering how the design should be modified to accommodate existing manufacturing capabilities and cost considerations. This process could result in several iterations of design changes. In previous eras, since

FIGURE 7.2   Industrial and manufacturing engineering—the bridge between design engineering and manufacturing.

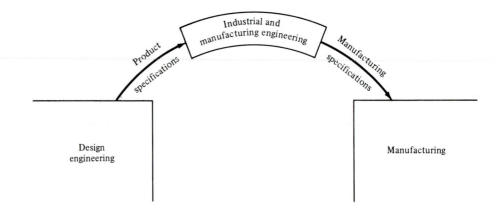

product technology was of paramount importance to management, manufacturing was often forced to accept designs that were very difficult and expensive to make. This is changing in many firms where the importance of manufacturing capabilities is understood. Today, design, industrial, and manufacturing engineering are working together as the product is designed. This type of cooperative effort is called concurrent engineering. As a result, the time to make the first product has been significantly reduced, and quality has been dramatically improved.

As the product is designed, process plans can be developed, new manufacturing equipment can be acquired (if needed), tooling and fixtures can be designed and acquired, etc. In addition, industrial and manufacturing engineering can work with design engineering to define families of parts utilizing group technology concepts. Using these families of parts, fewer new parts have to be designed. Also, industrial engineering can design the manufacturing facility so that material handling is minimized and equipment can be organized into cells. Industrial and manufacturing engineering also work with manufacturing operations and production planning and control to design management information systems needed to manage manufacturing. In addition, industrial engineering and manufacturing engineering are responsible for designing and integrating shop-floor automation and for applying new technologies. Some aspects of a new product design may require new manufacturing technologies to make the product competitive. It is important to note that design engineering should try to design a product that utilizes existing manufacturing capabilities. However, competitive market conditions may dictate a new design which requires that new manufacturing capabilities be acquired.

For a firm to be successful in the highly competitive world markets of today, all functions of the company must work together. Production control is the functional organization that translates management's strategic plans into manufacturing plans for producing parts in the right quantities, at the right time, at the lowest possible cost, and with the highest possible quality. Achieving these objectives is not easy. In the process of striving to attain these objectives, production control is responsible for large amounts of capital invested in material, inventory, equipment, and facilities. Consequently, once a product design is released by engineering to manufacturing, the production control function has a key role in assuring the profitability of a firm. This realization alone is enough to compel anyone designing an integrated CAD/CAM system to include production control within that system.

Types of information involved in this integration include bills of materials, process plans, product demands, inventory levels, inventory goals, part effectivities, machine processing rates, setup times, tooling requirements, work-in-process inventories, scrap rates, vendor delivery performance, and manufacturing lead times. This list is a small sample of the information that a production control function utilizes. Also, this information comes from many functional organizations within a firm. For instance, a bill of materials (BOM) defines all the parts that are used to make a product (see Figure 7.3); design engineering compiles the bill of materials. A BOM is often depicted in a product-structure tree format (see Figure 7.4). Process plans are developed by manufacturing engineering. Inventory levels

Level	Part number	Quantity	Lead time	Description
0	AA	1	2	Fictitious assembly
1	X	1	1	Inner assembly
2	K	1	1	Pin
3	R	6 in.	3	412 stainless steel, 0.250-in. diameter (purchased)
2	L	1	2	Bearing assembly
3	S	4	1	Ball bearings (purchased)
3	T	1	2	0.5-in. bearing holder (purchased)
1	Y	1	1	Outer assembly
2	S	8	1	Ball bearings (purchased)
2	M	1	2	1-in. bearing holder (purchased)
1	Z	1	1	Fastener (purchased)

FIGURE 7.3   Bill of materials for fictitious assembly.

are reported by inventory control, and inventory goals are a reflection of company strategic goals involving customer service and profits. Part effectivities denote when part design changes become effective; these are specified by design engineering. This description is representative of the many types of diverse data from several different organizations that must be brought together to perform the pro-

FIGURE 7.4   Producture-structure tree for fictitious assembly.

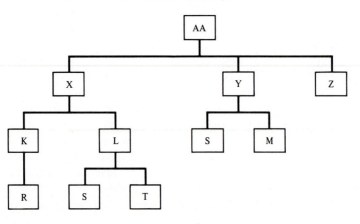

duction control function. Because part inventories and demands involve large amounts of data, computer systems have long been utilized to aid in managing a production control function. However, few of these systems have been integrated with other CAD/CAM systems. In today's competitive environment, however, this integration cannot be ignored.

Manufacturing operations is directly responsible for making the product. This can vary from a very integrated process where everything is made, beginning with raw materials, to a firm that assembles the final product from piece parts that are purchased from other firms. Within manufacturing operations, the importance of quality, on-time shipment, and cost dominate most considerations. As noted earlier, in recent years the importance of quality has been significantly elevated. Somewhat associated is the importance of the workers and their skills.

7.3 OVERVIEW OF MANUFACTURING ENGINEERING

Manufacturing engineering plays a key role in translating new product specifications from design engineering into process plans that manufacturing then uses to produce the product. Figure 7.5 shows the manufacturing engineering–related flow of information that occurs in a typical firm.

As the product is being designed, manufacturing engineering (manufacturing technical evaluation) works with design engineering to ascertain if the design can

FIGURE 7.5   Information flow in manufacturing engineering.

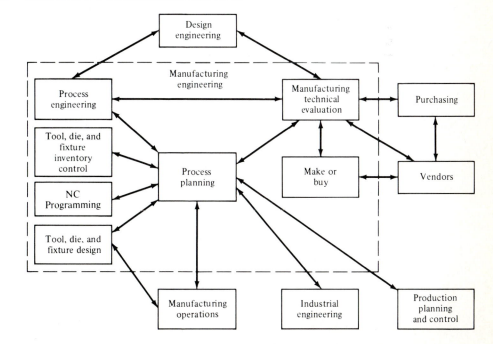

be manufactured and at what cost. Producibility guidelines are used in this analysis. Tolerances, materials, clearances, and assembly times are particularly important factors in this evaluation because they directly affect the producibility. Cost estimates are equally important. If a new process is needed, which might arise if existing processes are deemed too expensive or incapable of producing the desired product, process engineering would be asked to develop a new process. Another alternative would be to change the product design. Choosing the best alternative may be very difficult because it is based on many conflicting objectives (market share, customer specifications, cost, feasibility, timeliness, etc.).

Linear regression is a statistical technique that can be applied to developing cost estimates, especially when group technology has been utilized to designate families of parts. This statistical technique entails fitting a line through a series of points so that the sum of the squared deviations from the line is minimized. The equation for a straight line is $y = a + bx$, where a is the intercept and b is the slope. By applying the linear regression technique to a set of data, it is possible to calculate values for a and b. Then, for some value of x (the independent variable), a value for y (the dependent variable) can be calculated.

Assume that a family of gears has been designated. These gears are made out of stainless steel and have the same quality specifications, AGMA 12. Using linear regression, we will develop a relationship between the diameter of the gear (the independent variable) and the cost of making the gear (the dependent variable). Table 7.3 lists the data we will use. The objective is to fit a line between the points so that the sum of errors squared is minimized (see Figure 7.6). Because of random variations or other factors, a particular data element may not reside on the

TABLE 7.3
Ring gear manufacturing costs

Diameter, in.	Manufacturing cost, dollars
2.0	60.00
2.0	70.00
3.5	65.00
3.5	73.00
5.0	93.00
7.5	86.00
7.5	113.00
10.0	115.00
11.75	121.00
12.0	97.00
12.25	135.00
14.0	150.00
14.25	123.00
16.0	138.00
16.0	157.00
17.0	155.00

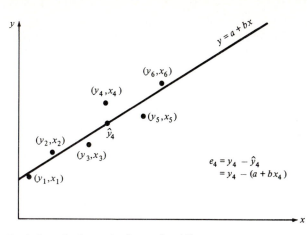

FIGURE 7.6   Deviation of a data point from a fitted line.

line. The error is

$$\sum_{i=1}^{N} e_i^2 = \sum_{i=1}^{N} [y_i - (a + bx_i)]^2 \qquad (7.1)$$

By differentiating Eq. (7.1) with respect to a and b, setting the results equal to 0, and then solving for a and b, we get

$$a = \bar{y} - b\bar{x} \qquad (7.2)$$

$$b = \frac{N \sum_{i=1}^{N} x_i y_i - \sum_{i=1}^{N} x_i \sum_{i=1}^{N} y_i}{N \sum_{i=1}^{N} x_i^2 - \left(\sum_{i=1}^{N} x_i \right)^2} \qquad (7.3)$$

A line can be fit to a set of data by first using Eq. (7.3) to compute an estimate of b. Then this value is substituted into Eq. (7.2) to calculate the estimate of a. Figure 7.7 shows a line fit to the data in Table 7.3

After fitting a line through a set of data using the above equations, you may want to determine how good a predictor the equation will be. One measurement is the coefficient of correlation r, where

$$r = \frac{N \sum_{i=1}^{N} x_i y_i - \sum_{i=1}^{N} x_i \sum_{i=1}^{N} y_i}{\sqrt{\left[N \sum_{i=1}^{N} x_i^2 - \left(\sum_{i=1}^{N} x_i \right)^2 \right] \left[N \sum_{i=1}^{N} y_i^2 - \left(\sum_{i=1}^{N} x_i \right)^2 \right]}} \qquad (7.4)$$

A positive coefficient of correlation means that increases in the independent variable will result in increases in the dependent variable. If r is negative, the indepen-

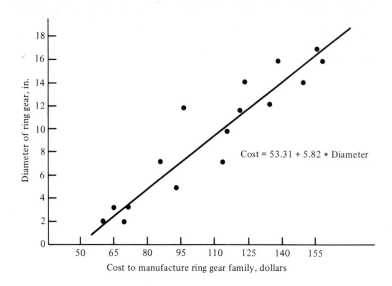

FIGURE 7.7   Diameter of ring gear versus manufacturing cost.

dent and dependent variables are inversely related. Also, if r is 0, there is no correlation between these variables.

The coefficient of determination, r^2, is another useful measure of the relationship between the variables. One interpretation of this measure is that it denotes the percentage of change in the dependent variable that is explained by changes in the regression line. Thus, a correlation of determination of 1.0 means that 100% of the changes are explained. This coefficient will have a value in the range of $0 \le r^2 \le 1$. The linear regression technique can be extended to equations that involve more than one independent variable. This statistical technique is very useful; it can be used to forecast product demand. This topic will be discussed later in the chapter.

Once a new part design is approved for production, the design is reviewed as to whether the part should be made or bought. At first glance, it appears relatively straightforward to determine if the lowest-cost alternative is to make or buy. Closer examination reveals many subtle factors that make this analysis difficult. For instance, when production is subcontracted to a vendor, control of quality and delivery is more difficult. Proprietary design and manufacturing processes are extremely difficult to protect at a vendor's location. Computing and tracking cost of vendor support is not easy; many hours of engineering and liaison assistance may be required, especially if complex parts are involved. Often indirect labor costs are allocated to ''in-house'' produced parts based on some factor such as direct labor hours. If production is sourced to vendors, there is less in-house production to absorb the indirect labor costs. Many other factors may need to be considered in a typical make-or-buy decision; this type of decision can be very complex.

Manufacturing engineering is responsible for preparing the process plans for parts that are to be made in-house. This function was covered in detail in an earlier chapter; consequently, it will not be discussed here. As the process plans are prepared, numerical control programs are developed and the necessary tools, dies, and fixtures are designed. After the process plans are approved, the new parts can be manufactured when production control deems appropriate.

7.4 OVERVIEW OF PRODUCTION CONTROL

Figure 7.8 shows the basic elements of the production control function. Demand forecasting provides an estimate of the demand for each type of product sold.

FIGURE 7.8   Production control information flow.

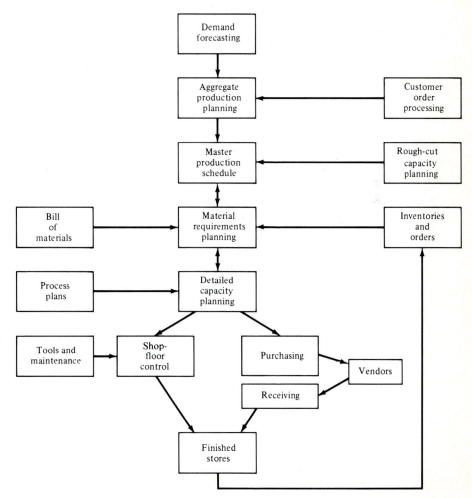

Long-term forecasting is done with enough lead time, usually 1 to 5 years, to adjust capacities which may involve constructing new buildings, buying new equipment, and hiring people. Short-term forecasting is done with enough lead time, usually 1 to 12 months, to manufacture the required products.

Once demand is known, an aggregate production plan can be prepared. Since demand and existing production capacity for some planning horizon, usually a year, will not normally be the same, some plan must be developed to rationalize these differences. That is the objective of the aggregate production plan. It is called an aggregate production plan because demand and the associated production requirements are represented in terms of some aggregate unit, such as direct labor hours. This aggregate measure is compared to the existing capacity, expressed in the aggregate units, for each interval, usually a month, in the planning horizon. At this point, the differences between demand and capacity are rationalized. Some techniques for doing this will be explained in this chapter.

After the aggregate plan has been developed, it must be disaggregated into a master production schedule. This is a plan that specifies how many of each type of end product must be produced in each period in the planning horizon. When this plan is finalized, all organizations should be in agreement as to what quantities of each will be available for sale in each interval in the planning horizon.

At this point the detailed production requirements can be planned. An end product usually consists of several detailed parts, normally represented as a bill of materials, and inventories usually exist for some if not all of these parts. Consequently, the master production plan cannot be used for the detailed production plan without some modification. The process of developing the detailed production plan is called *material requirements planning*. In this process, the bills of materials for each product to be produced are examined to ascertain what detailed parts are needed. Next, the requirements for each type of detailed part are determined. Then these requirements are reduced by the inventories on hand. The result represents the amount of each detailed part that must be produced in each planning interval in order to satisfy the master production schedule.

Often, the material requirements plan will exceed capacity for one or more of the detailed parts. This is determined by performing a detailed capacity analysis of the material requirements plan. Capacity limitations may be resolved by several means, such as working overtime or subcontracting. If the capacity limitations cannot be resolved, the material requirements plan and possibly the master production schedule must be modified. Consequently, developing a master production schedule and a material requirements plan can be an iterative process. Once these plans have been finalized, they can be initiated.

The material requirements plan is implemented by releasing production orders for the detailed parts that are to be manufactured, and by releasing purchase orders for the parts that are to be purchased. Releasing the orders does not assure that everything will be done as planned. In fact, Murphy's law (if something can go wrong, it will) must be considered. Consequently, the production control function must include shop-floor control and vendor-monitoring elements. These provide feedback on the progress of the individual production and purchase orders so that required corrective action can be performed in a timely manner. Once the detailed parts are manufactured or arrive from the vendors, these parts can be assembled and shipped to the customer.

Production control is sometimes described as the "heart" of manufacturing. For each planning period, this organization has the responsibility of finding answers to some extremely important questions, such as the following:

1. How much can we sell?
2. What parts are required to satisfy this forecast?
3. What do we have in inventory?
4. What do we have to make?
5. What manufacturing processes will be required?
6. What are the available capacities?
7. What must we procure?
8. What is the production lead time?

Often, answering these questions is not a trivial exercise. For example, Figure 7.9 shows some of the possible considerations in determining the production lead time. Many of these considerations introduce possibilities for substantial error in estimating the lead time. So, although performing the production control function may be difficult at times, it is an important part of any manufacturing company. We will now discuss in some detail each of the elements in Figure 7.8

7.4.1 Forecasting

If all future events were known, decision making that would result in achieving a desired objective would be relatively easy. Since this is not the case, decision makers attempt to make forecasts of future events with the intent of removing as much uncertainty as possible. The quality of management decisions often reflects the quality of the forecasts of key parameters, such as product demand, used to make these decisions. The time horizon—the number of time periods into the future—of the forecast will vary depending on the decisions that have to be made. For instance, it may take years to build a new plant, but a replacement order for a common part might be filled in days.

Because good forecasts are very important in decision making, considerable research has been devoted to developing accurate forecasting techniques. Even so, forecasting remains a combination of art and science. As might be expected, when the forecast time horizon becomes longer, the forecasting techniques become less reliable. Also, the forecasting technique that would provide the best result varies with the length of the time horizon.

Forecasting techniques can be grouped into three general types: qualitative, time series, and causal. *Qualitative techniques* apply structured methodologies to information and human judgment with the objective of developing an unbiased

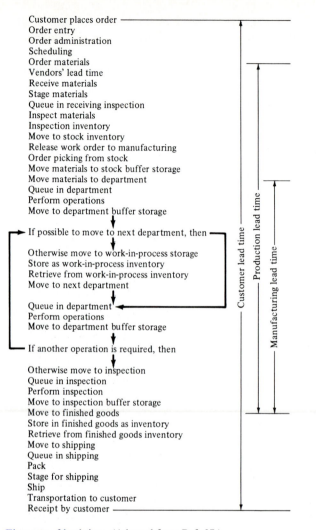

Customer places order
Order entry
Order administration
Scheduling
Order materials
Vendors' lead time
Receive materials
Stage materials
Queue in receiving inspection
Inspect materials
Inspection inventory
Move to stock inventory
Release work order to manufacturing
Order picking from stock
Move materials to stock buffer storage
Move materials to department
Queue in department
Perform operations
Move to department buffer storage

If possible to move to next department, then

Otherwise move to work-in-process storage
Store as work-in-process inventory
Retrieve from work-in-process inventory
Move to next department

Queue in department
Perform operations
Move to department buffer storage

If another operation is required, then

Otherwise move to inspection
Queue in inspection
Perform inspection
Move to inspection buffer storage
Move to finished goods
Store in finished goods as inventory
Retrieve from finished goods inventory
Move to shipping
Queue in shipping
Pack
Stage for shipping
Ship
Transportation to customer
Receipt by customer

Customer lead time

Production lead time

Manufacturing lead time

FIGURE 7.9   Elements of lead time. (Adapted from Ref. 27.)

and logical prediction of some factor. These techniques are used when quantitative techniques cannot be utilized because sufficient data are not available to develop quantitative relationships. The Delphi technique [19] is one of the better-known qualitative techniques. These techniques are often utilized to make predictions when several new technologies are involved and the time horizon is long, such as 5, 10, or more years. For example, the Delphi technique might be used to predict what the typical manufacturing environment might be in the year 2005 in the United States. The Delphi technique utilizes a panel of experts to respond to a series of questionnaires. Each expert is sent an initial questionnaire dealing with some subject, such as the use of group technology in a factory in the year 2005.

The results are used to produce the next questionnaire. Thus any new information generated is passed on to all of the experts.

A *time-series model* is a chronologically ordered set of data. Time-series analysis is the application of statistical techniques to a time series with the objective of making predictions. Regular variations may be hidden within a time series. If the interval of the variations approximates a year, the patterns are called *seasonal variations*. Otherwise, these patterns are called *cyclical variations*. In addition, a time series may be characterized by trends and growth rates of these trends. When time-series analysis is used for forecasting, the assumption is made that the historical data depicts the future. Consequently, time-series forecasting is usually not very accurate over the long term unless history is representative of the future. These techniques are often used in inventory control applications where product demand must be forecast for the next 3 to 6 months. Some basic time-series forecasting techniques will be presented later in this chapter.

A *causal model* often is a more reliable forecasting technique. Using statistical techniques, such as linear regression, it may be possible to define a relationship between ordered sets of data. For example, if you raise chickens to produce eggs, there should be a relationship between the number of eggs produced and the number of hens. A mathematical relationship can be developed between these factors, hens and eggs. If you want to predict how many eggs you might produce for the next 12 months, you would first predict how many hens you have and then calculate how many eggs they would produce. One problem with a causal model is that the independent variables, in this case the number of hens, must be predicted, which may be difficult to do with accuracy very far into the future. Also, the relationship between the dependent variable, the eggs in this case, and independent variables may change over time. Consequently, the functional relationship must be continually monitored and possibly changed.

One of the most widely used forecasting techniques is the moving average. If history is indicative of the future, this technique may be adequate for short-term forecasts. As the name implies, an average is computed from a selected number of historical data points. When a new average is computed, usually the most recent historical data points are used. In other words, a *moving average* is an average of the most recent N data points.

$$\hat{x}_t = \sum_{i=t-N+1}^{t} \frac{x_i}{N} \tag{7.5}$$

where $\hat{x}_t$ is the moving average estimate for period t, and x_i represents the datum point for period i. Recent history is being used to project the future, which seems logical. For this reason and because the technique is simple, moving averages are widely used for short-term forecasts.

One disadvantage of the moving-average technique is the data storage requirements. Also, each data element used in the moving-average computations is given equal weight. The exponential smoothing forecasting technique eliminates these two disadvantages while keeping most of the good attributes of a moving average. Equation (7.7) is the basic *single exponential smoothing model* (constant model):

$$\hat{F}_t = \alpha x_t + (1 - \alpha)\hat{F}_{t-1} \tag{7.6}$$

where $\hat{F}_t =$ is the forecast (smoothed statistics) for period t. In other words, $\hat{F}_t$ is the exponentially smoothed average of all previous data, and, for the constant model, it is the forecast for all future periods. The variable α, the smoothing constant, is constrained such that $0 \le \alpha \le 1$.

The constant model requires only three data elements, which, in a computerized system, minimizes data storage requirements. Consequently, exponential smoothing is often used in production control systems where demand for thousands of different items must be forecast. It is also interesting to note that varying the value of α has the effect of weighting the recent data relative to older data. For instance, if α has a value of 1, the forecast for all future periods is the same as the most recent observation. Usually, α will have a value between 0.1 and 0.3. Therefore, if the underlying process being modeled is level (constant), a "small" α will filter out most of the random variation while modifying the estimate of the constant level if it changes.

It is interesting to note that $\hat{F}_t$ is a weighted average because the weights sum to 1; proof of this observation will be left to an exercise. To satisfy the criterion of a weighted average, the weights must sum to 1.

Forecasts of differing time horizons are needed. A long-range forecast, 1 to 5 years, is needed to plan for long-lead-time items, such as buildings and some machine tools. An intermediate forecast, 3 to 12 months, is needed to plan for relatively long-lead-time production items, such as workforce sizing, equipment, etc. Some types of products, such as airplanes, that have a long production cycle require a longer intermediate forecast than 12 months. A short-range forecast of 1 week to 3 months is used to plan finished goods production.

7.4.2 Master Production Schedule

Typically the sales and marketing functions of a firm make demand forecasts, which are a combination of orders on hand and actual forecasts. These estimates must be reconciled with manufacturing constraints, such as capacity and availability of materials and labor. Also, since a firm may make and market thousands of finished goods, these forecasts and reconciliations are usually performed at some manageable aggregate product level. Consequently, this phase of production planning is sometimes called *aggregate planning*. Typically an aggregate plan is composed of 12 one-month intervals. Each interval denotes a planned level of production, required workforce, and anticipated inventory levels. Table 7.4 illustrates an aggregate plan.

One of the major uses of an aggregate plan is to "level" the production schedule so that production costs are minimized. Product demand varies from month to month. These variations in demand make it difficult to stabilize the workforce. However, by anticipating the high- and low-demand periods, the

TABLE 7.4
Example aggregate plan*

	Demand ($1000)	Cum demand	Work days	Work-force	Regular production	Excess production	Inventory ($ sales)
January	5,000	5,000	19	400	6,992	1,992	1,992
February	4,000	9,000	19	400	6,992	2,992	4,984
March	6,000	15,000	25	400	9,200	3,200	8,184
April	8,000	23,000	20	400	7,360	−640	7,544
May	8,500	31,500	20	400	7,360	−1,140	6,404
June	7,500	39,000	24	400	8,832	1,332	7,736
July	8,000	47,000	19	400	6,992	−1,008	6,728
August	9,000	56,000	20	400	7,360	−1,640	5,088
September	7,500	63,500	24	400	8,832	1,332	6,420
October	8,000	71,500	20	400	7,360	−640	5,780
November	10,000	81,500	18	400	6,624	−3,376	2,404
December	11,000	92,500	22	400	8,096	−2,904	−500

* Each direct labor hour produces $115 in sales. Objective is to maintain a level workforce.

workforce can be stabilized if inventory is accumulated during the low-demand periods for utilization during the high-demand periods. However, carrying inventory is expensive. Consequently, this cost must be reconciled with the costs of not carrying the inventory, which are the costs of hiring and firing of the workforce, overtime, and subcontracting to vendors.

Several mathematical programming formulations of aggregate models have been developed. One linear programming formulation to minimize total costs follows.

Minimize:

$$Z = \sum_{t=1}^{N} C_p P_t + C_r R_t + C_o O_t + C_i I_t + C_s C_t + C_h H_t + C_l L_t \qquad (7.7)$$

subject to:

$$I_t - S_t = I_{t-1} - S_{t-1} + P_t - F_t \qquad \text{for } t = 1, 2, \ldots, N \qquad (7.8)$$

$$R_t = R_{t-1} + H_t - L_t \qquad \text{for } t = 1, 2, \ldots, N \qquad (7.9)$$

$$O_t - U_t = K P_t - R_t \qquad \text{for } t = 1, 2, \ldots, N \qquad (7.10)$$

$$P_t, R_t, O_t, I_t, S_t, H_t, L_t, U_t, Z_0 \geq 0 \qquad \text{for } t = 1, 2, \ldots, N \qquad (7.11)$$

where P_t = production units scheduled in period t
C_p = cost per unit of production excluding labor
R_t = regular-time labor hours available in period t
C_r = regular-time labor cost per hour
O_t = overtime labor hours scheduled in period t

C_o = overtime labor cost per hour
I_t = inventory in stock at end of period t
C_i = carrying cost per unit of inventory
S_t = stockout quantity at end of period t
C_s = cost per unit of stockout
H_t = new hires in hours for period t
C_h = new hire cost per hour
L_t = layoffs in hours for period t
C_l = layoff cost per hour
F_t = forecast of unit demand for period t
K = labor hours required to produce one unit
N = number of periods in planning horizon

Although many models have been developed, few successful applications have been documented. This means that aggregate planning is performed using tableau and graphical methods. Spreadsheets are often used to try to minimize costs or maximize profits.

Since the aggregate plan does not denote specific products, this plan must be disaggregated into something that designates quantities of finished goods by time period. The result of this disaggregation is a master production schedule.

The master production schedule has several important uses:

1. Marketing, engineering, manufacturing, and finance use it to coordinate their activities.
2. Management uses it to plan and control workforce levels, plant facilities, equipment, materials, vendors, and costs.
3. Management uses it to specify what should be made.
4. Management uses it as a planning device to balance customer needs with plant capacities.

The master production schedule is not a demand or sales forecast, because the plant may not have the capacity to meet either forecast. Also, it is not a fabrication or assembly schedule, because inventories have not been considered. Sufficient inventory may exist to satisfy much of the master production schedule. So, although it provides a means of coordinating the activities of many of the functional departments within a firm, additional analysis must be performed before manufacturing has an executable plan.

7.4.3 Rough-Cut Capacity Planning

At this point, rough-cut capacity planning is performed to ascertain if the master production schedule is feasible. In rough-cut planning, some broad guidelines, such as direct labor for a specific production area, are used in this evaluation. These guidelines usually relate to a key resource. However, no attempt is made to determine what specific skills and equipment would be required. After a master production schedule is designated and exact production requirements are known, specific production capacity requirements must be determined using detailed capacity planning techniques. These techniques will now be discussed.

7.4.4 Material Requirements Planning

Up to this point in the production planning cycle, existing inventories have not been considered. Enough inventory may exist for a specific finished item to satisfy requirements; consequently, no production order or purchase order need be initiated for that item. Determining actual production requirements is difficult in the typical firm having thousands of finished goods assembled from many subassemblies and piece parts. Some of the components may be purchased, and many different lead times need to be considered. Fortunately, a technique called *material requirements planning* (MRP), combined with computer technology, has simplified this process. Because of the amount of detail involved, MRP was not viable until computers became practical.

MRP requires the following data:

1. Master production plan
2. On-hand inventories
3. Product structure (bill of materials)
4. Purchased or manufactured order status by item
5. Replenishment rules by item:
   a. Lead time
   b. Order quantity
   c. Scrap allowance
   d. Safety stock
   e. Etc.

Because so much data is required, a major MRP implementation problem is data accuracy.

We have already described the master production schedule. It represents everything the firm plans to ship: finished items, spares, etc. Manufacturing and procurement lead times must be considered; otherwise, the product cannot be shipped on time. The master production schedule represents the independent demand, because it is the combination of orders on the books and forecasts. However, end items are usually composed of several components; consequently, end-item demand creates a demand for these components. This latter demand is sometimes called *dependent demand*, since it is derived from the end-item requirements.

Accurate inventory data is vital, with accuracies of at least 95% being desirable. However, this level of accuracy is very difficult to attain. On-line data entry systems are an aid. These must be supplemented with employee education as well as motivational programs and cycle counting procedures. Using these procedures, the entire inventory should be audited over a time cycle such as 6 or 12 months. However, only part of the inventory is counted at one time. Some companies use a Pareto approach to minimize the time involved in performing an audit. In this case, the ''A'' items (the 20% of the items that compose 80% of the cost) may be counted individually every quarter, and the ''B'' items (the middle 30% of the items that account for 15% of the cost) may be counted every 12 months. The ''C'' items may be maintained in a two-bin system such that an order is automatically created when the first bin becomes empty. These items may be sampled every 12 months to ascertain that adequate inventories are maintained.

The bill of materials (BOM) designates what items and how many of each are used to make up a specific finished product. It is desirable to maintain a 99% level of accuracy for the BOM. The BOM is used to derive the dependent demands.

The status of an order may change after it is issued. Consequently, the status of previous orders must be considered each time the MRP analysis is performed, since the requirements calculated should represent the latest and most accurate information available.

Many types of replenishment rules are utilized within MRP, and most companies use more than one rule. The selection of what rules to use depends on the inventory policies of the individual firm, the value of the individual item, and management philosophy. Many research papers have been written on this subject. Consequently, an exhaustive list of rules and the advantages and disadvantages of each will not be attempted. The interested reader may consult Refs. 18 and 28. Examples of some of the replenishment rules are:

1. Fixed order quantity
2. Economic order quantity
3. Lot for lot
4. Fixed-period order quantity

As the name *fixed order quantity* suggests, inventory is replenished in a specified lot size. The exact quantity may be a function of many factors, such as equipment capacities, storage limitations, or shelf life.

The *economic order quantity* (EOQ) replenishment rule has been used in inventory management for several years. A mathematical expression for this rule is

$$EOQ = \frac{\sqrt{2C_p A}}{C_H} \qquad (7.12)$$

where C_p = order preparation cost
A = annual demand for an item
C_H = annual inventory carrying cost

The EOQ expression was derived using the assumption that a continuous, steady-rate demand exists. Since most demand does not satisfy this assumption, the term EOQ may be a misnomer.

Lot-for-lot order sizing is simple: The lot size is the same as the requirements for the period covered. This lot-sizing method minimizes inventory carrying costs.

Using the *fixed-period order quantity* rule, the user specifies how many periods a planned order should cover. The sum of the requirements during this span becomes the order size.

MRP logic takes the master production schedule (independent demand and spares) and breaks it down into dependent demands using the bill of materials. Some components may be used in different products or subassemblies; consequently, these demands must be consolidated. These consolidated demands are called *gross requirements*.

In the next step, inventory on hand is subtracted from gross requirements. At this time, any allowances for scrap and safety stock can be applied. Next, any purchased or manufacturing orders outstanding are subtracted, giving the *net requirements*.

Using order quantity rules, the order quantities are computed and then offset from the requirements date to allow for procurement and manufacturing lead times. Figure 7.10 gives the results of MRP computations for all the parts described in the BOM in Figure 7.3. Note that the MRP planning horizon must be long enough to include the manufacturing and supply lead times.

So far, the production planning process does not seem too difficult. However, most events do not seem to occur as planned, especially in a typical production environment which contains so many opportunities for variation from plan. Consequently, variations from previous plans must be considered before the current MRP cycle is completed. Because of these changes, orders may have to be expedited, delayed, increased, and/or decreased.

Because of the magnitude of computations involved, generative MRP systems (all computations are performed) are seldom run with a frequency less than 1 week. Some firms perform these computations only once a month. However, to facilitate more frequent analysis, *net-change MRP* has been developed to minimize the computations. Using this idea, MRP computations are performed only for products for which changes have occurred in the data used to make the last computations.

Firms that have successfully implemented MRP systems have derived many benefits (see Table 7.5). However, implementing an MRP system can be difficult. Most failures have been attributed to data integrity. Part routings and BOM data should be 99% accurate, and inventory data should be at least 95% accurate.

7.4.5 Capacity Planning

At this point the planned production orders from the MRP analysis represent a tentative manufacturing production plan. It is tentative because capacity may not

Part: AA, fictitious assembly
Lead time; 2 weeks

Week	1	2	3	4	5	6	7	8	9	10
Gross requirement	20	30	15	20	5	15	30	25	10	30
Scheduled receipt	25	10								
On hand (5)	10									
Net requirements			15	20	5	15	30	25	10	30
Planned order release	15	20	5	15	30	25	10	30		

Part: X, inner assembly (1 per assembly)
Lead time; 1 week

Week	1	2	3	4	5	6	7	8	9	10
Gross requirement	15	20	5	15	30	25	10	30		
Scheduled receipt	25									
On hand (10)	20									
Net requirement			5	15	30	25	10	30		
Planned order release		5	15	30	25	10	30			

Part: Y, outer assembly (1 per assembly)
Lead time; 1 week

Week	1	2	3	4	5	6	7	8	9	10
Gross requirement	15	20	5	15	30	25	10	30		
Scheduled receipt	40									
On hand (15)	40	20	15							
Net requirement				10	30	25	10	30		
Planned order release				30	25	10	30			

Part: Z, fastner (1 per assembly)
Lead time; 1 week

Week	1	2	3	4	5	6	7	8	9	10
Gross requirement	15	20	5	15	30	25	10	30		
Scheduled receipt	55									
On hand (5)	45	25	20	5						
Net requirement					25	25	10	30		
Planned order release				25	25	10	30			

Part: K, pin (1 per assembly)
Lead time; 1 week

Week	1	2	3	4	5	6	7	8	9	10
Gross requirement		5	15	30	25	10	30			
Scheduled receipt	15									
On hand (20)	35	30	15							
Net requirement				15	25	10	30			
Planned order release			15	25	10	30				

To part L

To part S & M

FIGURE 7.10   Example MRP releases.

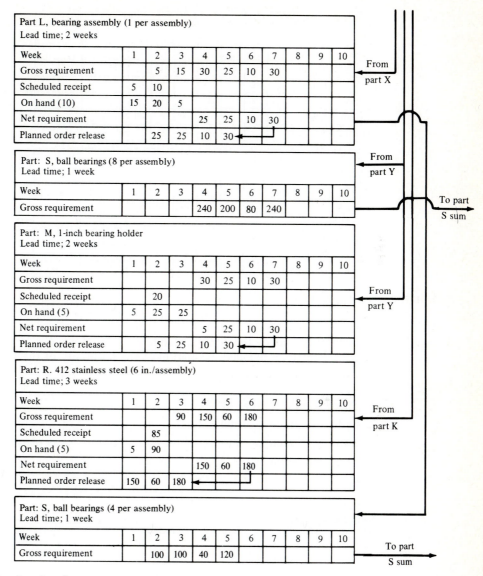

Part L, bearing assembly (1 per assembly)
Lead time; 2 weeks

Week	1	2	3	4	5	6	7	8	9	10
Gross requirement		5	15	30	25	10	30			
Scheduled receipt	5	10								
On hand (10)	15	20	5							
Net requirement				25	25	10	30			
Planned order release		25	25	10	30					

From part X

Part: S, ball bearings (8 per assembly)
Lead time; 1 week

Week	1	2	3	4	5	6	7	8	9	10
Gross requirement				240	200	80	240			

From part Y

Part: M, 1-inch bearing holder
Lead time; 2 weeks

Week	1	2	3	4	5	6	7	8	9	10
Gross requirement				30	25	10	30			
Scheduled receipt		20								
On hand (5)	5	25	25							
Net requirement				5	25	10	30			
Planned order release		5	25	10	30					

From part Y

Part: R. 412 stainless steel (6 in./assembly)
Lead time; 3 weeks

Week	1	2	3	4	5	6	7	8	9	10
Gross requirement			90	150	60	180				
Scheduled receipt		85								
On hand (5)	5	90								
Net requirement				150	60	180				
Planned order release	150	60	180							

From part K

To part S sum

Part: S, ball bearings (4 per assembly)
Lead time; 1 week

Week	1	2	3	4	5	6	7	8	9	10
Gross requirement		100	100	40	120					

To part S sum

FIGURE 7.10   (*continued*)

Part: S, ball bearings (sum of S requirements)
Lead time; 1 week

Week	1	2	3	4	5	6	7	8	9	10
Gross requirement		100	100	280	320	80	240			
Scheduled receipt	100									
On hand (150)	250	150	50							
Net requirement				230	320	80	240			
Planned order release			230	320	80	240				

Sum of S require-ments

Part: T, 0.5-inch bearing holder (1 per assembly)
Lead time; 1 week

Week	1	2	3	4	5	6	7	8	9	10
Gross requirement		25	25	10	30					
Scheduled receipt	20									
On hand (20)	40	15								
Net requirement			10	10	30					
Planned order release		10	10	30						

From part L

FIGURE 7.10   (*continued*)

exist to make all of the desired products. Therefore, the next step, capacity planning, entails a detailed evaluation of the feasibility of the MRP results. The following information is required:

1. Planned orders (from MRP)
2. Orders in process (order status)
3. Routings, including setup and run times (from process plans)
4. Available facilities

TABLE 7.5
Benefits of MRP

Improved level of customer service
Better production scheduling
Reduced inventory levels
Reduced component shortages
Reduced production lead times
Reduced manufacturing costs
Higher product quality
Less scrap and rework
Higher morale in production
Improved communication
Improved plant efficiency
Improved competitive position
Improved coordination with marketing and finance

5. Workforce availability

6. Subcontracting potential

In concept, capacity planning is simple. In reality, however, it is complex, because plant capacity is not a constant number: It changes continually as equipment fails and is repaired, as the workforce varies (hiring, firing, sickness, skills, etc.), and as the product mix varies. Examples of other complicating factors include:

1. Splitting of lots (batches) across identical machines

2. Splitting of lots to expedite a smaller quantity

3. Sequencing of lots to minimize setup time

4. Alternative routings that require different resources

5. Loading a facility by weight, volume, etc. (such as heat treating)

Most firms are continually struggling to estimate their current capacity.

The process of evaluating the capacity requirements of a tentative manufacturing plan begins with the due dates of each order. Using lead times, bills of materials, and routings, each order is back-scheduled from the due date through the required operations to determine when a particular workcenter will be utilized. Figure 7.11 was developed using the lead times and the bill of materials from Figure 7.3. Doing this for all orders and accumulating the results provide the capacity requirements. Up to this point, existing capacity limitations have not

FIGURE 7.11   Capacity requirements planning example.

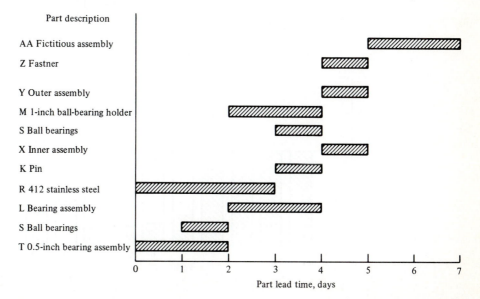

been considered; therefore, this portion of the analysis is sometimes called *infinite* capacity planning. When the capacity requirements are compared to what is available (*finite* capacity planning), workcenter overload conditions may be revealed (see Figure 7.12). Potential problems may be eliminated by shifting the sequence of orders, by expediting, by subcontracting, and/or by revising the manufacturing plan. Since the capacity and order base are constantly changing, some types of adjustment to the manufacturing plan are occurring daily.

7.4.6 Order Release

The process of notifying production or a supplier that work can begin on an order is called *order release*. Before a production order is released, material inventories are allocated to this order. Then, using the routing, a schedule is created and the appropriate shop records (paperwork or electronic media) are created to document the schedule and to notify the shop control system that the order has been

FIGURE 7.12   Workcenter load profiles: (*a*) infinite capacity planning versus (*b*) finite capacity planning.

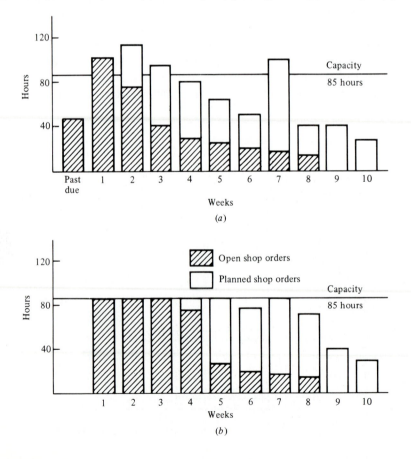

released to production. The order release system is an important element of a firm's inventory control procedures. If a manufacturing area has too much work-in-process inventory, reduction of order releases to this area will aid in reducing these inventories. Purchase order release depends primarily on the lead time of the supplier. This process is complicated by quantity discounts, quality considerations, and supplier reliability.

7.4.7 Shop-Floor Control

During the manufacturing process, many random events (machine failure, tool breakage, employee absence, shortage of materials, preempted order priorities, etc.) may occur that will affect the delivery of the product. Most of these random events are expected (only the timing of each event is unknown); consequently, production control relies on a good shop-floor control system to link higher-level planning, order release, detailed scheduling, and shop-floor data collection.

In order to keep the shop under control, production control and production supervisors rely on accurate and timely job order information, status of shop resources, and work status. As work orders progress through the shop, data are collected (possibly as frequently as at each operation) using sensors, bar-code readers, and/or manual data entry terminals. Figure 7.13 shows a data collection terminal that supports multiple types of inputs: bar-code wand or slot reader, magnetic-strip wand or slot reader, and keyboard.

Data collected from the shop floor is utilized in several ways, such as to develop new schedules, to provide order status visibility, and to generate mea-

FIGURE 7.13   Shop-floor data collection terminal. (Courtesy of IBM.)

sures of performance. The following order status information is representative of what a good shop-floor control system might provide:

1. Current locations of parts, tools, operators, etc.
2. Estimated completion date
3. Remaining operations
4. Times for all remaining operations
5. Starting batch size and current batch size
6. Job efficiency to date

This information should be on-line and readily available.

Shop-floor control has changed dramatically during the last 10 to 20 years, paralleling the development of automation and computer technology. Initially it involved primarily data collection of labor hours. This was expanded to include production lot tracking (location, number of good items, etc.). Today, machine monitoring and quality data are also collected, and data are also transmitted to the operators and to the machines. Thus, shop-floor systems are evolving toward a closed-loop control environment. An individual in customer service might interrogate an on-line computer system to determine the location of a particular production lot in order to answer an inquiry. The quality of a process might be monitored and plotted as production is performed. Numerical control programs might be loaded directly into a CNC machine. Consequently, many improvements have been made in the shop-floor control function, and many more will be made in the near future.

Managing the shop floor is an important and demanding function. Success requires planning, execution, feedback, and flexible response. It is not surprising that the results of a Yankee Group survey [24] reports that most U.S. industrial sites designate the plant floor as a primary area for investment: "More than 90% of survey respondents considered a plant floor or process monitoring and control application to be one of the most important projects." The plant-floor system can be viewed as a decision support system for factory-floor workers and management. This system should provide the ability to make "constant, incremental changes to processing." Figure 7.14 illustrates the foundations for a plant-floor/process control system. The technology is available today to implement the foundation depicted.

SHOP-FLOOR SCHEDULING. After a work order is released to production, each order must be assigned to the appropriate workcenter and then sequenced through the required operations. This process is called *scheduling*. The typical scheduling objectives are to schedule all work orders so that all due dates are satisfied, manufacturing through-put time is minimized, and work in process is minimized. Sometimes these are conflicting objectives; therefore, the scheduler must rationalize the workcenter assignments and the sequence of each order through each workcenter.

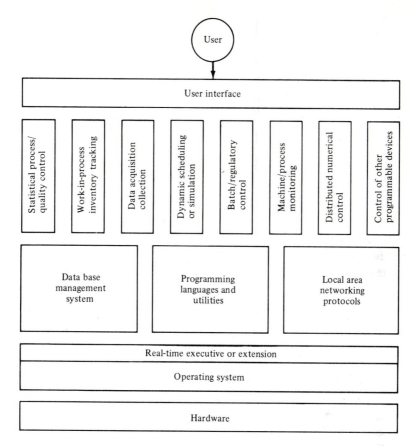

FIGURE 7.14   Foundation for a plant-floor planning and control system [24].

Work-order manufacturing through-put time is composed of the following elements:

1. Run time (operation or machine run time multiplied by batch size)
2. Setup time
3. Move time
4. Queue time (time waiting to be processed)

In a typical U.S. firm, queue time can account for 80% to 90% of the manufacturing through-put time. Consequently, scheduling is an important factor in the profitability of a firm.

Gantt charts are often used to assist in developing a schedule for a shop. Figure 7.15 shows a set of parts that must be processed through a series of operations. Figure 7.16 is a Gantt chart showing a schedule for completing the required work. No attempt has been made to minimize the makespan (the time to

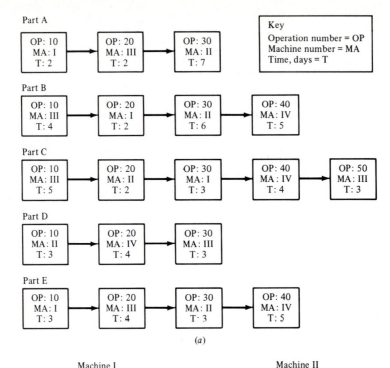

FIGURE 7.15 Part machining requirements.

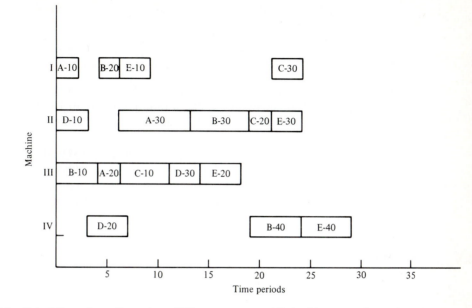

FIGURE 7.16 Scheduling using a Gantt chart. UCL = upper control limit; CL = centerline; LCL = lower control limit.

make all of the parts including machine idle time); however, the schedule is feasible.

Priority sequencing rules have been developed to assist the scheduler in determining what work order should be processed next at a workcenter. Some of the commonly used rules are:

1. *Earliest operation due date:* Process the order with the earliest due date.
2. *Order slack:* Sum the setup times and run times for all remaining operations on an order, subtract this from the time remaining until the order is due, and call the remainder slack. Work on the order with the least slack. This rule addresses the amount of remaining work.
3. *Slack per operation:* Divide the slack by the number of remaining operations. Work on the order with the smallest value. The premise for this rule is that a larger number of remaining operations increases the complexity of the scheduling task.
4. *Shortest operation first:* Work on the order with the shortest operation time at the particular workcenter. It has been proven that this rule will maximize the number of shop orders that go through a workcenter and will minimize the number waiting in queue. However, due dates are not explicitly addressed; consequently, some orders may be very late if this rule is followed without exception.

5. *Critical ratio:* Orders are sequenced based on the critical ratio, CR:

$$CR = \frac{\text{due date} - \text{now}}{\text{lead time remaining (including setup, run, move, and queue times)}} \tag{7.13}$$

A ratio of 1.0 means that the order is on time; a ratio of less than 1.0 indicates a behind-schedule condition; and a ratio greater than 1.0 indicates an ahead-of-schedule condition. So, orders with the smallest ratio should be scheduled first.

The critical ratio rule is often used in computerized shop-floor control systems.

Shop-floor control is an important function in today's competitive manufacturing environment. When a shop is well managed, manufacturing can be a competitive weapon that responds quickly to changing customer, market, and/or internal company needs.

7.4.8 Quality Assurance

As production proceeds, the quality of the product must be assured. As noted earlier, the subject of product quality has received a lot of emphasis in recent years. Quality is maintained in two ways: (1) each worker monitors his or her work; and (2) periodic inspections are performed by individuals from the quality control function. As the importance of producing only good products has become better understood, firms have emphasized that each worker should be responsible for checking his or her work. In some firms, whenever anyone spots a major defect, work stops until the cause is found. This philosophy will reduce, and even possibly eliminate, in-process inspection. The result is higher-quality products, lower through-put times, and lower costs. After assembly and final inspection, finished products are moved to shipping or to stores.

Inspection is necessary at several places in the manufacturing process, including:

1. Inspection of raw materials
2. Inspection of manufactured product:
   a. Preprocess
   b. In-process
   c. Postprocess
3. Inspection of production process parameters:
   a. Tools
   b. Fixtures
   c. Production machinery
4. Verification/calibration:
   a. Inspection fixtures
   b. Inspection gauges
   c. Inspection machinery

Ideally, suppliers should provide good products. If the receiving firm can depend on the quality of the materials, quality assurance will not have to inspect incoming product. If inspection is required, statistical sampling techniques are often used. These procedures are called *acceptance sampling* and are based on the premise that a random sample drawn from a batch is representative of the entire batch. Thus, only the sample is inspected in determining the quality of the entire batch.

There are two basic types of acceptance sampling: variables and attribute. *Variables inspection* is characterized by precise measurement of dimensions, weight, surface finish, etc. Each of these features can be described on a continuous scale. Measurement may involve precision instruments, closely calibrated, to ensure accuracy. *Attribute inspection*, in contrast, refers to evaluating whether a part's quality characteristics do or do not conform to specifications. There are two types of attributes: (1) those that cannot be measured, such as damage and missing parts; and (2) those that are not measured to save time and cost, such as using a "go/no-go" gauge to evaluate conformance to a hole size.

One objective of acceptance sampling is to assist in balancing the cost of production with the cost of shipping defective product. Ideally, no defective product should be shipped; however, this quality level will cost more than one that tolerates some defective products. Some product applications require that no defects be shipped. For example, a propulsion engine for a commercial airplane must meet rigid performance specifications because many lives may depend on that performance. Therefore, each jet engine part and assembly is carefully inspected during production, and each engine is tested before it is shipped. However, for another product, such as a ceiling fan, performance is not as critical; consequently, fan quality does not receive the same level of emphasis during production as the propulsion engine. Larger expenditures can be justified for propulsion engine quality than can be justified for ceiling fan quality because of the application. From this discussion, you can surmise that there are many types of sampling plans depending on the circumstances involved, such as costs and acceptable quality levels.

In many companies that stress the production of high-quality parts, workers inspect the parts during the production process to ensure that the process is under control. Some operations involve 100% inspection and some use random samples. Statistical process control (SPC) charts (see Figure 7.17) may be used to assist in monitoring the production process.

An SPC chart consists of a centerline (CL) that represents the expected or target value for a measured characteristic of the process or part and the control limits: an upper control limit (UCL) and a lower control limit (LCL). The control limits denote the extreme values of a measured characteristic that can occur while the process is in control. If the limits are exceeded, the process is no longer under control and something should be done. Otherwise, many, if not all, of the subsequent parts produced will have defects.

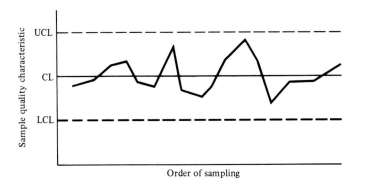

FIGURE 7.17 Typical statistical process control chart.

In recent years there has been growing recognition of the value of SPC charts. Montgomery [17] gives some reasons why control charts are used:

1. Control charts are a proven technique for improving productivity. A successful control chart program reduces scrap and rework, which are primary productivity killers in any operation. If scrap and rework are reduced, productivity increases, cost decreases, and production capacity (measured as the number of conforming parts per time period) increases.

2. Control charts are effective in preventing nonconformity. The control chart helps keep the process in control, which is consistent with the "do it right the first time" philosophy. It is rarely cheaper to sort out good units from bad ones later on than it is to build them correctly in the first place. If there is no effective process control, operators are being paid to make nonconforming product.

3. Control charts prevent unnecessary process adjustments. A control chart can distinguish between background noise and abnormal variation; no other device, including a human operator, is as effective in making this distinction. If operators adjust the process based on periodic tests unrelated to a control chart, they will often overreact to the background noise and make unneeded adjustments. These unnecessary adjustments can actually result in deterioration of process performance. In other words, the control chart is consistent with the "if it isn't broken, don't fix it" philosophy.

4. Control charts provide diagnostic information. Frequently, the pattern of points on the control chart will contain information of diagnostic value to an experienced operator or engineer. This information allows the implementation of a change that improves the performance of a process.

5. Control charts provide information about process capability. The control chart provides information about the value of important process parameters and their stability over time. This allows an estimate of process capability to be made. This information is of tremendous value to product and process designers.

Since quality is an extremely important consideration in global competitive markets, it is easy to understand why SPC and control charts are being implemented by many firms.

SPC can involve collecting a large amount of data combined with extensive data analysis and possibly graphical plots. This is an obvious computer application. Figure 7.18 shows some electronic data collection equipment that is available. Using devices such as these, data can be read directly into a computer or be retained by the collection device until a batch of data is loaded in a computer. The computer can then update quality assurance statistics and prepare desired plots. This is illustrative of computer applications in CAI (computer-aided inspection). Unfortunately, most of these applications are also good examples of islands of automation, because they are usually stand-alone systems.

The coordinate measuring machine (CMM) is another good example of CAI (see Figure 7.19). A CMM consists of a table which will hold a part in a fixed, known position, a movable head, a sensing probe mounted on the head, and a computerized control unit. Some CMMs also have a rotary fixture mounted on the table. The movable head has three axes of freedom corresponding to the x, y, and

FIGURE 7.18   Electronic quality assurance data collection equipment. (Courtesy of Dataputer.)

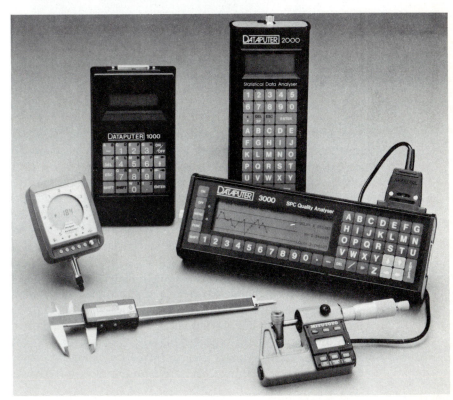

FIGURE 7.19   Coordinate measuring machine.

z coordinates. A CMM is a contact inspection device: The head moves the probe into contact with the surface of a part that is being measured. Very good accuracies can be achieved with machines of this type (up to 0.0001 in.). However, higher accuracies require a controlled environment. The CMM must be located on a pad that is isolated from vibrations, and temperature and humidity must be controlled within a relatively narrow range. A very accurate CMM facility can cost as much as a million dollars. For complex surfaces, however, these machines can dramatically cut inspection times and costs and improve inspection capability. One of these machines can accomplish what might otherwise require a multiplicity of instruments, machines, and a surface plate. A CMM can often check feature location, distances between features, and shape/form definition without repositioning the part. Also, measurement data can be automatically compared to design specifications, with any deviations being printed out and/or transferred to computer storage. A CMM is especially effective for the first article inspection of small-batch, complex parts that are made on expensive machines. Under these conditions, manufacturing costs are high and it is uneconomical to provide in-process gauging.

Programming a CMM can be very time-consuming. In general, a program written for one type of CMM will not run on another CMM type. Some vendors are providing off-line programming capabilities utilizing a CAD system. These programs require a preprocessor for each type of CMM, resulting in a proliferation of preprocessors. The Dimensional Measuring Interface Specification (DMIS) was developed to provide an interfacing standard for inspection data between CAD systems and computerized inspection equipment [3].

DIMS defines a high-level language with an APT-like vocabulary (APT is a popular NC machine programming language which will be discussed in Chapter 9) that establishes a neutral format for exchanging inspection programs and data between automated equipment, such as CAD systems, CMMs, and quality information systems. This standard is an important development in achieving a CIM environment.

Computer-aided testing (CAT) is another area where computers have been used extensively in quality assurance. We noted earlier that today's market includes a wide variety of products; some are very complex. Consequently, testing these products can be very difficult and time-consuming. For example, a propulsion engine for an airplane would not be shipped to a customer without undergoing some very exhaustive testing. This is typically done in a computer-aided test cell where many parameters are monitored and analyzed. Often these cells are also good examples of islands of automation.

Computer applications within quality assurance functions are proliferating. It is clear that management has accepted the premise that ''quality pays for itself''; in fact, it can result in lower costs. High product quality is a significant competitive advantage because in today's environment the customer is aware of the product quality of competing firms. Table 7.6 lists some of the effects of poor quality. Although this list is not exhaustive, it is clear that the current emphasis on quality is warranted.

Some firms have attempted to ''inspect in'' quality. However, this will only identify defects; it will not reduce or eliminate the cause of the defects. The result is that production costs will increase as more inspectors are hired and more

TABLE 7.6
Effects of poor quality

Low customer satisfaction
Decline in market share
Increased inspection
Low morale
Material waste
More scrap
Increased amounts of rework
More work-in-process inventory
Reduced plant capacity
More process bottlenecks
Increased uncertainty

defects are identified and routed to rework operations, or the product is scrapped. Instead, the goal should be to design and manufacture quality into the product. We should strive to achieve zero defects.

TOTAL QUALITY MANAGEMENT/CONTINUOUS IMPROVEMENTS. The quest for quality cannot stop when the initial product is designed and successfully manufactured. Rather, throughout a product's life cycle, efforts should be made to continually improve quality. Some term this process *continuous improvement* or *total quality management* (TQM). When quality is viewed in this broader context, we begin to realize that a quality assessment could be performed on all tasks performed. Within an organization, TQM is the effort to achieve and maintain an environment that supports excellence in whatever an employee does. Externally, TQM is being recognized for quality of products and services. The Japanese have demonstrated that this management philosophy can be very powerful.

In the 1950s, product made in Japan was thought to be junk or cheap. Today, just the opposite is associated with Japanese products; consequently, consumers are willing to pay more for many Japanese products. One reason attributed to this almost unbelievable change is that everyone in these firms is concerned with the quality of the tasks they perform. So the quest for quality improvement does not stop after the initial parts were designed and produced. Instead, there is a continuous effort to make incremental improvements in the product design, manufacturing processes, and service provided to the customer.

Although TQM is addressed in this text under the heading of quality assurance, it is really a management philosophy. At this time many individuals are proclaiming this philosophy. Some, such as W. Edwards Deming and Joseph Juran, were pioneers in developing these concepts. This philosophy is becoming so pervasive that some, such as P. B. Crosby, are terming it the "quality revolution" [8]. Although there are variations in the concepts presented and the recommended ways for achieving the desired goal, the fundamental philosophy is the same: We must continually improve. The customer is becoming the predominant consideration as we strive to make goods or provide services that meet or exceed the customer's expectations. So it is not a matter of which individual's philosophy is right or better; rather we should support efforts to adopt the underlying philosophy.

Sink [21] proposed five major quality checkpoints as an aid to understanding the concepts of TQM:

1. *Selection and management of upstream systems* (i.e., internal and external suppliers and vendors). Emphasize the development and maintenance of standards, specifications, and requirements; open communication with the organizations from which your organization receives inputs; explicit and clear communication; and cooperation and coordination.

2. *Incoming quality assurance.* Assure that all inputs (material, labor, energy, capital, and data/information) received are on time and are what was specified, expected, and needed.

3. *In-process quality management and assurance.* Assure that value-adding processes are effective, efficient, and build quality into the goods and/or services. Continuous improvement should be emphasized.

4. *Outgoing quality assurance.* Assure that your organization is producing the desired goods and/or services on time and within specifications.

5. *Management of downstream systems* (i.e., customers, end users, and other people or organizations that affect your organization). Proactive assurance that your organization is meeting or exceeding your customers' needs, specifications, requirements, wants, desires, and expectations.

If an organization manages these five checkpoints successfully, it will have successfully adopted the concepts of TQM.

It was noted earlier that several people, such as Deming and Juran, have devoted most of their careers to the concept of TQM. However, even these individuals do not totally agree on the fundamental concepts of TQM. Recognizing this fact, Sink [21] selected two TQM plans that were developed to improve quality. From these plans he extracted the following list of TQM foundational concepts:

1. TQM is customer-oriented.

2. TQM involves a long-term commitment to the continuous improvement of all processes.

3. TQM success demands top management leadership and continuous involvement.

4. Products and services are the result of processes, and all processes are subject to inherent variation.

5. Much of the knowledge needed to improve a process resides in the workforce and with the customer(s); however, only management can make many of the necessary changes happen.

6. Responsibility for establishment and improvement of processes lies with management.

7. Managers are responsible for the quality of supplier products and services received.

8. TQM is a strategy for continuously improving performance at every level and in all areas.

9. TQM aims at achieving one broad, unending objective: continuous improvement of products and services.

10. Successful TQM implementation depends on establishing a nurturing, encour-

aging environment, a disciplined organizational goal-setting methodology, and a formal, structured process-improvement methodology.

Sink notes that this list may not be comprehensive or undisputable; however, it does represent the efforts of experts to put a TQM plan on paper.

From this discussion, we can see that TQM is a management philosophy that concentrates on the fundamental processes in a company. As experience is accumulated with TQM, these ideas will be enhanced. The result may be called something else; however, evidence at this time verifies that TQM has had a very significant impact on firms that have implemented these concepts successfully.

7.4.9 Manufacturing Planning and Control Systems

The use of computerized manufacturing planning and control has become commonplace for both large and small firms. One reason is the many "off-the-shelf" software systems which can be purchased that will run on a variety of computers. The Manufacturing Accounting and Production Information Control System (MAPICS) from IBM is widely used and is representative of what is available.

MAPICS consists of 19 interrelated applications (see Table 7.7). Although this type of system involves many thousands of lines of computer statements, versions can be purchased that run on microcomputers as well as versions that run on mainframes. Typical computer constraints are the number and type of applications, the amount of data that will be stored, and the number of on-line terminals.

TABLE 7.7
MAPICS application modules

Accounts payable
Accounts receivable
Capacity requirements planning
Cross-application support
Data collection system support
Financial analysis
Forecasting
General ledger
Inventory management
Inventory management for process
Location/lot management
Master production schedule planning
Material requirements planning
Order entry and invoicing
Payroll
Product data management
Production control and costing
Purchasing
Sales analysis

Source: From Ref.11.

We will look briefly at eight of the principal applications modules in MAPICS from the viewpoint of the major reports and on-line displays. Demand forecasts for end items and aggregate groups (families) are made by the forecasting module using statistical techniques. The major reports are:

1. *Forecast detail report:* Shows forecast performance data and 1-year forecasts in units for individual items.
2. *Forecast summary report:* Shows aggregate forecasts in units and standard costs for each product line or value class.
3. *Inventory summary report:* Shows a summary of the cost associated with the calculated safety stock and reorder point.
4. *Seasonal profile report:* Details group and item seasonal parameters and shows them graphically for visual evaluation of seasonal patterns.

The *product data management module* is used to manage the bills of materials, manufacturing routings, workcenter information (description of machines and other facilities), and the item master. The *item master* contains detailed information about each part, such as part number, description, standard cost, part type, and drawing number. The major displays and reports are:

1. *Bill of materials:* Several formats of BOM can be displayed.
2. *Item where-used:* Shows all the assemblies or products using a particular component.
3. *Costing:* Shows actual costs at current or standard cost values.
4. *Product feature/options:* Shows all the options for all the features available for a product.
5. *Routings:* Shows routings for a part.

The *master production schedule planning module* develops production plans and resource requirements for groups of parts. A plan can be reviewed and changed on-line. Some of the major displays and reports are:

1. *Display/maintain production families:* Lets you assign master scheduled end items to production families. Also, master production schedules can be reviewed by family.
2. *Display/maintain family operating plans:* Lets you review and change a family's desired level of production.
3. *Display/maintain item trial plans:* Permits you to set and adjust production levels for items in each family.
4. *Family plan inquiry:* Assists in evaluating family production plans.
5. *Resource requirements analysis:* Lets you review the quantities and dollar values of critical resources that will be needed for each planning period.

6. *Display/maintain master schedule:* Lets you review and change the master production schedule for any master scheduled item.

7. *Available to promise:* Calculates and displays current quantities of an item that will be available for sales in each planning period.

The *material requirements planning module* converts the master schedule into a manufacturing plan. This plan can be revised. Either the generative or net-change mode of performing the MRP computations can be utilized. Some of the major displays and reports are:

1. *Master-level item requirements versus forecast of customer orders:* Displays the requirements for the master-level items versus the sales demand.

2. *Material requirements plan:* Assists the planner in reviewing the planned orders.

3. *Order recommendation:* Shows the planner which orders should be released, rescheduled, or canceled.

4. *Purchase planning report:* Projects material needs into the future for all items purchased from a vendor.

You can plan and control production using the *production control and costing module*. When manufacturing orders are released, the shop packet can be printed with component material lists and routings. Detailed records for open orders are maintained. When shop-floor activity is reported, it is edited against and posted to these records. Some of the major displays and reports are:

1. *Shop packet:* A set of information that is moved around the shop floor with a batch of parts. This information specifies the work-order number, beginning quantity, routing, processing information, due date, etc.

2. *Order status:* Displays status (location) of an order.

3. *Workcenter status:* Displays the open orders at the workcenter and those coming into the workcenter.

4. *Critical order list:* Identifies orders that may need special attention, such as expediting to meet due dates.

5. *Worklist:* Shows orders by priority for each workcenter.

6. *Workcenter analysis report:* Provides summary information on utilization, efficiency, queue size, and output for each workcenter.

The feasibility of the manufacturing plan can be evaluated using the *capacity requirements planning module*. The workcenter and time periods with an overload or underload condition can be identified. The major displays and reports are:

1. *Workcenter load analysis:* Shows each workcenter's accumulated workload by period and compares this to available capacity.

2. *Workcenter load analysis detail:* Shows the planner the operations that have contributed to the workcenter overload.

3. *Workcenter variable capacity:* Displays the temporary capacity changes in the manufacturing facility.

The last module we will discuss is the *purchasing module*, which supports planning and control of orders released to vendors. The major reports are:

1. *Open requisition analysis:* Shows backlogs.

2. *Dock-to-stock transaction audit:* Shows all transactions related to a particular purchase order accepted by the system.

3. *Purchase order status:* Provides current status of any purchase order in the system.

4. *Prioritized dock-to-stock worklist:* Tracks the progress of purchase order receiving activities.

This completes our brief overview of MAPICS. The above modules are representative of many similar systems that are available.

7.5 CELLULAR MANUFACTURING

7.5.1 Overview

As manufacturing competition has increased, firms have sought any advantage. Many firms are realizing that cellular manufacturing can provide significant advantages: reduced through-put, reduced work-in-process inventory, reduced setup time, reduced materials handling, improved quality, higher morale, etc. (see Chapter 5). Consequently, cellular manufacturing concepts are being adopted at an increasing rate.

Cellular manufacturing is the organization of manufacturing machines and people into groups responsible for producing a family of parts. A *real cell* is a fixed, physical group of machines and people (see Figure 5.26 in Chapter 5). A *virtual cell*, in contrast, is a group of machines and people that are not located in one physical location; some of these resources might be shared with or assigned to other cells as requirements change. A flexible manufacturing system, which is a special type of cell, consists of [7]:

1. Two or more work stations with computer-controlled machine tools

2. An automated materials handling system for transporting work-in-process inventory

3. Mechanisms for transferring the work-in-process inventory between the transportation systems and the machine tools

4. Storage by an automated storage and retrieval system of work-in-process inventory and sometimes tooling

5. Central computer control of the entire process.

The first step in migrating to a cellular manufacturing environment is the implementation of group technology concepts so that families of parts can be identified. Once families of parts are identified, potential cells can be designed.

The next step is selecting machines for the cell, followed by arranging the cell (allocation of equipment if it is a virtual cell) and assigning the operators. The capacity of a cell is usually dictated by one or two bottleneck operations. Consequently, the other operations need only produce at a pace equal to these bottleneck operations. Therefore, something less than the fastest available machines may be selected for some of the operations.

In selecting machines, process engineers should prepare a set of functional specifications that can be compared with the capabilities of machines being considered. If the family of parts to be made in the cell is large, 20% of the parts that make up 80% of the load should be used to select the machinery. Otherwise, the amount of detail may be overwhelming. If some of the remaining 80% of the parts cannot be processed by the equipment selected, additional equipment that might have low utilization can be added to the cell. Otherwise, some operations for these parts might be performed by equipment located elsewhere.

The cell design is also dependent on the size and shape of the parts and interdependent setup times. Materials handling considerations are also dependent on these factors and the numbers of components. If large volumes of components are involved, automated handling alternatives should be considered. Materials handling should not be the limiting factor. The capacity of a cell should be determined by the key machine(s)—the one(s) with the longest cycle times.

A cell should be laid out so that an operator can tend two or more machines at once. Ideally, each operator should have the skills to operate all of the equipment in the cell. A Gantt chart can be used to evaluate alternative operator assignments.

A completed cell design can be thoroughly evaluated using computer simulation. This evaluation should include a sensitivity analysis of the important design parameters. For instance, changes in production rates, number of parts in a family, and technology should be considered. Simulation can be very useful in the process of cell design.

Cells simplify scheduling because they can be scheduled as a single entity. Consequently, the cell capacity must be at least as great as the required production rate. This concept is very important: Underutilized equipment may be required to achieve it. Since setup time can vary considerably from one part type to another, the sequencing of the parts may be important. After the parts have been sequenced, however, they should flow through the cell in a FIFO (first-in, first-out) manner. Once a batch of parts is started, it should be completed. Splitting batches complicates control.

Many cells involve little or no automation; control is provided by the operators. As a cell is refined, automation can be added with less risk.

Cells eliminate materials handling, since most part movement is internal to the cell. Because these moves consist of short moves of small quantities of parts, material tracking and control is eliminated. Batches of parts are delivered to the cell and may not be accumulated until just prior to leaving the cell. Outside the cell, moves of batches of parts are longer and less frequent; consequently, materials handling is reduced and more efficient.

Associated tooling and gauges should be located at the cell. This will result in better tool and gauge control, which will result in reduced tool inventories and a higher conformance to gauge specifications.

Improved quality can be expected from cellular manufacturing. There are several reasons. After parts are grouped into a family to be processed within a cell, the next step is to thoroughly analyze the various operations involved. Fixturing and tooling are also evaluated. The result is an excellent understanding of the processes involved and a refinement of tooling and fixturing. Also, the number of different types of tools and fixtures is usually reduced, as are setup times.

A reduction in setup times usually means that the setups have been simplified. Once setup times are reduced, shorter through-put times are a consequence. These shorter setup times encourage small batch sizes, resulting in a more flexible manufacturing facility. Cells also facilitate "pipelining" of batches (passing parts directly from one operation to another without accumulating the entire batch), which reduces through-put times.

Shorter through-put times and smaller batch sizes aid in improving quality, since the feedback time of quality problems is reduced. These is a closer connection between defect occurrence and detection, which aids in determining the appropriate corrective action. Reduced materials handling results in less defects. Also, cells increase worker knowledge of the parts they process; consequently, self-inspection is more effective.

Since cell operators are expected to operate several different machines, inspect their own work, and perform some machine maintenance, it is reasonable to expect increased morale and higher levels of motivation. Operators should be encouraged to help each other when problems are encountered. The members of a cell become a team.

7.5.2   Hierarchical Manufacturing Control Model

As modern manufacturing organizations migrate toward cellular manufacturing, there is a proliferation of associated computer software and hardware. Cellular manufacturing facilitates the use of new manufacturing philosophies, such as JIT (just-in-time). However, cellular manufacturing progress has been hindered by the lack of standards; consequently, the many different types of machines, computers, controllers, and associated software have been difficult to interface and integrate.

Recognizing this problem, the National Bureau of Standards (NBS) has developed a generic architecture for real-time production control [13]. Figure 7.20 shows the control levels in this architecture. The proposed model must exhibit flexibility and modularity so that additions, deletions, and substitutions of software and hardware may occur at all levels. An Automated Manufacturing Re-

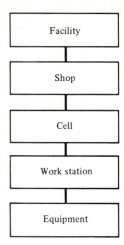

FIGURE 7.20   Generic control levels for real-time production control.

search Facility (AMRF) has been developed to demonstrate these concepts. This real-time production control system [13] has been:

1. Partitioned into a hierarchy in which the control processes are isolated by function and communicate via standard interfaces
2. Designed to respond in real time to performance data derived from machines equipped with sensors
3. Implemented in a distributed computer environment using recent advances in software engineering, microcomputers, and artificial intelligence programming techniques

This design requires a very reliable management information system that can provide data in a timely manner.

Figure 7.21 shows the hierarchy control structure used in the AMRF facility. Flow of control is strictly vertical and between adjacent neighbors; however, data are shared across one or more levels. Each control module decomposes an input command from its supervisor into: (1) procedures to be executed at that level; (2) subcommands to be issued to one or more subordinate modules; and (3) status feedback sent back to the supervisor. This decomposition process is repeated until a sequence of primitive actions is generated. Status data are provided by each subordinate to its supervisor to close the control loop and to support adaptive actions.

FACILITY CONTROL SYSTEMS. This highest level of control is broken down into three subsystems: manufacturing engineering, information management, and production management. The planning horizon within these areas can be several months to several years.

Manufacturing engineering functions have already been described in detail; they remain the same in this hierarchical control system. The information management subsystems provide data and information system support to the user. Applications are very broad, such as cost estimating, order entry, billing, payroll, etc. Using the production management subsystems, major projects and production resources can be managed.

SHOP CONTROL SYSTEM. Production jobs and associated resources are coordinated at the shop level. Consequently, the planning horizon is several weeks to several months. The control system is composed of two modules: a task manager and a resource manager.

The task manager schedules jobs and equipment maintenance and tracks equipment utilization. This person is also responsible for capacity planning, combining orders into batches, activating and deactivating virtual cells, assigning and releasing orders to cells, and tracking individual orders to completion. Cells are organized around group technology codes or support functions. Virtual cells are activated on the basis of production requirements. Cell controllers are designed to manage particular part families.

The resource manager allocates the various resources (work stations, tooling, materials, etc.) to cell-level control systems for particular jobs. Inventories of raw stock, work in process, tools, and replacement parts are also monitored. Since virtual cells can be altered dynamically, the resource manager must be able to respond quickly to these changes. Cells bid for resources on the basis of job priority, costs, etc.

FIGURE 7.21   Hierarchy control structure. (From Ref. 13.)

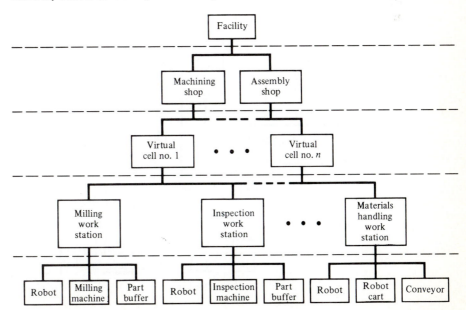

CELL CONTROL SYSTEM. Batch jobs are sequenced through work stations of a cell, and support services such as materials handling are supervised at this level. The planning horizon can be several hours to several weeks. As noted earlier, these are virtual cells which permit the time sharing of work stations. This design brings some of the efficiencies of a flow shop to small-batch production. Modules within the cell control system perform task decomposition, analyze resource requirements, report job progress and system status to shop control, make dynamic batch routing decisions, schedule operations at assigned work stations, dispatch tasks to work stations, and monitor the progress of those tasks.

WORK STATION CONTROL SYSTEM. The work station control level manages the activities of small integrated physical groupings of shop-floor equipment. The planning horizon may be several minutes to several hours. A typical work station is composed of a robot, a machine tool, a material storage buffer, and a control computer (see Figure 7.22).

A machining work station processes trays of parts delivered by the materials handling system. The controller sequences the equipment-level subsystems through job setup, part fixturing, cutting processes, chip removal, in-process inspection, job takedown, and cleanup. Since the cells are virtual, the cell-to-work station control interface is designed to be independent of the work station type.

FIGURE 7.22   Typical work station control architecture. (From Ref. 13.)

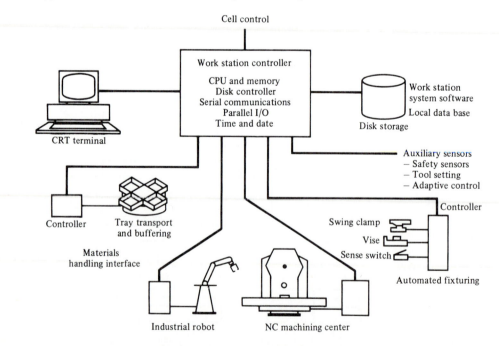

EQUIPMENT CONTROL SYSTEM. Equipment control systems are designed to interface between the work station control system and the vendor-supplied equipment controllers which are provided on robots, NC machine tools, coordinated measuring machines, delivery systems, and storage/retrieval devices. The equipment control system translates the commands from the work station control level into a sequence of simple tasks that can be understood by the vendor-supplied controller. This control system also monitors the execution of these tasks using sensors attached to the equipment.

GENERIC CONTROL MODULE. Every control module in the hierarchy control structure operates in essentially the same way: Input commands from a higher level are decomposed, status feedback from a lower level is processed, and new commands and status information are output. The developers of this control system refer to this mode of operation as reaction. As the control module is developed further and sophistication is increased, three additional levels of intelligence are planned: planning, optimization, and learning.

INFORMATION MANAGEMENT SYSTEM. The information management system is composed of two subsystems: data administration and network communication. The data administration system provides a uniform method of access to data for

FIGURE 7.23   Example factory network architecture. (From Ref. 13.)

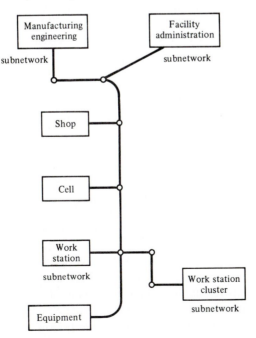

all the control modules. It is composed of data dictionaries, commercial data base management systems, physical data storage devices, logical-view processors, data-manipulation language translators, and interface software. Operation of this subsystem is transparent to the user.

The network communication system includes the hardware and software required to move information within this hierarchical control system in a timely manner. The final architecture contains a broad-band factory network with several subnetworks (see Figure 7.23). The network conforms to OSI (Open Systems Interface) international standards. The OSI computer networking standards are being developed to simplify the network of dissimilar computer systems. These standards will be discussed in greater detail in Chapter 12.

The experience obtained in developing this prototype hierarchical control system will be very beneficial to the many companies considering an integrated environment. It should provide an important part of the foundation for developing standard interfaces and control modules.

7.6 JIT MANUFACTURING PHILOSOPHY

Just-in-time (JIT) as an operations management philosophy has received an increasing amount of support in recent years. Originally, JIT was interpreted to mean that the right number of parts would be delivered to the right shop-floor operation at the right time. As a result, work-in-process inventories would be minimized. However, implementing this philosophy requires that most manufacturing practices (inflexible setup times, large lot sizes, inefficient plant layouts, poor quality, etc.) be evaluated. If they are not, the JIT implementation will fail. JIT, when applied vigorously, will result in a concerted effort to eliminate anything that does not add value to the product.

Implementing JIT requires that a pull system of production be utilized. In a pull system, product is made to order; in a push system, by contrast, product is made to stock. In the pull system, when a worker runs low on parts, a request is passed back to previous operations for replacement parts. This procedure is repeated throughout the manufacturing operations. As a result, lot sizes must be small and setup time must be short. Also, materials handling must be minimized and quality must be very good. Short materials handling moves and short setup times support the conversion to cellular manufacturing that is occurring at many firms.

JIT exerts pressure on most other organizations within a firm. Engineering will strive to design or utilize existing parts within existing families. Relationships with suppliers will be closely evaluated to ensure timely deliveries of high-quality parts. Most firms that have implemented JIT have significantly reduced the number of suppliers. Often, long-term relationships are established that contribute to improved quality. This results in formal agreements to encourage the supplier to be more innovative in reducing costs and improving quality. Schedules for suppliers must be frozen within some reasonable time period; otherwise, the supplier will have to build inventory to meet the customer's demands.

The benefits that firms have achieved from successfully implementing JIT concepts have been very significant. Table 7.8 indicates what some firms have reported. JIT is based on the premise of continuous improvement; consequently, you should note in Table 7.8 that JIT-associated benefits increased over the years.

Other organizations, such as plant engineering, must also change the way they do business. If a machine fails, production is disrupted. Consequently, preventive maintenance and a spare parts inventory are critical, as is timely response to a machine failure. Plant engineering records assume a special significance.

There has been some controversy regarding whether MRP and JIT can be integrated. Actually, the two systems are complementary because MRP is a planning system and JIT is an execution system. Care must be used in defining where each system starts and ends. JIT eliminates MRP execution control such as job tickets for end-of-the-line counts of units finished, work hours consumed, and materials used. As the JIT system pulls parts through the production process, product status information can then be compared to the plan generated by the MRP system.

Relative to JIT, MRP systems operate with a large amount of "fat." Lead times are much longer for purchased and manufactured parts. In fact, these lead times are fixed for a given part and are utilized in determining when a batch of parts should be released to fabrication or assembly. The MRP algorithm typically utilizes other inventory-creating factors, such as safety stock, lot-sizing rules, and scrap factors. Planning time periods are weeks or months instead of shifts or days. Also, the MRP planning process is seldom performed more frequently than at weekly intervals. From this brief comparison, it can be seen that, relative to MRP, JIT can result in much shorter manufacturing cycle times. However, many other manufacturing operating practices must also be changed if this is to happen.

TABLE 7.8
Reported JIT benefits

	Improvement	
	First year	**Cumulative 3–5 years**
Manufacturing cycle times	30–40%	80–90%
Inventories:		
Raw materials	10–30%	35–70%
Work in process	30–50%	70–90%
Finished goods	25–60%	60–90%
Labor costs:		
Direct	3–20%	10–50%
Indirect	3–20%	20–60%
Space	25–50%	40–80%
Quality costs	10–30%	25–60%
Material costs	2–10%	5–25%

Source: From Ref. 9.

Most firms in the United Stats still operate in a continual expedite mode. For many years the employee has been trained to respond to problems—so much so that it is not the schedule that counts, but the expeditor's hot list. Transitioning to JIT is difficult. The manager of manufacturing must understand the importance of having the right systems, accurate data, educated users, and an equitable measurement system. Then, all involved persons must continue to strive to excel at their jobs.

Buker [5] recommends 10 principles to be followed when attempting a transition to a JIT environment:

1. Management education and leadership.
2. Worker involvement programs.
3. Total quality control from design through shipping.
4. Simplification of product design.
5. Reduction of inventory levels.
6. Production in small lots and reduction of setup times.
7. Improving plant layout.
8. JIT purchasing agreements with vendors. Purchase in small lots, reduce number of suppliers, develop long-term relationships, and encourage supplier involvement and support.
9. Total preventive maintenance.
10. Solving and preventing problems, eliminating waste, and striving for continual improvement.

This set of principles is very similar to the concepts of total quality management (TQM). In fact, the JIT concepts can be characterized as TQM focused on manufacturing planning and control.

7.7 INTEGRATION OF CAD/CAM REQUIRES MRP II

The heart of an integrated manufacturing environment is manufacturing resources planning (MRP II), because an excellent business plan cannot be achieved without an excellent supporting manufacturing resources plan. Consequently, integrative CAD/CAM must address how MRP II can be included in these integration efforts.

Some data elements are common to both CAD and MRP II systems. For instance, bills of materials and some part specifications and costs are utilized in both types of systems. As design engineers develop a new design, they may want to interrogate the MRP II data base several different times. For instance, they may want to extract parts with similar group technology codes, or evaluate existing inventories. In addition, during manufacturing planning and processing there is a periodic need to utilize data residing in the CAD data base. For instance, process planners need to review geometric information, retrieve part numbers and drawing revision numbers, and review processing specifications. Production control analysts need access to the bills of materials, and shop-floor operators need access to processing specifications and geometric data. Data residing at other CAM islands of automation is also useful to people throughout a firm. Therefore, an obvious objective is to integrate this data and, as the need arises, provide the desired information in a timely manner. Although substantial progress has been made, fully achieving this objective is not an easy accomplishment. This subject will be discussed in some detail in Chapter 12.

7.8 SUMMARY

Manufacturing planning and control is a broad subject area. It is a very important part of any firm, especially in the rapidly changing, globally competitive environment of today. Computers have been applied in a variety of ways in manufacturing; most of these applications, however, are examples of islands of automation. This is not necessarily bad, because, in the transition to an integrated environment, these islands must first be developed. Then integration can be accomplished.

More advances have been made in manufacturing planning control during the last decade than in any previous decade. Our increased understanding of the importance of manufacturing will assure that we continue to concentrate significant efforts in this area. Also, increased understanding of what has to be done will reinforce these efforts. So, during the next decade, the rate of change in manufacturing planning and control should accelerate.

EXERCISES

To answer some of these questions, the student will have to utilize other references besides this text.

1. Wickham Skinner's book [22], *Manufacturing: The Formidable Competitive Weapon*, is an important contribution to the literature that has been written on the importance of manufacturing. The following questions are based on the material in this book:
   (a) In Chapter 5, Skinner list some 25 trade-off decisions in manufacturing, such as "span of process: make or buy." Select five of these trade-offs and explain when you would make a specific decision, such as when you would make a part.
   (b) Explain what is meant by a "focused factory" and why such a factory might be better.
   (c) One of the suggested ways to design a production system or to analyze an existing system is to perform a manufacturing audit. What are the eight "cardinal sins" that this audit might reveal?
2. It is not unusual for a company to have several bills of materials for one product.
   (a) Should a company have only one bill of materials for a product? Explain your answer.
   (b) Describe at least three different bills of materials for one product.
3. The term "kanban" is often associated with JIT. Explain:
   (a) What does kanban mean?
   (b) Can the JIT philosophy be implemented without kanban?

4. W. Edwards Deming is renowned for his pioneer work in the area of quality improvement. He recommends 14 points for achieving a cultural change for quality. What are these points?

5. Krajewski et al. [15] compare three types of production systems: kanban, MRP, and ROP (reorder point). Review this article and summarize the authors' conclusions.

6. Quality, cost, and schedule are often listed as the three most important objectives for a manager of manufacturing. Rank these objectives in order of importance and justify your ranking.

7. Speciality Manufacturing specializes in making ring gears. Data from six representative jobs is given below. If the selling price of a new gear is established as 1.5 × manufacturing cost, what selling price would you suggest for a new gear that is 30 in. in diameter?

Job	Diameter (in.)	Manufacturing cost
1	12	$15.25
2	14	20.10
3	12	12.50
4	16	32.50
5	25	77.25
6	47	121.75
7	64	205.25

8. Forecast the number of students that will be enrolled in your department for each of the next 5 years.

9. Show that the coefficients α and $1 - \alpha$ in Eq. (7.6) sum to 1 over time.

10. Explain why the master production schedule is not a forecast.

11. Figure 7.24 denotes processing times for parts from family A. The parts are to flow through a cell in the order that the machines are listed in Figure 7.24. Assume that a

FIGURE 7.24   Cycle times for family A parts.

Part	Machines				Total cycle time (sec)
	1	2	3	4	
A1	22	42	59	18	141
A2	17	36	48	18	119
A3	28	45	70	18	161
A4	15	39	55	18	127

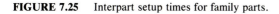

	A1	A2	A3	A4
A1	–	30	45	40
A2	20	–	15	10
A3	10	15	–	10
A4	5	10	15	–

From

FIGURE 7.25   Interpart setup times for family parts.

demand rate of 70 parts per hour is expected. You are to design a cell by specifying the number of required machines and operators. Also, denote how you would locate the machines.

12. Figure 7.25 specifies interpart setup times for part family A described in Figure 7.24 (Exercise 11). In what order would you schedule parts in this family through a cell?

13. Using the following, data develop an aggregate plan that results in a level work force:

Period	Unit demand	Work days
1	55	19
2	50	19
3	60	25
4	55	20
5	50	20
6	65	24
7	45	19
8	45	20
9	55	24
10	90	20
11	150	18
12	170	22

Assume that no overtime is allowed, a work day is 40 hours, each unit requires 60 direct labor hours, and the starting finished goods inventory is 10 units.

14. The master schedule for part A-10 is as follows:

Time period	1	2	3	4	5	6	7	8	9	10	11	12
Gross requirements		2	1	3	3	6	2	6	4	0	9	6	6

The bill of materials for this part is specified in Figure 7.26. Determine the lot-for-lot order releases for A-10, A-20, A-21, and A-30. Assume not-scheduled order receipts for any time.

15. Using the data in Figure 7.14, develop a schedule that minimizes the makespan for this set of parts.

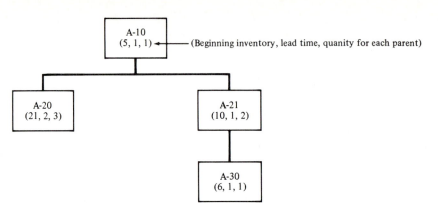

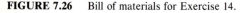

FIGURE 7.26   Bill of materials for Exercise 14.

16. Using current literature, prepare a report describing how you would implement an SPC program in a high-volume discrete-parts manufacturing company.

17. Using current literature, prepare a report describing the status of cell control systems.

18. Assume that you have given the task of selecting a capacity planning system for a low-volume discrete-parts manufacturing company with sales of $100 million a year. Using current literature, prepare a report with your recommendations.

REFERENCES AND SUGGESTED READING

1. "Back to Basics," *Business Week*, June 16, 1989, p. 17.
2. Bedworth, D. D., and J. E. Bailey: *Integrated Production Control Systems*, 2d ed., John Wiley, New York, 1987.
3. Brown, C. W.: "DIMS—An Overview," Allied Signal, Kansas City, Mo., 1988.
4. Brown, K. H.: "Higher Quality Helps Boost U.S. Products," *The Wall Street Journal*, January 11, 1988, p. 1.
5. Buker, D. W.: "10 Principles to JIT Advancement," *Manufacturing Systems*, March 1988, p. 55.
6. Business Education Associates: "Winning the Implementation Game: The Determinant Factors for a Successful MRP Implementation," Business Education Associates, Saratoga Springs, N.Y., 1987.
7. Clark, K. E.: "Cell Control: The Missing Link to Factory Integration," *1989 International Industrial Engineering Conference Proceedings*, Toronto, May 1989, pp. 641–646.
8. Crosby, P. B.: "How Goes the Quality Revolution," *Quality and Productivity Management*, vol. 7, no. 2, 1989, pp. 3–6.
9. Datapro Research Corporation: "How U.S. Manufacturing Can Thrive," *Management and Planning Industry Briefs*, March 1987.
10. Greene, T. J., and C. M. Cleary: "Is Cellular Manufacturing Right for You?" *1985 Annual International Industrial Engineering Conference Proceedings*, Chicago, December 1985, pp. 181–190.
11. Groover, M. P., and E. W. Zimmers, Jr.: *CAD/CAM: Computeraided Design and Manufacturing*, Prentice-Hall, Englewood Cliffs, N.J., 1984.
12. IBM Corp.: *Introducing Advanced Manufacturing Applications, MAPICS II*, IBM, Atlanta, Ga., 1985.

13. Jones, A. T., and C. R. McLean: "A Proposed Hierarchial Control Model for Automated Manufacturing Systems," *Journal of Manufacturing Systems*, vol. 5, no. 1, 1986, pp. 15–25.

14. Kinney, H. D., Jr., and L. F. McGinnis: "Design and Control of Manufacturing Cells," *Industrial Engineering*, October 1987, pp. 28–38.

15. Krajewski, L. J., B. E. King, L. P. Ritzman, and D. S. Wong: "Kanban, MRP, and Shaping the Manufacturing Environment," *Management Science*, vol. 33, no. 1, January 1987, pp. 39–57.

16. Manufacturing Studies Board, National Research Council: *Toward a New Era in U.S. Manufacturing: The Need for a National Vision*, National Academy Press, Washington, D.C., 1986.

17. Montgomery, D. C.: *Introduction to Statistical Quality Control*, John Wiley, New York, 1985.

18. Orlicky, J.: *Material Requirements Planning*, McGraw-Hill, New York, 1975.

19. Sackman, H.: *Delphi Critique*, Lexington Books, D. C. Heath, Lexington, Mass., 1975.

20. Schonberger, R. J.: "Frugal Manufacturing," *Harvard Business Review*, September–October 1987, pp. 95–100.

21. Sink, D. S.: "TQM: The Next Frontier or Just Another Bandwagon to Jump on?" *Quality and Productivity Management*, vol. 7, no. 2, 1989, pp. 6–21.

22. Skinner, W.: *Manufacturing: The Formidable Competitive Weapon*, John Wiley, New York, 1985.

23. Stoner, D. L., K. J. Tice, and J. E. Ashton: "Simple and Effective Cellular Approach to a Job Shop Machine Shop," *Manufacturing Review*, vol. 2, no. 2, June 1989, pp. 119–128.

24. "Support for the Plant Floor," *Manufacturing Engineering*, March 1989, pp. 29–30.

25. "The Computer, Where Is It Now and Where Is It Going?" *Modern Machine Shop 1989 NC/CIM Guidebook*, Modern Machine Shop, Cincinnati, Ohio, 1989, pp. 12–32.

26. "The Payoff from Teamwork," *Business Week*, July 10, 1989, pp. 56–62.

27. Tompkins, J. A.: *Winning Manufacturing*, Industrial Engineering and Management Press, Norcross, Ga., 1989.

28. Vollmann, T. E., W. L. Berry, and D. C. Whyback: *Manufacturing Planning and Control Systems*, Richard D. Irwin, Homewood, Ill., 1984.

29. Wheelwright, S. C., and R. H. Hayes: "Competing through Manufacturing," *Harvard Business Review*, January–February 1985, pp. 99–109.

CHAPTER
8

MANUFACTURING CONTROL— COMPUTER CONTROL

Now we are on the threshold of a new kind of revolution, . . . vastly increasing man's "thinking" capabilities of planning, analyzing, computing, controlling.

Robert Ledley in 1960 [6]

Digital computer control of industrial processes became a reality in the late 1950s, although the processes controlled were almost all of the continuous manufacturing category: chemical and refinery processes, paper and steel production, and cement manufacturing, to name just a few. The controlling computers in those initial applications were what we would probably today call a mainframe or maxicomputer, though those early computers were really "maxi" only in terms of physical size. Usable memory was at most one-quarter of the memory size in today's smaller personal computers, and there was virtually no backup memory in terms of tape or disk. Price was certainly "maxi," with computers costing hundreds of thousands of dollars, while programming was thrown in as a way for vendors to "buy into the market." Microcomputers nowadays have larger capacity, are much faster in operation, and can provide control capability for $10,000 or less.

A schematic for a generalized control computer is given in Figure 8.1. Information is captured from a physical process (such as cement kiln rotation speed, machine tool load current, robot gripper status, automatically guided vehicle speed) in digital format (on/off, open/closed) or analog format (voltage). Analog

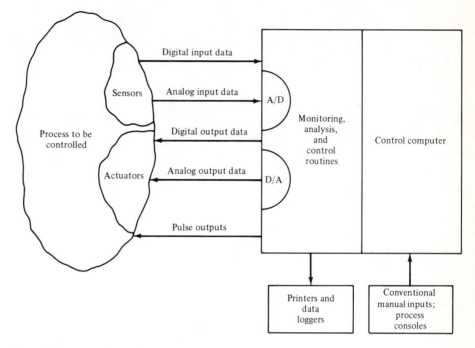

FIGURE 8.1   Control computer schematic.

inputs have to be converted to a digital representation (A/D). Because the process is operating in real time, the data input has to be in real time, responding to problems in perhaps fractions of a second. After analysis of the data and possible use of control algorithms, signals might be sent to the process for adjustment, shutting down, and so on. In addition to digital and analog outputs, pulse outputs often drive stepping motors, frequently used for machine tools and other equipment. In addition to communicating directly with some process, the control computer has conventional computer input/output equipment for operator/management communication. The fact that a control computer has to communicate with a physical process, often under severe timing constraints, probably constitutes the main characteristic that distinguishes computer controllers from general-purpose computers. It is the process interface with timing restrictions that makes an understanding of control computers so important to the CAD/CAM specialist.

As will be seen in later sections, a digital computer used to control a certain process (such a computer will be called a *control computer*) is no different from a conventional digital computer. The person responsible for developing such systems has to be very knowledgeable about the process to be controlled, including process timing constraints, the different ways in which sensors can capture process information and then transmit that information to the computer, data analysis procedures for control purposes, and ways to get controlling signals from the computer back to the process being controlled.

FIGURE 8.2   IBM PC controlling a laboratory experiment. (Courtesy of Industrial Engineering Controls Laboratory, Arizona State University.)

FIGURE 8.3   Square D programmable controller. (Courtesy of Industrial Engineering Controls Laboratory, Arizona State University.)

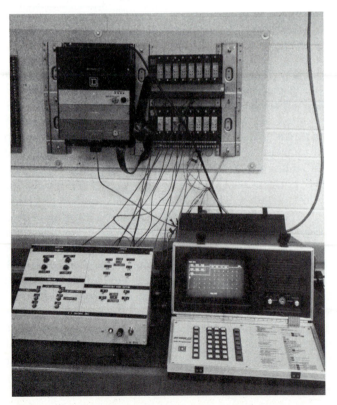

The major thrust of this chapter will be to show how control computers can input, analyze, and transmit process data within severe timing constraints. Inherent in this discussion will be the fact that many things have to occur concurrently in a physical process, and so approaches to concurrent task control will also be presented. A specialized form of controlling "computer" will be considered after the basic control computer material has been introduced: *programmable controllers*, which have, in the past, been somewhat specialized controllers of on/off processes (traffic lights, motor control, machine control, robot actuator control, conveyor control, etc.). Programmable controllers are fast approaching the capabilities of the more general control computers, as will be shown when they are discussed, and even though they are often thought of as having different control characteristics from a digital computer, they really do not. Thus their inclusion within a chapter on control computers is not only logical, it is almost mandatory. Figure 8.2 shows an IBM personal computer for controlling a process, while Figure 8.3 similarly depicts a Square D programmable controller.

8.1   THE ROLE OF COMPUTER CONTROL
IN CAD/CAM INTEGRATION

As has been suggested, a *hierarchy* of computers that encompasses all the computer functions needed for manufacturing operations, such as information capture, decision making, and dissemination of results from decision making, is highly desirable, regardless of whether automation is prevalent at the shop-floor level or not. If the functions of CAD or CAM that have been discussed in previous chapters are implemented in a manufacturing plant, then the hierarchical structure becomes an absolute necessity. Such a structure might be as shown in Figure 8.4.

The hierarchical levels are akin to the span of control in a management hierarchy, in which a chief executive officer of a plant should have not more than five to eight people (say, vice presidents) reporting directly to her; similarly, each vice president should have a restricted number of people, perhaps department heads, reporting to him. At the bottom of the management tree we might see eight to ten line workers reporting to a foreman. In this way each manager in the organization has a tractable number of people to manage and control.

The hierarchical computer system has a tractable number of computers "reporting" or being "controlled" by a parent computer. At the bottom of the computer hierarchical tree lie those computers that are actually doing equipment monitoring/control. Multiple computers at the control level constitute a *distributed* system of computers in which somewhat autonomous control is distributed among the computers. In many cases these computers will be integral components of specific vendor-supplied equipment: computer numerical control machines, controllers for specific robots, materials handling control systems, automatic warehouse controllers, computer-aided testing equipment where the computer might provide test signals from which some reaction is monitored for in-control conditions, a people-protection system to prevent robots from operating if a human penetrates within a certain distance from the robot, as well as many other

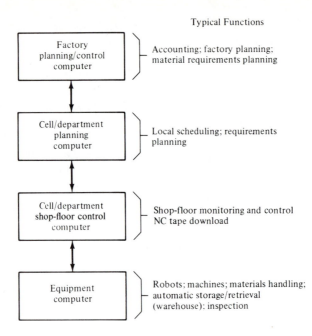

Typical Functions

Factory planning/control computer — Accounting; factory planning; material requirements planning

Cell/department planning computer — Local scheduling; requirements planning

Cell/department shop-floor control computer — Shop-floor monitoring and control NC tape download

Equipment computer — Robots; machines; materials handling; automatic storage/retrieval (warehouse); inspection

FIGURE 8.4   Representative manufacturing hierarchical computer system.

possibilities. All of these cases have one thing in common: A computer, probably a microcomputer, generates signals in real time to direct the equipment to accomplish tasks in some optimum fashion. In order to do this the computer must be able to gather information from the process (Where is an automatic guided vehicle at the present time?), analyze the data (What is needed to get the vehicle to the correct location at the correct time?), and transmit information to accomplish the required control (speed up the vehicle by transmitting a signal while setting inductive switching equipment to send the vehicle to the correct location). A single microcomputer might monitor, analyze, and effect actuation of many sensors at the same time, thus leading to a multitasking, real-time computer operating system philosophy. An expert in CAD/CAM systems should *at least* be conversant with the concepts of real-time control computers and their cousins, the programmable controllers, since, in terms of factory automation, the computers at the lower end of the hierarchy are just as critical to the success of the entire operation as are the computers at the top end.

Knowing the characteristics of such computers will allow the overall control system to be designed so that problems do not occur because one control computer cannot handle all the processes it is supposed to control or a vendor recommends a particular operating system that is really not suited to a particular manufacturing system. This chapter is designed to introduce the concepts of real-time computer control (and the use of programmable controllers) to the CAD/CAM professional who needs an overview of this very important topic. The material will be presented from a microcomputer point of view, but it is just as appropriate to control by minicomputer.

8.2   KEY DEFINITIONS

Analog I/O   Input or output of an analog signal from or to a physical process. An analog signal in this chapter will refer to a voltage that is analogous to some process characteristic.

Analog-to-digital (A/D) conversion   The act of converting an analog signal (voltage) to a digital format so that the value can be manipulated by a digital computer. The converted value is frequently called a digital *count*.

Bit   A single digit in the binary number system. Comes from *bi*nary digi*t*.

Continuous process   A process without discrete breaks in the production characteristics. Typical would be paper production, petrochemical production, steel rolling, cement production, and so on.

Continuous signal   A set of data from a process that has no discrete breaks, such as a continuous voltage signal.

Control computer   In this chapter, a digital computer used in the control of a physical process; for example, a manufacturing operation, or control of an airplane landing on an aircraft carrier deck, can be accomplished through the use of a control computer.

Controlled process   A physical situation, such as a machine tool, a robot, and so on, that is automatically controlled in some manner.

Conversion count   The digital value that results from an analog-to-digital conversion. Frequently, the digital value that is input to a digital-to-analog convertor is called a *count*.

Digital I/O   The input or output of binary-type digital data to or from a physical process by a control computer. Binary data might be data from an on/off device or from an overheat or not-overheat sensor, for example.

Digital-to-analog (D/A) conversion   The conversion of a digital number (count) that represents an analog value (voltage) to an output voltage form, possibly for actuation of a control instrument in a process.

Discrete process   A physical process that operates in discrete entities rather than as a continuous process. Typical is the production of parts such as appliances, airplanes, or computer boards.

Hierarchical computer system   A group of computers arranged in a vertical communication fashion similar to that of a management organization system. The higher the computer in the hierarchy, the higher the level of planning accomplished. The lowest level in a manufacturing computer hierarchy is made up of the control computers that actually control equipment.

Input/output board   In this chapter, a board in a microcomputer that facilitates the process input/output programming needed for a control computer.

Ladder diagram   A form of programming structure originally devised to allow programmable controller programming to be ''user-friendly'' for shop-floor technicians.

Mainframe computer   Generally, the plant's prime data base computer. At one time mainframe computer referred to one that was huge in terms of cost and size. Nowadays, a minicomputer might well be the mainframe computer. In terms of a manufacturing hierarchy of computers, the mainframe is generally the computer that is highest in the hierarchy.

Masking   A software process that allows some facility for bit manipulation to occur in high-level programming languages (say, BASIC) so that digital inputs can be decoded in an efficient manner.

Microcomputer   A computer that is small in size and in cost is a simplistic definition of a microcomputer. Control computers in this chapter will be discussed in terms of microcomputers.

Minicomputer   This is very hazy to define. At one time it referred to a computer much cheaper than a mainframe computer (perhaps $35,000 to $50,000) with some reduced operating characteristics. Minicomputers today may be very powerful and expensive but in smaller physical packages than originally. Many control computers are classified as minicomputers.

Multitasking operating system   A computer operating system that allows software tasks to compete for the central processing system on the basis of user-specified priorities and timing constraints.

Port   An interface device that allows the computer central processing unit to communicate with the outside world. Input data comes from the controlled process through an input port to the control computer. Similarly, controlling data is sent to the process through an output port.

Priority interrupt   A technique whereby tasks to be accomplished by a computer may interrupt currently running tasks that are of a lower priority than the interrupting task. Branching from and returning to the interrupted task is accomplished automatically based on parameters set by the programmer, including possible times for interrupts to occur.

Programmable controller (PC)   A special form of control computer that operates primarily on cyclic control of a process. The program is completely reviewed on a repetitive cycle and actions are taken depending on logical relationships between input devices.

Programmable logic controller (PLC)   Another name for programmable controller.

Programmable relay controller (PRC)   Original name for a programmable controller.

Radix   The base of a number system. For the decimal system this would be 10 (the number of unique digits in the system). Most common in computer work are the binary (radix 2), octal (radix 4), and hexadecimal (radix 16) number systems.

Real-time control   This can be a very misleading and misunderstood term. In reality, everything operates in real time. Real-time control here refers to the fact that reactions to events in a process have to occur with relation to our normal time: immediately, every 10 s, at the beginning of each shift, and so on.

Register A grouping of binary digits (bits) in terms of the number of bytes (groups of 8 bits) that a particular computer can manipulate. The register is a *hardware* grouping of bits. For a microcomputer this is usually 8, 16, or even 32 bits.

8.3 BACKGROUND OF COMPUTER CONTROL

We mentioned in the introduction to this chapter that the late 1950s saw computer control of industrial processes become a reality, but that these early controlling computers were of large physical size, small memory capacity, and required huge dollar outlays. The late 1960s saw a move to minicomputers where "mini" was a function not only of reduced price (perhaps $35,000 to $50,000) but often a commensurate reduction in computer speed, at least in comparison with the mainframe computers on the market at that time. The cost reduction allowed the number of control computer installations to increase dramatically and even allowed for a measure of decentralization, where one computer might control only part of a complex industrial process. A disadvantage of the early control computer applications, in which one computer controlled an entire plant, was the possibility of plant shutdown because of computer failure. These early applications had to allow for parallel *complete* manual control capability to protect against computer failure, though it should be mentioned that industrial control computers had amazingly high reliability statistics, even in the early 1960s. Having multiple computers control somewhat autonomous segments of the process not only protected against complete plant shutdown in the event of computer failure, it also was a precursor of the hierarchical/distributed system for manufacturing computers discussed in Section 8.1. A major benefit for a system of distributed control computers has to be protection against the possibility of computer failure causing complete plant failure. One suggestion often heard is to put as many equipment-related computer functions (scheduling, controlling) as far down the hierarchy as possible in order to minimize the possibility of machine downtime.

It is very difficult today to categorize computers into a "maxi" or "mini" ranking, since many minis, for example, are used as so-called mainframe computers (and often called superminis). The advent of the microcomputer has created an even more interesting categorization situation, though the microcomputer is probably easier to classify than the other two categories. Cassell [2] states that the difference between the categories of computers is a matter of scale, with the boundaries between them often being indistinct. Microcomputers, though, can have all the characteristics of the other categories of computer, including the ability to control physical processes within real-time constraints, with considerable advantages accruing from their reduced cost and reduced physical size. Some disadvantages can be expected in order to achieve reductions in cost and size, but these are fast diminishing with continuing rapid advances in microcomputer technology. From a real-time, control point of view, these disadvantages have related primarily to speed of computation and ease of real-time programming, and neither seems to be a major problem at the present time since a large amount of application programming is feasible using a high-level language such as "C," Fortran, or BASIC rather than requiring the use of assembly-language programming. Speed of computation seems to be improving almost daily, judging by advertisements in trade journals. This chapter will assume that a high-level language is feasible for control computer application and that control will be accomplished through a microcomputer.

The cost/capability advantages inherent in microcomputers have made many industrial control applications possible, just as the minicomputer opened up control possibilities that could not be justified using the early mainframe computers. The manufacturing hierarchical/distributed computer structure is predicated on having microcomputers at the lowest level of the hierarchy—controlling "things" at the shop-floor level. Restricting the discussion to "microcomputers" should not disturb the reader who is interested in using minicomputers for control. The material is just as applicable to any category of controlling computer; in fact, the computer historian can show that current computer control technology is heavily dependent on capabilities developed for the early 1960s computers, particularly General Electric's GE/PAC 4000 series. Typical capabilities of that series included multilevel priority interrupts, microcoding possibilities, and Fortran/assembly-language interfacing.

Specific information that will be helpful for the analyst learning about applications of microcomputers for controlling industrial processes relates to number systems and data handling material. Input/output of physical process data may be binary or on/off-type information. Such data is handled more efficiently if several sensors or actuators can have their status inputs or required state outputs manipulated at the same time. The binary number system is a logical way to handle this. Another form of sensor data is analog, where, for example, valve position might be represented by a voltage signal. Analog input data has to be transformed to a digital format for computer analysis, and actuators triggered by analog signals need the digital results determined by computer analysis converted to analog output. The accuracy of the analog signal analysis will be proportional to the maximum integer number that the convertor is capable of handling. The remainder of this section will be concerned with those topics relating to the previous material: number systems, number handling, and analog/digital (A/D) and digital/analog (D/A) conversion.

8.3.1 Number Systems

The number system we are most familiar with is the decimal system, which has 10 unique possible values for each digit or position in the number (0 through 9). The base of this system, or *radix*, is 10. The term *radix* will be utilized instead of "base" throughout the remainder of this material. As an aside, it has been suggested that a radix of 12 might have been more useful to humanity, since 12 is evenly divisible by 2, 3, 4, and 6, while 10 is divisible evenly only by 2 and 5. Be that as it may, we only have 10 fingers or toes and so we also use a radix of 10. If we did use a radix of 12, we would need 12 unique symbols to represent each digit in a number, say, α and β in addition to the 10 normally accepted symbols 0 through 9.

A widely accepted radix within computer usage is the hexadecimal system, where the radix is 16. Here we use the symbols A through F in addition to 0 through 9. The larger the radix, the more compact a particular number representation may be. For example, 126 with a radix of 10 is equivalent to a hexadecimal value of 7E. These will be represented as $(126)_{10}$ and $(7E)_{16}$, respectively. The most common radix in computer usage is of course 2, with only the symbols 0 and 1 being allowed for each digit. This is called the *binary* system. In terms of sensor data that is binary in nature (on/off, open/closed, start/stop), the value 1 might represent one state—say, open—while 0 represents closed. For example, $(1101)_2$ might tell us that of four relays being monitored by a computer, three have open contacts and one has closed contacts. It is somewhat cumbersome in high-level programming languages to utilize binary numbers, it being far easier to use the equivalent decimal, octal, or hexadecimal representation. Hexadecimal and octal conversion can usually be handled with a programming command in a language such as BASIC, and so only conversion to binary will be considered in the subsequent material. Also, as far as binary sensor data is concerned, only integer numbers need be considered. Fractional conversion, handled differently from integer conversion, will be considered, but only for later discussion of analog data analysis.

Consider an integer number with radix p, say Np, that is to be converted to a number with radix q. The general equation for a number with radix p is:

$$Np = d_n p^n + d_{n-1} p^{n-1} + \cdots + d_i p^i + \cdots + d_1 p^1 + d_0 p^0 \tag{8.1}$$

where the integer coefficients d_i represent values for the ith digit in the number Np and $0 \le d_i < (p - 1)$. Also, d_0 is the lowest-order digit in the number.

An example in the decimal system would be

$$(136)_{10} = (1)(10)^2 + (3)(10)^1 + (6)(10)^0$$

Now, if Np is converted to a number with radix q, the same value is:

$$Np = D_m q^m + D_{m-1} q^m + \cdots + D_j q^j + \cdots + D_1 q^1 + D_0 q^0 \tag{8.2}$$

where D_j is the new coefficient for the jth digit.

To find out what these new digit values are, Np in Eq. (8.2) is divided recursively by the new radix (q) in the following manner:

$$\frac{Np}{q} = \underbrace{\frac{D_0}{q}}_{\text{fraction} = F_0} + \underbrace{D_m q^{m-1} + D_{m-1} q^{m-2} + \cdots + D_1 q^0}_{\text{integer} = I_0}$$

Since D_0/q is a fraction, it follows that D_0 is an integer remainder from the division that has a value greater than 0 and less than $q - 1$, a requirement for a digit value with radix q.

Now, if we divide I_0 by q, the second step in the recursive procedure, the result is:

$$\frac{I_0}{q} = \underbrace{\frac{D_1}{q}}_{\text{fraction} = F_1} + \underbrace{D_m q^{m-2} + D_{m-1} q^{m-3} + \cdots + D_2 q^0}_{\text{integer} = I_1}$$

and D_1 is the second digit in the number with radix q. Recursive division of the integer I_1 is next made to get the third digit, and this general procedure continues until all digits have been found.

Example. Convert $(367)_{10}$ to binary.

$2\ \underline{|367}$
$\downarrow$
$2\ \underline{|183}$ remainder 1 ← (lowest-order digit value)
$\downarrow$
$2\ \underline{|91}$ remainder 1
$\downarrow$
$2\ \underline{|45}$ remainder 1
$\downarrow$
$2\ \underline{|22}$ remainder 1
$\downarrow$
$2\ \underline{|11}$ remainder 0
$\downarrow$
$2\ \underline{|5}$ remainder 1
$\downarrow$
$2\ \underline{|2}$ remainder 1
$\downarrow$
$2\ \underline{|1}$ remainder 0
$\downarrow$
0 remainder 1 ← (highest-order digit value)

The result is

$$(367)_{10} = (101101111)_2$$

(highest-order   (lowest-order
digit value)   digit value)

How can it be determined if the above value is correct? One way is to convert the binary number to a decimal number using Eq. (8.1):

$$N_{10} = (1)(2)^8 + (0)(2)^7 + (1)(2)^6 + (1)(2)^5$$
$$+ (0)(2)^4 + (1)(2)^3 + (1)(2)^2 + (1)(2)^1$$
$$+ (1)(2)^0$$
$$= 256 + 64 + 32 + 0 + 8 + 4 + 2 + 1$$

which does give $(367)_{10}$. If the decimal number is to be converted to any other radix, then the previous procedure is followed except that division or multiplication is accomplished with the new radix, not with 2. Table 8.1 gives binary, octal, and hexadecimal equivalents for the first 16 integer numbers.

It is interesting to note that if we have a *binary* number, there is a trick to finding equivalent numbers whose radixes are even powers of 2 (e.g., binary, octal, and hexadecimal). Suppose that the power of 2 is k, where k is 1 for binary, 3 for octal, and 4 for hexadecimal. If the binary number's digits are partitioned in groups of k digits per group from the lowest-order digit, the values from Table 8.1 can be used to convert to the other radixes. For example, conversion to hexadecimal would require partitioning the binary numbers into groups of four digits. If the binary number is $(11011011101)_2$, we get:

$$(110 \quad 1101 \quad 1101)_2$$

$$(6 \quad\quad D \quad\quad D)_{16}$$

Similarly, conversion to octal would be:

$$(11 \quad 011 \quad 011 \quad 101)_2$$

$$(3 \quad 3 \quad 3 \quad 5)_8$$

If the computer can make the conversion from decimal to octal, or from decimal to hexadecimal, then it is a simple matter to convert the result to binary using Table 8.1 values in a manner similar to that shown above.

Now let us return to the original question of how to check the status of binary sensors. If it is necessary to check to see if the relay status is $(1100)_2$, a

TABLE 8.1
Integer number equivalence with radixes of 10, 2, 8, and 16

Decimal	Binary	Octal	Hexadecimal
0	0	0	0
1	1	1	1
2	10	2	2
3	11	3	3
4	100	4	4
5	101	5	5
6	110	6	6
7	111	7	7
8	1000	10	8
9	1001	11	9
10	1010	12	A
11	1011	13	B
12	1100	14	C
13	1101	15	D
14	1110	16	E
15	1111	17	F

logic check could be inserted in a program, possibly:

$$\text{IF (RELAY = 12) THEN 100 ELSE 80}$$

The value 12 is, of course, the decimal equivalent of the given binary word. If the particular configuration does give 12, then a branch is taken to statement 100. If not, then the branch is to statement 80. Unfortunately, the combination $(1100)_2$ is only one of 16 possible combinations of status for the four relays. If different actions must be taken depending on which of the 16 combinations is currently in place, then 16 IF statements could be employed to determine the correct decision. But if 16 relays are involved, then 65,536 IF statements will be required, a ridiculous situation. Section 8.4 will show how such a problem can be greatly simplified even when a high-level programming language is employed.

Equation (8.1) gave the general formulation for an *integer* number. A general fractional number can be represented by

$$Np = d_{-1}p^{-1} + d_{-2}p^{-2} + d_{-3}p^{-3} + \cdots \tag{8.3}$$

Combining Eqs. (8.1) and (8.3), the number $(136.83)_{10}$ can be written as

$$(136.83)_{10} = \underbrace{(1)(10)^2 + (3)(10)^1 + (6)(10)^0}_{\text{integer}}$$

$$\underbrace{+ (8)(10)^{-1} + (3)(10)^{-2}}_{\text{fraction}}$$

A question germain to later analog conversion material is how the decimal fraction can be converted to binary. *Integer* radix conversion requires a recursive *division* process. Similarly, *fraction* radix conversion needs a recursive *multiplication* approach. If a radix of q is needed for the number Np in Eq. (8.3), the general result is

$$Np = D_{-1}q^{-1} + D_{-2}q^{-2} + D_{-3}q^{-3} + \cdots \tag{8.4}$$

Multiplying Eq. (8.4) by q gives

$$(Np)(q) = \underbrace{D_{-1}}_{\text{integer} = I_0} \quad \underbrace{+ D_{-2}q^{-1} + D_{-3}q^{-2} + \cdots}_{\text{fraction} = F_0}$$

D_{-1} is the lowest-order digit in the new radix system and is an integer overflow from the multiplication. Multiplying F_0 by q gives

$$(F_0)(q) = \underbrace{D_{-2}}_{\text{integer} = I_1} \quad \underbrace{+ D_{-3}q^{-1} + \cdots}_{\text{fraction} = F_1}$$

F_1 is then multiplied by q to get D_{-3}, and so on.

Example. Convert $(0.83)_{10}$ to binary.

Step 1: $(0.83)(2) = 1.66$
The integer 1 is the highest-order binary digit.
Step 2: $(0.66)(2) = 1.32$
The integer 1 is the next binary digit.
Step 3: $(0.32)(2) = 0.64$
The integer 0 is the next binary digit.

Continuing in this fashion yields

$$(0.83)_{10} = (0.110101)_2$$

The reader who verifies this will find that when to stop the recursive multiplication may not be clear. In the example above, six recursions were accomplished. A rule of thumb is to determine 3 bits (binary digits) for *each* fractional position in the decimal number and then find one additional bit. This gives 7 bits in the example. It will be found that the seventh bit is 0 and so is not needed. To go from fractional binary to decimal requires Eq. (8.4):

$$N_{10} = (1)(2)^{-1} + (1)(2)^{-2} + (0)(2)^{-3} + (1)(2)^{-4} + (0)(2)^{-5} + (1)(2)^{-6}$$

$$= 0.5 + 0.25 + 0 + 0.0625 + 0 + 0.015625$$

$$= (0.83)_{10}$$

With integer conversion, grouping binary data in threes will allow the octal number to be found, while grouping in fours realizes the hexadecimal number (looking up the octal digit values in Table 8.1). Grouping is always done from the decimal point. The same holds for fractional radix conversion if the radix is a power of 2. For example:

$$(. \underbrace{1\ 1\ 0}\quad \underbrace{1\ 0\ 1})_2$$
$$(.\quad 6\quad\quad 5\ )_8$$

To go to decimal from $(.65)_8$ requires

$$(0.68)_8 = (6)(8)^{-1} + (5)(8)^{-2} = (0.83)_{10}$$

For a hexadecimal equivalence:

$$(. \underbrace{1\ 1\ 0\ 1}\quad \underbrace{0\ 1\ 0\ 0})_2$$
$$(.\quad D\quad\quad 4\ )_{16}$$

One final point that will be useful in later sections is the following: Dividing a decimal integer by numbers that are powers of 10 just requires the decimal point to be shifted to the left by the power that 10 is raised to. Dividing by 1000 requires the decimal point to be moved three places to the left. A similar situation exists for dividing binary numbers by numbers that are powers of 2 (2, 4, 8, 16, etc.). For example, dividing $(1001.)_2$ by 4 requires a shift of two places to the left, since $2^2 = 4$.

$$4 \lfloor (1001.)_2 = 4 \lfloor (9.)_{10}$$

$$(10.01)_2 = (2.25)_{10}$$

The reader can easily verify that the decimal conversions are correct.

8.3.2 Data Handling with Microcomputers

It has been shown that digits in the binary system, 0s and 1s, can be used to represent sensor binary status as well as required actuator binary condition. Each binary digit (bit) in a status word can represent the status or condition for one physical device. For efficient manipulation of data it is beneficial to input or output the binary information for several devices at one time. A limitation on the number of devices that can be handled by one input or output is the number of bits in the microcomputer's word size.

A *word* for a microcomputer has usually been 8, 16, or 32 bits in length, with 16 or 32 bits now being the most prevalent. The word size (sometimes called the *register size*, because a register is the hardware that handles a word of data) governs the accuracy the computer can maintain. For example, the maximum integer number that can be handled by a 16-bit register is $[2^{(15)} - 1]$. For an N-bit integer this would be $(2^N - 1)$. Using information in Table 8.1, this can be readily seen:

A 1-bit word has a maximum value of 1, or $(1)_2$, which is $2^1 - 1$.
A 2-bit word has a maximum value of 3, or $(11)_2$, which is $2^2 - 1$.
A 3-bit word has a maximum value of 7, or $(111)_2$, which is $2^3 - 1$, and so on.

A *byte* of information refers to the number of bits required to hold one alphabetic character of data. Eight bits constitute a byte, which explains why register sizes are frequently 8, 16, or 32 bits in length. Depending on the configuration of the microcomputer, the number of binary sensor devices that can be input or output simultaneously is a function of the word length and number of bytes. These binary sensor inputs/outputs will be called digital I/Os in the next section, in accordance with usual practice. A common setup for 16-bit-register machines is to use one byte for eight digital inputs and the second byte for eight digital outputs. However, input/output is handled through ports in the computer, and I/O ports can be set up for input or output as the user sees fit. A system that has one byte for input and one byte for output may also be set up for 16 bits of input *or* 16 bits of output. Of course, the physical device wired to specific connector pins has to be in accordance with whether the device is inputting data to the computer or receiving information from the computer. This will dictate how the ports are set up for input/output. Actual handling of digital I/Os within a control program will be discussed in Section 8.4.

Analog data, such as a voltage input or output, has to be in digital form within a computer program. So analog input information has to be converted to a digital format, and digital results must then be converted to an analog output format in some manner. Analog-to-digital (A/D) and digital-to-analog (D/A) convertors are used for this purpose.

A/D CONVERSION. A/D convertors are hardware devices that take a voltage within some stipulated range and linearly convert that voltage to a numerical value. The numerical value depends on the size of the conversion register, and the size of the conversion register dictates the accuracy of conversion. For example, common register sizes for convertors are 10, 12, or 16 bits, with convertor cost being a function of the register size.

The maximum conversion number (often called the conversion count) for a 10-bit register is $2^{10} - 1$, or 1023. The number of unique possible results from conversion is, of course, 1024, since zero is a possible conversion value. Now, suppose that the analog input range to the convertor is 0 to 10 V DC. A 10-V signal will have a converted count of 1023, and a 0-V signal will have a resultant count of 0. Any voltage between 0 and 10 V will have a count that is a linear interpolation between 0 and 1023. Similarly, an input analog range of -10 V to $+10$ V will also have a linearly interpolated count between 0 and 1023.

This relation can be expressed in equation form as

$$\text{Count} = \frac{(\text{input voltage} - \text{minimum voltage})\,(1023)}{\text{voltage range}} \tag{8.5}$$

Thus, if the input voltage is 3 V when the range is -10 to $+10$ V (a 20-V range), the conversion count is:

$$\text{Count} = \frac{[3 - (-10)]\,(1023)}{20} = 665$$

Of course, if the register size is 16 bits, then the value of 1023 must be replaced by 65,535.

The *resolution* of the convertor is a function of how small a change in voltage can be detected in the conversion to a digital number. For an N-bit convertor, the resolution is:

$$\text{Resolution} = \frac{\text{voltage range}}{2^N} \tag{8.6}$$

Resolution as a function of the convertor register size is:

	10-V range	20-V range
8-bit register:	0.0391 V	0.781 V
12-bit register:	0.0024 V	0.0488 V
16-bit register:	0.0002 V	0.0003 V

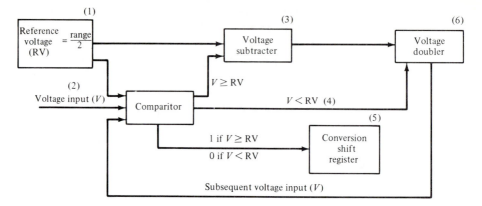

FIGURE 8.5 Logic diagram for A/D conversion.

A variety of hardware approaches are available for accomplishing analog/digital conversion and, of course, the hardware is part of the control computer configuration. One approach to accomplishing the conversion is the *doubler* circuit, which has appeared in many variants over the years. The logic diagram for this method is given in Figure 8.5. Steps in the A/D conversion process (corresponding to numbers in parentheses in the figure) are as follows:

1. A reference voltage (RV) equal to half the A/D convertor input range is provided. If the range is 10 V (0 to +10 V), then RV = 5.
2. The voltage to be converted is input to a comparator which has an output based on the comparison of the input voltage (V) to the reference voltage (RV).
3. If $V \geq$ RV, then the voltage subtracter computes ($V - $RV) and this replaces the previous cycle's value for V.
4. If $V < $RV, then the subtraction process is bypassed.
5. Based on the comparison result, a 0 or a 1 is shifted into a conversion register. If $V \geq$ RV, then a 1 is shifted in; if $V < $RV, then a 0 is shifted in. The highest-order bit is shifted in first.
6. Subsequent voltage inputs to the comparitor are the doubled values of V, *not* the *initial* input voltage. This procedure cycles as many times as there are bits in the conversion register, with the lowest-order bit being the last to be shifted in.

> **Example.** Suppose that the input voltage is 3.5 V with a convertor that has an input range from 0 to 10 V (RV = 5). Also, assume that we have a 4-bit convertor.
>
> Cycle 1: V of 3.5 < RV of 5; 0 is the *highest-order bit*. Subtraction is bypassed and second cycle $V = (2)(3.5)$.
>
> Cycle 2: V of 7.0 > RV of 5; 1 is the next bit in the conversion register. Subtraction gives $V - $RV = 2, and the doubler forms the next V of $(2)(2) = 4$.

Cycle 3: V of $4.0 <$ RV of 5; 0 is the third bit in the conversion register. Subtraction is bypassed and the next V is calculated by the doubler to be $(2)(4) = 8$.

Cycle 4: V of $8.0 >$ RV of 5, so the final bit shifted into the 4-bit conversion register is 1, giving a final result of $(0101)_2$.

Since the maximum value for a 4-bit register is $(15)_{10}$, the resultant voltage calculated in a computer routine using the converted count is:

$$\text{Volts} = \frac{(\text{conversion count})(\text{voltage range})}{(\text{count range})} \tag{8.7}$$

Giving

$$\text{Volts} = \frac{(5)(10)}{(15)} = 3.33 \text{ V}$$

This is 0.17 V less than the true input voltage of 3.5 V. The interested reader may verify that a 7-bit register gives a value of 3.465 V and a 10-bit register gives a conversion of 3.4995 V, thus showing the advantage of a larger-size conversion register (a fact demonstrated earlier when convertor resolution as a function of register size was demonstrated).

D/A CONVERSION. Given a calculated voltage (a *digital* value in the computer) that is to be converted to an equivalent analog signal for actuating a control device, it is necessary to determine the conversion count for output by modifying Eq. (8.7):

$$\text{Conversion count} = \frac{(\text{volts})(\text{count range})}{(\text{voltage range})} \tag{8.8}$$

As will be seen in Section 8.4, this is now output to a D/A convertor, which provides a voltage output proportional to the count value.

Example. Suppose that the D/A convertor is a 4-bit device (of course, 10- and 12-bit convertors are more realistic from the viewpoint of both accuracy and real applications) and the voltage output from the convertor is linear in the range 0 to $+10$ V. A control algorithm determines that a 7.4 V signal has to be output to actuate a specific process device. The conversion count from Eq. (8.8) is:

$$\text{Conversion count} = \frac{(7.4)(15)}{10} = (11)_{10} = (1011)_2 \tag{8.9}$$

The convertor might have four electronic relays set according to the 4-bit result just determined.

A typical D/A resistive circuit is shown in Figure 8.6.[1] This convertor is a constant-current source *(I)* device, where I is switched into the circuit by electronic

---

[1] Readers who are not familiar with circuit analysis may skip this discussion, as it is not necessary for an understanding of the related computer control material to be given in Section 8.4.

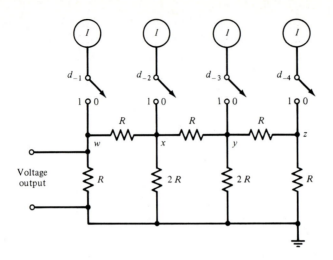

FIGURE 8.6   D/A conversion circuit.

switches marked on the figure by d_{-1} through d_{-4}. A closed switch corresponds to a binary 1 and an open switch relates to a binary 0. The negative subscripts on d indicate that the resultant binary number is a *fraction* rather than an integer. This fraction is

$$\text{Number} = (2^{-1})(d_{-1}) + (2^{-2})(d_{-2}) + (2^{-3})(d_{-3}) + (2^{-4})(d_{-4})$$

An analysis of the circuit shows that resistance to ground to the *right* of nodes z, y, x, and w is R, $2R$, $2R$, and $2R$, respectively (remember $2R$ in parallel with $2R$ gives R). Similarly, resistance to ground to the *left* of nodes w, x, y, and z is also R, $2R$, $2R$, and $2R$, respectively. From this, it can be seen that if the electronic switch d_{-1} is closed, then ($\frac{2}{3}$) I will flow through the output resistor. If d_{-2} is closed, its contribution of I splits three ways at node x so that ($\frac{1}{3}$) I flows through the output resistor. Similar analysis shows that closing d_{-3} or d_{-4} will allow ($\frac{1}{6}$) I or ($\frac{1}{12}$) I to flow through the output resistor. Therefore, the total current flowing through the output resistor is:

$$\text{Current} = (I)(R)[(d_{-1})(\tfrac{2}{3}) + (d_{-2})(\tfrac{1}{3}) + (d_{-3})(\tfrac{1}{6}) + (d_{-4})(\tfrac{1}{12})] \qquad (8.10)$$

The output voltage is found by rewriting Eq. (8.10) and knowing that (R) (current) is voltage.

$$\text{Volts} = (\tfrac{4}{3})\,(I)(R)(\underbrace{d_{-1}2^{-1} + d_{-2}2^{-2} + d_{-3}2^{-3} + d_{-4}2^{-4}}_{\text{Binary fraction}}) \qquad (8.11)$$

Assuming that all switches are closed (all $d_{-i} = 1$), then the binary fraction for 4 bits is

$$(1)(0.5) + (1)(0.25) + (1)(0.125) + (1)(0.0625) = 0.9375$$

For a 0- to +10-V D/A convertor, the maximum output should of course be 10 V. Therefore, I and R in Eq. 8.11 can be chosen so that

$$(\tfrac{4}{3}) \, (I)(R)(0.9375) = 10$$

For this case Eq. (8.11) reduces to:

$$\text{Volts} = (10.67)(d_{-1}2^{-1} + d_{-2}2^{-2} + d_{-3}2^{-3} + d_{-4}2^{-4}) \qquad (8.12)$$

Now, the perceptive reader will wonder what is the relationship between the *fractional* bit configuration in Eq. (8.12) and the *integer* count determined earlier in Eq. (8.9). The latter value was found to be $(1011)_2$ for a voltage requirement of 7.4 V. When integer and fractional binary numbers were discussed earlier, it was shown that dividing a binary integer by 2 raised to a power can be accomplished by shifting the decimal point a number of places to the left equal to that power. For a 4-bit convertor there are 2^4 (or 16) possible conversion combinations. A *fractional* conversion count is the *integer* value divided by 16. Since 16 is 2 raised to a power of 4, the fractional value is equal to the integer value with the decimal point moved four places to the left. It turns out, rather obviously perhaps, that for the unique situation where an N-bit integer is divided by 2^N, the integer bit sequence is the *same* as the fractional bit sequence:

$$
\begin{array}{cccc}
d_3 & d_2 & d_1 & d_0 \qquad \text{(integer sequence)} \\
\downarrow & \downarrow & \downarrow & \downarrow \\
1 & 0 & 1 & 1 \\
d_{-1} & d_{-2} & d_{-3} & d_{-4} \qquad \text{(fractional sequence)} \\
\nearrow & \nearrow & \nearrow & \nearrow \\
.1 & 0 & 1 & 1
\end{array}
$$

This says that the integer count value that has the lowest-order integer bit (d_0) forces closure of electronic switch (d_{-4}) when d_0 is 1. The same holds true for d_1, d_2, and d_3, causing the open/closure of switches d_{-3}, d_{-2}, and d_{-1}, respectively.

Substituting in Eq. (8.12) gives the actual voltage output:

$$\text{Volts} = (10.67)[(1)(0.5) + (0)(0.25) + (1)(0.125) + (1)(0.0625)] = 7.33 \text{ V}$$

This is very close to 7.4 V. It should be remembered that a 4-bit register has only 16 possible combinations, and for a 10-V range the precision is just 0.625 V. We were fortunate in that our example conversion is very close to one of the 16 combinations. A calculated value of 7.1 V would still realize a D/A conversion output accuracy of 7.33 V. Thus, more register bits will improve the conversion accuracy.

This concludes the material that will serve as background to the control computer programming-oriented material in the next section. There we will see how a computer can actually interface with a manufacturing process in a real-time control mode.

8.4 SOME COMPUTER CONTROL PROGRAMMING CONCEPTS

It was shown earlier that control digital computers are really no different from general-purpose digital computers. The input of data from a process can be interpreted in a manner analogous to addressing conventional peripherals such as keypads for data entry. Similarly, data to be output to actuate some process controller can be thought of as output to a device similar to a printer or screen. Granted, process input data are contact closure status values as well as analog information rather than the conventional alphanumeric data expected from a console or keypad. Output data to the process has the same characteristics as input process data.

The process that a control computer is monitoring and actuating dictates certain possible differences in the control computer, such as that output response to input data frequently has to be extremely fast, possibly in small fractions of a second. Tasks being controlled, by the very nature of the physical process, have to operate simultaneously, necessitating an operating system that can handle a real-time, multitasking environment. In line with all these aspects of computer control, this section will show through a subset of the BASIC language how a microcomputer might be interfaced with a process, specifically with digital and analog I/O. It should be realized that control can be effected more easily with a language such as "C." BASIC is currently more widely understood and so will be used in this material as the demonstrating language.

8.4.1 Digital Input

Examples of digital input include relays being open or closed to indicate that a bearing is hot through a bimetal switch or that a part is present at a robot unloading station. Many industrial alarm situations or condition status notifications fall into the digital input classification. As discussed in Section 8.3, such information can be efficiently represented in binary format. For example, an input of $(11011010)_2$ might indicate that, of eight motors being interrogated, five are running—the five with a "1" input. Of course, the external relays have to be wired correctly to an input port at the back of the computer so that the programmer knows what bit position relates to which motor. If a control action has to be taken given the particular bit combination $(11011010)_2$, a test to see whether the binary number is $(218)_{10}$ can be made and program branching taken depending on the result. Unfortunately, if we only want to test one motor's status—say, the one represented by the left-hand bit—then there are 128 binary combinations that might have a "1" in the left-hand position. This is a simple matter at the assembly-language programming level, but most applications-oriented people find higher-level programming to be far more efficient in terms of program development. How to simplify this bit-checking problem with a higher-level language leads to the concept of masking.

Many vendors have developed real-time, process input/output boards for personal computers to simplify process I/O programming. For example, Data

Translation [4] markets a board which has macro subroutines programmed on the board in assembly language. The user may call these subroutines from a high-level program, perhaps using BASIC or "C" with an IBM PC, and all multiplexing and signal conditioning is handled through the subroutines. The Data Translation philosophy will be used to exemplify how a computer can communicate with a physical process, since the CALL statements are very similar to those of many other real-time computer systems. While it cannot be said that they are truly generic, a standard being developed for real-time Fortran has a similar CALL structure [8].

For digital input, a CALL statement can be inserted in a program somewhat as follows:

```
CALL DINP(INPORT, IMASK, IVAL)
```

where the first three symbolic names have been set to integer values prior to the CALL according to vendor instructions:

DINP is a symbolic name for the starting address of the subroutine digital input.

INPORT is the port number for digital input (the port to which process sensors are wired).

IMASK is a *mask* value which controls which bits are being interrogated.

IVAL is the digital input value resulting from the CALL. IVAL shows the status of the desired sensors.

The mask value determines which bits can be monitored through the input call. For example, if the input port is set to receive a single-byte input, 8 bits, then the mask is an 8-bit number where a 0 forces the equivalent process input to be 0 and a 1 in a mask position allows the process input to be the true input value, either 0 or 1. A 0 in a mask position says that we are *masking out* a certain bit. The advantage of masking can be shown with a simple example. Suppose that we have bimetal strip switches checking bearing temperatures on two motors; they send a 1 to the computer when they are closed, indicating an alarm state. Further, the two switches in question are connected to the inport port's bits 3 and 5. It should be remembered from the discussion of integer conversion that the lowest-order integer bit is bit 0. If we want a test to see if either bit is a 1 in the input value, IVAL, all bits are masked with the exception of bits 3 and 5, giving

$$IMASK = (00101000)_2 = (40)_{10}$$

The value 40 is, of course, $2^3 + 2^5$. Now, if the bimetal strip switches are both open but the other six sensors connected to INPORT are sending 1, the value of IVAL will be $(0)_{10}$, because the six 1s will be masked out. It is a simple matter to check the status of bits 3 and 5 as shown in Figure 8.7[2] Of course, if we wanted to

---

[2] BASIC is used in some examples, but readers familiar only with some other language should be able to follow the discussion.

		Comments
10	DEFINT D, I	Force all symbolics whose names start with D or I to an integer value.
20	DINP=63	Location of assembly-language digital input program set to vendor's specification.
30	INPORT=1	Computer port for digital input set to 1.

300	IMASK=40	Set input mask to mask out all bits except bits 3 and 5.
310	CALL DINP(INPORT,IMASK,IVAL)	Input status for bits 3 and 5.
320	IF IVAL=32 THEN 500 ELSE 330	If bit 5 is a "1," go to statement 500; otherwise, go to next line.
330	IF IVAL=8 THEN 600 ELSE 340	If bit 3 is a "1," go to statement 600; otherwise, continue with normal operation.

500 (Take whatever action is needed to shut off bit 5 motor.)

600 (Take whatever action is needed to shut off bit 3 motor.)

FIGURE 8.7   Segments of a BASICA program to monitor two motors.

check if both bearings are overheating, then line 320 in Figure 8.7 could be:

$$320 \ \text{IF IVAL} = 40 \ \text{THEN} \ \langle \text{take action} \rangle$$

8.4.2 Digital Output

Digital output of binary information (on/off, start/stop) can be handled in a similar way to input:

$$\text{CALL DOUTP(OUTPORT, OMASK, OVAL)}$$

where

DOUTP is a symbolic name for the starting address of the subroutine digital output.

OUTPORT is the port number for digital output.

OMASK is the output mask.

OVAL is the output digital value to actuate devices in the process.

The mask value is *extremely* important in digital output. Suppose, for example, that we are controlling a model train and we can start and stop the train through output bit 0. Also, we can switch track segments with bits 1 and 2. Further, a stop/go signal is controlled with bit 4. The train is started with a bit 0 value of 1 and stopped with a 0. The signal is green if bit 4 is a 1 and red if it is 0. If we want to switch the track segments by outputting a 0 in bits 1 and 2 but the

Comments

```
10   DEFINT D,O                              Force all symbolics whose names start with D or O to an integer value.
20   DOUTP=66                                Location of assembly-language digital output program set to vendor's specifications.
30   OUTPORT=0                               Computer port for digital output is set to 0.
     .
     .
     .
300  OMASK=6                                 Mask out all bits except bits 1 and 2.
310  OVAL=0                                  The output value is set to some number that contains 0s in bit positions 1 and 2.
320  CALL DOUTP(OUTPORT,OMASK,OVAL)          Output to switch the tracks.
330      (Continue with the program.)
```

FIGURE 8.8   Segments of a BASICA program to output 0s to bits 1 and 2.

particular subroutine doing this does not determine the train start/stop status or the signal red/green condition, what should the output be to bits 0 and 4? The output mask is used to *mask out unknown* bit conditions so that, regardless of what is placed in those bits, a 0 or a 1, *no change* will be made to the outputs corresponding to those masked bit positions. Bits are masked out by having 0s in the corresponding output mask positions. Logically, the output mask would have only 1s in the output value's (OVAL) bit positions that we really can control—in this example, bits 1 and 2. Possible OVAL and OMASK values that would work are:

$$OMASK = (00000110)_2 = (6)_{10}$$

$$OVAL = (00000000)_2 = (0)_{10}$$

Actually, with a decimal mask of 6, *any* OVAL would work that has 0s in bit positions 1 and 2. Figure 8.8 shows how BASICA program segments might handle this case. If the switch controlled by bit 2 is to receive a 1 and the other switch is to remain unchanged, then the OMASK could still be 6 and the OVAL would logically be $(00000100)_2$, or $(4)_{10}$. Of course, having an OMASK value of $(4)_{10}$ would also work.

8.4.3   Analog Input

The *concepts* of analog/digital (A/D) and digital/analog (D/A) conversion were presented in Section 8.3. Analog input/output (with conversion) can be handled in a manner similar to that just presented for digital input/output. A CALL for analog input could be:

```
CALL ANINP(CHAN, GAIN, ANIN)
```

where

ANINP is a symbolic name for the starting address of the subroutine analog input.

CHAN is the analog input channel number through which an analog input is brought.

GAIN is a "programmed" gain whereby the input voltage is amplified prior to conversion.

ANIN is the conversion value (or count value).

Since an analog conversion will create a value much larger than a bit representation as received in digital inputs, each sensor in the process sending out an analog signal has to be wired into a separate set of screw connections leading to the convertor. Each set is commonly called an input *channel*. A small system might have 16 input channels, say, channel numbers 0 through 15. A programmable gain can be useful if input voltages are very small values. Amplifying the input voltage by some known gain prior to conversion may help in the accuracy of conversion. We will assume for our purposes that GAIN has possible values of 1, 2, 4, and 8.

The conversion value, ANIN, is a function of the number of bits in the convertor (as was shown in Section 8.3). A common size is a 12-bit convertor, which will allow a conversion range of 0 to 4095. This means that if we have a 0- to 10-V convertor, the digital conversion value will be linearly interpolated in the range 0 to 4095. A 10-V input with a GAIN of 1 will realize an ANIN value (see previous CALL structure) of 4095, a 0-V input will give a 0 conversion, and a 5-V input will result in a conversion of 2048. The use of GAIN will *amplify* the input voltage *before* conversion, so the 5-V input with a programmed GAIN of 2 will give a conversion of 4095 rather than the 2048 without GAIN. Obviously, if the program is going to print out the equivalent in volts, then the GAIN has to be taken into account. A *general* voltage equation to determine volts from the conversion count is

$$\text{Volts} = \frac{\left[\dfrac{(\text{conversion count})(\text{voltage range})}{(\text{conversion range})} + \text{low voltage}\right]}{\text{gain}} \tag{8.13}$$

The low voltage allows for a bipolar convertor—say, ±10 V—to be applicable.

Figure 8.9 gives a BASICA program structure that allows analog input with variable gain control.

8.4.4 Analog Output

An analog output CALL might look like the following:

```
CALL ANOUTP(CHANO,AVOLT)
```

where

ANOUTP is a symbolic name for the starting address of the subroutine analog output.

CHANO is the analog output channel number through which the voltage is to be sent to the process being controlled.

AVOLT is the output voltage desired in count (0–4095) form.

Comments

10	DEFINT A	Force all symbolics whose names start with A to an integer value.
20	ANINP=3	Location of assembly-language analog input program set to vendor's specifications.
30	ACHANI=5	Set analog input channel to number 5.

.
.
.

100	AGAIN=1	Set gain to a value of 1.
110	CALL ANINP(ACHANI,AGAIN,ANIN)	Perform one analog conversion through channel 5.
120	IF (ANIN > 2048) THEN 200	If the input is in the upper range, convert to volts.
130	IF (AGAIN=8) THEN 200	If the conversion is in the lower end of the range and gain is maximum, convert to volts.
140	AGAIN=AGAIN*2	Double the gain if the conversion is in the bottom half of the range and gain is less than 8.
150	GOTO 110	Do another conversion with the higher gain value.

.
.
.

200	VOLTS=(((ANIN*20)/4095)–10)/AGAIN	Convert conversion count to volts. Voltage range is 20 (±10).
210	(Continue with the program.)	

FIGURE 8.9   Segments of a BASICA program that allow for analog input with programmable gain.

It will be noticed immediately that there is no GAIN in the output, and a little reflection will reveal that output GAIN is meaningless. A D/A convertor—say, a 12-bit convertor—will allow a value between 0 and 4095 to be converted to a voltage between −10 and +10V (assuming a bipolar convertor). Any gain required would be an integral component of the convertor itself. Figure 8.10 shows typical BASICA commands used in accomplishing an analog output with associated D/A conversion assuming that the required output voltage is 5.2 V. Equation (8.13) gave the equation for finding VOLTS given a conversion count. Line 100 in Figure 8.10 rearranges Eq. (8.13) to allow a required conversion count to be found given a specific voltage:

$$AVOLT = \frac{(\text{required volts} - \text{low voltage})(\text{conversion range})}{(\text{voltage range})} \tag{8.14}$$

Since we have a required voltage of 5.2, a conversion range of 4095, a voltage range of 20 (±10 V), and a low voltage of −10 V, the output digital count (AVOLT) is

$$AVOLT = \frac{[5.2 - (-10)](4095)}{20} = 3112$$

8.4.5   A Final Example

Combining all the I/O concepts covered to this point leads to the example shown in Figure 8.11. The situation programmed is as follows:

1. Check to see if digital input bits 4 and 5 are both 1. If not, repeat digital input until they are. The digital input CALL is put in a subroutine (as are all other CALLS) so that calls can be made in several sections of a program.

10	DEFINT A,C	Force all symbolics whose names start with A or C to an integer value.
20	ANOUTP=24	Location of assembly-language analog output program set to vendor's specifications.
30	CHANO=1	Set analog output channel to number 1.

.
.
.

100	AVOLT=((5.2+10)*4095)/20	Calculate digital count assuming that output voltage is to be 5.2 V with 12-bit bipolar converter.
110	CALL ANOUTP(CHANO,AVOLT)	

.
.
.

(Continue with the program.)

FIGURE 8.10   Segments of a BASICA program to do analog output.

FIGURE 8.11   Segments of a BASIC program doing digital and analog I/O.

Comments

10	DEFINT A,C,D,I,O	Lines 10–70 are initialization lines—done only once in program.
20	DINP=63	Lines 20–50 identify assembly-language routine locations according to vendor instructions.
30	DOUTP=66	
40	ANINP=3	
50	ANOUTP=24	
60	INPORT=1	Lines 60–70 set digital input and output ports.
70	OUTPORT=0	

.
.
.

100	IMASK=48	Mask out all input bits except bits 4 and 5.
110	GOSUB 1000	Go to digital input subroutine to do one input.
120	IF IVAL=48 THEN 200 ELSE 110	If input bits 4 and 5 are 1s, do digital output; otherwise, repeat digital input.

.
.
.

200	OMASK=192	Mask out all output bits except bits 6 and 7.
210	OVAL=192	Go to digital output subroutine to do one output.
220	GOSUB 2000	After digital output, do analog input.
230	GOTO 300	

.
.
.

300	AGAIN=1	Set GAIN to 1.
310	ACHANI=10	Set analog input channel to number 10.
320	GOSUB 3000	Go to analog input subroutine to do one input.
330	IF ANIN > 2457 THEN 400 ELSE 320	If input voltage > 2 V, do analog output; otherwise, repeat analog input.

.
.
.

400	CHANO=1	Set analog output channel to number 1.
410	AVOLT=ANIN	Set analog output count to same value as was input earlier.
420	GOSUB 4000	Go to analog output subroutine to do one output.
430	END	Program run complete.

.
.
.

1000	'DIGITAL INPUT SUBROUTINE	
1010	CALL DINP(INPORT,IMASK,IVAL)	Perform digital input—port and mask must be set by caller.
1020	RETURN	Return to statement number after the one that requested the subroutine.
2000	'DIGITAL OUTPUT SUBROUTINE	
2010	CALL DOUTP(OUTPORT,OMASK,OVAL)	Perform digital output—port, mask, and output must be set by caller.
2020	RETURN	
3000	'ANALOG INPUT SUBROUTINE	
3010	CALL ANINP(ACHANI,AGAIN,ANIN)	Perform analog input—channel and gain must be set by caller.
3020	RETURN	
4000	'ANALOG OUTPUT SUBROUTINE	
4010	CALL ANOUTP(CHANO,AVOLT)	Perform analog output—channel must be set by caller.
4020	RETURN	

2. If input bits 4 and 5 are set, output 1s through digital-out bits 6 and 7.

3. Perform analog input until the input voltage is greater than 2 V (programmable gain is not used in this example). The reader can confirm that an input count of 2457 does equal 2 V for a ± 10-V, 12-bit convertor.

4. Once the input voltage is greater than 2.0 V, that same voltage is to be output through a digital-to-analog conversion. Since gain is not a factor, the input conversion will be the value required for output (assuming the same ± 10-V, 12-bit convertor constraints), so the output count is set to the input value count before actually doing the D/A conversion and output.

The objective of this section has been to show how a high-level language can be used to assist in computer communication with a physical process—namely, digital and analog input/output. It should be apparent that the examples we have looked at are lacking a few things:

1. Timing constraints have not been considered. What if digital inputs have to be accomplished every 10 ms?

2. Waiting for an analog value to reach a specific value will use valuable computer time. There should be some way to allow tasks to occur on some priority basis. The next section will briefly introduce ways to handle timing constraints as well as how multiple tasks might be handled in some priority fashion.

8.5 TIMING, INTERRUPTS, AND MULTITASKING

As just mentioned, computer control requires some way to ensure that all tasks can be handled within quite severe timing constraints: a cement kiln might require that all sensors be polled every 0.5 s in addition to cement blending and analysis being accomplished; a conveyor sorting system might require that part sensing be accomplished at least every 0.1 s to ensure that high-speed movement does not preclude parts from being tracked; an automated guided vehicle system might require checking equipment status every 2 s. If tens or hundreds of such tasks must be handled by one computer, some form of prioritization must be established to allow tasks to be controlled and monitored in an optimum fashion. This section will look at ways to alleviate some of these problems: timing, priority interrupts, and multitasking.

8.5.1 Timing

The earliest form of timing available to digital computers in process control was digital clocks, where an *external* digital clock could be interrogated just as any other form of sensor might be checked. Hours, minutes, and seconds could be input in binary-coded decimal (BCD) and decoded to determine what actions should be taken depending on the time input. Programs had to be written in such a way that the clock's time would be input at least once a second. This led to the concept of an *executive control routine*, where a task scheduler and time decoder would be entered many times from any routine. This was cumbersome, inaccurate, and time-consuming in terms of programmer effort. Many improvements have been made in recent years.

The IBM PC (and compatibles) BASICA language has a TIMER function that allows time increments of about 0.05 s to be utilized. If we have the following command in a BASICA program,

```
10 XX = TIMER
```

XX will hold the time in seconds since the computer was turned on or since midnight. If we wish to delay for 0.35 s, a program segment might be:

```
10 AA = TIMER
20 IF TIMER-AA <0.35 THEN 20 ELSE 30
30 (Continue after delay)
```

Use of this TIMER function can assist in quite large real-time control problems in academic activities. More complex control problems require more advanced means for taking care of timing *as well* as task scheduling. Priority interrupts and multitasking operating systems are two such approaches.

8.5.2 Priority Interrupts

Priority interrupts have been available for computer systems for more than 20 years. For example, typewriters used to be used for printing with computers. Let us say that the word "PRI" is to be typed. While the letter "P" is being typed, the computer should be able to do other tasks, such as a calculation, during the tenth-of-a-second or so it takes to type "P." Once the "P" bar returns to its home position the calculation can be interrupted by the computer operating system, all register values can be automatically saved, and program control can branch to printing, where printing of "R" can be initiated. After initiation, program control can go back to the calculation, which could resume where it had left off. This ability to *interrupt* programs in *progress* in order to tackle other programs led to the term *priority interrupt* (PI).

Groover [5] suggests priority interrupt assignments might be as follows:

Priority level	Assignment
1 (lowest)	Operator-initiated interrupt
2	System-generated interrupt
3	Timer interrupt
4	Process command passing
5 (highest)	Process-generated interrupt

TABLE 8.2
Priority interrupt data

Task	Priority (3 is highest)	Expected task duration	Interrupt time (from time 0)
A	2	5 s	0 s
B	1	10	2
C	3	6	3
D	1	3	6
E	2	8	10

In a multilevel priority interrupt system, only the highest-level tasks are guaranteed to be completed without interruption. Lowest-level priority tasks can be interrupted by any higher-level task, so that with a five-level scheme:

Tasks at level 1 can be interrupted by tasks at levels 2 through 5.

Tasks at level 2 can be interrupted by tasks at levels 3 through 5.

Tasks at level 3 can be interrupted by tasks at levels 4 through 5, and so on.

Suppose we have five tasks that can be initiated by priority interrupt. We have a three-level interrupt system with the interrupts for the tasks and task data as given in Table 8.2.

Figure 8.12 shows schematically when the five tasks will start and be completed. Task A starts at time 0 but is interrupted by higher-priority task C at time 3. Task C goes to completion at time 9 since it is a highest-level task. Even though tasks B and D could start by time 9, task A will resume at time 9 because it has a priority equal to or greater than these tasks. Task A completes at time 11 and is followed by Task E, since E has a higher priority than task B or D. Task E completes at time 19 and is followed by task B and then D. Tasks B and D have the same priority, but task B was initiated first.

FIGURE 8.12 Task start/stop times under priority interrupt.

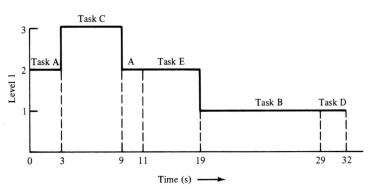

There is one problem with interrupt operation: Careful planning is needed to ensure that all tasks will be completed within some reasonable period of time. If a large number of tasks are specified for the highest-level priority interrupt, some lower-level tasks might not be completed.

8.5.3 Real-Time, Multitasking Operating Systems

Another method of handling task sequencing and time constraints that is becoming quite common, even in microcomputer systems, is to utilize a multitasking operating system. A multitasking operating system allows tasks to compete for the computer's central processing unit on the basis of priorities and the status of the tasks. Tasks may be thought of as subroutines but they are really independent, small programs that run in the CPU according to the criteria just mentioned.

A typical task might be a digital input routine. Another might be decoding of digital inputs, while a third might be process outputs based on the status of the inputs. The operating system will ensure that the digital input decoding task will not compete for the CPU until the input task has received process inputs. Similarly, the process output task will not compete for the CPU until the inputs have been decoded.

Multitasking allows the one thing that computer scientists generally abhor: It allows the computer user to tap into the computer's operating system (under protective constraints, however). Tasks are moved by the operating system into "states" depending on event occurrences (time, physical change, etc.). A typical state-space diagram showing the possible state transitions is given in Figure 8.13. The possible states in which a task might find itself are as follows.

1. *Nonexistent*. An operating system can, in general, maintain only a limited number of tasks. The user has to activate tasks by some CALL command (say, in Fortran, but the language is immaterial, the concept is not). The Data General Real-Time Input/Output System [3], developed in the early 1970s, has proven over time to be an excellent multitasking system. In fact, a European group is developing a set of standards for industrial real-time Fortran [8] which is based on the Data General System. Sample commands will be given to demonstrate multitasking from these sources, though it should be realized that these represent only 3 to 5% of such commands. A possible CALL to identify a task is

        CALL ITASK (taskname, task ID, error, priority)

   where

   taskname is the name of the task to be identified.
   task ID is a number corresponding to that task.

error is a location where errors could be decoded (another task has the same name, etc.).

priority is task priority, just as with priority interrupt.

Program sequence in a MAIN program might be

```
100 CALL ITASK(DIGIN, 65, $300, 2)
            110    (continue with program)
             .
             .
             .
            300    (decode error)
```

2. *Dormant.* Tasks that have been identified but, because running criteria have not yet been met, are not yet competing for the CPU.

3. *Pending.* Tasks that are waiting for the CPU in a queue to get to the CPU, based on task priority.

4. *Running.* A task that has control of the CPU is in the *running* state. It will be assumed that once a task has the CPU, it will be run to completion. Thus, task

FIGURE 8.13   A typical multitasking state-space diagram.

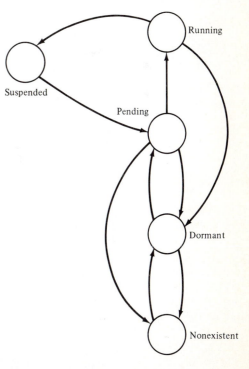

run times should be short—input one analog signal or group of signals, for example.

5. *Suspended.* Tasks that are run on a cyclic basis are *suspended* until the next cycle has been completed. Also, tasks may suspend themselves or may be suspended by other tasks, waiting for some time to occur to reactivate the task.

TYPICAL CALLS. To exemplify the power of a multitasking system, a few CALLs will now be presented:

1. Suspend a task:
   a. CALL SUSP—Suspends current task.
   b. CALL HOLD (task ID)—Suspends another task.
   c. CALL WAIT (number of time units, units), where units might be

$$0 = \text{milliseconds}$$
$$2 = \text{sixtieths of a second}$$
$$3 = \text{seconds}$$
$$4 = \text{minutes}$$
$$5 = \text{hours}$$

So, if after a digital input we want to delay a half-second before the next input, we could use

$$\text{CALL WAIT}(30,2)$$

Such a CALL is certainly far superior to the TIMER function in BASICA, introduced earlier.

2. Start a suspended task:

$$\text{CALL START (task ID, number of time units, units)}$$

If the task identified earlier as DIGIN is to be started immediately, we might have

$$\text{CALL START}(65,0,0)$$

3. Put a task into nonexistent state:
   a. CALL KILL—Takes current task out of system.
   b. CALL KILL (task ID)—Takes another task out of system.
   Frequently, a main program will be used to identify tasks and then KILL itself as all other tasks can operate according to their identified priorities and internal characteristics.
4. Bit testing (functions):
   a. BTEST (digital input word, bit)—Tests if a particular bit of a digital input word is 1 or 0. Returns a "true" state of 1. For example, if the digital input

word is DIGIN1, we might have

$$100 \text{ IF BTEST (DIGIN1, 4) THEN 200 ELSE 400}$$

If bit 4 is a 1, then the program would branch to line 200. If bit 4 is a 0, the program will go to line 400. The reader who has done much process control using a high-level language will realize the benefits of such a command.

b. ISET (register word, bit)—Sets the bit position in the word to 1.

c. IBCL (register word, bit)—Sets the bit to 0.

A possible structure for a multitasking program might be as exemplified in Figure 8.14. Common blocks have been ignored, but it should be apparent that

FIGURE 8.14 Hypothetical multitasking program.

		Comments
10	TASK MAIN	Main task will be used to identify other tasks in the system.
20	CALL ITASK(ONE,45,$60,3)	Start task identifications.
30	CALL ITASK(TWO,46,$60,2)	Priority 3 is highest; priority 1 is lowest.
40	CALL ITASK(THR,47,$60,1)	
50	GOTO 100	Tasks identified.
60	(DECODE ERROR)	Error decoding and action.
70	(PRINT ERROR MESSAGE)	
80	(EITHER STOP, WAIT FOR OPERATOR CHANGE, OR CONTINUE.)	
	.	
	.	
	.	
100	CALL HOLD(46)	Suspend tasks 2 and 3—to be
110	CALL HOLD(47)	started as needed by tasks 1 and 2.
120	CALL KILL	Remove main task from the system.
130	END	
	.	
	.	
	.	
10	TASK ONE	Assuming that a priority of 3 is highest, this will be next task in CPU.
20	CALL DIGIN(PORT,MASK,INVAL)	Do a digital input.
30	CALL START (46,0,0)	Make sure task 2 will decode inputs.
40	CALL WAIT(30,3)	Suspend this task (1) for 30 s.
50	END	
	.	
	.	
	.	
10	TASK TWO	
20	IF BTEST(INVAL,6) THEN 30 ELSE 50	Test if bit 6 in digital input is 1.
30	ISET(OUTVAL,5)	If yes, set output value's bit 5 to 1.
40	GOTO 60	
50	OUTVAL=0	If no, all output bits set to 0.
60	CALL START(47,0,0)	Start task 3 immediately for outputs.
70	CALL SUSP	Task 2 suspends itself.
80	END	
	.	
	.	
	.	
10	TASK THR	
20	CALL DIGOUT(PORTO,MASKO,OUTVAL)	Task 3 outputs digital status.
30	CALL SUSP	Then suspends itself.
40	END	

information common to tasks will have to be identified. Assuming that a language is available that has all the CALLs given in Figure 8.14, the three tasks could cycle indefinitely doing input, decode, and output.

The advantages of a multitasking scheme might not seem apparent unless some experience has been gained using one. Tasks can be written independent of other tasks as easy-to-debug programs. The operating system takes care of task sequencing; the user just specifies requirements that must be satisfied prior to the task running in the CPU. The only disadvantage might be some difficulty in debugging—where was the program when problems occurred? Judicious use of print statements in each task will help in this regard. More and more microcomputers are appearing with multitasking operation systems, some process control–oriented. The industrial real-time standard, mentioned earlier, might well be a reality within a few years for such languages as Fortran, Pascal, or ''C.'' Tools such as priority interrupts and multitasking operating systems can only serve to aid the analyst working with computer control. In turn, this will be a major aid to computer-aided manufacturing.

8.6   PROGRAMMABLE CONTROLLERS

A special case of digital computer control is programmable controllers. Because the method for programming a programmable controller has historically been predicated on boolean logic, another name frequently encountered is *programmable logic controller* (PLC). In their earliest development, programmable controllers used relay interfaces rather than software interfaces, and so at one time the term *programmable relay controller* (PRC) was quite common. The most general acronym today for programmable controllers, especially on the shop floor, is PC, which creates some problems with the now-accepted designation for a personal computer. The acronym PLC will be utilized in this section to refer to a *programmable (logic) controller*.

In the majority of applications, PLCs utilize digital inputs and control processes with digital outputs. The interface between inputs and outputs is called a *soft* interface, to denote that software links the input states to desired output states. Rapid advances in microchips have enabled PLCs to take on functions that are far more broad than just digital input and output. Ports are often available for analog inputs and outputs as well as for printing alarm conditions or management reports. Analog I/O is most frequently used for proportional/integral/differential (PID) control, a control methodology originating with the process industries but applicable to internal machine tool controlling (see Chapter 11).

A programmable controller is a ''cyclic'' controller in that it goes through all the inputs, analysis, and outputs on a repetitive basis. The length of time for a cycle is a function of the number of ''rungs'' in a ladder diagram. The PLC cycles through this ladder diagram from top to bottom. Typical applications that are cyclic include traffic lights, machine operations, train monitoring, factory cell operation (robot, machine, materials handling, inspection), and so on.

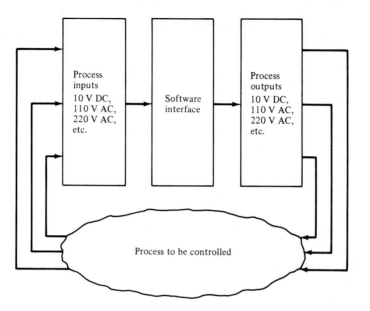

FIGURE 8.15   Programmable controller modular structure.

The PLC uses boolean logic in a software mode to tie physical inputs to physical outputs, thus eliminating hardwiring that could be expensive to accomplish or change. The relationship among inputs, the software interface, and outputs is shown in Figure 8.15. An on/off input device furnishes one of many possible input voltages. A PLC input module sets software "contacts" according to the input states. The software interface evaluates specified combinations of the software "contacts" and, through output modules, allows an output voltage (not necessarily the same as the input voltages) to be sent to the process when programmed characteristics are satisfied. The software interface also allows counting and timing functions as well as mathematical operations and printing of status information.

Some typical physical input (on/off) devices are shown in Figure 8.16. A limit switch is opened or closed by some object striking the switch, such as a part on a conveyor opening a limit switch which stops the conveyor until the part is removed by a robot. A pushbutton changes state momentarily when the button is depressed; an emergency stop button on an assembly line is an example. A relay contact is one that closes or opens upon energization of a solenoid. A temperature-activated relay might be triggered by a bimetal switch. The relay stays open or closed as long as an over-temperature condition exists. A normally open (NO) switch is just what it says: There is no continuity in the usual situation. A NO limit switch allows continuity when pressed.

Many other devices could be specified, including a photoelectric switch that sends a signal depending on whether or not the light source is broken by some

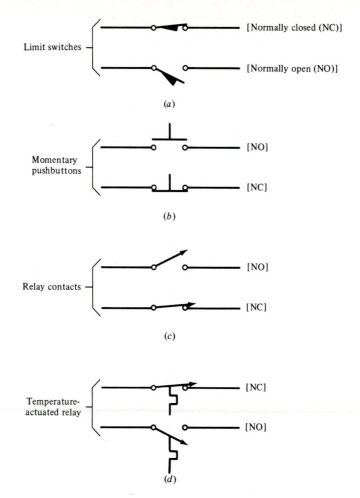

FIGURE 8.16 Typical on/off input devices.

intruder (part or human). However, all devices are characterized by the fact that when they are closed, they allow current to flow; and when they are open, current does not flow. Whatever the device, *all* can be represented by schematic contacts:

Physical input devices will be represented in the PLC interface by such schematic contacts.

One problem with a programmable controller is that although it will know when a voltage is *received* from a physical input device, it cannot know if the

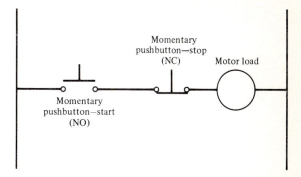

FIGURE 8.17   Single-rung-motor starter circuit.

device transmitting the voltage is normally open (NO) or normally closed (NC). Obviously, a relatively complicated programming structure would allow this, but it is not necessary. A simple example will demonstrate.

Suppose that we have a physical motor starter circuit as shown in Figure 8.17. The starter circuit is drawn as a single-rung ladder diagram. Interpretation of the rung tells us that activation of the output device (motor load) will occur when all input devices on the rung provide continuity from left to right to the output device (solenoid coil). When the start pushbutton is depressed and the stop button is *not* depressed, the motor should run. There is an obvious fallacy to this: Releasing the momentary start button will stop the motor! Another path to the motor load is needed to keep the motor running. If the output solenoid will close a relay when activated, then that relay can be used to keep the motor running, as shown in Figure 8.18. When the start button is depressed, the motor will start and the relay contact parallel to the start button will close. When the start button is

FIGURE 8.18   Corrected single-rung-motor starter circuit.

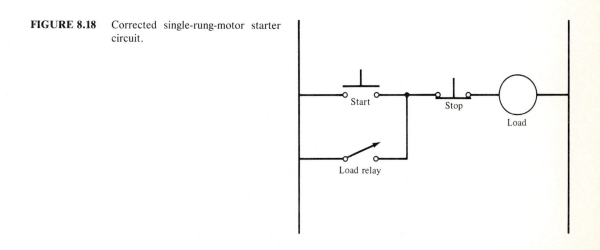

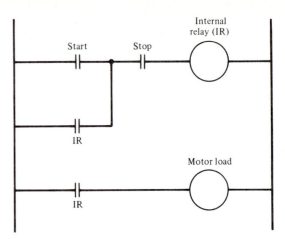

FIGURE 8.19 A two-rung software ladder diagram for the physical setup of Figure 8.18.

released, continuity will flow through the relay contact to keep the motor running until the stop button is depressed. This will stop the motor, thus opening the relay contact. We still have a *one-rung* ladder diagram in that all inputs feed to a single output device. In terms of boolean Logic, the motor will run if:

(START depressed OR IR relay closed) AND (STOP closed)

Now, how can this be programmed on the PLC software interface using only the contact symbols ‖ and ⫫? To get away from the problem of not knowing whether the device is NO or NC, it will be assumed that the software contact will *change state* (from ‖ to ⫫, or from ⫫ to ‖) when the physical input device being represented sends a voltage to the PLC. Figure 8.19 shows this for the previous physical relay diagram, Figure 8.18. When the start button (NO) is depressed, voltage is sent to the PLC for the short instant of depression time and so the software contact changes state from ‖ to ⫫ for this same period of time. Assuming that the stop button (NC) is not depressed, the voltage from the stop button changes the stop contact's state from ‖ to ⫫, allowing continuity to a software internal relay. This is called *activation* of the internal relay. Software contacts can be activated by this internal relay; such a contact is shown parallel to the start contact. Any time a software internal relay is activated, similarly labeled contacts will change their status (‖ to ⫫, or ⫫ to ‖). The internal relay contact provides continuity to the internal relay until the stop button is depressed. The second rung in Figure 8.19 allows the internal relay contact actually to energize the motor load; internal relays are purely internal—there is no output to the process.

Several points can be made concerning this simple example:

1. *Input* software contacts can be *repeated* in a ladder diagram.
2. Only *one output* should occur in any rung, and a *specific* output (say, motor AA-1) should appear only once in a diagram.

3. Currents do *not* actually flow in the *software interface*. Logical bits are set (say, 1 for IR when ⟋⟍ and 0 when ‖) when the internal relay is "energized" ("energized" means that if the start and stop bits are both 1, then the internal relay status is also 1). Output voltages to the process are allowed when output "load" status is 1.

4. The diagrams could be developed in many ways. The convention used in this section conforms with the Square D convention [11]. We could as easily have used a Texas Instruments approach or an Allen-Bradley convention. Typical controllers have a portable screen with a keypad that allows rung development using symbols as just seen in Figure 8.19. A typical screen and keypad are shown in Figure 8.20.

5. Ladder diagramming should be simple for technicians familiar with relay logic and symbols. The development of the ladder approach came about for this reason: to alleviate fear of programming for shop-floor people. Microchip advances are leading to even more powerful ways for programming PLCs.

It should be apparent that the momentary pushbutton for motor start could physically have been a normally closed switch (though that might be illogical from

FIGURE 8.20   Texas Instruments screen and keypad. Typical ladder diagram video programming units. (Courtesy of Texas Instruments.)

a systems design viewpoint). If that were the case, then a voltage would normally be input to the PLC when *start* is *not* required. The software start contact would now be programmed as ⫪, so that when the voltage is received by the PLC (all the time except when the button is depressed), the contact changes state to ‖. When start is pressed, this contact momentarily reverts to ⫪.

Further points concerning programming ladder diagrams (software) include the following.

1. Input contacts are programmed as *open* contacts (‖) if *logical continuity* is required when *voltage comes to the PLC* from external on/off devices.
2. Input contacts are programmed as *open* contacts (‖) if *logical continuity* is to be *broken* when *no* voltage comes to the PLC from external on/off devices.
3. If condition 1 or 2 is not correct, program the contact as ⫪.
4. Physical input devices are wired to *numbered* input modules, and programming of software contacts uses the same numbers. We will assume that input devices are numbered 101, 102, . . . , 116. Similarly, output devices will be assumed to be connected to output module contacts 201, 202, . . . , 216. Software internal relays will be numbered R301, R302, . . . , R316.

FIGURE 8.21 Configuration for PLC Example 1. S_1 and S_2 are track switch sensors that send a voltage to the PLC if they are switched to the inner loop and that send no voltage if they are switched to the outer loop.

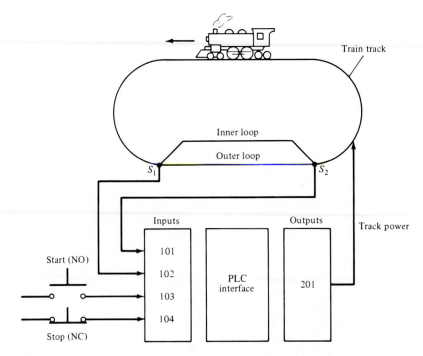

To introduce some PLC capabilities, four examples will be given. The first will consider only digital input/output. The second will add counting capability, and the third will introduce timing possibilities. A fourth example will show how timers might be used in a robot alarming situation.

PLC EXAMPLE 1. The Arizona State University Industrial Controls Laboratory has a small model railroad that is controllable both by computer or by a Square D programmable controller. The track arrangement is shown in Figure 8.21, with only the sensors/actuators needed for this first example being included.

In this example, the train is started with a NO momentary pushbutton (103). However, if the track switches are not both in the *same* position (inner or outer loops), the train should not start until the switches are manually put in the same position (this could be done automatically by the PLC, of course). While the train is running, it is stopped by the PLC if the same switches are ever moved manually to a situation where they are not both at the inner or both at the outer loop position (or the train would derail). Finally, an NC momentary pushbutton allows the train to be stopped by an operator if needed. A ladder diagram to allow this simple control is given in Figure 8.22.

One thing should be clear to the reader at this point: Many different solutions to the same problem are feasible. Figure 8.23 gives an alternative solution to this first PLC example.

PLC EXAMPLE 2. Now suppose that the track arrangement is modified by adding two proximity sensors in the track that will send a voltage to the PLC when the train passes over them and will send no voltage when the train is *not* over the

FIGURE 8.22 Ladder diagram for PLC Example 1.

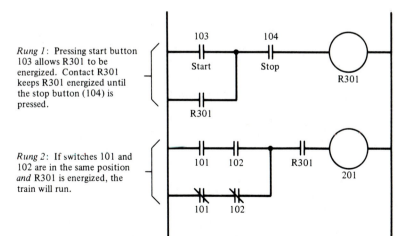

Rung 1: Pressing start button 103 allows R301 to be energized. Contact R301 keeps R301 energized until the stop button (104) is pressed.

Rung 2: If switches 101 and 102 are in the same position *and* R301 is energized, the train will run.

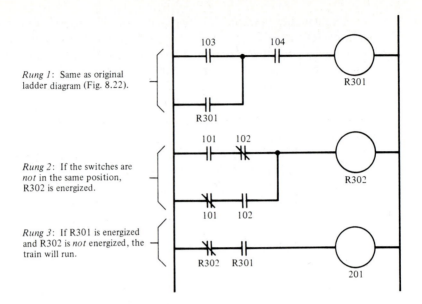

Rung 1: Same as original
ladder diagram (Fig. 8.22).

Rung 2: If the switches are
not in the same position,
R302 is energized.

Rung 3: If R301 is energized
and R302 is *not* energized, the
train will run.

FIGURE 8.23 Alternative solution to PLC Example 1.

sensor. This is shown in Figure 8.24 with sensors A (105) and B (106). As a point
of interest, these sensors are magnetically operated relays that are activated by a
magnet located at the front of the train's engine. The PLC is to control the train as
in PLC Example 1 with the addition of stopping the train when it has gone over
sensor A four times in a row without crossing sensor B *or* sensor B four times in a
row without crossing sensor A.

Needless to say, some way of counting is needed. With every counter, four
functions are desirable:

1. Some identification number is needed. A PLC might have as many as 50 or 100
   counters. We will use C1 through C16 to identify counters.
2. When should the counter count, and does it count up or down?

FIGURE 8.24 Addition of two sensors for PLC Example 2.

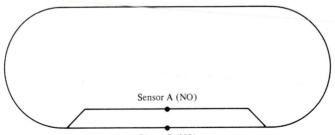

Sensor A (NO)

Sensor B (NO)

3. When should action be taken based on a counter's value? This is a *decode* function.

4. When should a counter be reset to some value (usually 0), and when should it be allowed to count? This is called *reset/enable*.

We will assume that at least three rungs are needed for each counter operation: an identification/count rung, a reset/enable rung, and a decode rung. Figure 8.25 shows such a counter structure. The Square D counters can count to 999, so if the counter is reset to 0 and a subtraction is made before an addition, the result is 999. It is possible to reset counters to values other than 0, but the reader should obtain PLC manuals if problems more complex than those given in this section are to be set up.

The ladder diagram for the second train example is given in Figure 8.26. The rung logic should be checked carefully to ensure understanding of the result. For example, at rungs 4 and 5, the sensor A counter is reset when sensor B is crossed

FIGURE 8.25   Typical counter ladder logic.

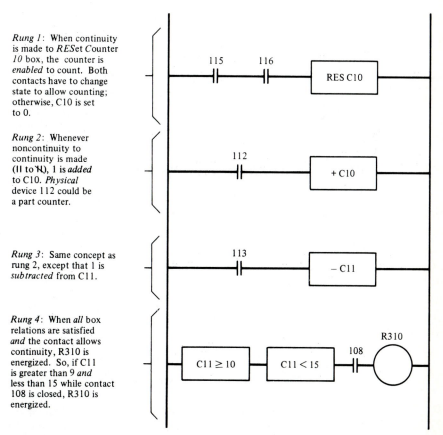

Rung 1: When continuity is made to *RES*et *C*ounter *10* box, the counter is *enabled* to count. Both contacts have to change state to allow counting; otherwise, C10 is set to 0.

Rung 2: Whenever noncontinuity to continuity is made (ǀǀ to ᴎ), 1 is *added* to C10. *Physical* device 112 could be a part counter.

Rung 3: Same concept as rung 2, except that 1 is *subtracted* from C11.

Rung 4: When *all* box relations are satisfied *and* the contact allows continuity, R310 is energized. So, if C11 is greater than 9 *and* less than 15 while contact 108 is closed, R310 is energized.

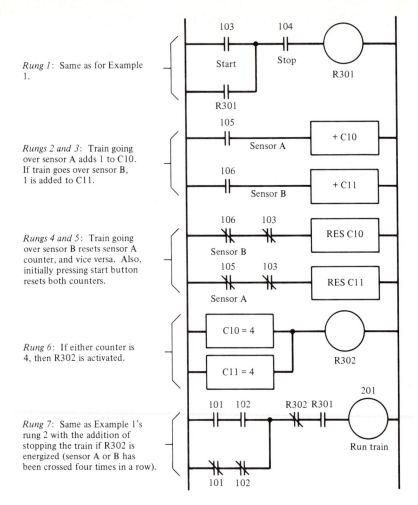

Rung 1: Same as for Example 1.

Rungs 2 and 3: Train going over sensor A adds 1 to C10. If train goes over sensor B, 1 is added to C11.

Rungs 4 and 5: Train going over sensor B resets sensor A counter, and vice versa. Also, initially pressing start button resets both counters.

Rung 6: If either counter is 4, then R302 is activated.

Rung 7: Same as Example 1's rung 2 with the addition of stopping the train if R302 is energized (sensor A or B has been crossed four times in a row).

FIGURE 8.26   Ladder diagram for PLC Example 2.

to ensure that decode is accomplished only if a sensor is crossed four times consecutively. For the same rungs, when the start button (NO) is pressed, contacts 103 (⊣⊢) are opened momentarily, allowing the counters to be initially reset. The *count* and *reset* boxes should be interpreted as *outputs*—only one (+C11, for example), per ladder diagram.

PLC EXAMPLE 3. In addition to the functions taken care of in PLC Examples 1 and 2, now assume that if the train takes longer than 25 s to go around the track, it is assumed that derailment has occurred and that train power should be cut off. A timer is now required to check the time between sensors A to B, A to A, B to A, or B to B, since we do not know when the train will take the inner or outer loop.

Timers can be treated exactly like counters. In fact, the Square D controller uses counters and timers interchangeably; a timer will *count* in tenths of a second or in seconds (with a maximum time of 99.9 or 999 s, respectively). As with counters, we will need an identification rung, a reset/enable rung, and a decode rung. Such a general situation is shown in Figure 8.27. If a contact is put in front of rung 1, then when that contact is open (‖), the timer is *not* identified. It is identified only when *all* inputs allow continuity. For example, we could have

T10 would be defined only if all three inputs allow continuity: C10 greater than a count of 15, T5 less than 3 s (if defined as timing in second increments), and contact 113 allowing continuity (⫫). The ladder diagram to allow train derailment to be caught is given in Figure 8.28.

PLC EXAMPLE 4. Suppose that a robot-assembly area is protected by a gate system as shown in Figure 8.29. Opening a gate will open an NC limit switch, thus

FIGURE 8.27   Typical timer rung structure.

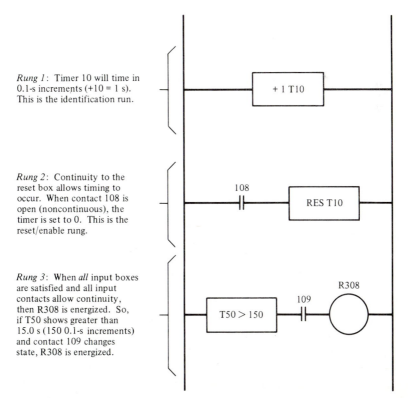

Rung 1: Timer 10 will time in 0.1-s increments (+10 = 1 s). This is the identification run.

Rung 2: Continuity to the reset box allows timing to occur. When contact 108 is open (noncontinuous), the timer is set to 0. This is the reset/enable rung.

Rung 3: When *all* input boxes are satisfied and all input contacts allow continuity, then R308 is energized. So, if T50 shows greater than 15.0 s (150 0.1-s increments) and contact 109 changes state, R308 is energized.

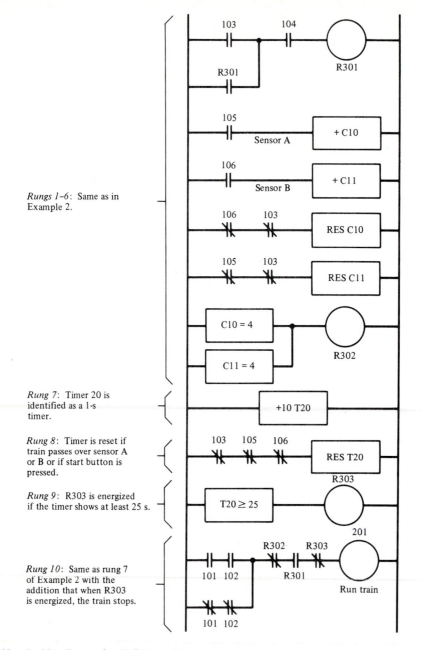

Rungs 1–6: Same as in Example 2.

Rung 7: Timer 20 is identified as a 1-s timer.

Rung 8: Timer is reset if train passes over sensor A or B or if start button is pressed.

Rung 9: R303 is energized if the timer shows at least 25 s.

Rung 10: Same as rung 7 of Example 2 with the addition that when R303 is energized, the train stops.

FIGURE 8.28 Ladder diagram for PLC Example 3.

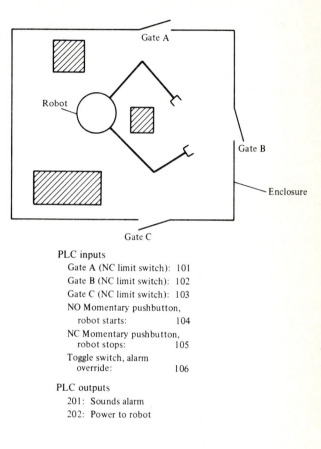

PLC inputs
Gate A (NC limit switch): 101
Gate B (NC limit switch): 102
Gate C (NC limit switch): 103
NO Momentary pushbutton,
    robot starts: 104
NC Momentary pushbutton,
    robot stops: 105
Toggle switch, alarm
    override: 106

PLC outputs
201: Sounds alarm
202: Power to robot

FIGURE 8.29   Robot-assembly problem.

sending no voltage to the PLC. When any gate is opened, an alarm is to be sounded (10 s on and 5 s off). Even though the gate is closed (the person opening the gate might still be *inside* the fence), the alarm is to continue until a toggle switch cuts it off (this is also an override switch so that maintenance can be accomplished on the robot without an alarm going off). A possible diagram solution is given in Figure 8.30.

8.6.1   The Future of PLCs

This section has only *introduced* the capabilities of PLCs. We have assumed that the PLC program is in the form of a ladder diagram and that all rungs within the diagram are scanned once per program cycle. Also, we assumed only digital I/O, although analog I/O is found in more and more applications. The PLC software interface allows changing of input to output relationships (sensor to actuator) *without* having to change expensive hardwiring between equipment. The ladder diagram concept makes programming easy to accomplish and easy to change.

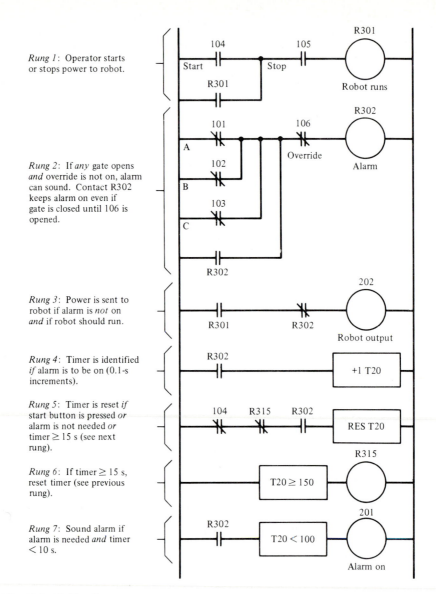

Rung 1: Operator starts or stops power to robot.

Rung 2: If *any* gate opens *and* override is not on, alarm can sound. Contact R302 keeps alarm on even if gate is closed until 106 is opened.

Rung 3: Power is sent to robot if alarm is *not* on *and* if robot should run.

Rung 4: Timer is identified *if* alarm is to be on (0.1-s increments).

Rung 5: Timer is reset *if* start button is pressed *or* alarm is not needed *or* timer $\geq$ 15 s (see next rung).

Rung 6: If timer $\geq$ 15 s, reset timer (see previous rung).

Rung 7: Sound alarm if alarm is needed *and* timer < 10 s.

FIGURE 8.30   Robot ladder diagram.

Also, adding rungs allows addition of parallel control functions if necessary, assuming that the basic program cycle is satisfactory for all functions accomplished within the total ladder diagram. Extra rungs will, of course, increase the cycle time, though this increase will generally be less than a millisecond for the worst-case application. It should be realized that PLCs today have much greater capability than presented in the preceding material. The concept of ladder diagramming and its role in cyclic controllers is the basic knowledge required for an analyst to tackle more recent innovations in PLCs, possibly including the use of a programming language more akin to high-level BASIC than the programming of a ladder diagram on a CRT. The high-level language will still utilize the boolean logic relationships among input information in order to determine output requirements.

What, then, is the future for PLCs? Jesse Quatse presented some possibilities in an article in *Control Engineering* [9].

- *Operating system.* An operating system will be available that requires real-time and ''slow''-time services. In fact, this is available in some recent PLC releases. Rungs within the ladder diagram may require widely differing response times; the operating system of the future will allow this to be realized in an efficient manner.

- *Languages.* Even though ladder diagram logic may be user-friendly for the technician, it might not be for the engineer who has been brought up with high-level computer languages as an integral part of the engineer's toolkit. In fact, senior-level students tend to have a hard time realizing the benefits that accrue from the ladder diagram concept. Quatse gives typical examples of how a high-level language might work:

    START T6 might initialize and start a timer numbered 6.

    IF T6, DONE THEN would be one form of testing the timer state.

- *Network compatability.* Programmable controllers will become MAP (manufacturing automation protocol, discussed in Chapter 12) nodes in the future, thus facilitating communication among all controlled equipment in a facility as well as integrating the control structure itself.

- *Programming center.* Most PLCs today are still programmed through portable hand-held pendants or small CRTs. The trend for the future has microcomputers (predominantly IBM PLCs or compatibles) emulating the programmable controller to allow the analyst to write the program and debug it before downloading to the controller itself. Anyone who has tried to debug a large ladder diagram program will see the advantages of this programming tool.

Even though it might seem that the PLC and computer capabilities in CAM are merging into one device, a microcomputer with a real-time operating system, there still exists a need for a relatively inexpensive cyclic controller in almost countless manufacturing applications. The PLC fills that need. The ''future'' capabilities of PLCs will make them more user-friendly, faster, and more integrative devices. They will be an indispensable component of the computer-aided manufacturing facility for a long time to come.

8.7 SUMMARY

The digital computer permeates all components of computer-integrated manufacturing operations. At the top of the manufacturing hierarchical structure we find overall operation planning. At the bottom is physical control: machine tool control, conveyor control, robot control, and so on. This chapter has discussed some fundamentals of computer control that fall at the bottom of this computer hierarchy.

Inherent in any form of computer control is the sequence of determining process status by data input followed by data analysis and then data output to the process. Data input considered was digital input for binary status definition as well as analog input for voltage signals. Similarly, outputs to the process that were considered were digital output and analog output. A quite common output that was not covered is pulse output, where voltages are output at a very fast rate for control of such devices as stepping motors (see Chapter 11). A programmable controller, which is itself a simplified control computer, can be used to generate pulsed outputs if needed. Programmable controllers, in this chapter, were treated as digital input/output controllers with timing and counting capability, though it should be realized that PLCs can be used with analog inputs and outputs, especially in process control functions.

In addition to process communication, the ability to handle control with extremely tight *timing* constraints is a feature that frequently distinguishes control computers from general-purpose computers. These times might well be in the millisecond (or even microsecond) category. To aid in the complex programming requirements forced by real-time control applications, aids such as multilevel priority interrupt as well as multitasking operating systems are recommended. Real-time multitasking systems are becoming more and more available to the CAM application engineer, even with microcomputer systems. However, a word of caution: We have just touched on an introduction to control computers (and programmable controllers), so the applications engineer is well advised to delve further into the literature in these areas. The result will be well worth the effort expended.

EXERCISES

1. Within a field in which you are interested (manufacturing engineering, industrial engineering, airline operations, etc.), perform a literature search to determine how well control computers have been applied in that field. Suggest one or two new applications.

2. Perform a literature search to determine recent advances in programmable controllers, especially with relation to control of manufacturing operations. For each application that you report on, comment on the feasibility of microcomputers doing the control and why programmable controllers might be better.

3. Realizing that you are probably not a pilot, and if you are a pilot that you have never landed an airplane on an aircraft carrier, suggest how a control computer might be used in the automatic landing of a plane on a carrier.

4. Perform the following integer number conversions:
   (a) $(316)_{10}$ to octal
   (b) $(316)_{10}$ to binary
   (c) $(3574)_8$ to decimal
   (d) $(1101100111101)_2$ to decimal

> (e) $(453)_{10}$ to a number with a radix of 6
>
> (f) $(4DEF)_{16}$ to decimal

5. Perform the following fractional number conversions:

    (a) $(0.4632)_{10}$ to octal

    (b) $(0.4632)_{10}$ to binary

    (c) $(0.357)_8$ to decimal

    (d) $(0.110111011)_2$ to decimal

    (e) $(.35)_{10}$ to hexadecimal

6. Using Table 8.1 only, perform the following integer number conversions:

    (a) $(325)_8$ to binary

    (b) $(101110110111)_2$ to hexadecimal

    (c) $(5FD3)_{16}$ to octal

7. Using Table 8.1 only, perform the following fractional number conversions:

    (a) $(0.C3D)_{16}$ to binary

    (b) $(0.6423)_8$ to binary

    (c) $(0.1101101101)_2$ to hexadecimal

8. Using the philosophy of Exercises 6 and 7, convert $(32123.2103)_4$ to binary, octal, and hexadecimal.

9. What is the resolution of a 4-bit A/D convertor if the convertor can handle an input voltage with a range of +20 to −20 V? What if we had a 20-bit convertor?

10. Using the A/D convertor logic given in Figure 8.5, determine the digital count resulting from conversion if the input voltage is 11.3 V, the input range is 0 to 20 V, and the convertor register is 10 bits.

11. Using the information in Exercise 10, determine the percentage error in voltage conversion as a function of conversion register size (assume that this size can be 10, 12, 14, 16, 18, or 20 bits). For each register size, first determine the conversion count and then use Eq. (8.7) to find the equivalent voltage.

12. Assume that we have a resistive D/A convertor (similar to Figure 8.6) that has electronic switch positions d_{-1} through d_{-8}. What would be a logical numeric value for $(I)(R)$ in Eq. (8.11)? (The output voltage range is 0 to 10 V.)

13. Digital input by computer assumed an input CALL of

```
CALL DINP(INPORT, IMASK, IVAL)
```

We have a 16-bit input register and want to check if bits 3 and 13 are *both* 1s. Develop an IMASK value to allow this and show how one IF statement could then check if both bits are set to 1.

14. Suppose that we have an 8-bit input register that is used in monitoring eight on/off functions on a machine tool (bearing hot, motor on, access door open, etc.). Using the DINP CALL in a subroutine, write a simplified program to allow the eight functions to be checked.

15. Suppose you are an analyst who is designing a microcomputer application for monitoring patient status in a critical-care recovery room. Suggest possible digital input characteristics that might measure patient status (i.e., breathing or not breathing). What type of sensor might capture this information?

16. A computer program has to send digital outputs through bit positions 3, 6, and 7 in an 8-bit output word depending on analyzed conditions. The remaining five bits are con-

trolled by another subroutine. Suggest possible output mask values that might be feasible for use with a DOUTP CALL. Explain when these mask values might be used.

17. Write a simplified computer program to allow digital input and output to occur in the following manner with 8-bit input and output registers:
    (a) If input bits 3 and 5 are both 1s, output a 1 through bit 6.
    (b) If input bits 1, 6, and 7 are all 1s, output 1s through bits 5 and 7, and a 0 through bit 6.
    (c) If *all* input bits are 0s, output 0 to bits 5, 6, and 7.
    In your program, use one subroutine for digital input and another subroutine for digital output.

18. A microcomputer is to be used for safety monitoring and control in a home. Suggest typical digital output applications that might be feasible and actuators that might be used in these applications.

19. Write a variable-gain analog input program to sequentially input values through input channels 3 and 6 every 0.2 s (i.e., input two values, take action, and repeat 0.2 s later). The A/D convertor handles input voltages between 0 and +10 V and is an 8-bit convertor. If the input voltage for either channel is less than 3 V, then increase the gain and input the signal again (feasible gains are 1, 2, and 4). Convert the input count to true volts and print the result (use something like: 110 PRINT RESULT). If the amplified voltage is *never* greater than 3 V for *any* channel, print an error message and stop.

20. A digital input value (8-bit input register) is to be used as the count value for a D/A conversion. The convertor is an 8-bit register and the output voltage is in the range 0 to 10 V.
    (a) Suppose that input bits 0, 3, 6, and 7 are 1s and the rest are 0s. What would be the output voltage?
    (b) Write a simplified program to allow output of any voltage in the range, depending on the digital input values.

21. A timer with a programmable controller can only time to 999 s if in the 1-s-increment mode. Draw a ladder diagram to show how a 30-min delay might be allowed in a particular program.

22. Three momentary pushbuttons (NO) on a milling machine console are connected to a programmable controller. Switch functions are as follows:

    • When switch A is pressed, it indicates start of a batch run.
    • Switch B, when pressed, indicates completion of a part.
    • The end of a batch run (several parts) is indicated by pressing switch C.

    Draw a ladder diagram to allow the controller to:
    (a) Count the parts during a batch run.
    (b) Stop the machine at the end of a batch run.
    (c) Time the length of the batch run (assume that it is never longer than 999 s).

23. A building has five entry doors, each with a security relay attached that is closed when the door is closed. The five relays are inputs to a programmable controller. If a door is opened at night, a noise alarm is to sound—cyclic for 5 s on and 3 s off. The alarm can be turned off only by a security guard who is notified of a problem by a light that comes on in her office (the light is on constantly until the alarm is turned off by the guard). Draw a ladder diagram to handle this situation.

24. A control computer (CC) and a programmable controller (PLC) are in a hierarchical structure that allows the following:

   (a) If the CC sends out a 1 through digital output bit 0, the PLC is to start a model train. The digital output closes a relay that allows a voltage to go to the PLC input.

   (b) If the train takes more than 35 s to go through a complete track loop, the PLC is to notify the CC by a voltage output that closes a relay which is "read" by the CC through digital input bit 3 (which is 1 if the PLC sends an output).

   (c) If the CC is notified that the train has spent more than 35 s in a loop, the CC stops the train and prints some appropriate message.

   (d) If loop times are always less than 35 s, the PLC stops the train after 15 loops.

   Draw a PLC ladder diagram to handle the PLC position of control and develop a simplified computer program to handle the CC functions.

25. A programmable controller is to control the north–south (NS) red–green lights for an intersection traffic light as follows:

   (a) If NS is red and one or fewer cars arrive in the *north* lanes, red is maintained for 20 s (obviously, green in east–west).

   (b) A proximity sensor allows the PLC to count the cars arriving in the *north* lane only. If two or more cars arrive while the light is red, then it is immediately changed to green *if* it has been red for at least 10 s. Otherwise, after 20 s red, it is changed to green. Draw a ladder diagram to handle this situation.

26. A multilevel priority interrupt system has three interrupt levels—A, B, and C—with C being the lowest priority and A the highest. Three tasks are affected by the interrupt system:

Task	Task duration (s)	Priority
1	6	C
2	2	B
3	5	A

   If task 3 is started by an interrupt every 8 s and task 2 is similarly started every 10 s, is there any limitation on how often task 1 might have interrupted *starts*? Show all your work developed in arriving at your answer.

27. A multilevel interrupt system has the same interrupt levels as given in Exercise 26. Now we have seven tasks affected:

Task	Task duration (s)	Time of interrupt	Priority
1	6	3	B
2	2	6	C
3	5	15	A
4	8	12	B
5	10	8	B
6	4	22	C
7	7	18	A

   If times of interrupt are after *start* of a computer run, determine the start and complete time for each task relative to the computer run start time.

28. A single-level priority interrupt system says that all tasks have the same priority. What would be the start and completion times for the seven tasks in Exercise 27 if we now have a single-level interrupt system?

REFERENCES AND SUGGESTED READING

1. Asfahl, C. R.: *Robots and Manufacturing Automation*, John Wiley, New York, 1985.
2. Cassell, D. A.: *Microcomputers and Modern Control Engineering*, Reston Publishing, Reston, Va., 1983.
3. Data General: *Real Time Input/Output System User's Manual*, 093-00095-02, Data General, Southboro, Mass., May 1975.
4. Data Translation, *User Manual for PCLAB*, Data Translation, Marlborough, Mass., 1983.
5. Groover, M. P.: *Automation, Production Systems, and Computer-Integrated Manufacturing*, Prentice-Hall, Englewood Cliffs, N.J., 1987.
6. Ledley, R. S.: *Digital Computer and Control Engineering*, McGraw-Hill, New York, 1960.
7. Mellichamp, D. R., D. D. Bedworth, O. Petterson, P. Rony, L. Bezanson, W. Higgins, and G. Korn: "Real-Time Computing and the Engineering Support System," *IEEE Micro*, vol. 5, no. 5, October 1985.
8. Petterson, O.: "Advanced Real-Time Operating System ARTOS," Internal Working Paper, Chemical and Nuclear Engineering Department, University of California at Santa Barbara, February 1985.
9. Quatse, Jesse T.: "Programmable Controllers of the Future," *Control Engineering*, vol. 33, no. 1, January 1986.
10. Rembold, U., M. K. Seth, and J. B. Weinstein: *Computers in Manufacturing*, Marcel Dekker, New York, 1977.
11. Square D Company: *SY/MAX 20 Programmable Controller Technical Manual* (Bulletin M-553), Square D Company, Milwaukee, Wisc., April 1981.
12. Vail, P. S.: *Computer-Integrated Manufacturing*, PWS-Kent Publishing, Boston, Mass., 1988.

MANUFACTURING CONTROL— NUMERICAL CONTROL

Mr. Wilkinson has bored us several cylinders almost without error; that of 50 inches diameter, which we have put up at Tipton, does not err the thickness of an old shilling in any part.

Mathew Boulton (in 1776) [quoted in 12]

The introductory quotation refers to a cylinder-boring device invented in 1774 by John Wilkinson. This has been claimed to be the first machine tool in the modern sense of the word, and it was Wilkinson's invention that allowed James Watt to build a full-scale steam engine [12]. In fact, the introductory quotation was made by Watt's partner, Mathew Boulton.

Machining within a tolerance of "the thickness of an old shilling" would be a pretty sad state of affairs today, but the passage of 200 years has obviously seen tremendous advances in all facets of human endeavor. The most dramatic advances in automation of machine tool operation have occurred only within the last 30 to 40 years; without these advances the concept of computer-integrated manufacturing would be just a glimmer in a futurist's eye. In fact, not only was the advent of numerical control in the 1950s the first major move into what is now computer-aided manufacturing, the development of the processing language APT, which facilitated programming of NC tools, could well be considered the precursor of today's modern computer graphics and computer-aided design.

Numerical control refers to the use of coded numerical information in the automatic control of equipment positioning. For machine tools this might refer to the motion of the cutting tool or the movement of the part being formed against a rotating tool as well as changing cutting tools. Positioning and inserting electronic components in a circuit board line may also be accomplished by numerical control. Another example is the process of laying composite material to form lightweight alternatives to machined metal parts. Even though the emphasis in this chapter will be on numerical control for machine tool operation, it should be realized that the application spectrum is far broader.

There are several acronyms for numerical control, but the one most commonly used, not very surprisingly, is NC. Others are CNC, for computer numerical control, and DNC, which frequently has two meanings: direct numerical control or distributed numerical control. The CNC and DNC concepts will be deferred until Sections 9.5 and 9.6; the initial emphasis will be on general NC concepts: the role of NC in integration, typical NC equipment, functions, and advantages, as well as a computer language for facilitating NC.

9.1   THE ROLE OF NC IN INTEGRATION

The NC process can be presented as in Figure 9.1. First, an engineering design of a part is developed. This is often received by the manufacturing planner in the form of a blueprint. A manufacturing plan for the part production is then made. Typical information includes machine(s) that will be used to produce the part, sequence of operations that will need to be accomplished on the machine(s) to produce the part, time estimates for setup and production so that schedules can be developed, and tooling and raw material requirements. An NC analyst then programs the necessary geometry and motion statements that will be processed by a general-purpose NC language such as APT (Automatically Programmed Tools) with resultant cutter location data (CLDATA) being developed. The CLDATA is further processed with a *postprocessor* to tailor the information to the characteristics of a specific machine tool, since machine tools are not generic with regard to dynamic characteristics of the machine. Typical characteristics include control aspects such as interpolation and coolant flow that augment the CLDATA, which is basically just a sequence of coordinate information about the cutting tool location. Finally, numerical instructions are input to the machine tool, commonly through a tape medium.

Figure 9.2 shows how NC might fit into an integrated CAD/CAM structure. The left-hand path will be considered first. Boxes I through V represent an integration methodology developed at Arizona State University by Li [8]. This methodology was tested using IBM's CADAM design package with United Technologies' CMPP generative process planning software.

First, an engineering design is accomplished with a CAD package. Most design packages can put the design information into the standard IGES (Initial Graphics Exchange Specification) file format [5]. This allows a generic *feature-recognition algorithm* to be developed. The purpose of the algorithm is to translate CAD language information into CAPP information, as shown in Figure 9.3.

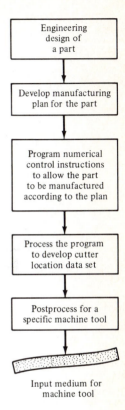

FIGURE 9.1   The basic NC process.

The generic part definition data structure (GPDS) takes the CAPP information as well as all other data necessary to describe a part completely (general information, coordinate dimensions and tolerances, geometric tolerances, and raw material information) and puts it into a neutral file structure accessible by any CAPP language.

The GPDS information goes into the CAPP input data file using the specific CAPP input formats. A process plan is then realized, as discussed in Chapter 6.

The CAPP route allows one to process a part through several machines, a common cell manufacturing situation. The right-hand path in Figure 9.2 shows how single-machine production could be handled. Many design packages have an associated NC processor that can evolve CLDATA for the part and put the CLDATA into an IGES structure. This can then be postprocessed and downloaded directly to the machine tool. In fact, even though it might not be the most judicious use of computer facilities, it is feasible to go directly from the CAD computer to the machine tool. Section 9.4 will discuss computer languages for NC in light of this integration framework.

Figure 9.4 shows a schematic of a manufacturing cell layout for automatic production of spur gears. While this example was utilized in Chapter 5, it can summarize the integrative role of NC.

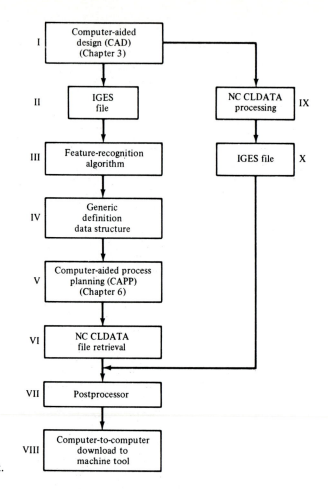

FIGURE 9.2 NC in a CAD/CAM environment.

FIGURE 9.3 CADD-to-CAPP language conversion.

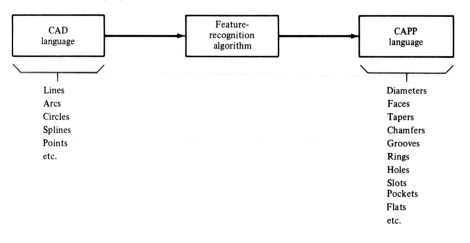

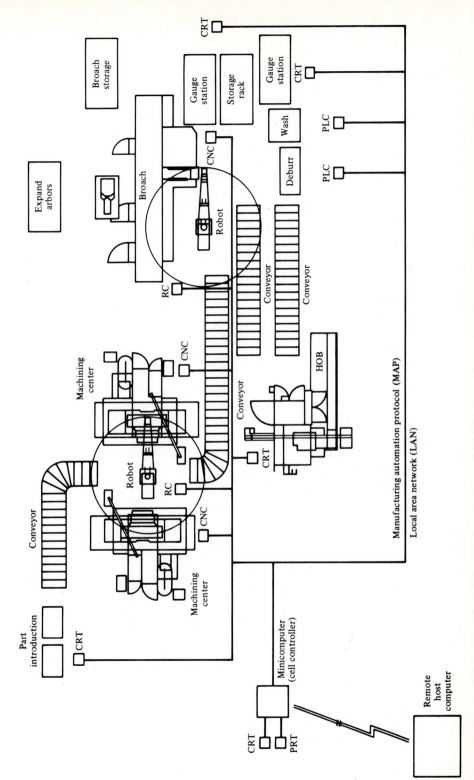

FIGURE 9.4   Typical manufacturing cell layout. (Courtesy of Garrett Engine Division of Allied-Signal Corporation.)

On a shift or daily basis, the remote host computer downloads to the cell controller numerical control–processed data for the parts to be processed for that time period. When each part (or batch of parts) is ready to be processed, the cell controller downloads the appropriate NC information to the manufacturing equipment's memory unit. As will be discussed in Section 9.3, the manufacturing planner and machine operator will still be a vital cog in the NC process, even though it is *theoretically* possible to go directly from CAD to the machine tool. In general, NC programming by an analyst will be a major requirement in the NC process for many years to come. Upon completion of the part(s), setup will be accomplished for the next part and the appropriate NC information will be downloaded. Only those programs needed for a logical production time (shift, day) will be held by the cell controller at any one time. Without such a manufacturing philosophy, complete cell automation would not be feasible.

9.2 KEY DEFINITIONS

Several terms used in this chapter are specific to numerical control computer processors and will be fully defined as they are introduced. This section introduces more generic terms relative to numerical control.

Adaptive Control   Control of a process that responds to changes in the manufacturing operating conditions. Typical conditions include tool dulling, temperature increases, and vibration changes.

APT   A software compiler (Automatically Programmed Tool) for simplifying numerical control programming. APT is the most widely used processor for this purpose.

Canned cycle   A series of instructions prestored in a machine's computer memory that perform a series of machine operations. Typically associated with *computer* numerical control (see CNC).

CL data   Cutter location data that is generated by a numerical control computer processor. CL data represents position coordinates for the tool performing the cutting.

CNC   An acronym for *computer numerical control*, where a computer (usually a micro) is part of the controller and is dedicated to a single CNC machine.

Continuous cutting   Machining process in which removal of material is continuous rather than at discrete points. Milling is a typical continuous process.

Controller (MCU)   Refers to the actual device for controlling the motions of a machine tool. The controller may or may not include a dedicated computer.

Direct numerical control   See DNC.

Distributed numerical control   See DNC.

DNC   There are two meanings for this acronym. *Direct* numerical control refers to the original time-sharing use of a mainframe computer for transmitting motion information to the machines being controlled. The more common application is now *distributed* numerical control, in which a host computer can transmit entire part motion sequences to the machine tool's CNC memory.

Hardwired control Refers to a controller that has control effected by somewhat inflexible electronic components as contrasted to a controller that is programmable.

Interpolated motion Controller function that translates motion between two points into linear, circular, parabolic, and maybe cubic polynomial shapes.

Macro A set of computer instructions that can be efficiently represented by a single command; for example, the drilling of several points.

Numerical control The use of numerical data for the control of operations such as complex machining. The numerical data is generated through a computer and is input to the machine by a tape medium (conventional NC and CNC) or directly by computer (CNC or DNC).

Part definition In this chapter, the machining analogy of design geometry.

Point-to-point Machining operations that are performed at discrete points; typical are drilling, boring, tapping.

Postprocessing The computer procedure that translates generic CL data for a specific machine tool.

Processing language A computer language that greatly facilitates the generation of CL data for a complex part; for example, see APT.

Retrofit In this chapter, upgrading a conventional numerical control machine to CNC or DNC.

9.3 NC OPERATION AND EQUIPMENT

Before talking about computer programming languages that make part development easier for the manufacturing planner, it will be well to spend a little time on the operation of the NC function as well as introduce typical NC hardware. Following the subsequent section on NC software, some further comments will be made relative to such hardware topics as NC controls.

Even though numerical control evolved through metal-cutting operations, the NC concept is now applied to a much broader spectrum: grinding, sheet metal press operations, welding, flame cutting, tube bending, riveting, assembling, inspection, wire cabling, and others. This chapter emphasizes metal cutting, but it should be recognized that the *concepts* are just as valid for all applications.

It is very important to realize that NC is not a machining method [10]. Rather, it is a concept of machine control. Even with computer-aided process planning or even the potential direct interface of a solid model design to machine tool cutting, there will still be a role for the NC analyst and machine operator in most NC applications. While there are some relatively rare examples of complete manufacturing facilities that produce varieties of parts automatically, full production and assembly automation of a complex product is a remote dream even to the most optimistic. Consider, for example, the manufacture of a turbine engine. One such facility has some 14,000 shop orders in its manufacturing lines at any time competing for about 1000 machines and other resources. Further, each part may require up to 50 operations on a variety of the machines, the sequence of which is determined by the manufacturing requirements for the part. Group technology and computer-aided process planning can aid in the automation of segments but not all of such a facility. It therefore seems logical at this time to comment on the analyst's and operator's roles in NC.

9.3.1 The Analyst's Role in NC

The manufacturing analyst determines how a part can be produced in an economical manner consistent with customer due dates and other requirements. As discussed earlier in Chapter 6, this information involves translating geometric design information into a sequence of machining operations while determining tooling requirements for the machines. In addition, cutting feeds and speeds have to be specified, with resultant time requirements being evolved for each production step. Granted, modern CAD packages often have an ancillary NC processor that allows generation of cutter location data *directly* from the design. The analyst still has to ensure that correct tooling will be available and that the assigned machines, sequence of machines, and machining parameters are correct.

Automatic generation of CL data is not prevalent in most production situations. Even if it is, as with some CAD packages, the NC analyst usually has to stipulate machine actions such as cutting paths, coolant flow characteristics that depend on the stage of machining, and so on. These may be programmed into an NC machine through the machine control unit (MCU), which converts the instructions into the required machine tool actions [16]. An NC software compiler can be used to generate CL data and associated machining parameter conditions efficiently. Such a compiler is APT (Automatic Programmed Tool), which will be discussed in some detail in Section 9.4. It might be mentioned at this time that APT is the predominant NC compiler that is ancillary to many CAD packages.

9.3.2 The Operator's Role in NC[1]

The operator's duties can be divided into three specific areas:

1. Setup and program input
2. Running and monitoring
3. Editing, adjusting, and diagnostics

Setup and program input includes such functions as:

1. Placing work-holding fixtures on the work table
2. Assembling perishable tooling (drills, mills, reamers, etc.) into tool holders and then loading them into the NC machine's tool magazine.

---

[1] Appreciation is tendered to Randy Wiemer of the Garrett Turbine Engine Company for this and other segments of this chapter.

3. Jogging the machine table so that the spindle is aligned with the fixture. In turn, this permits the part program to be "aligned" with the part in the fixture.

4. Reading the program into the machine control unit, usually with 1-in.-wide tape. When there is a direct computer-to-machine interface (see DNC in Section 9.3.5), the operator is responsible for setup and subsequent notification to the computer facility that the machine is ready to receive the machining operations.

5. Dry-running a noncutting tryout cycle at a controlled feed rate (without a part in the fixture) to debug the program. If no motion errors are detected, the part can be loaded and machining can commence. Often the operator may be considered to be on setup until the first part is verified by the inspection department.

Running and monitoring operator tasks include such functions as the following.

1. During the actual running of the part, the operator often *senses* problems through sight and sound.

2. Other problems may be determined through use of the machine's control unit CRT screen. The screen often has the ability to show the part program as well as to display such events as the current axis locations (absolute and relative) as well as the distance remaining before the current commanded position is attained. Other monitoring devices register the feed/speed settings as well as the load on the spindle. This latter information is often used to monitor tool life, since a dull or damaged cutting tool requires considerable horsepower consumption to drive through the material being cut.

3. Alarms and error codes may also appear on the screen, allowing immediate corrective action to be feasible.

4. If, for some reason, the operator must override the NC controls, a freeze-hold button can be pressed to freeze all axis movement. In the worst case, a large emergency stop button can be pressed to shut everything down.

Editing and adjusting is the third operator function to be considered. *Flexibility* is a major advantage for NC machines. On many such machines the operator can make "offset" adjustments to the tooling and fixture. For example, the operator can "offset" the tool down using a tool compensation feature if the cutting tool does not remove sufficient stock on the finish cut. Similarly, cutter compensation permits economic utilization of undersized or reground cutters.

The operator may be allowed to make some changes to the program displayed on the CRT screen. Usually these changes will be limited to adjusting feeds and speeds, but they could include extensive tool reprogramming to correct for tools that are causing problems.

The editing capability can be abused by the operator, needlessly delaying a production run while editing to satisfy operator preference (operators often have their favorite feeds and speeds at which they like to run). On the other hand, time spent in eliminating wasted motions and maximizing feeds and speeds can be time well spent, especially when long production runs are required.

The remainder of this section will give information relative to the hardware characteristics of NC. It might at first reading seem to be a little disjointed. However, for the reader wanting an introduction to NC concepts these are all topics that should be considered. Some topics will be introduced in this section and then discussed further in the subsequent controls section. They are introduced at this time so that the discussion of software will be more meaningful.

9.3.3 Axes in NC Operations

The standard six axes of motion for NC are shown in Figure 9.5. One's right hand is used to depict the positive direction for the main x, y, and z axes. The other three prime axes of motion are rotation about the x, y, and z axes. It is somewhat unusual to have a machine with as many axes of motion as six, though NC standards have defined 14 possible axes of motion (no machine will ever have all 14). Three examples of axis specification for specific equipment are given in Figure 9.6.

9.3.4 Types of NC Systems

The two major types of NC systems are point-to-point (PTP) and contouring. PTP is used when the *path* of the tool *relative* to the workpiece is not important, since

FIGURE 9.5   The six major axes for NC. (From Modern Machine Shop's *1986 NC/CAM GUIDEBOOK* [10], with permission.)

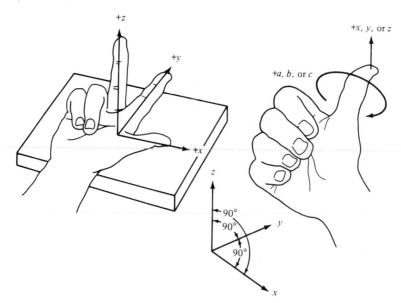

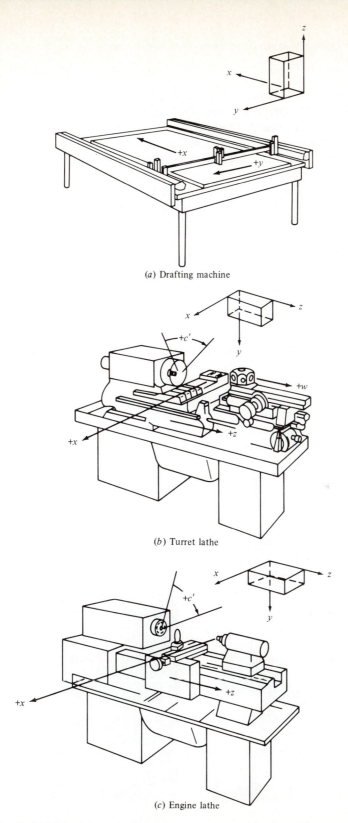

(a) Drafting machine

(b) Turret lathe

(c) Engine lathe

FIGURE 9.6 Typical NC equipment with respective motion axes. (From Modern Machine Shop's *1986 NC/CAM GUIDEBOOK* [10], with permission.)

the tool is not in contact with the work while traveling from one point to another. Typical PTP examples include drilling, punching, tapping, and component insertion in a circuit board. The specific *x, y* coordinates are of extreme importance in drilling, for example, but how to get from one hole to another is not so important other than that the travel distance should be held to a minimum to maximize production rates.

Contouring systems, often called *continuous paths*, are found on most NC lathes and milling machines when the motion of the tool relative to the part being machined is critical. Complicated shapes can be machined as the MCU controls the acceleration and deceleration of each machine axis involved in order to produce the required programmed shape. As an example, all that is normally required for the programming of an arc are the coordinate locations of the endpoints, the radius of the circle, and the coordinates of the circle's center.

9.3.5 NC/CNC/DNC

Before giving some specific examples of NC machine tools, it will be helpful to discuss briefly the terms NC, CNC, and DNC. Also, some advantages will accrue if the terms are introduced before discussing some of the software aspects of NC.

The original numerical control machines were referred to as NC machines. These had "hardwired" control, whereby control functions such as motion interpolation were accomplished through the use of inflexible physical electronic components. Punched-paper-tape readers constituted the way to get the motion instructions to the controller, with no memory capability to allow the program to be resident within the MCU after loading.

FIGURE 9.7 CNC lathe. (Courtesy of the Garrett Engine Division of Allied Signal Corporation.)

FIGURE 9.8   CNC machining center. (Courtesy of the Garrett Engine Division of Allied Signal Corporation.)

FIGURE 9.9   Tool changer mechanism for CNC machining center. (Courtesy of the Garrett Engine Division of Allied Signal Corporation.)

CNC, computer numerical control, refers to a machine that has a microprocessor as an integral component of the MCU. A program can thus be loaded into the MCU so that dependency on the tape reader for each part of a batch is eliminated. The microprocessor also facilitates editing of the program by the operator as well as providing diagnostic analysis with resulting information being sent to the operator. Motion interpolation is handled by "softwired" control capability allowed by the microprocessor, a much more flexible system than with the original conventional NC's "hardwired" control. A further advantage is the availability of preprogrammed cycles of machining commands, usually called *canned cycles*. The programming of such functions as hole tapping can be greatly simplified through the use of these cycles.

The development of CNC over many years, coupled with the development of local area networks, has evolved into the modern concept of distributed numerical control (DNC). An off-site host computer holds tool motion commands for all the parts to be produced in the DNC facility. This information is downloaded only when the machine that will do the work has been determined. Control information is thus *distributed* to machines in the most efficient manner possible. Obviously, the DNC concept is an extremely important key to computer integration of manufacturing.

FIGURE 9.10   NC point-to-point assembly system. (Courtesy of the Micro-Relle Company.)

9.3.6 Some Equipment Examples

Figure 9.7 shows a CNC lathe with the operator control panel and screen display in the right-center of the picture. The 10-tool turret uses five inside-diameter (I.D.) tools and five outside-diameter (O.D.) tools. The CNC controls handle two axes of motion, x and z.

A machining center is shown in Figure 9.8. A machining center can handle a wide variety of operations by using a tool changer mechanism (see Figure 9.9). Some tool changers can handle more than 100 different tools; the one pictured has a 35-tool magazine. Recent advances in technology have seen the development of cameras that can be mounted in a magazine slot. The camera can then be brought to the part, in the same manner as a tool would be brought to the part, and used to

FIGURE 9.11 Twin-spindle CNC vertical mill. (Courtesy of the Garrett Engine Division of Allied Signal Corporation.)

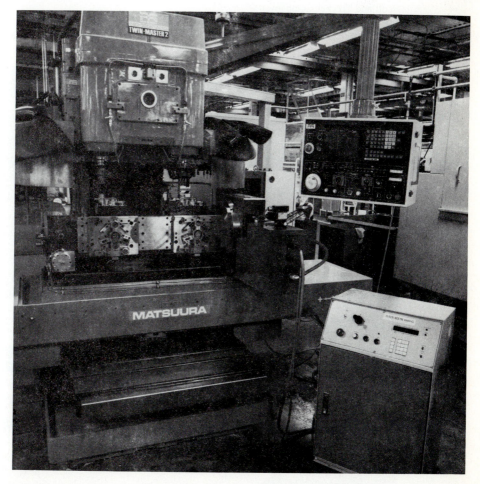

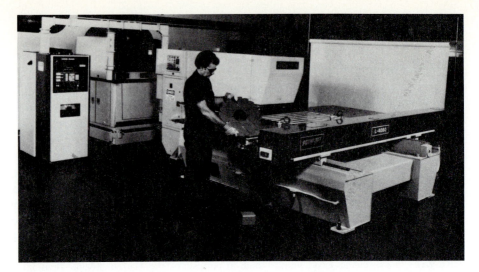

FIGURE 9.12 NC laser cutting machine. (Courtesy of the Wiedermann Division of Warner and Swasey, a Scott and Trecker Company.)

FIGURE 9.13 Digitizer to input information to the laser cutter in Figure 9.12. (Courtesy of the Wiedermann Division of Warner and Swasey, a Scott and Trecker Company.)

verify part position as well as machining correctness through visual analysis. The vertical machining center (Figure 9.8) performs such operations as milling, drilling, boring, facing, reaming, threading, and tapping. The machine shown is a four-axis machine (*x, y, z,* and a removable rotary table *a* axis).

Figure 9.10 shows a point-to-point assembly operation. Integrated circuits, diodes, resistors, and capacitors are automatically positioned on a ceramic substrate. This particular assembly is utilized in a heart pacemaker.

A twin-spindle vertical CNC machining center is shown in Figure 9.11. Two complex parts, seen in the picture, are produced simultaneously. The rotary table allows a good deal of flexibility in the manufacturing operations that can be handled.

A truly flexible machine is the laser cutting machine shown in Figure 9.12. This machine cuts complex shapes out of metal at speeds up to 500 in. per minute. An optional electronic tablet and pen (Figure 9.13) allows drawings, blueprints, and even parts to be converted directly into machine programs.

Obviously, the spectrum of application for NC equipment is much broader than it is feasible to show in a page or two. Whatever the application, however, the role in computer-aided manufacturing is quite clear. Now it is time to consider some of the software aspects of NC.

9.4 NC PROGRAMMING

Many NC processing languages have been developed since the advent of NC. Koren [6] discusses 15 such languages, and Luggen [9] suggests that APT, the language that has the largest vocabulary of the general processor languages, was also the first to be developed. Many variations of APT have been developed, and most mainframe, full-spectrum computer-aided design packages have an APT-like processor that can evolve cutter location data (CL data) directly from the design geometry. For this reason, the APT structure will be used in this section to exemplify NC programming. It is stressed, however, that this structure is only one of many available to the user.

Our APT processor will follow IBM's APT-AC (Automatically Programmed Tool—Advanced Contouring) organization philosophy [4][2] This organization contains the part definition, the machining specification, and the machining plan.

The *part definition* contains the elements of geometry that make up the physical description of the part to be made. Since the purpose of any NC language is to transform *design* specifications into a *machining plan*, it follows that geometry elements are *not* a complete design specification of the part. Rather, they define key characteristics that are germane to the manufacturing plan—for example, a machining reference point and center points of holes to be drilled.

---

[2] Permission to use this philosophy and much of the example material was graciously granted by the IBM Corporation. The APT command formats to be given are generic to most APT systems.

The *machining specification* realizes that the characteristics of each machining process will vary, and so the machining environment has to be specified. Some discussions classify the machining specification into *auxillary* and *postprocessor* statements.

The *machining plan* relates to how tool *motions* are controlled in order to remove necessary material from the part. These motions are controlled with relation to the part geometry specified in the part description. These three APT organizational groupings will now be discussed in more detail to show how simple part processing might be accomplished.

9.4.1 Part Definition (Geometry)

Part geometry will be defined in the cartesian coordinate system as shown in Figure 9.14. We will be concerned with axis direction when motion commands are considered, but some geometry statements require axis notation. APT-AC has 21 geometry types that can be used in defining part geometry. To exemplify the language, we will utilize the six types listed in Table 9.1.

The general format for geometric statements is

$$\langle symbol \rangle = \text{geometric type/definitional modifiers}$$

For example, a point at coordinate location $x = 3$, $y = 4$, $z = 5$ might be defined by

$$\text{PTA} = \text{POINT/3,4,5}$$

FIGURE 9.14 Conventional cartesian coordinate system.

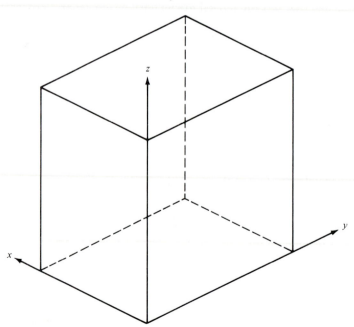

TABLE 9.1
Typical APT geometry vocabulary words

Geometric type	APT vocabulary word
Point	POINT
Line	LINE
Plane	PLANE
Circle	CIRCLE
Pattern (of parts)	PATERN
Cylinder	CYLNDR

To simplify the programmer's job while providing a great deal of flexibility, there are many ways to define some of the geometry types. Some of these will be given for the six geometry types listed in Table 9.1.

POINT (APT word is POINT).

1. *Define in terms of cartesian coordinates.* This was just exemplified by

$$PTA = POINT/3,4,5$$

2. *Define at intersection of two lines.* If two lines have been defined as LIN1 and LIN2, then a point can be defined at their intersection by

$$PTB = POINT/INTOF,LIN1,LIN2$$

as shown in Figure 9.15*a*.

3. *Define at intersection of a line and a circle.* One of four modifiers (YSMALL, YLARGE, XSMALL, XLARGE) is used to distinguish between the two possible points created when a line goes *through* a circle. The modifier gives the relative *x*- or *y*-axis position of the point in relationship to the other possible point, as given in Figure 9.15*b*.

$$PTD = POINT/YSMALL,INTOF,LIN3,C1$$
$$PTD = POINT/XSMALL,INTOF,LIN3,C1$$
$$PTC = POINT/YLARGE,INTOF,LIN3,C1$$
$$PTC = POINT/XLARGE,INTOF,LIN3,C1$$

4. *Define at intersection of two circles.* In a fashion similar to the line/circle intersection, we get the definitions for Figure 9.15*c*:

$$PTE = POINT/YLARGE,INTOF,C1,C2$$
$$PTE = POINT/XLARGE,INTOF,C1,C2$$
$$PTF = POINT/YSMALL,INTOF,C1,C2$$
$$PTF = POINT/XSMALL,INTOF,C1,C2$$

5. *Defined at the center of a circle.* From Figure 9.15*d*, we get:

$$PT7 = POINT/CENTER,C6$$

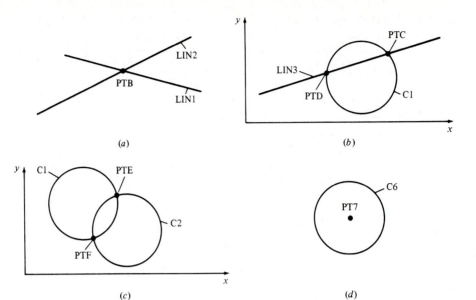

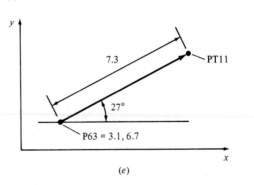

FIGURE 9.15 Some APT point definitions: (*a*) intersection of two lines; (*b*) intersection of lines and circles; (*c*) intersection of two circles; (*d*) center point of a circle; (*e*) defined by distance from reference point (RADIUS) and angle to *x* axis.

6. *Defined by a reference point, radius, and angle.* The point PT11, in Figure 9.15*e*, is defined by:

$$PT11 = POINT/3.1,6.7,RADIUS,7.3,ATANGLE,27$$

If the reference point had been defined earlier in the *x*–*y* plane at 3.1, 6.7 with the symbol P63, we could have

$$PT11 = POINT/P63,RADIUS,7.3,ATANGLE,27$$

It should be apparent that defining a pattern of points for NC drilling in an aircraft's fuselage baffle would be extremely tedious to accomplish in the ways just given. The PATERN vocabulary word facilitates multiple-point definition (as would loop definitions, which will be treated when motion is presented).

PATTERN (APT word is PATERN).

1. A *linear pattern of points* may be generated in terms of a starting point ⟨start⟩, ending point ⟨end⟩, and the total number of points ⟨n⟩ needed:

$$\langle Symbol \rangle = PATERN/LINEAR, \langle start \rangle, \langle end \rangle, \langle n \rangle$$

Using this command, $n - 2$ points are generated in an equidistant fashion *between* ⟨start⟩ and ⟨end⟩. The points in Figure 9.16a would be generated by

$$PATG = PATERN/LINEAR, P16, PT3, 6$$

It will be apparent that the only points that have a symbolic name in the linear pattern are P16 and PT3. If it is desired to give a symbolic to one of the inner

FIGURE 9.16 Forming patterns of points: (*a*) linear pattern generation from two points; (*b*) grid pattern formed by COPY.

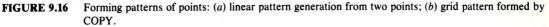

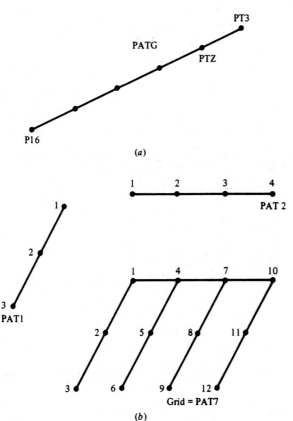

points, this can be handled by

$$PTZ = POINT/PAT6,5$$

which gives the fifth point from the beginning ⟨start⟩ the name PTZ. This is shown in Figure 9.16a.

It should be mentioned at this time that the inner points do *not* have to be named in order for drilling to be accomplished at those points.

2. A *grid of points* may be specified using the COPY modifier:

$$⟨symbol⟩ = PATERN/COPY,PAT1,ON,PAT2$$

Figure 9.16b shows how PAT1 may be attached to PAT2 (since no further modifier is given, it is assumed that the first point generated in PAT1 will be attached to each point, *in sequences*, on PAT2). Figure 9.16b has numbers above each point that indicate the sequence from

$$PAT7 = PATERN/COPY,PAT1,ON,PAT2$$

If the grid of points represents the centers of 12 circles to be drilled, it should be apparent that drilling in the sequence indicated in Figure 9.16b may not minimize travel (remember that, of the 12 points, we have named only the original start and end points for PAT1 and PAT2, which are now points, 1, 3, and 10).

The modifiers SAME and UNLIKE can be used to define the sequence for a grid of points in many ways.

The modifier SAME placed *after* the pattern designator will force that pattern's sequence numbers to follow their original sequence. If the modifier is omitted, as in

$$PAT7 = PATERN/COPY,PAT1,ON,PAT2$$

it is assumed that SAME refers to the first pattern symbolic—PAT1 in this case. So Figure 9.16b has the four sequences for PAT1 in monotonic order: (1, 2, 3), (4, 5, 6), (7, 8, 9), (10, 11, 12). However, if we have SAME modifying the second pattern, as in

$$PAT8 = PATERN/COPY,PAT1,ON,PAT2,SAME$$

then the sequence of points for the three PAT2s is (1, 2, 3, 4), (5, 6, 7, 8), (9, 10, 11, 12), as shown in Figure 9.17a.

This would probably not be optimum as far as motion is concerned in going to each point. The UNLIKE modifier can be of assistance in obtaining other sequences. UNLIKE says that the sequence of points will be reversed on the second cycle from that of the first and that the third will be the same as the first (or reversed from the second), and so on. Figure 9.17b shows this situation for

$$PAT11 = PATERN/COPY,PAT1,ON,PAT2,UNLIKE$$

Similarly, Figure 9.17c gives the sequence for

$$PAT12 = PATERN/COPY,PAT1,UNLIKE,ON,PAT2$$

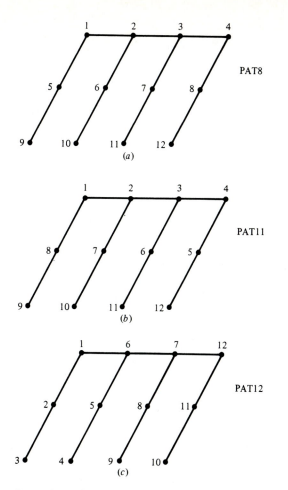

FIGURE 9.17 Generating points in different sequences: (*a*) grid by PATERN/COPY,PAT1,ON,PAT2,SAME; (*b*) grid by PATERN/COPY,PAT1,ON,PAT2,UNLIKE; (*c*) grid by PATERN/COPY,PAT1,-UNLIKE,ON,PAT2.

Many variations of the PATERN command exist which will not be covered in this material. Attach points for the patterns can be changed in the COPY modifier, and PATERNs can be other than linear, such as circular, for example. Also, it is possible to generate sequences which are not equidistant in separation or to eliminate unwanted points.

LINE (APT word is LINE). APT-AC has 27 different ways in which a line may be defined. Figure 9.18 gives three somewhat obvious ways of definition. Figure 9.18*d* needs a little clarification. The point may be a symbolic from a previously defined point or the coordinates for that point. If the angle is clockwise from an axis, it is *negative*; if counterclockwise, it is *positive*.

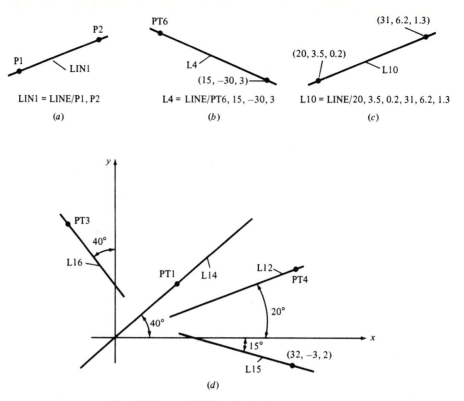

L12 = LINE/PT4, ATANGLE, 20, XAXIS
L14 = LINE/PT1, ATANGLE, 40   (*x* axis is default)
L15 = LINE/32, −3, 2, ATANGLE, −15, XAXIS
L16 = LINE/PT3, ATANGLE, 40, YAXIS

FIGURE 9.18   Some APT line definitions: (*a*) line defined in terms of two predefined points; (*b*) line defined in terms of one predefined point and another point's coordinates; (*c*) line defined in terms of two points' coordinates; (*d*) line defined in terms of a point and at an angle to the *x* or *y* axis.

Lines relative to circles. A line can be defined in many ways relative to a circle (the circle has to have been defined earlier and some examples for this will be given shortly).

Figure 9.19 shows how a line might be defined in terms of a single circle and a point. The lines are *tangent* to the circle, and the LEFT or RIGHT modifier indicates whether the line is at the left or right tangent point, depending on how one looks at the circle from the point.

In a similar manner, Figure 9.20 shows how a line can be defined tangent to two circles. The position modifiers assume a direction looking *from* the first circle defined (C3 in the definition of L6) *to* the second circle defined (C4 in the definition of L6).

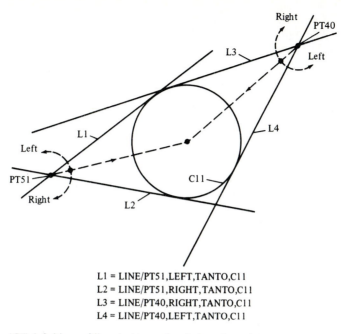

L1 = LINE/PT51,LEFT,TANTO,C11
L2 = LINE/PT51,RIGHT,TANTO,C11
L3 = LINE/PT40,RIGHT,TANTO,C11
L4 = LINE/PT40,LEFT,TANTO,C11

FIGURE 9.19   APT definitions of lines in terms of a circle and a point.

FIGURE 9.20   Defining a line in terms of two circles.

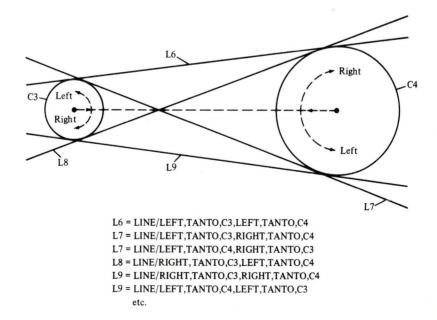

L6 = LINE/LEFT,TANTO,C3,LEFT,TANTO,C4
L7 = LINE/LEFT,TANTO,C3,RIGHT,TANTO,C4
L7 = LINE/LEFT,TANTO,C4,RIGHT,TANTO,C3
L8 = LINE/RIGHT,TANTO,C3,LEFT,TANTO,C4
L9 = LINE/RIGHT,TANTO,C3,RIGHT,TANTO,C4
L9 = LINE/LEFT,TANTO,C4,LEFT,TANTO,C3
        etc.

Line in terms of a previously defined line and point. Figure 9.21 shows how a line can be defined in terms of a point through which it goes, as well as a line to which it is either perpendicular or parallel.

Line in terms of intersection of planes. Assuming that two intersecting planes have been defined, a line can be defined at their intersection, as shown in Figure 9.22.

PLANE (APT word is PLANE). Three possible ways to define a plane are given in Figure 9.23. The most common is in terms of three points, where the three points may *not* lie in a straight line. A second method is in terms of being parallel to an already-defined plane and a point in the new plane. The third is in terms of being parallel to a previously defined plane with an axis modifier (YSMALL in the figure) giving the relative position of the *new* plane with respect to the old one.

FIGURE 9.21   Some other APT line definitions: (*a*) LN3 = LINE/PNT6,PARLEL,LN15; (*b*) LN4 = LINE/PNT5,PERPTO,LN13.

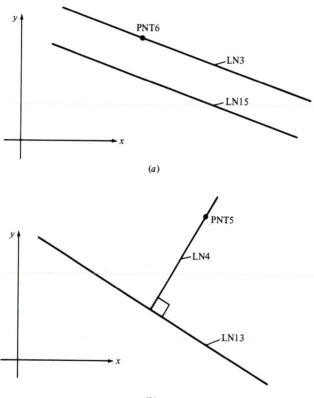

(*a*)

(*b*)

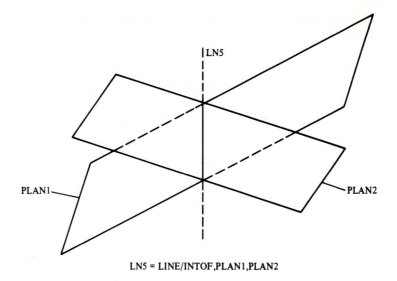

LN5 = LINE/INTOF,PLAN1,PLAN2

FIGURE 9.22   A line in terms of planes.

FIGURE 9.23   Some APT plane definitions.

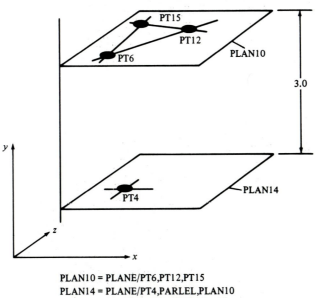

PLAN10 = PLANE/PT6,PT12,PT15
PLAN14 = PLANE/PT4,PARLEL,PLAN10
PLAN14 = PLANE/PARLEL,PLAN10,YSMALL,3.0

CIRCLE (APT word is CIRCLE). Figure 9.24 indicates how a circle can be defined in terms of points and/or a line. First (Figure 9.24a), a circle can be represented by its center point and radius. The center point comes first in the definition and may be in terms of a symbolic or actual coordinates.

Figures 9.24b and 9.24c show how a circle can be defined in terms of its center point and a tangent line or point on the circle.

Finally, with regard to circle definitions, a circle can be defined in terms of two previously defined lines as shown in Figure 9.25. The axis modifiers indicate the relationship of the circle's center point to the tangent point of the line and circle. Some possible definitions are:

C3 = CIRCLE/YLARGE,LN6,XLARGE,LN4,RADIUS,2.0
C3 = CIRCLE/XLARGE,LN6,YSMALL,LN4,RADIUS,2.0
C1 = CIRCLE/YLARGE,LN6,YLARGE,LN4,RADIUS,3.0
C2 = CIRCLE/XSMALL,LN6,XSMALL,LN4,RADIUS,1.5
C2 = CIRCLE/YLARGE,LN4,YSMALL,LN6,RADIUS,1.5

CYLINDER (APT word is CYLNDR). One of several ways to define a cylinder is as tangent to two planes. As can be seen in Figure 9.26, there are four possible ways

FIGURE 9.24 Some APT circle definitions: (a) C1 = CIRCLE/3,6,5,4.3, C1 = CIRCLE/CENTER,PT3,RA-DIUS,4.3; (b) C3 = CIRCLE/CENTER,PT6,TANTO,LN4; (c) C7 = CIRCLE/CENTER,PT8,PT5.

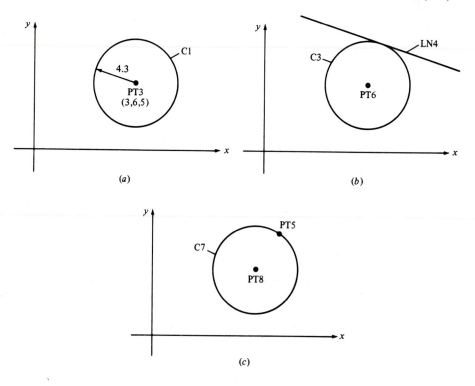

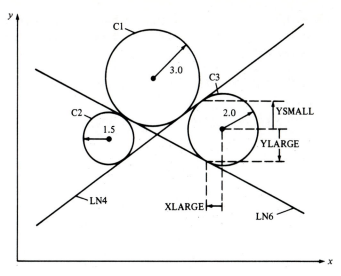

FIGURE 9.25 Circles defined in terms of two lines.

this may happen, with only one of these being pictured in the illustration. The definition structure becomes a little cumbersome because of this:

⟨symbolic⟩ = CYLNDR/⟨axis modifier⟩,TANTO,⟨first plane⟩,⟨axis modifier⟩,
TANTO,⟨second plane⟩,RADIUS,⟨radius value⟩

FIGURE 9.26 An APT definition of a cylinder.

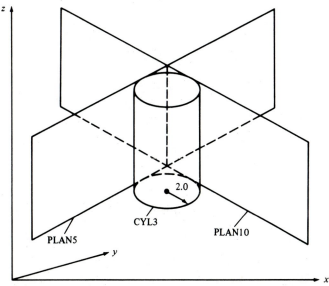

The axis modifier depends on the relationship of the cylinder center point to the tangent point of the plane the modifier precedes. Examples for the cylinder in Figure 9.26 include:

CYL3 = CYLNDR/XLARGE,TANTO,PLAN5,YSMALL,

TANTO,PLAN10,RADIUS,2.0

CYL3 = CYLNDR/XSMALL,TANTO,PLAN10,XLARGE,

TANTO,PLAN5,RADIUS,2.0

Geometry example. In order to tie together some of the geometry concepts presented so far, consider the top view of a plate as shown in Figure 9.27. The outer shape of this plate is to be milled and the grid of holes drilled (we will see how this can be done in Section 9.4.2). In order to do this it is necessary to define the geometry of the part—its outer shape and the locations of the holes.

We will assume a reference point at the top left corner, shown as (4,5,0) on Figure 9.28. Symbolic names have been attached to three holes as well as the lines and the circle. From the design geometry and symbolic names we can now completely define the geometry of the plate shown in Figure 9.29.

FIGURE 9.27   Example part.

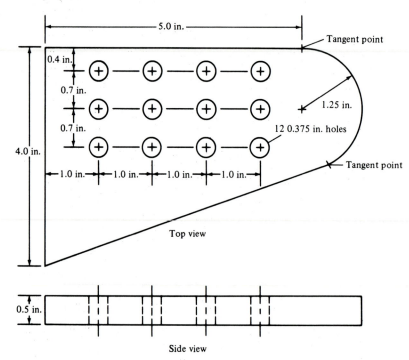

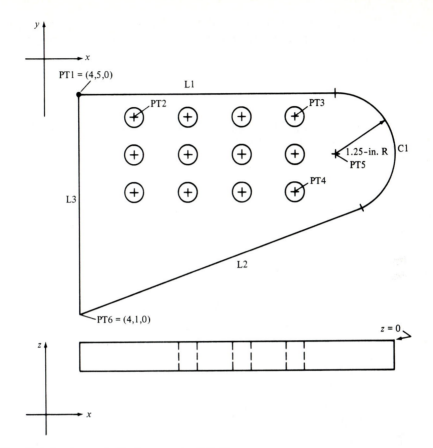

FIGURE 9.28   Part showing symbolics for geometry definition.

FIGURE 9.29   Geometry definition for example part.

```
  PT1 = POINT/4,5,0
  PT2 = POINT/5,4.6,0
  PT3 = POINT/8,4.6,0
  PT4 = POINT/8,3.2,0
  PT5 = POINT/9,3.75,0
   C1 = CIRCLE/CENTER,PT5,RADIUS,1.25
  PT6 = POINT/4,1,0
   L1 = LINE/PT1,LEFT,TANTO,C1
   L3 = LINE/PT1,PT6
   L2 = LINE/PT6,RIGHT,TANTO,C1
PLAN1 = PLANE/PT1,PT2,PT3
PLAN2 = PLANE/PARLEL,PLAN1,ZSMALL,0.5
 PTN1 = PATERN/LINEAR,PT2,PT3,4
 PTN2 = PATERN/LINEAR,PT3,PT4,3
 PTN3 = PATERN/COPY,PTN2,UNLIKE,ON,PTN1
```

The two planes are given to define the top and bottom surfaces of the part. The ⟨UNLIKE⟩ modifier in PTN3 forces the following sequence of point generation:

1	6	7	12
2	5	8	11
3	4	9	10

Therefore, if the same tool can do the part milling as well as drilling (an end mill), we can start and finish milling at PT1 and then optimize drill travel time. Now let us see how APT can control this type of motion.

9.4.2 The Machining Plan

There are three prime operating categories for numerical control:

1. Point-to-point
2. Straight cut
3. Contouring

Point-to-point refers to operations that require fast movement to a point followed by a manufacturing operation at that point. Such operations include drilling and punching. Point-to-point is analogous to pick-and-place in robotic applications. *Straight cut* indicates motion along only a major axis; sawing is an application. *Contouring* refers to the complex, continuous removal of material in an application such as turbine-blade machining.

APT has certain motion commands that relate to point-to-point and others used primarily for contouring. We will look first at the point-to-point commands and then at contouring, though in most systems the two can be interleaved.

POINT-TO-POINT (PTP). The three commands usually associated with PTP are:

FROM/⟨point location⟩
GOTO/⟨point location⟩
GODLTA/⟨coordinate increments⟩

The ⟨*point location*⟩ may be given in terms of the x, y, and z coordinates, or it may be a previously defined symbolic as described in the last section. FROM/ denotes that the point location is a starting point for the tool, with the end of the tool being at that point. Motion from the starting point to the desired location is straight-line.

GOTO/ refers to a rapid, straight-line move to the point location indicated— say, a point at which a drilling operation is to occur. GODLTA/ commands that the tool be moved an incremental distance from the current position. For example,

$$GODLTA/0,0,-0.5$$

will cause the tool to be moved half an inch in the negative z direction (assuming we are not working in the metric system). GODLTA motions are straight-line motions.

Figure 9.30 shows a triangular plate which has to have three 0.3-in.-diameter holes drilled in it. The home point, P0, has a 0.1 value for z to allow for clearance of the tool when it approaches the part (the top surface has $z = 0$). Also, the three points for the circle centers will be given z-axis values of 0.1 for the same reason:

$$P1 = POINT/1.0,2.7,0.1$$

$$P2 = POINT/2.0,2.7,0.1$$

$$P3 = POINT/1.0,2.0,0.1$$

FIGURE 9.30   Part for point-to-point example.

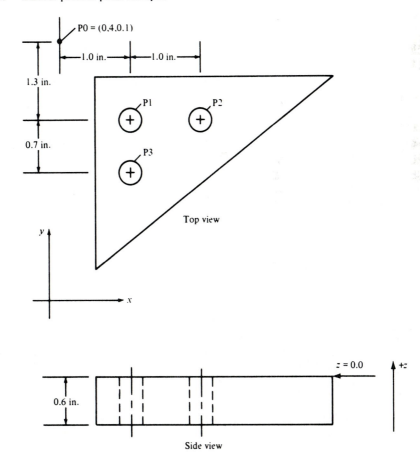

The motion statements to allow drilling through the 0.6-in.-deep plate could be:

```
FROM/P0
GOTO/P1
GODLTA/0, 0, -0.8
GODLTA/0, 0, +0.8
GOTO/P2
GODLTA/0, 0, -0.8
GODLTA/0, 0, +0.8
GOTO/P3
GODLTA/0, 0, -0.8
GODLTA/0, 0, +0.8
GOTO/P0
```

The depth of motion is set to 0.8 to allow for the tool being 0.1 in. above the part at the start of drilling as well as to ensure that the tool clears the bottom of the part.

Three conclusions can readily be realized from this simple example:

1. The use of looping or a subroutine would greatly facilitate point-to-point motion with many destination points.
2. While the PATERN/ command given in the previous section allows easy definition of many points, access to these points by a GOTO/ is not realistic since a symbolic name will, in general, not be available for all points.
3. Depending on the part material and operation involved, certain auxiliary commands will be necessary to control such functions as spindle speed (revolutions per minute for drilling) and feed rate (inches per minute).

Auxiliary commands (machining specification) will be introduced in Section 9.4.3 to fill the void realized in conclusion 3. The concept of "canned cycle" programs can greatly simplify the problem raised by conclusion 2. Such programs will be introduced in conjunction with CNC concepts in Section 9.5. However, at this time the use of *looping* and *macros* will be presented to show how such techniques can simplify APT programming. Both approaches can be applied to contouring as well as to point-to-point operations.

MACROS. In its simplest form, a *macro* is a single computer instruction that stands for a given sequence of instructions [14]. If the sequence of instructions is used many times in different sections of a program, the macro instruction can replace the sequence in each program section. Such a sequence of instructions might be

$$\text{GODLTA}/0, 0, -0.8$$
$$\text{GODLTA}/0, 0, +0.8$$

This is obviously the most simple case, as only *two* instructions will be replaced by *one* macro instruction. The APT format for a macro is:

⟨name⟩ = MACRO/⟨possible parameters⟩⟨sequence of instructions⟩
TERMAC

For our simple macro we might have

$$\text{DELTA} = \text{MACRO}$$
$$\text{GODLTA}/0, 0, -0.8$$
$$\text{GODLTA}/0, 0, +0.8$$
$$\text{TERMAC}$$

The macro can be used any time in the APT program (*after* the MACRO is listed) by inserting

CALL macro name (, list of parameters)

In our case this would be

CALL DELTA

The original APT program for drilling the three holes in Figure 9.30 might be written as shown in Figure 9.31. The interested reader should trace through the program to be sure that the holes will be drilled correctly.

LOOPING. The use of loops in APT allows one to position to a large number of points with efficient programming. The APT structure for a loop is:

RESERV/⟨array name⟩, ⟨maximum number of variables in array⟩
LOOPST
      ⟨loop instructions⟩
LOOPND

The RESERV statement reserves space for a variable array if one is needed. The LOOPST and LOOPND are self-explanatory: They define the start and end of the

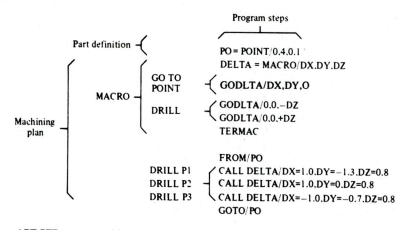

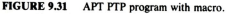

FIGURE 9.31 APT PTP program with macro.

loop. A couple of useful commands are IF and JUMPTO:

1. If ⟨numeric value or equation⟩ NEG, ZER, POS says jump to NEG, ZER, or POS depending on the numeric result being negative, zero, or positive.
2. JUMPTO/⟨symbolic location⟩.

Suppose we want to assign symbolic locations for the 12 circle centers shown in Figure 9.32. An APT loop to allow the 12 points to be defined might be:

```
        RESERV/PNT, 12
        LOOPST
        STARTY = 5.0
           J = 1
     5     I = 1
        STARTX = 3.0
    10  PNT(I) = POINT/STARTX, STARTY, 1
        IF (I-4) 15, 20, 20
    15  I = I + 1
        STARTX = STARTX + 2.0
        JUMPTO/10
    20  IF (J - 3) 25, 30, 30
    25  J = J + 1
        STARTY = STARTY + 1.0
        JUMPTO/5
    30  LOOPND
```

A MACRO may not be inserted between LOOPST and LOOPND, but MACRO CALLs can be (also, MACRO CALLs can be inserted within another MACRO).

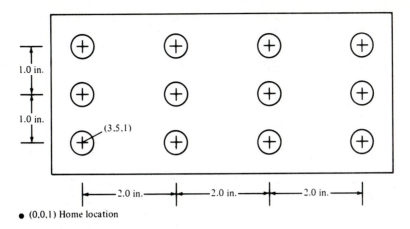

● (0,0,1) Home location

FIGURE 9.32   Part for looping example.

As an example, assume that we need to drill the 12 holes in Figure 9.32 to a depth of a 0.5 in. A possible APT program to accomplish this is shown in Figure 9.33. The 12 points are now *not* set up as an array, as this is not needed: The macro CALL is made as soon as a point is defined. If we *were* using the array values, PNT(I) from the original loop, we might have the MACRO CALL:

$$\texttt{CALL DRILL/PT = PNT(I), DZ = 0.7}$$

Obviously, an index I will have to be set sequentially from a value of 1 to a value of 12.

FIGURE 9.33   Sample APT program with loop and macro.

```
        DRILL = MACRO/PT.DZ
        GOTO/PT
        GODLTA/0,0,–DZ
        GODLTA/0,0,DZ
        TERMAC
        FROM/0,0,1
        LOOPST
        STARTY=5
        J=1
   5    I=1
        STARTX=3.0
  10    PNT=POINT/STARTX,STARTY,1
        CALL DRILL/PT=PNT.DZ=0.7
        IF (I–4) 15,20,20
  15    STARTX=STARTX+2.0
        JUMPTO/10
  20    IF (J–3) 25,30,30
  25    J=J+1
        STARTY=STARTY+1.0
        JUMPTO/5
  30    LOOPND
```

The point-to-point operations can be further simplified using the PATERN/ command to define sets of points and the CYCLE/ command to perform manufacturing operations at those points. For example, if we have a pattern of points already defined as PTN3, we might have

```
CYCLE/DRILL⟨, list of parameters⟩
GOTO/PTN3
CYCLE/OFF
```

The list of parameters would include such items as depth of drill (z), feed rate (f), spindle speed (i), clearance above part during move (c), and so on.

The same approach can be used with points:

```
CYCLE/DRILL, z, f, i, c
GOTO/P1
GOTO/P2
GOTO/P3
CYCLE/OFF
```

CYCLE/ is usually used for z-axis motions such as boring, drilling, and countersinking. It can be used for others, though, such a pocket milling.

CONTOURING. Needless to say, computer processing for continuous removal of metal in three-dimensional space is far more complex than point-to-point. A brief overview can give only a cursory introduction to the topic; the interested reader should investigate the technical literature for detailed information.

The point-to-point commands GOTO/, FROM/, and GODLTA/ all force straight-line motion from one position to the next. Contouring is of course not restricted to such gross limitations on motion. As a result, APT requires that three surfaces be defined at all times to control tool movement:

Part surface: The surface on which the *end* of the tool is riding. Frequently this is a hypothetical surface such as a plane lying $\frac{1}{4}$ in. below the surface of the part.

Drive surface: The surface against which the *edge* of the tool rides; typically this is the actual part edge being cut by a milling tool.

Check surface: A surface at which the current tool motion is to stop.

All three surfaces are shown schematically in Figure 9.34; a more realistic depiction of contouring motion with respect to the three surfaces is shown in Figure 9.35.

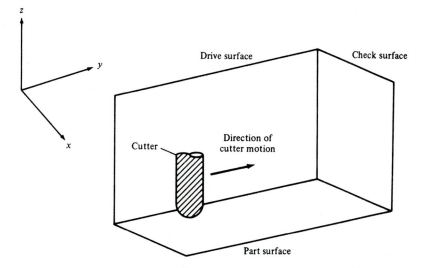

FIGURE 9.34   The three definition surfaces for contouring. [Reprinted by permission from IBM Corporation, *System/ 370 Automatically Programmed Tool—Advanced Contouring Numerical Control Processor: Program Reference Manual* (Program Number 6740-M53), 4th ed., September 1986.]

FIGURE 9.35   Contouring with the three APT surfaces. [Reprinted by permission from IBM Corporation, *System/ 370 Automatically Programmed Tool—Advanced Contouring Numerical Control Processor: Program Reference Manual* (Program Number 6740-M53), 4th ed., September 1986.]

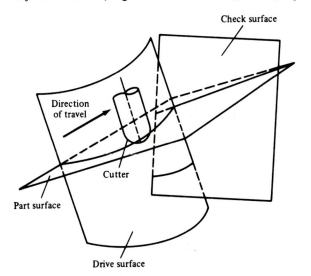

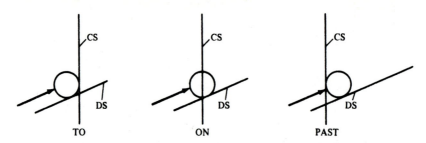

FIGURE 9.36   Use of the TO, ON, and PAST modifiers. [Reprinted by permission from IBM Corporation, *System/370 Automatically Programmed Tool—Advanced Contouring Numerical Control Processor: Program Reference Manual* (Program Number 6740-M53), 4th ed., September 1986.]

FIGURE 9.37   Complete GO/ command example for contouring. [Reprinted by permission from IBM Corporation, *System/370 Automatically Programmed Tool—Advanced Contouring Numerical Control Processor: Program Reference Manual* (Program Number 6740-M53), 4th ed., September 1986.]

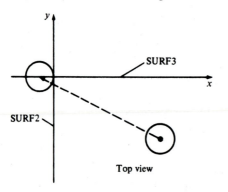

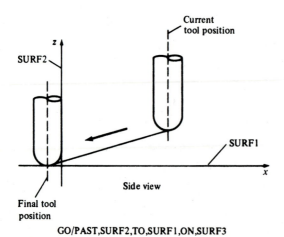

GO/PAST,SURF2,TO,SURF1,ON,SURF3

In order to position the cutting tool with respect to the surface, a *GO surface* command is needed:

$$\text{GO/} \begin{bmatrix} \text{TO} \\ \text{PAST} \\ \text{ON} \end{bmatrix}, \text{ DRIVE SURFACE, } \begin{bmatrix} \text{TO} \\ \text{PAST} \\ \text{ON} \end{bmatrix}, \text{ PART SURFACE, } \begin{bmatrix} \text{TO} \\ \text{PAST} \\ \text{ON} \\ \text{TANTO} \end{bmatrix}, \text{ CHECK SURFACE}$$

We will assume that all three surfaces have to be defined, though this is not always the case. The TO, PAST, ON, and TANTO modifiers indicate the desired location of the cutter with respect to the particular surface. For example, a tool moving to the check surface (CS) along the drive surface (DS) has three possible ending locations as shown in Figure 9.36. A complete GO surface command is shown in Figure 9.37.

The *TANTO* modifier can be used only in conjunction with a *check* surface. Figure 9.38 shows the TANTO relationships.

A: GO/TO,L1,TO,PL2,TANTO,C1
B: GO/PAST,L1,TO,PL2,TANTO,C1

FIGURE 9.38   Use of the TANTO modifier.

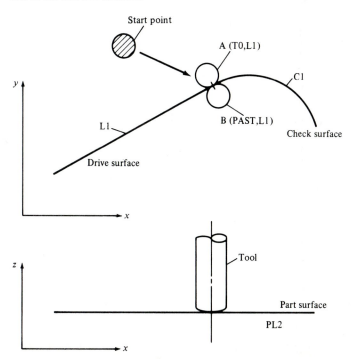

As already mentioned, only a few APT statements out of many have been given in this introduction to the subject. Modifiers are needed, for example, to indicate whether the tool is *offset* from the part so that the tool edge is cutting or the tool point is *on* the part.

So far we have considered orienting the tool only to its initial position with respect to the three control surfaces. Now we need to consider motion commands that are relative to the previous motion direction:

GOLFT/	Move left along the drive surface
GORGT/	Move right along the drive surface
GOUP/	Move up along the drive surface
GODOWN/	Move down along the drive surface
GOFWD/	Move forward from a tangent position
GOBACK/	Move backward from a tangent position

As an example, consider the tool path required to mill the edge of the plate shown in Figure 9.39. The tool path required is

$$\text{START} \rightarrow A \rightarrow B \rightarrow C \rightarrow D \rightarrow E \rightarrow \text{START}$$

Assume that the lines and point indicated in Figure 9.39 have been identified correctly and that we have a hypothetical plane (PL1) for the part surface that lies just below the bottom of the plate, possible APT motion commands are:

```
FROM/START

GO/TO, L1, TO, PL1, ON, L3

GORGT/L1, TANTO, C1

GOFWD/C1, TANTO, L2

GOFWD/L2, PAST, L3

GOLFT/L3, PAST, L1

GOTO/START
```

Let us clarify some important points about this example. The motion commands contain the drive surface and the check surface:

```
GORGT/⟨drive surface⟩, ⟨check surface⟩
```

The part surface is required only if a change is required in the current part surface. Also, a GOBACK would be feasible only from tangent point B or C. And, if

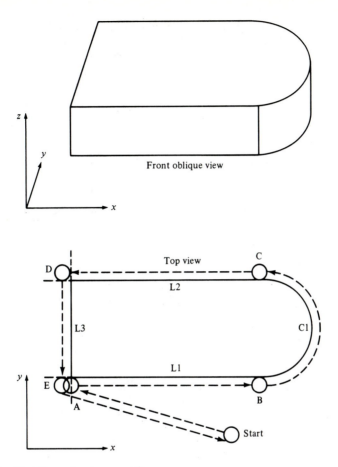

Front oblique view

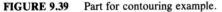

FIGURE 9.39   Part for contouring example.

GOUP or GODOWN is required (a z-axis change), then a new part surface definition would also be required.

Several questions should still be answered relating to tolerances, coolant flow, and use of MACRO in contouring, rough and finish machining, and so on. To tie things together, we will look first at machining specifications (Section 9.4.3) and then consider an example that illustrates all the concepts presented so far (Section 9.4.4).

9.4.3   Machining Specifications

While a great deal of standardization has been accomplished through the use of processors such as APT, such standardization is not as apparent in the machine tool hardware. As a result, an APT program has to be *postprocessed* for a particular machine tool. Typical postprocessor commands listed by Groover [3] include:

```
MACHIN/
COOLNT/
FEDRAT/
SPINDL/
TURRET/
END
```

It is possible to include these in MACRO specifications.

MACHIN/ is used to specify the machine tool and call the postprocessor for that tool:

```
MACHIN/DRILL,3
```

might specify the third NC drill in the shop.

COOLNT/ allows the coolant fluid to be turned on or off; typical modifiers include:

```
COOLNT/MIST
COOLNT/FLOOD
COOLNT/OFF
```

FEDRAT/ specifies the feed rate for moving the tool along the part surface in inches per minute:

```
FEDRAT/4.5
```

SPINDL/ gives the spindle rotation speed in revolutions per minute:

```
SPINDL/850
```

TURRET/ can be used to call a specific tool from an automatic tool changer:

```
TURRET/11
```

Modifications allow for a tool identification to be specified as well as a tool holder location.

END inserted in the program forces a machine tool stop so that the operator can manually perform an inspection or perhaps change a tool.

Other APT statements, called auxiliary by Groover, are typified by the following factors.

TOLERANCE SETTING. Nonlinear motion is accomplished in straight-line segments, and INTOL/ and OUTTOL/ statements dictate the number of straight-line segments to be generated. This is shown in Figure 9.40. The actual tolerance—say, a thousandth of an inch—is inserted in the statement:

$$INTOL/.0015$$

$$OUTTOL/.001$$

FIGURE 9.40 Examples of INTOL and OUTTOL specifications: (*a*) OUTTOL specification only; (*b*) INTOL specification only; (*c*) use of both INTOL and OUTTOL.

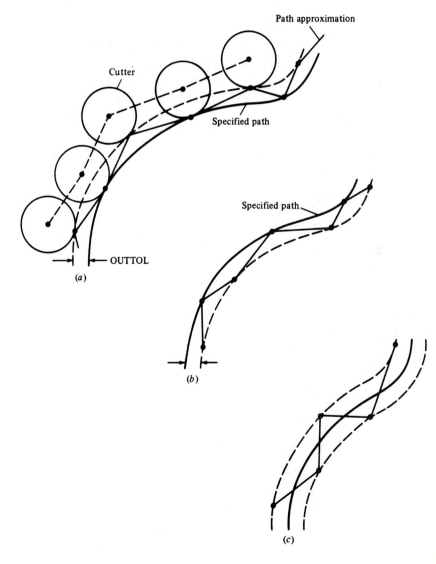

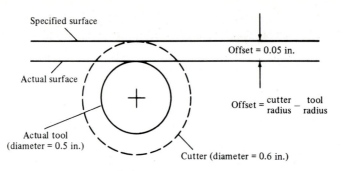

FIGURE 9.41   Use of CUTTER/ for offset.

PARTNO identifies the part program and is inserted at the start of the program.

CLPRINT indicates that a cutter location printout is desired.

CUTTER/ specifies a cutter diameter for offset (rough versus finish cutting). If a milling cutter is 0.5 in. in diameter and we have

$$\text{CUTTER/0.6}$$

then the tool will be offset from the finish cut by 0.05 in. as shown in Figure 9.41.

FINI specifies the end of the program.

9.4.4 APT Contouring Example

Now let us look at a contouring example that uses a MACRO as well as some postprocessing statements: Consider the part shown in Figure 9.42. The periphery

FIGURE 9.42   A simple example for APT contouring.

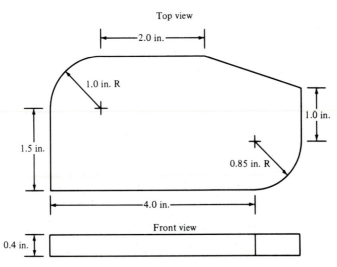

of this part is to be milled in two passes. The first pass will be a rough cut to 0.01 in. of the final geometry specification, and the second pass will be to the final periphery specifications.

Some pertinent information relative to machining is as follows:

	Spindle speed	Feed rate	Coolant	Turret location for cutting tool
PASS 1	600	3.0	Full	4
PASS 2	900	2.0	Full	6

Figure 9.43 identifies the geometric entities so that the APT geometry data can be written. It is assumed that the "home" position for the tool is at

$$P0 = (0,0,1.1)$$

Also, the lower left-hand corner of the part is at

$$P1 = (1,1,0.5).$$

The z value of 0.0 is defined in Figure 9.43 to allow the bottom of the milling tool to be below the bottom of the part when edge milling takes place.

FIGURE 9.43   Part definition information.

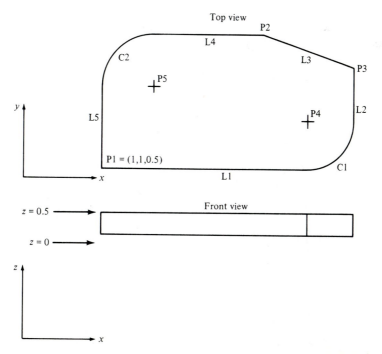

Figure 9.44 gives a possible APT program for the example using information given earlier. An outer tolerance for the curved sections has been set abritrarily at 0.0015; no inner tolerance is specified. A few points are in order concerning this program:

1. PL2 is defined a being $\frac{1}{2}$ in. *below* PL1 (through the modifier YSMALL), and this will form the *drive* surface for position.

2. The two motion commands,

GOLFT/L3,PAST,L4

GOFWD/L4,TANTO,C2

FIGURE 9.44 APT program for sample part.

```
PARTNO    P1534                                        ⎫ Machining
          MACHIN/MILL,4                                ⎬ specifications
          CLPRINT                                      ⎭
          OUTTOL/0.0015
P0        = POINT/0,0,1.1                              ⎫
P1        = POINT/1,1,0.5                              ⎪
P2        = POINT/4,3.5,0.5                            ⎪
P3        = POINT/5.85,2.85,0.5                        ⎪
PL1       = PLANE/P1,P2,P3                             ⎪
PL2       = PLANE/PARLEL,PL1,YSMALL,0.5                ⎪ Part
P4        = POINT/5,1.85,0.5                           ⎬ definition
P5        = POINT/2,2.5,0.5                            ⎪ (geometry)
C1        = CIRCLE/CENTER,P4,RADIUS,0.85               ⎪
C2        = CIRCLE/CENTER,P5,RADIUS,1.0                ⎪
L1        = LINE/P1,RIGHT,TANTO,C1                     ⎪
L2        = LINE/P3,LEFT,TANTO,C1                      ⎪
L3        = LINE/P2,P3                                 ⎪
L4        = LINE/P2,RIGHT,TANTO,C2                     ⎪
L5        = LINE/P1,LEFT,TANTO,C2                      ⎭
MILLS     = MACRO/CUT,SSP,FRT,CLT                      ⎫
          CUTTER/CUT                                   ⎪
          FEDRAT/FRT                                   ⎪
          SPINDL/SSP                                   ⎪
          COOLNT/CLT                                   ⎪
          FROM/P0                                      ⎪
          GO/TO,L1,TO,PL2,TO,L5                        ⎪
          GORGT/L1,TANTO,C1                            ⎪ Macro
          GOFWD/C1,TANTO,L2                            ⎬ for
          GOFWD/L2,PAST,L3                             ⎪ milling
          GOLFT/L3,PAST,L4                             ⎪
          GOFWD/L4,TANTO,C2                            ⎪
          GOFWD/C2,TANTO,L5                            ⎪
          GOFWD/L5,PAST,L1                             ⎪
          COOLNT/OFF                                   ⎪
          GOTO/P0                                      ⎪
          TERMAC                                       ⎭
          TURRET/4                                     ⎫
          CALL/MILLS,CUT=0.52,SSP=600,FRT=3.0,CLT=FULL ⎪ Machining
          TURRET/6                                     ⎬ specifications
          CALL/MILLS,CUT=0.5,SSP=900,FRT=2.0,CLT=FULL  ⎪ and
          SPINDL/0                                     ⎪ machining
          END                                          ⎪ plan
          FINI                                         ⎭
```

might seem inconsistent. The GOFWD along the drive surface L4 is used instead of GOLFT because the angle change is less than 45°. If L4 had required a drive parallel to the *y* axis to the right, then we would have used GORGT/L4, ...

3. The check surface modifiers (TO, ON, PAST, TANTO) are determined based on assuming that the check and drive surfaces are not truncated to form the part outline. For example,

GOFWD/L3,PAST,L4,

allows complete milling of the part's L3 as shown below:

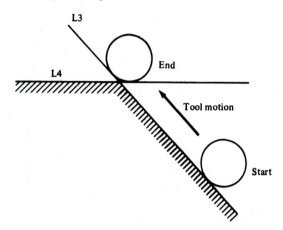

4. The APT statements given are *representative* of those found in many APT processors and in general conform to ANSI APT standards. However, variations will occur; for example, tool information with respect to a tool changer will most often include a changer location *and* a tool identification.

In line with item 4, the purpose of this section has been to give an *example* of NC APT processing. Over 200 statements are available; we have viewed only a small subset of these. The next section will consider briefly the CNC and DNC aspects of numerical control, from both hardware and software points of view.

9.5 COMPUTER NUMERICAL CONTROL (CNC)

Processors such as APT have eliminated much of the drudgery in preparing to machine complex parts. It should not be assumed, however, that such processors are the last word in human/machine tool interfaces. On the other hand, it should not be assumed that other recent developments have eliminated the need for APT. The fact that an APT text was released in 1987 illustrates this point [7]. As just indicated, however, APT is not the last word in this area. A case in point is computer numerical control (CNC).

CNC evolved in the 1970s when the inception of minicomputers provided the possibility of economically allowing one minicomputer to augment the control of one machine tool. More recent developments in microprocessors have made economic viability even more realistic.

The original numerical control used hardwired controls, whereby electronic components provided the control capability (pulses to drive the axes, for example). CNC incorporates a mini- or microcomputer (nowadays it will be a micro) into the control unit to provide *softwired* systems.

The dedication of a single computer to a single machine tool can provide many benefits [6, 9, 13].

1. Softwiring is more *flexible* than hardwiring. An analogy from Chapter 8 is the advantage of a programmable controller over a control system with fixed wiring going from sensors, through some logic relay circuitry, to activator components.

2. The dedicated computer's memory allows a program tape to be read into memory, with production of subsequent parts being sequenced from the *memory* program. This saves tape read-in time as well as eliminates potential errors inherent in punched tape being used in harsh environments.

3. Along with point 2, *edit* capability allows override/correction of tape instructions from the CNC console.

4. *Displays* on the CRT screen show pertinent information to the operator regarding current machining conditions.

5. *Canned cycles* allow quite complex machining operations to be accomplished with very few programming steps.

The development of "canned cycle" programs stored in the resident computer has meant that small machine shops can purchase numerical control equipment without needing extensive programming capability, as would be necessary to use a full APT processor.

9.5.1 Canned Cycles

In many ways, canned cycle programming seems a regression to the earliest form of NC programming, when manual programming had to define *blocks* of tape data with somewhat cumbersome word structures. An NC processor then used the block information from the geometry, motion, and auxiliary statements.

The tape block structure was not introduced in the earlier sections of this chapter because it was deemed to be more beneficial to emphasize the geometry and motion structure from APT in a general introduction. Also, APT is a processor that is interfaced by many CAD systems.

Blocks of data have an alphabetic code letter followed by a numerical modifier. For example, N denotes a block sequence number, say, N0000. G81, as another example, is the standard code for a drilling cycle.

For drills and machining centers, canned cycles are involved principally for z-axis machining motions (drill, bore, tap, etc.). The cycle is initiated by a G preparatory code, with the complete operation being encompassed in a block of data. As Childs shows [1], if two holes are to be drilled at points C and E (Figure 9.45), then the two blocks might be:

N__G81 X__Y__C__D__F__EOB

N__X__Y__EOB

The dashes after the alphabetic letters indicate that numeric information needs to be inserted:

N__	Gives the block number
G81	Is the canned cycle for drilling; the program would be filed in the CNC memory by this designation
X__ Y__	Rapid traverse of drill to these coordinates
C__	Rapid Z traverse to C coordinate before start of machining
D__	Depth for drilling at a given feed rate
F__	Specified feed rate
EOB	End-of-block designator

This example is a little simplistic, since parameters might include inch/metric, repeats per hole, retraction characteristics, and so on.

The block information can be programmed directly into memory through the CNC console (see Section 9.3) or via tape input. The significant point is that the canned cycle programs are stored in memory and require little programming skill.

FIGURE 9.45   Canned cycle structure for drilling. (From Ref. 1, with permission.)

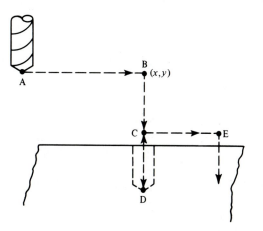

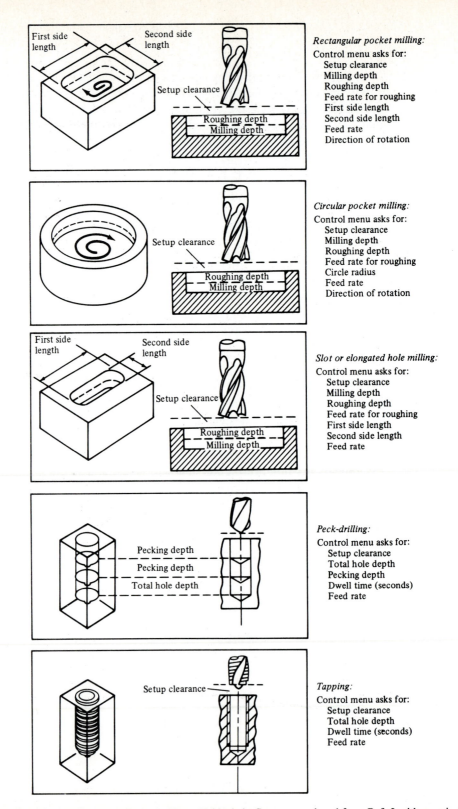

FIGURE 9.46 Menu-driven canned cycles. [From Heidenhain Corp., reproduced from Ref. 2 with permission.]

9.5.2 Further CNC Possibilities

Figure 9.46 shows how a computer can be used to excellent advantage in canned cycles to get away from the traditional code words such as G, N, X, Y, etc. The operator is asked to input certain machining parameters through an interactive CRT session. This will increase the speed of program data input as well as eliminate many potential errors.

One point should be made concerning CNC: what to do with all the traditional existing NC equipment. Gayman [2] suggests that it is possible to retrofit CNC to older NC equipment, with some retrofits costing around $10,000 for control update with maybe $5,000 to $30,000 additional for a total retrofit. Assuming that the original equipment is mechanically sound, it is suggested that a technically up-to-date piece of equipment might be obtained for less than 30% of the original cost.

Childs [1] lists some advantages to CNC over NC:

1. Increased accuracy and speed of controls
2. Built-in machining aids (cutter compensation, circular interpolation, etc.)
3. Programmable machine/control interface permitting closer mating of machine tool and control
4. Program editing and visualization aids
5. Input ports to punched tape, magnetic tape, diskettes, and even distributed numerical control (DNC)

9.6 DISTRIBUTED NUMERICAL CONTROL (DNC)

The first DNC (direct numerical control) systems came on the market in the middle 1960s. In the earliest form, one computer controlled one tool—say, a punch press. An objective was to bypass the tape process. Because of the price of the controlling computer, a mainframe, control costs were exhorbitant. This cost was reduced when several machines were controlled by one computer in a time-share mode, but the pioneer systems saw the machine tools being ineffective when the computer failed or was in maintenance. Thus, when CNC became available in the 1970s, the original DNC concept went into disfavor.

DNC now has a little different connotation than the original direct numerical control systems. In fact, the meaning of the acronym has taken on another context: *distributed* numerical control. With local computer capability from CNC, it is now possible to download NC program control information *directly* to the machine tool's local memory, thus allowing the machine to work independently from the host computer.

A greatly simplified schematic for a DNC system is shown in Figure 9.47. The NC programs for all the parts to be produced under distributed control are

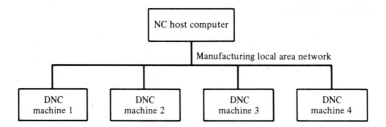

FIGURE 9.47 Simplified DNC schematic.

stored in the host computer system—probably on magnetic tapes with large systems. The programs are *not* stored in *postprocessed* format, because it will probably not be known in advance on what machine the part is to be produced. Once a machine tool has been fixtured for a part, the setup operator notifies the host computer facility that the machine is ready. The NC program is then postprocessed for the particular machine and downloaded to the machine tool through the machine shop's network communication system—say, MAP (manufacturing automation protocol) or maybe ETHERNET. The operator verifies that part production is satisfactory and proceeds to set up another machine.

This same philosophy is key to manufacturing cell operation and flexible manufacturing production (see Chapter 12). Earlier, Figure 9.4 showed a schematic for cell production and discussed how NC plays an integral role in the effective operation of the cell. The cell controller can store a day's schedule of NC programs, downloading to the machine tools when needed. Now a *hierarchical* and *distributed* information system is apparent.

9.6.1 CLDATA

The NC data is usually stored at the host computer site as CLDATA (cutter location data). If an APT processor is used, CLDATA is the result of the compilation process and the actual tool location coordinates are generated for a particular shape. For example, Figure 9.48 shows a portion of the CLDATA output generated by IBM's APT-AC, the processor discussed earlier in this chapter. The values of the *x, y,* and *z* coordinates through which the tool will pass are listed under the word "CIRCLE." The interesting thing about this data is that *no* APT code was written. Rather, a geometric shape was designed using IBM's CADAM[3] package, and the CLDATA was generated through an APT-AC interface. Feed rate, cutter information, and so on, were all input through menu-screen interfacing. The possibility for automatic design to manufacturing is now a reality.

---

[3] *Computer-Aided Design And Manufacturing.*

0011 CUTTER/	0.5000	0.0	
0011 MACHTOL/	0.0010	0.0010	
0011 FEDRAT/	60.0000		
0011 FROM/	POINT		
	−1.0781	−2.3291	3.0000
0012 GOTO/	POINT		
	0.0	−1.0000	3.0000
0014 TURRET/	12.0000		
0015 MACHTOL/	0.0002	0.0002	
0017 LOADTL/	1.000D-04		
0017 FEDRAT/	45.0000		
0017 GOTO/	POINT		

FILE: APT   OUTPUT   A1

	0.0	−1.2500	−0.1000	
0018 SUBFACE			CIRCLE	DS(IMP-TO)
	0.0	0.0	−0.1000	
	0.0	0.0	1.0000	1.2500
0018 GO—/	CIRCLE			
	0.0981	−1.2461	−0.1000	
	0.1955	−1.2346	−0.1000	
	0.2918	−1.2155	−0.1000	
	0.3863	−1.1888	−0.1000	
	0.4784	−1.1548	−0.1000	
	0.5675	−1.1138	−0.1000	
	0.6531	−1.0658	−0.1000	
	0.7347	−1.0113	−0.1000	
	0.8118	−0.9505	−0.1000	
	0.8839	−0.8839	−0.1000	
	0.9505	−0.8118	−0.1000	
	1.0113	−0.7347	−0.1000	
	1.0658	−0.6531	−0.1000	
	1.1138	−0.5675	−0.1000	
	1.1548	−0.4784	−0.1000	
	1.1888	−0.3863	−0.1000	
	1.2155	−0.2918	−0.1000	
	1.2346	−0.1955	−0.1000	
	1.2461	−0.0981	−0.1000	
	1.2500	4.073D−07	−0.1000	
	1.2461	0.0981	−0.1000	
	1.2346	0.1955	−0.1000	
	1.2155	0.2918	−0.1000	
	1.1888	0.3863	−0.1000	
	1.1548	0.4784	−0.1000	
	1.1138	0.5675	−0.1000	
	1.0658	0.6531	−0.1000	
	1.0113	0.7347	−0.1000	
	0.9505	0.8118	−0.1000	
	0.8839	0.8839	−0.1000	
	0.8118	0.9505	−0.1000	
0018 GO—/	CIRCLE			
	0.7347	1.0113	−0.1000	
	0.6531	1.0658	−0.1000	
	0.5675	1.1138	−0.1000	
	0.4784	1.1548	−0.1000	
	0.3863	1.1888	−0.1000	
	0.2918	1.2155	−0.1000	
	0.1955	1.2346	0.1000	
	0.0981	1.2461	−0.1000	
	3.924D−07	1.2500	−0.1000	
0019 GOTO/	POINT			
	−3.2500	1.2500	−0.1000	
0020 GOTO/	POINT			
	−3.2500	−1.2500	−0.1000	
0021 GOTO/	POINT			
	0.0	−1.2500	−0.1000	
0022 GOTO/	POINT			
	−1.0781	−2.3291	3.0000	

FIGURE 9.48   CLDATA generated directly from IBM's CADAM package through APT-AC (generated at Arizona State University.)

9.6.2   Retrofitting to DNC

According to Weck [16], there are three main ways to connect a DNC system directly to the host computer:

1. Conventional numerical control with behind-the-tape (BTR) connection
2. Backup controls (machine tool controller, MTC)
3. Programmable numerical controls (CNC) with BTR operation

The MTC system replaces a conventional NC controller and allows direct softwired interface to the host computer. According to Groover [3], such systems have an advantage in flexibility, since the control functions themselves can be changed to make improvements without rewiring. Groover points out, however, that CNC with BTR allows similar advantages.

If a machine tool is to be retrofitted for DNC, the BTR scheme has many advantages, not the least of which is cost. With BTR, information is downloaded to the machine tool in the same format as it would have been input by the reader. As BTR implies, the reading function is bypassed.

9.7   CONTROLS IN NC

Many texts are devoted in their entirety to numerical control, and a large portion of those texts is usually devoted to controls in NC. In a text with a broad CAD/CAM scope it is not possible to cover all NC topics. However, it will be worthwhile to briefly introduce some controls information. This will be limited to levels of control, differences between point-to-point and contouring control, adaptive control, and position interpolation.

9.7.1   Open-Loop and Closed-Loop Control

The early NC control systems were open-loop systems. That is, it was assumed that a sequence of pulses would always drive stepping motors to the correct position. No sensing of position feeding back to the controller was made. In fact, even some of the early DNC systems were open-loop.

A closed-loop system senses axis position and feeds this back to the NC controller for corrective action. Figure 9.49 shows several levels of closed-loop control as well as open-loop control. According to Gayman [2], machines are only as accurate as the feedback device.

In addition to using position measurement to improve accuracy, dimensional gauging can be utilized for on-line testing purposes. Sheffield [15], in a very interesting review of 66 centuries of measurement, discusses physical probe devices as well as optical gauging. Electronic probes actually touch the material in detecting deflection of a stylus caused by the workpiece. Optical gauging allows sensing without touching the part. Scanning lasers, according to Sheffield, can have resolutions of ten-millionths of an inch. A beam of light scans the measure-

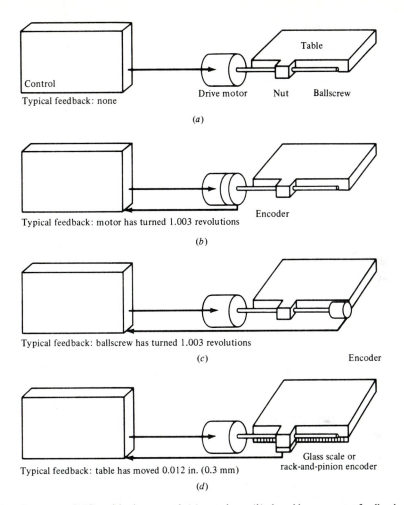

Table

Control
Typical feedback: none

Drive motor   Nut   Ballscrew

(*a*)

Encoder
Typical feedback: motor has turned 1.003 revolutions

(*b*)

Typical feedback: ballscrew has turned 1.003 revolutions

(*c*)   Encoder

Glass scale or
rack-and-pinion encoder
Typical feedback: table has moved 0.012 in. (0.3 mm)

(*d*)

FIGURE 9.49   Spectrum of NC positioning control: (*a*) open loop; (*b*) closed loop—motor feedback; (*c*) closed loop—ballscrew feedback; (*d*) closed loop—work table position feedback. (From Ref. 2, with permission.)

ment area at a constant speed. The object being measured interrupts the beam for a period of time which is proportional to the diameter or thickness of the part being measured.

A linear array of light-sensitive diodes comprises a nonmoving sensing system in which a source of light on one side of the part to be measured is detected by the matrix of diodes on the opposite side. Sheffield indicates that dimensions to fifty-millionths of an inch or less are feasible with this system.

Television cameras are also being used in systems which represent an inspector's eyes. Prototype systems now mount a camera in a tool-changer turret so that the camera can be brought in to sense part characteristics in the same manner as a tool is brought to the workpiece.

9.7.2 Point-to-Point Control

Even though point-to-point and contouring operations are usually available on modern CNC machines, many systems that are only point-to-point are still in operation. Koren [6] shows that these systems generally use position down counters on each axis of motion. The x and y axes are driven at full speed until the correct count is reached. Overshoot is common, however, so that corrective back-tracking is required. This is not feasible in contouring, where excess material may be removed with the overshoot.

9.7.3 Contouring Control

Koren [6] gives the typical closed-loop system contouring shown in Figure 9.50. There are two feedback loops in the system. The tachometer converts motor speed to voltage, and this value is then compared to the voltage analog of the required speed. Thus, instant correction is maintained for motor speed. The outer feedback loop has a transducer which transmits both position and velocity information for the particular axis. This information is compared to required conditions and instantaneous correction is effected. The reference signal is generated as a function of feed rate specifications.

9.7.4 Adaptive Control

Feeds and speeds set by the manufacturing technologist for part production are generally specified on optimum conditions—a sharp tool, for example. As the tool dulls, drive amperage, for example, may increase. Monitoring this information can allow for an algorithm to reduce the feed rate to compensate for the dulling of the tool. Typical sensed information can be torque, temperature, vibration, and part size. Sensors might be placed on the spindle, tool, and workpiece. Sensing information and feeding it back to an on-line adaptive control algorithm can result in optimum cutting conditions. In fact, a machining center equipped with tool-changing capability might be able to choose tools and machining parameters to allow cost optimization to be an objective commensurate with design parameter maintenance.

FIGURE 9.50   Contouring closed-loop control schematic. (From Ref. 6, with permission.)

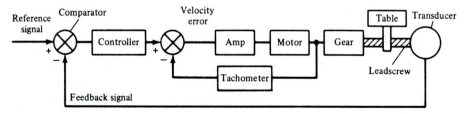

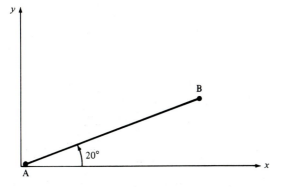

FIGURE 9.51   Linear interpolation for move from A to B.

9.7.5   Interpolation

In order to track a given shape, it is necessary to interpolate position points. The two common types of interpolation are linear and circular. One of the major differences between hardwired and softwired controllers is how they accomplish this interpolation: One (softwired) is flexible and the other (hardwired) is not.

Linear interpolation refers to how the controller moves machine components (tool, table) in a supposedly straight line. Assuming that the axis drive speeds can be driven at different rates (which they can), the function of linear interpolation is to drive the axis motors at the proper speed. A move from A to B in Figure 9.51 would require the x motor to drive faster. Using the tangent of 20° (0.364), it follows that the y drive speed would have to be 36.4% of that of the x drive.

Circular interpolation allows the program to specify that a circular path is to be taken between two points, say, A and B. CNC CYCLE G codes define, for example, whether the interpolated motion is to be clockwise (G03) or counterclockwise (G02). Several algorithms are available to approximate circular motion with small linear segments. Other types of interpolation are also available, including parabolic and polynomial fits higher than linear (cubic). Obviously, the interpolation scheme will have a great bearing on position accuracy.

9.8   CONCLUDING COMMENTS

It is apparent that any "factory of the future" will have to consider numerical control. The concept of *distributed* numerical control implies such an automated philosophy. Numerical control is not just for large, defense-type industries. Small machine shops can afford a CNC machine with its concomitant simplified programming possibilities. Also, at the other end of the spectrum, programming can be eliminated completely in some applications where a computer-aided design package has an interfaced NC processor. A case in point is IBM's CADAM design package, which interfaces directly with the APT-AC processor to produce CLDATA directly from the design geometry.

9.8.1 When to Use NC

Numerical control is best suited to medium-sized batch production, where the number of similar parts to be produced in the same batch run might be 50 to 75 or fewer. In fact, in flexible manufacturing systems (see Chapter 12), batch sizes are approaching one with rapid setup and automated part movement between machines and at a machine.

The concept of group technology was introduced in Chapter 5 as a possible way to group machines and parts into families for production. An NC machine will have limited cutting capabilities (part sizes, rotational characteristics, cutting characteristics, etc.). Group technology can be used to sort parts by machine capability, so that each machine can operate on a group of parts with similar geometric characteristics, similar materials (to simplify tooling and possible chip sorting/removal). Logically, complex part shapes with similar characteristics and low to medium batch sizes are candidates for NC.

9.8.2 Advantages of NC

Many advantages have been alluded to in the prior discussions. A formalization of some advantages to be gained from NC utilization is as follows.

1. Many parts cannot be machined in any other way because of material, fixturing, and complexity problems.
2. Engineering design changes can be easily implemented. A program change is easier than conventional change implementations—blueprints, operation plans, operator interpretation, and so on.
3. Cell and higher-order automation approaches cannot be attained without CNC or DNC.
4. One operator can service a group of machines, setting up one machine while the others are operating.
5. Many sources report that machine tools may be cutting only 20% of the time. Childs [1] says that this 80% nonproductive time might be attributed to:
   a. Handling and setup of the workpiece, cutting tools, and fixtures
   b. Constant review of blueprints and process sheets
   c. Checking part tolerances between operations
   d. Operator fatigue
   e. Trips to the tool crib.
   NC can aid in many of these areas. Childs indicates that an NC machine can be cutting 60 to 70% of the time. Flexible manufacturing systems claim even higher figures as a result of flexible fixturing (fixtures can be controlled through pneumatic or hydraulic controls), off-line palletizing, and so on.

6. Accuracy and repeatability in production can be improved significantly. Childs discusses machines at the Lawrence Livermore Laboratory that will hold one ten-millionth of an inch in positioning accuracy [1]. Working systems are now available with spatial accuracies of one ten-thousandths of an inch. This is quite phenomenal when the operating conditions, especially temperature, are considered.
7. Because of the improved cutting time factor, capital investment in machine tools can be reduced dramatically.
8. With reduction in the number of machines comes a related reduction in floor space required, a major cost of production.
9. As discussed in Chapter 7, the production planning and control function is the heart of the production operating system. Automating the production system allows the production control system to be simplified (though it also has to be computerized). Savings in planning time, inventory costs, and so on, will accrue.

These are just a few of the advantages of NC. Integrating CAD with CAM would not even be feasible without NC. In fact, as was pointed out at the beginning of this chapter, both CAD and CAM had their roots in NC development and the associated APT-like processors.

EXERCISES

1. Review the literature to determine the approximate percentage of manufacturing plants in the United States that utilize some form of NC.
2. Search the literature to find a noncutting example for numerical control (inspection, assembly, composite material laying, etc.). Review the article to point out benefits from such an application.
3. Review library material to find an article that covers a DNC application. What are the economic benefits of such an application?
4. Consider the six points in Figure 9.52a.
   (a) Define the point locations using APT's conventional POINT/ statements. Assume that P1 is at the *x, y, z* coordinates (0,0,0).
   (b) Define the point locations using an APT LOOP set of statements.
5. Define the six point locations in Figure 9.52a using a PATERN/ statement. Assume that P1 is at the *x, y, z* coordinates (0,0,0).
6. Define the 24 point locations in Figure 9.52b using PATERN/ and PATERN/COPY statements. Do this so that the points are sequenced in the following manner:

1	5	9	13	17	21
2	6	10	14	18	22
3	7	11	15	19	23
4	8	12	16	20	24

Assume that P1 is at the *x, y, z* coordinates (0,0,0).

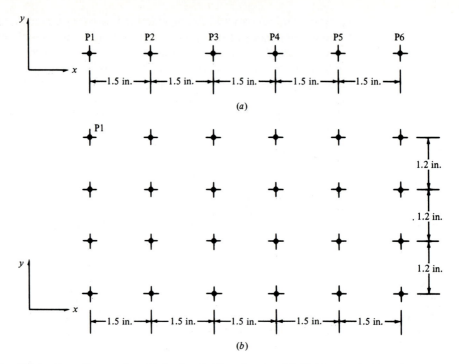

FIGURE 9.52 Point patterns for exercises: (a) 6 points for Exercises 4 and 5; (b) 24 points for Exercises 6, 7, 8, 9, 11, 12, and 13.

7. Repeat Exercise 6 so that the sequence is

4	5	12	13	20	21
3	6	11	14	19	22
2	7	10	15	18	23
1	8	9	16	17	24

8. Repeat Exercise 6 so that the sequence is

6	5	4	3	2	1
7	8	9	10	11	12
18	17	16	15	14	13
19	20	21	22	23	24

9. Use an APT LOOP sequence of statements to generate the pattern of points in Figure 9.52b. The points should be identified as array variables: P1, in the figure, has coordinates of 0,0,0.

10. Use a PATERN/COPY statement to generate the coordinates in Figure 9.17 in the following sequence:

4	3	2	1
8	7	6	5
12	11	10	9

Note: You will have to redefine PAT1 and PAT2 in Figure 9.16.

11. Write the APT machining plan (motion) statements to allow drilling of the holes in Figure 9.52*b* assuming that the points were defined as array variables (see Exercise 9). The depth of drill is to be 0.22 in. and the *z*-coordinate value of 0 is to be 0.1 in. above the top surface of the part. Also, the tool's home location is $(-1, -1, 0.2)$. Be sure to use a MACRO in your solution.

12. Put all the APT-AC part definition, machining plan, and machining specification statements into one program to allow the part in Figure 9.52*b* to be drilled. Assume the conditions given in Exercise 11 with the following additional data:

   - The cutting tool is in turret 6.
   - Coolant is to be on FULL during the actual drilling.
   - Spindle speed is 1200 rpm.
   - Feed rate is 9 in./min.
   - You are *not* to use a CYCLE/ command in your solution.

FIGURE 9.53   Part for peripheral milling.

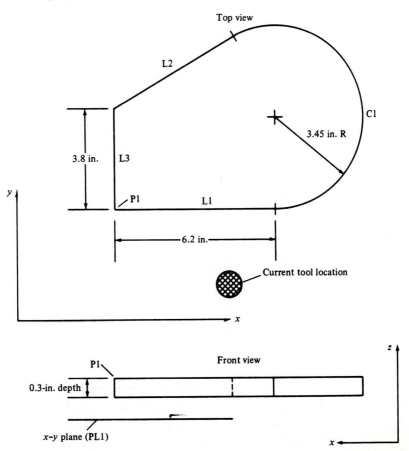

13. Repeat Exercise 12, except this time you may use a CYCLE/ command in your solution.

14. Consider Figure 9.53. Sketch where the tool will end after each of the following contouring initialization statements:
    (a) GO/TO,L1,TO,PL1,PAST,L3
    (b) GO/TO,L2,TO,PL1,TO,L3
    (c) GO/PAST,L1,TO,PL1,TANTO,C1
    (d) GO/PAST,L3,TO,PL1,PAST,L2
    (e) GO/TO,L2,TO,PL1,TANTO,C1

15. The outside periphery of the plate in Figure 9.53 has to be milled in three cutting passes. The first pass will be a rough cut to within 0.05 in. of the finish dimensions. The second cut is to be within 0.02 in. of the finish surface, and the final cut will be to the required specifications. The end-mill's diameter is 0.32 in. Specify the necessary values for DIA in CUTTER/DIA.

16. Specify the APT part definition statements needed to define the part in Figure 9.53 assuming that the outer periphery is to be milled. P1 is at the x, y, z location (3.2,2.2,0.0), and the plane PL1 has a z coordinate of -0.5.

17. Write the machining plan statements (motion) to allow the part in Figure 9.53 to be milled in accordance with the conditions given in Exercises 15 and 16. A MACRO should be incorporated in your solution for obvious programming efficiency, and the current tool position should be assumed to be at the x, y, z coordinates (7.0,1.0,2.0).

18. The part in Figure 9.53 now has two slots to be cut as indicated in Figure 9.54. Write a set of APT statements to handle the cutting of the two slots using an appropriate slotting MACRO. Each slot should be cut with only one pass of the cutting tool.

FIGURE 9.54   Additional slotting information for part in Figure 9.53.

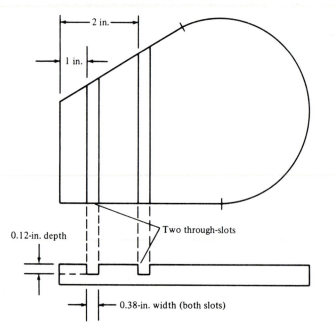

19. The slotting and milling MACROs developed in the two previous exercises should now be incorporated into an APT program for a single machine that can handle both operations. Parameters are as follows:

- Postprocessor machine designation: MILL,3.
- Milling cutter is in turret holder 4.
- Slotting cutter is in turret holder 6.
- Milling spindle speed is 920 rpm.
- Slotting spindle speed is 780 rpm.
- Milling feed rate is 4.8 in./min.
- Slotting feed rate is 3.1 in./min.
- Coolant is to be on FULL for both operations.

List all the APT statements needed to handle the required work. Include the part definition statements in your list.

20. A baffle is to have 62 holes drilled according to the pattern in Figure 9.55. The holes are to be bored after all 62 have been drilled. The distance between any two adjacent holes along the y axis is 1.2 in. and along the x axis is 1.6 in. Write an efficient APT program to allow the part definition information to be specified, followed by point-to-point drilling and boring. You may assume reference points, feed rates, spindle speeds, turret locations, etc., as you see fit.

21. Figure 9.56 shows a plate that has to have a slot milled, a hole drilled, and the outside periphery milled. Assume that there are three different tools in the machine's tool changer for these purposes. Specify a complete APT program (part definition, machin-

FIGURE 9.55 Grid of holes in a baffle.

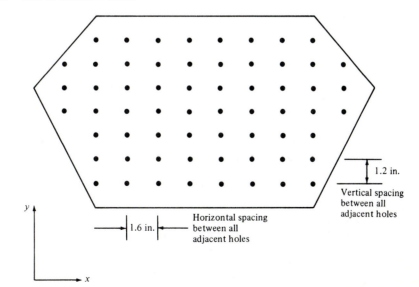

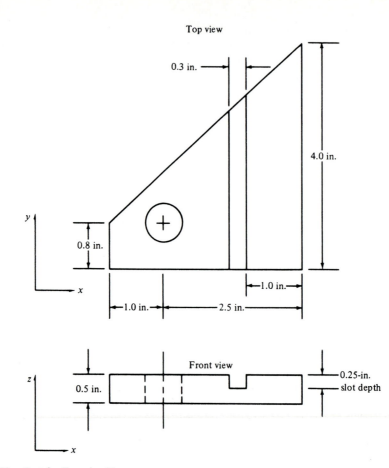

FIGURE 9.56 Part for Exercise 21.

ing specification, and machining plan) to allow the required work to be accomplished. Specify your own information for missing items such as spindle speeds.

22. Specify the part definition data for the plate shown in Figure 9.57 so that the outer periphery of the plate can be milled. All circle/line intersections are tangential.

23. Write statements to allow the outer periphery milling for the part in Figure 9.57. Make necessary assumptions concerning tool home location and so on.

24. Locate and review an article that discusses the potential for computer vision in machining operations. Pay particular attention to any comments that concern the future potential for computer vision in automated machining operations.

25. It is interesting to note in technological development how much of what is often thought of as a current innovation has its roots quite a long way back. Review the literature to determine how much of what is thought of as modern APT really came from a European extension of APT, called EXAPT I, some 20 years ago. A useful reference would be the text *Numerical Control User's Handbook*, edited by W. H. P. Leslie (McGraw-Hill, New York, 1970).

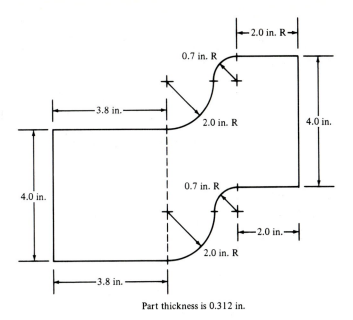

Part thickness is 0.312 in.

FIGURE 9.57   Part for Exercises 22 and 23.

26. One of the ultimate manufacturing automation developments for discrete-item production is supposed to be the concept of flexible manufacturing systems. Review a recent article that discusses a flexible manufacturing system application to determine the need for NC-type operations in such a system.

27. Chapter 12 discusses a manufacturing protocol for networking equipment on the shop floor to allow effective communication to, from, and between equipment. Find an article that discusses the MAP protocol in relationship to DNC-type operations. What type of information, other than the APT-type control information, would be needed to transmit to and from machine tools in a completely automated facility, such as a flexible manufacturing system?

REFERENCES AND SUGGESTED READING

1. Childs, J. J.: *Principles of Numerical Control*, 3rd ed., Industrial Press, New York, 1982.
2. Gayman, David J.: "Calibration, Compensation, and CNC," *Manufacturing Engineering*, January 1987.
3. Groover, M. P.: *Automation, Production Systems, and Computer-Integrated Manufacturing*, Prentice-Hall, Englewood Cliffs, N. J., 1987.
4. IBM Corporation: *System/370 Automatically Programmed Tool—Advanced Contouring Numerical Control Processor: Program Reference Manual* (Program Number 6740-M53), 4th ed., IBM Corporation, White Plains, NY, September 1986.
5. IBM Corporation: *Initial Graphics Exchange Specification (ICES) Processor, Program Reference and Operations Manual* (Program Number 5668-904), IBM Corporation, New York, April 1985.
6. Koren, Y.: *Computer Control of Manufacturing Systems*, McGraw-Hill, New York, 1983.

7. Kral, Irvin H.: *Numerical Control Programming in APT*, Prentice-Hall, Englewood Cliffs, N. J., 1987.
8. Li, Rong-Kwei: "A Conceptual Framework for the Integration of Computer-Aided Design and Computer-Aided Manufacturing," unpublished Ph.D. dissertation, Department of Industrial and Management Systems Engineering, Arizona State University, Tempe, Arizona, December 1986.
9. Luggen, W.: *Fundamentals of Numerical Control*, Delmar Publishers, Albany, N.Y., 1984.
10. Modern Machine Shop: *1986 NC/CAM Guidebook*, Gardner Publications, Cincinnati, Ohio, 1986.
11. Niebel, B. W., A. B. Draper, and R. A. Wysk: *Modern Manufacturing Process Engineering*, McGraw-Hill, New York, 1989.
12. Pease, W.: "An Automated Machine Tool," *Scientific American*, September 1952.
13. Puszta, J., and M. Salva: *Computer Numerical Control*, Reston Publishing, Reston, Va., 1983.
14. Ralston, A., Ed.: *Encyclopedia of Computer Science and Engineering*, Van Nostrand Reinhold, New York, 1983.
15. Sheffield Measurement Division: *66 Centuries of Measurement*, The Sheffield Measurement Division of the Cross Corporation, Dayton, Ohio, October 1984.
16. Weck, Manfred: *Handbook of Machine Tools, Volume 3: Automation and Controls*, John Wiley, New York 1984.

CHAPTER 10

ROBOTICS

Whatever the intentions of their creators, robots are always going to be compared with men in terms of their attributes and general behavior

Joseph F. Engleberger [7]

The popular notion of a robot looking and acting like a human being is obviously a myth imposed by early science fiction movies and even more recent cinema fare (for example, R2D2 in *Star Wars*). Pragmatically, robots installed in manufacturing environments have to earn their keep. If a robot accomplishes work that could be performed by a human, then a careful justification for use of the robot must be made in terms of productivity, operational costs, flexibility in terms of changeover to new tasks, ability to operate in hazardous environments, and the social implications relative to those persons displaced by the robot.

There are many examples of successful robot installations. *Time* magazine [16] cites a few of these:

- Robots perform more than 98% of the spot welding on Ford's Taurus and Sable cars.
- A robot drills 550 holes in the vertical tail fins of an F-16 fighter in 3 hours at General Dynamics. It used to take 24 worker-hours to do the job manually.
- Robots insert disk drives into personal computers and snap keys onto electronic typewriter keyboards.

Unfortunately, there are similar examples of installations that were not so successful. Again, *Time* gives some examples:

- Robots skipping key welds in an automobile production line led to the recall of some 30,000 vehicles.
- Campbell Soup donated a robot to a university after thousands of dollars of damage were caused because the robot dropped 50-lb cases of soup when defective cases were encountered. The robot was replaced by three humans!

A cautionary tone has been injected into these introductory comments to force the realization that significant evaluation has to be made *before* deciding if a robot installation is feasible and, given that it is, significant care needs to be taken in determining what robot to utilize. Concerning this last point, *Time* [16] points out that the number of manufacturers making robots in 1987 was 300, down from 328 in 1986. Another source [17] lists addresses for 237 robot manufacturers world-wide. Of these, 70 are in the United States, 59 are in Japan, and 15 are in West Germany. Obviously, a wide spectrum of robots is available for any potential application; care *must* be taken to ensure that the choice is correct.

In spite of the comments made so far, it can safely be said that the technology encompassing the field of robotics is extremely exciting and innovative. Vision and position sensing, "hand" tactile sensing, dexterous linkages, efficient power supplies, extremely user-friendly programming languages, and innovative control methodologies are just a few of the technological areas within the field of robotics. Further, there are many computer-aided manufacturing applications that would not be feasible without some aid from robotics.

The aim of this chapter is to give an overview of robotics as the field relates to CAM. The breadth of potential applications is as broad as the imagination of the CAM analyst, and so a complete review of such applications is neither feasible nor desirable. The reader wanting a more in-depth coverage of the *management* implications of robots might want to obtain a copy of Maus and Allsup's book, *Robotics: A Manager's Guide* [17]. A more technical review can be found in *Industrial Robotics* by Groover et al. [12]. A seminal work that is still excellent reading is the ICAM Robotics Guide [29]. It is rare to find a current technical robotics book that does not cite this guide regarding robot classification as well as accuracy and repeatability measures.

So far, no definition of a robot has been offered. We will try not to get bogged down in the philosophical arguments that abound on this subject. Rather, we will accept the oft-quoted Robot Institute of America definition [13]:

> . . . a programmable multifunction manipulator designed to move and manipulate material, parts, tools, or specialized devices through variable programmed motions for the performance of a variety of specified tasks.

The most common tasks are often those that require demeaning or drudging effort, frequently on a repeating basis, as well as tasks performed in hazardous environments. Whether or not the definition affects the discrepancy in the reported number of Japanese robot installations relative to U.S. installations (118,800 in Japan versus 25,000 in the United States in 1987 [16]) will be discussed in Section 10.3.

10.1 THE ROLE OF ROBOTICS IN CAD/CAM INTEGRATION

Industrial robots account for only 2% of the $24 billion factory-automation business, according to *Time* magazine [16]. What, then, is the role of robotics in CAD/CAM integration? *Time* also says that many sectors of manufacturing have been drastically altered by the intrusion of robots. The U.S. automobile industry in 1987 had half of the American robots in use! When one considers the spectrum of CAM, the automobile assembly industry has to be a major component. However, it is obviously not the only component.

A *Manufacturing Engineering* magazine survey [9] estimates that by 1995, 70% of all robots purchased in the United States will be for manufacturing cells or flexible manufacturing systems. As has been pointed out many times in this text, these two approaches are key to computer-aided manufacturing in the so-called job-shop environment.

The *Manufacturing Engineering* survey, accomplished over several years, also included projections on applications for robots, as shown in Table 10.1. The most interesting trends are in the decline of spot welding (automobile production lines!) and the increase in assembly and inspection applications.

Within the context of manufacturing cells and flexible manufacturing systems, there is no doubt that the robot can play a major role in integrating the grouped equipment in a materials handling role. Further, the robot is probably the most flexible means for automatically inserting parts into and removing them from

TABLE 10.1
U.S. robot sales by application

	1985	1990	1995
Machine tending	16%	15%	15%
Material transfer (excluding machine tending)	16	15	15
Spot welding	26	15	10
Arc welding	10	10	9
Spray painting/coating	10	10	7
Processing (drilling, grinding, etc.)	5	7	7
Electronics assembly	6	12	14
Other assembly	5	8	12
Inspection	5	7	10
Other	1	1	1
	100%	100%	100%

Source: Reproduced with permission from *Manufacturing Engineering* [Ref. 9].

a machine tool. Many examples exist where complex, specialized mechanisms are used for this purpose, but they are *not flexible*.

Continuing further, the major manufacturer of flexible manufacturing equipment in the United States utilizes a robot for loading tools in the correct sequence into a machining center's tool-changer magazine. Needless to say, tool unloading is accomplished simultaneously. Figure 10.1 depicts such an application.

An even stronger statement regarding the role for robots in cell manufacturing is given in *Control Engineering* [18]. In a list of requirements for an Air Force [29] cell development, the first was that "the work cell must employ a currently available industrial robot."

When one looks at manufacturing it is apparent that many operations are repetitive with short operation times. Others have to be accomplished in hazardous environments. Parts have to be moved between, into, and from machines according to parameters dictated by the machine and the part. If these functions are to be automated within a computer-aided manufacturing environment, it is apparent that robots will play a key role in accomplishing the automation.

FIGURE 10.1   Robot performing tool-loading function. (Courtesy of the Roberts Corp., Scott and Trecker, Milwaukee, Wisconsin.)

Finally, one other comment must be made in relation to robotics and both CAD and CAM. Use of robots—say, in assembly—will dictate that the *design* of a product consider the fact that a robot is to be used in the product's assembly. For example, positioning a vacuum cleaner motor in a housing will be greatly facilitated if the motor and housing can be made somewhat conical in shape, thus allowing the motor to be slipped into the housing without excess positioning tolerance requirements being required. In light of this, Section 10.7 will address accuracy constraints of robots in relation to their linkage capabilities.

10.2 KEY DEFINITIONS

Specialized terms will be defined when those terms first arise in succeeding sections of this chapter. More general terms will be defined now to provide a basic foundation for the general reader of this material.

Anthropomorphic Refers to a robot configuration that resembles the configuration of a human body. The robot configuration is frequently referred to as *articulated*.

Articulated See *Anthropomorphic*.

End effector The robot equivalent of a human hand and the tooling that the hand grips in order to perform a particular task.

Gripper An end effector used for holding on to a particular object. Grippers may or may not have force control capability.

Materials handling The function of moving parts, materials, tooling, and other equipment from one location to another. This is a major task for a robot.

Robot (See Robot Institute of America definition in Section 10.3.1)

Robot, assembly A robot designed primarily to perform assembly tasks. This frequently requires high accuracy capability combined with dexterous movement ability and force sensor control.

Robot, computerized servo A robot powered by DC servo motors that has teach and movement-optimization capability.

Robot, pick-and-place A robot used for moving an object from one location to another, as in assembly or pallet-loading operations. Specific contouring capability is not required.

Robot, sensing A robot that is instrumented to allow abilities similar to human sensing capabilities: touch, vision, voice recognition, etc.

Robot, teach A robot that can be "taught" positions by an operator in manual mode. These positions can then be accessed by the controlling program.

Robotics The general science of robots, including design and application.

Simulation The ability to emulate physical characteristics of a robot. This is usually accomplished using a digital computer with graphics capability, and so the term "digital simulation" arises.

Tactile sensing The capability of a robot to emulate the touch or feel capability of a human. This requires complex gripper instrumentation.

Teach pendant   A hand-held device with which it is possible to position a robot manually. Buttons on the pendant allow specific positions to be recorded into the robot's computer memory.

Welding, adaptive   The capability of a robot to adapt to environmental conditions while performing arc welding. This generally entails a vision system to evaluate the varying conditions.

Welding, arc   The joining of two pieces of metal with a continuous seam evolved with a robot end effector that has the ability to melt a wire with a welding gun.

Welding, spot   The fusion of two pieces of metal through the application of an electric current through a specially designed end effector.

10.3   CHARACTERIZATION OF ROBOTS

The number of combinations of different makes of robots and their potential applications is enormous. Deciding on a particular robot for a specific task would be a monumental task indeed if there were not some way to characterize robots into groups, somewhat in the manner that group technology is used to group machines into cells for specified tasks. Grouping of robots, in this section, will be according to two prime criteria: (1) motion ability and (2) drive power requirements. Criteria relating to cost, control, accuracy, and application can all be functions of the two prime criteria. Before discussing motion and power characteristics, it is advisable to comment briefly on the definitional aspects relating to "What is a robot?"

10.3.1   What Is a Robot?

The Robot Institute of America (RIA) definition of a robot will be repeated at this time to facilitate further discussion:

> . . . a programmable multifunction manipulator designed to move and manipulate material, parts, tools, or specialized devices through variable programmed motions for the performance of a variety of specified tasks. [13]

First of all, the robot is *programmable*, which indicates the need for a controlling computer. Second, it is a *multifunction* device, which eliminates from being a robot something like a cam-operated "pusher" linkage designed to push parts off a conveyor at some specified cyclic rate. The Japanese Industrial Robot Association (JIRA) defines, as one of six types of robot, a *fixed-sequence robot*, which is a "manipulator" that repetitively performs successive steps of a given operation according to a predetermined sequence, condition, and position, and where set information cannot be easily changed [13]. The "pusher" linkage could be construed as a robot according to the Japanese definition. In fact, an even more eye-opening JIRA robot definition is "a transport arm controlled directly by an operator with or without power assistance, depending on the application'' [13]. Now we see one reason for the number of Japanese robot installations being so disproportionate to those in the United States.

To get back to the point at hand, the Robot Institute of America robot definition, a third observation is that in order to manipulate an item through variable motions, it is necessary that the robot be able to *grip* that item. A hand, or *gripper*, can range from a vacuum pick-up device to a complex, multifingered appendage with tactile sensing ability (see Chapter 11).

A common form of gripper has ''bang-bang'' control. Simply put, this means that the gripper ''bangs'' open or ''bangs'' shut, but has no ability to control the force of closure from one application to another other than by adjusting the power source—say, the pneumatic pressure.

An alternative kind of gripper is shown in Figure 10.2. The Skinner hand is a three-fingered design which can allow a wide variety of holding approaches.

Lastly, using the RIA definition, a *manipulator* is the mechanism (linkage) that allows something to be moved along a path or to a particular point. The interface between the manipulator and the object to be moved or worked upon is called an end effector. Thus the end effector may be a gripper, or it might be a tool

FIGURE 10.2   The Skinner hand. (Reproduced with permission from Ref. 29.)

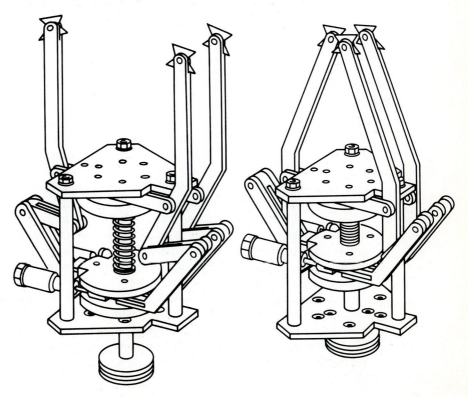

FIGURE 10.3 Spot-welding tooling mounted on robot tool-changing mechanism. (Courtesy of Applied Robotics, Latham, N.Y.)

mounted on the manipulator. Figure 10.3 shows an end effector used for spot-welding operations.

10.3.2 Robot Motions

A robot's motion capability is a function of the axes it is possible to control. This is directly analogous to numerical control motion control discussed in Chapter 9. The four most common robot configurations are shown in Figure 10.4.

The joint arm configuration is often called an anthropomorphic or articulated arrangement. It is the kind most often seen in television commercials, since the configuration allows for extremely dexterous positioning such as is required for automobile spot welding or painting.

The cylindrical arm arrangement is limited to applications where there are no obstructions in the work area, whereas the articulated robot might be programmed to avoid such obstructions. Moving objects from one place to another (pick-and-place), on a cyclic basic, is an example for a possible cylindrical configuration task.

A variant of the cylindrical arm configuration is the spherical arrangement, often classified as a polar configuration. It should be realized that the work *volume* allowed by the robot configuration is of paramount importance in any application. The difference between the cylindrical and spherical volumes can be seen in Figure 10.5.

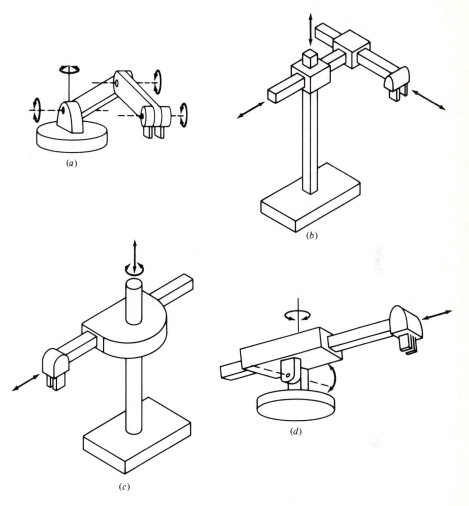

FIGURE 10.4   Common robot configurations: (*a*) jointed arm; (*b*) cartesian or *x–y–z* arm; (*c*) cylindrical arm; (*d*) spherical. (Reproduced with permission from Ref. 29.)

The last manipulator configuration is the cartesian (or rectangular) arrangement. This is probably the most limited configuration for a robot. Some part assembly operations are accomplished with the cartesian-configurated robot.

Care must be taken to ensure that a robot can reach a particular position in its particular envelope, even if that envelope seems to be sufficient. Figure 10.6 shows a work volume that can be reached by the robot, but for which spot welding at a point at the inner top left of the envelope is not possible (the robot can obviously reach the *outside* of that point). In a reverse vein, extended grippers can be used to extend a work volume if necessary, as shown in Figure 10.7.

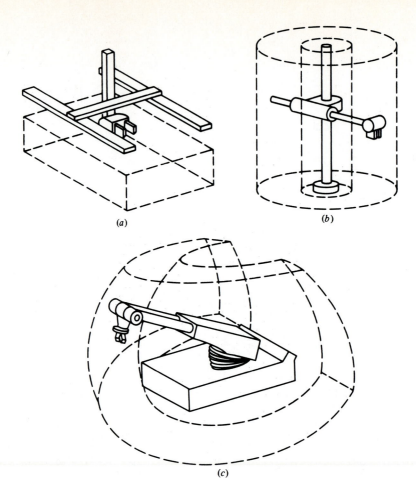

(a)

(b)

(c)

FIGURE 10.5 Typical robot work volumes: (*a*) rectangular; (*b*) cylindrical; (*c*) spherical. (Reproduced with permission from Ref. 29.)

FIGURE 10.6 Satisfactory work volume that is not feasible. (Reproduced with permission from Hinson, R.: "Knowing Work Envelopes Helps in Evaluating Robots," *Industrial Engineering*, July 1983.)

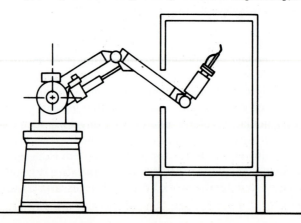

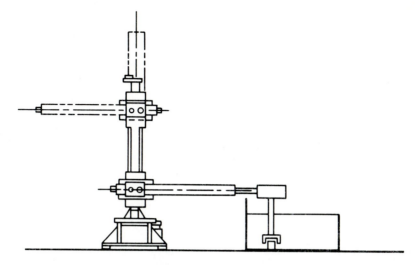

FIGURE 10.7   Using an extended gripper to enlarge the work volume (Reproduced with permission from Hinson, R.: "Knowing Work Envelopes Helps in Evaluating Robots," *Industrial Engineering*, July 1983.)

END-EFFECTOR MOTIONS. The jointed arm configuration shown in Figure 10.4 gave four axes of motion (or degrees of freedom, DOF). This configuration is obviously trying to mimic human motions and, as such, should provide further wrist/hand motions. The joint possibilities of the PUMA 560 robot are shown in Figure 10.8. The four points corresponding to the more generic arrangement in Figure 10.4 are the waist, shoulder, elbow, and wrist bends. The wrist rotation and flange joints give two additional DOFs. Actual gripper closure operation would be a seventh DOF.

It is one thing to specify a robot's configuration, but it is another to achieve that configuration with desired accuracy specifications. This is analogous to the numerical control problem of maintaining spatial accuracy within a hostile environment. The articulated robot configuration presents significant accuracy problems depending on the particular position of the robot with a specified load. Section 10.4 discusses accuracy and repeatability measures with a look at the kinematic problems associated with specific configurations.

10.3.3   Robot Drive Power

The three prime power sources used to drive a robot have been (1) pneumatic, (2) hydraulic, or (3) electrical. There is no doubt that the trend for the future is the electrical drive. This is exemplified in an article in *Time* magazine [16] that comments on the early American standard robots: large, powerful, hydraulic robots ranging in price from $30,000 to $200,000 each. The hydraulic power allowed for heavy payloads but was subject to accuracy and breakdown problems. The Japanese, on the other hand, tended to develop smaller, electrically powered robots

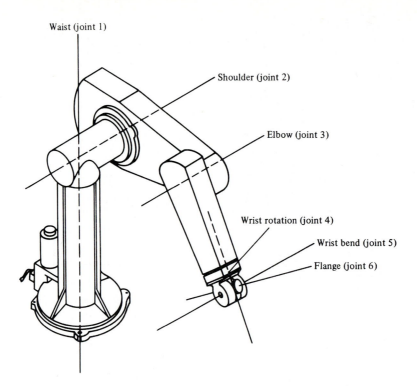

Waist (joint 1)

Shoulder (joint 2)

Elbow (joint 3)

Wrist rotation (joint 4)

Wrist bend (joint 5)

Flange (joint 6)

FIGURE 10.8 PUMA 560 joint configuration. (Reprinted with permission from *PUMA Mark II Robot 500 Series Equipment and Programming Manual*, 398 Pl, Unimation/Westinghouse, Danbury, Conn., 1983.)

that were far less expensive ($5,000 to $40,000). The smaller robots are more accurate in tasks requiring delicate manipulation, such as spot welding.

For heavy loads, hydraulic power is the most effective. With the trend toward highly flexible robots that can be reprogrammed quickly for relatively small applications, the desirability of hydraulic power diminishes. Further, hydraulic equipment tends to be noisy, dirty, and space-intrusive.

An alternative source of power that has just not caught on is pneumatic power. Pneumatics, like hydraulics, requires a cylinder to effect motion. Pneumatic cylinder drives have tended to be hard to control—they are very good for on/off cyclic control but not for delicate positional control. As a result, even though the main power drive for a robot is electric, it is often found that the gripper open/close control utilizes pneumatic power, as shown in Figure 10.9.

One reason that pneumatic power *has* been considered for robot drives is *cost*. One manufacturer specified a 50-lb load capacity with a concomitant cost less than $10,000. The resultant robot was very jerky, and the desired accuracies were very difficult to achieve.

Electrical power tends to be quiet, small on space utilization, with excellent accuracy potential for relatively light loads. *Stepping motors* have been used for

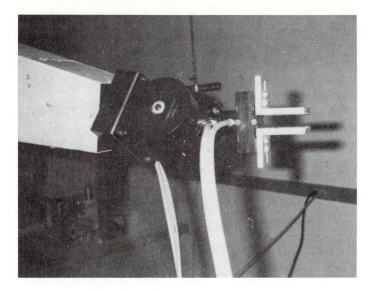

FIGURE 10.9   Gripper controlled by pneumatic power. (Courtesy of Unimation/Westinghouse, Danbury, Conn.)

the drive power because they can be effectively controlled by a series of DC pulses, thus allowing for a logical computer drive. The trend now is for DC servomotors to be the main drive source. As discussed in the next chapter, servomotors have a winding that is driven by the *error* realized when there is a difference between a desired position and the true position. The *larger* the error, the *faster* the motor attempts to correct error. Also, when the error is effectively zero, the motor stops. This makes the servomotor ideal for robot and numerical control positioning.

Some work is continuing on the application of AC motors to close-tolerance positioning. The advantage of the DC motor is that speed is controlled by voltage changes, whereas AC motor speed is effected by frequency changes. Frequency control is difficult and expensive to effect.

10.3.4   Types of Robots

The motion and power classification of robots leads to characterization by *type*. This allows some comments to be made regarding *relative* costs. Absolute costs will not be suggested because of rapid technological and market changes, but the reader can easily review current costs by pursuing the plethora of available magazines relating to robotics applications.

PICK-AND-PLACE ROBOTS. The cheapest robot is one designed to move something from one place to another. The numerical control analogy is of course point-to-point operation. Some of the so-called table-top robots fall in this category and could cost less than $5000.

COMPUTERIZED SERVO ROBOTS. A computerized servo robot functions something like contouring with numerical control, in that movement is optimized. Programming can be by "teach" mode (see Section 10.9.1 for further information) or by high-level language. The range of costs might well be three or more times those of pick-and-place robots.

SENSING ROBOTS. In a sensing robot the computerized servo operation is augmented by sensors that try to emulate human sensory perceptions: vision and feel (tactile), for example. Costs can be considerable, though vision capability is becoming much more affordable and realistic in operation.

ASSEMBLY ROBOTS. Some progress has been made in simulating a human being's approach to assembly: two parallel articulated configurations. Other assembly robots have used the cartesian coordinate approach or a single articulated robot. We are on the threshold of *complex* assembly applications. Costs can be very expensive.

10.4 ROBOT MOTIONS

10.4.1 Introduction

Robot motions range in complexity from simple 1-DOF pick-and-place robots with one rotational joint to seven- and eight-link robots with combinations of cylindrical, spherical, and translational motions. The performance of the robot is, of course, affected by the complexity of the robot linkages. One would expect a 1-DOF robot to be easily controlled by a simple electronic controller, and to be fast and accurate. Larger robots with more links require more sophisticated controllers, and the accuracies in position at the gripper are harder to achieve considering that the accuracy at the tip is influenced by the inaccuracies of all of the intermediate links.

Because of this relationship and because we should have an idea of the accuracies and the performance in general of a multitude of robot types, it is necessary to discuss the kinematics or motion of a robot as a system of independent links moving in a coordinated manner. Our goal here is to mathematically model the linkage motion of the robot to predict, among other things, the position and orientation of the end effector, the accuracy of positioning, and the repeatability. A controller should have the ability to move the individual links and predict the location of the gripper. Conversely, the controller should, given a position and orientation of the gripper, be able to calculate the exact linkage motions required to reach that condition. Other possible goals would be to also predict the dynamics of the arm, that is, the motor torques required to move the gripper to a new location at a given speed and the dynamic response of the robot to torque or force inputs.

Robot dynamics is beyond the scope of this text. We will, however, model robot *kinematics* by applying to the individual robot links similar transformations to those in Chapter 3. In addition, we will look at some examples of kinematics models of theoretical and existing robot arms.

10.4.2 Kinematic Link Chains

A 2-DOF ARM. As a simple first example of robot kinematics solutions, consider the 2-DOF arm in Figure 10.10, which contains two rotational links. This is a planar mechanism, because the axes of link rotation force the links to lie in one plane. We can define two vectors, **r1** and **r2** for links 1 and 2:

$$\mathbf{r_1} = [L_1 \cos(\theta_1),\ L_1 \sin(\theta_1)]$$
$$\mathbf{r_2} = [L_2 \cos(\theta_1 + \theta_2),\ L_2 \sin(\theta_1 + \theta_2)]$$

Adding these vectors gives the point at the robot gripper in terms of the universal coordinate system at the robot base:

$$x = L_1 \cos(\theta_1) + L_2 \cos(\theta_1 + \theta_2)$$
$$y = L_1 \sin(\theta_1) + L_2 \sin(\theta_1 + \theta_2)$$

This is called the forward transform for the 2-DOF robot described above. It gives the gripper position in terms of the link variables, in this case link angles. This is a valuable solution if we would like to know where the gripper is, especially if we are teaching the robot a point by controlling the angles of the joints.

On the other hand, we would also like to give the robot a point in space on which to place the gripper. In fact, we would like to give the controller both a gripper position and an orientation [six values: $(x, y, z$, roll, pitch, yaw)]. We call this the reverse solution: Calculate the joint angles (or other joint parameters) to

FIGURE 10.10   A 2-DOF planar robot.

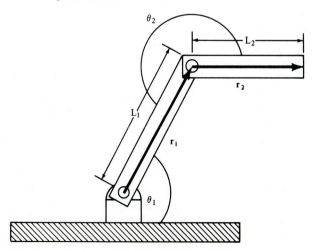

put the gripper in the defined position and orientation. For this simple 2-DOF robot, the reverse solution can be derived by solving the above equations for θ^1 and θ^2. The result (see Ref. 12 for the solution) is:

$$\cos \theta_2 = \frac{x^2 + y^2 - L_1^2 - L_2^2}{2L_1L_2} \tag{10.1}$$

$$\tan \theta_2 = \frac{[y(L_1 + L_2 \cos \theta_2) - xL_2 \sin \theta_2]}{[x(L_1 + L_2 \cos \theta_2) + yL_2 \sin \theta_2]} \tag{10.2}$$

Now, knowing the lengths L_1 and L_2 and the x and y desired location of the gripper, we can calculate angles θ_1 and θ_2. As it turns out with this and other robots having an elbow-type joint, there are two solutions: one with the elbow bent up and one with the elbow bent down. These two solutions are evident if you examine Eq. (10.1), which will yield both positive and negative values.

Textbooks on robotics devote quite a bit of time and space to discussing the forward and backward solutions to many robots. Simple configurations such as the 2-DOF example above are straightforward to solve. The challenge is greater when the complexity of the robot approaches 6 DOF and the joints are not limited to planar motion. In cases such as these, the kinematics equation analysis gives way to the use of 4×4 homogeneous transformations. Before rediscussing how these transformations developed in Chapter 3 apply to robotics, we first need to establish the concept of coordinate reference frames to apply to the many robot links.

10.4.3 Link Geometries

We will begin by describing the mathematical modeling of location and orientation. We will first assume that a universal coordinate system has been established from which to measure points in space and orientation angles. A point P, shown in Figure 10.11, in the coordinate system can be represented by the vector

$$\mathbf{P} = \begin{bmatrix} p_x \\ p_y \\ p_z \end{bmatrix}$$

where P_x, P_y, and P_z are x, y, and z coordinates of the point P.

A rigid-body location can be described by giving the point P of a location on the body in the universal coordinate system. This reference point P is usually at the origin of a coordinate system fixed to the body, that is, the body reference frame (Figure 10.12).

The orientation of the body reference frame can be described by a three-dimensional rotation matrix as defined in Chapter 3. This matrix is specified by the rotations of the body reference frame around the x, y, and z axes of the universal coordinate system. The combination of the point P and the three rotations precisely determines the rigid-body location and orientation.

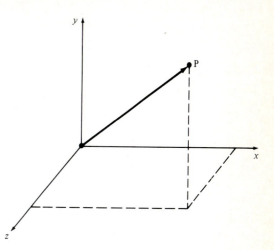

FIGURE 10.11   Representing a point in space.

10.4.4   Frame of Reference

A robot of several links is typically defined according to each link's position and orientation compared to other links in the robot. The links form a kinematic chain (Figure 10.13). Link 0 is the ground link and has its position and orientation defined in relation to the universal coordinate system. The remaining links are defined hierarchically: Link 1 is defined in relation to link 0, link 2 in relation to link 1, etc. Each link possesses its own coordinate system axes, which are usually aligned with the link geometry in some way. A prismatic link may have a set of axes whose origin is at one of the ends (called its base), and the axes may be lined up such that the z axis is along the major length of the link (Figure 10.14).

FIGURE 10.12   World and local frames of reference.

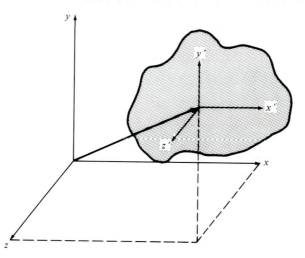

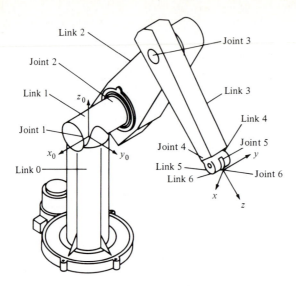

FIGURE 10.13   A robot as a kinematic chain. (From Ref. 26, p. 68.)

l0.4.5 Orientation

We now need to digress a moment to review motion in three dimensions. To describe the position of a rigid body—say, a pencil—in space, six coordinates are required; x, y, and z give the location in space of a point on the pencil—for

FIGURE 10.14   The coordinate system of a prismatic link.

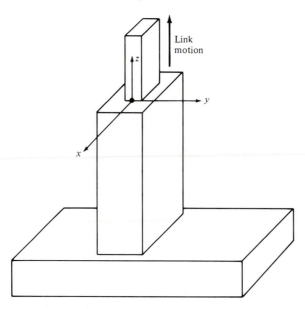

example, its sharpened tip. This becomes the origin of a reference coordinate system. To define the axes of this coordinate system we need the three orientation angles roll, pitch, and yaw, which define the attitude of the pencil. *Roll* is defined as rotation around the long axis, *pitch* is the "up-and-down" angle, and *yaw* is the "right-and-left" angle. The definitions of "up and down" and "right and left," of course, depend on the viewer. But these three direction axes—the long axis, the axis around which the pencil can be moved up and down, and the axis around which the pencil can be oriented right and left—define the z, x, and y reference coordinate system we can use to describe the pencil location and orientation. The six coordinates to define the pencil's location and orientation exactly are the x, y, and z of the local axis origin and the roll, pitch, and yaw angles around the axes. As an example, imagine that we rotate the pencil to give it 45° of yaw and then 20° of pitch. The resulting pencil with respect to the reference coordinate system is shown in Figure 10.15.

10.4.6   Changing Frames of Reference

Assume that we have a robot as shown in Figure 10.16 [23], which corresponds to the geometry of a robot called the Stanford Manipulator, one of the first academically developed and analyzed manipulators. It can be seen that there are six motions: rotation of the entire robot about the axis z_0, rotation of link 1 around the axis z_1, translational motion along z_2, and the motions at the wrist: rotations about z_3, z_4, and z_5. The combinations of motions (six in all) allow the robot gripper to be placed in many positions.

For the Stanford arm and, in fact for robots in general, the first reference coordinate system is attached to the ground and is used to measure the location and orientation of the first or ground link, labeled in Figure 10.16. The origin of the

FIGURE 10.15   An object exhibiting pitch and yaw.

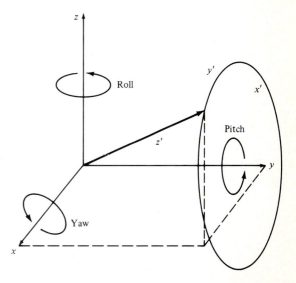

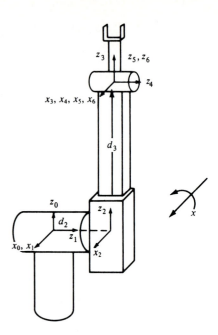

FIGURE 10.16 The Stanford Manipulator. (From Ref. 23, p. 56.)

first or ground coordinate system is labeled x_0, y_0, z_0. The rotation of the robot takes place first about this z_0 axis. The coordinate system attached to link 1 is labeled x_1, y_1, z_1, and the axes are so oriented that the motion of link 2 takes place around the z_1 axis. The coordinate system axes for each link are defined similarly. Can you see that the origins for the coordinate systems for links 4, 5, and 6 (that is, x_3, y_3, z_3 through x_5, y_5, z_5) are coincident? But their z axes are not all aligned. This tells us that link 4 rotates about z_3 in a roll-type motion, link 5 rotates about z_4 in a pitch-type motion, and link 6 (the gripper) rotates about z_5 in a roll-type motion.

All of the above links move rotationally about an axis. The odd link, link 3, moves in a translational motion along z_2. It telescopes to extend or shrink its reach.

Now, the above is just a descriptive treatment of robot motions. We could instead have developed the theory that justifies these coordinate systems, but in a text such as this, it is important to provide a quick practical introduction to how one describes robot motion. This is important, because the sections in the latter part of this chapter deal with measuring the accuracy and repeatability of the robot. To gain an appreciation for the problems of obtaining this accuracy, it is important to realize that robot motion is a combination of as many as six (some-times even more) separate links. And each of these links is controlled by a sepa-rate software/hardware system. Programming a robot to reach a certain point with a specific orientation requires knowledge about each of the link motions and their coordinate systems so that the robot controller can calculate the rotation and/or the translation of each link motion to achieve the desired gripper position. The controller does this mathematically, using the concept of homogeneous transformations which we discussed in Chapter 3.

The corresponding translation, rotation, and scaling matrices are identical to those developed in Chapter 3:

$$\text{Trans}(a,b,c) = \begin{bmatrix} 1 & 0 & 0 & a \\ 0 & 1 & 0 & b \\ 0 & 0 & 1 & c \\ 0 & 0 & 0 & 1 \end{bmatrix}$$

$$\text{Rot}(x,\theta) = \begin{bmatrix} 1 & 0 & 0 & 0 \\ 0 & \cos\theta & -\sin\theta & 0 \\ 0 & \sin\theta & \cos\theta & 0 \\ 0 & 0 & 0 & 1 \end{bmatrix}$$

and so on.

10.4.7 Forward Transformation (Six Degrees of Freedom)

We can define the links of the Stanford Manipulator hierarchically by starting with the ground link with coordinate system $[x_0, y_0, z_0]$ and defining the position and orientation of link 1 with coordinate system $[x_1, y_1, z_1]$. We can see that the position of the origin $[x_1, y_1, z_1]$ is right on top of the origin $[x_0, y_0, z_0]$ but rotated so that the z_1 axis is aligned with the $-y_0$ axis. To move from the origin of 0 to the origin of 1, we must rotate $-90°$ around the x axis. This gives us merely the $[x_1, y_1, z_1]$ origin of the link 1 at rest with no motion. Of course, link 1 may rotate about its z_1 axis an amount θ_1. This motion can be represented by a rotation matrix about z_1 and must be compounded with the $-90°$ rotation around x_0. The compound effect of these two motions can be included by multiplying the individual matrices for each motion to arrive at $\mathbf{A}_1$, where $\mathbf{A}_1 = \text{Rot}(z,\theta_1) * \text{Rot}(x,-90)$. The order of the matrix multiplication is important and should be done in the order shown as in Chapter 3. The matrices should go from right to left in the order performed. The resulting $\mathbf{A}_1$ matrix is:

$$\mathbf{A}_1 = \text{Rot}(z,\theta) * \text{Rot}(x,-90) = \begin{bmatrix} \cos\theta_1 & -\sin\theta_1 & 0 & 0 \\ \sin\theta_1 & \cos\theta_1 & 0 & 0 \\ 0 & 0 & 1 & 0 \\ 0 & 0 & 0 & 1 \end{bmatrix} * \begin{bmatrix} 1 & 0 & 0 & 0 \\ 0 & 0 & 1 & 0 \\ 0 & -1 & 0 & 0 \\ 0 & 0 & 0 & 1 \end{bmatrix}$$

$$= \begin{bmatrix} \cos\theta_1 & 0 & -\sin\theta_1 & 0 \\ \sin\theta_1 & 0 & \cos\theta_1 & 0 \\ 0 & -1 & 0 & 0 \\ 0 & 0 & 0 & 1 \end{bmatrix}$$

Similar $\mathbf{A}_i$ matrices can be calculated for each **Link**$_i$ with respect to the preceding **Link**$_{i-1}$ until a total of six $\mathbf{A}$ matrices are derived. Each matrix gives the transformation to go from the origin of **Link**$_{i-1}$ to **Link**$_i$. The total transformation

$\mathbf{T}_6$, then, to go from the ground origin to the gripper origin is the product of all the $\mathbf{A}_i$ matrices.

$$\mathbf{T}_6 = \prod_{t=1}^{5} \mathbf{A}_i$$

The resulting $\mathbf{T}_6$ matrix can be very complex and is certainly a nonlinear function full of $\sin \theta_i$ and $\cos \theta_i$ terms. Examples of the $\mathbf{T}_6$ matrix can be found in Ref. 23.

The significance of the $\mathbf{T}_6$ matrix is that it calculates the gripper position and orientation based on the individual joint displacements (rotation angles or prismatic translations). It can be used to predict a location of the gripper given joint displacements. Additionally, however, the user may (and, in fact, typically does) specify a gripper location and orientation and expect the robot controller to calculate the joint parameters to reach that gripper position. This is the inverse problem of the $\mathbf{T}_6$ matrix and thus is called the *inverse transformation*.

10.4.8 Solving for Joint Angles

The matrix $\mathbf{T}_6$ contains variables for all joint angles in the form

$$\mathbf{T}_6 = \begin{bmatrix} n_x & o_x & a_x & p_x \\ n_y & o_y & a_y & p_y \\ n_z & o_z & a_z & p_z \\ 0 & 0 & 0 & 1 \end{bmatrix}$$

where the vector (n_x, n_y, n_z) is the normal vector of the gripper coordinate axes (Figure 10.17), (a_x, a_y, a_z) is the gripper approach vector, (o_x, o_y, o_z) is the

FIGURE 10.17 The local coordinate system for the end effector. (From Ref. 23, p. 42.)

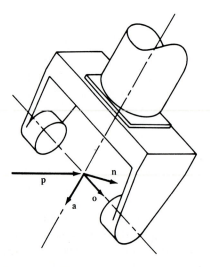

orientation vector, and (p_x, p_y, p_z) is the position vector of the gripper origin with respect to the ground origin. The normal vector is superfluous and can be found by taking the cross-product of the orientation and approach vectors. All of these vector components together with the last-row elements exist in the $\mathbf{T}_6$ matrix as functions of joint angles:

$$[\mathbf{V}_x, \mathbf{V}_y, \mathbf{V}_z] = f(\theta_i, \ i = 0, \ \text{number of joint angles})$$

where $\mathbf{V}$ is either **a, o, n,** or **p.**

The matrix $\mathbf{T}_6$ is a system of 16 simultaneous equations, some of which are redundant, yielding six independent equations to calculate the six independent joint variables. In other words, given a gripper location as (p_x, p_y, p_z), (a_x, a_y, a_z), and (o_x, o_y, o_z), the individual components of the $\mathbf{T}_6$ matrix can be used to find all six joint angles. The procedure is beyond the scope of this text, but an example can be found in Ref. 23.

10.5 WORKSPACE DESCRIPTIONS

Finding the joint angles required to position an end effector in a specific position and orientation is one way to program a robot. The end-effector location is determined by the application the robot is required to perform. For example, in a paint-spraying application, it may be necessary to program the specific path of the paint gun (the end effector) in terms of *x, y, z* coordinates and roll, pitch, and yaw angles and have the computer then determine the angular input to each joint to reach the required positions. It is possible that the robot will not be able to reach all the positions in the painting path. The computer controller will tell the user which positions are reachable and which are impossible.

The set of all possible positions attainable by a robot end effector is called the *robot workspace* or *work volume*. A subset of this is the set of all points that can be reached by a fixed orientation of the robotic hand. This subset can be exemplified by a robot which is required to maintain a tangency to a surface as it traverses the surface. To do this it would be necessary that the robot be able to move continuously with specified orientations through a sequence of points. Just because the robot can move its end effector through a series of points does not necessarily mean that it can do so with a specific orientation. It may need to reorient the end effector as it moves because of geometric or range-of-motion limits.

The workspace is defined by the geometry of the links and the range of motion of each of the joints. In a simple case as shown in Figure 10.18, a 2-DOF planar manipulator workspace can be determined by rotating each of the joints throughout their ranges of motion simultaneously. Assuming that joint 2 can rotate from $-60°$ to $+60°$ and that joint 1 can rotate a full 360°, the workspace of the robot would be defined as the shaded area in Figure 10.18. Notice that the end effector can reach most of the area contained inside the outer circle, but that there is an unreachable region because of the limited motion of link 2. This limited type of workspace is typical of most robots. This limit is usually due to limitations in

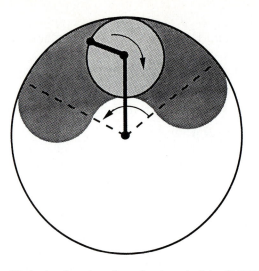

FIGURE 10.18   Workspace for a two-dimensional manipulator (2 DOF).

link mobility but can also be due partially to the load carried by the gripper. It may be that the fully extended arm is not capable of supporting large loads. The movement created about the links may be larger than the link actuators can handle. The workspace, then, is dependent on both the range of motion and the load being carried.

Some example workspaces for standard configurations were shown earlier in Figure 10.5. When determining the workspace required for a specific application, it should be noted which regions it is necessary to reach. Robots vary greatly in their work envelopes, and it is usually true that the robot geometry is chosen to satisfy a certain application such as spray painting or pick-and-place or to give a maximum work volume. Some commercial robot work volumes are shown in Figure 10.19. Notice that these volumes are merely the set of points that are reachable by the end effector independent of workload or orientation. If these latter two constraints are added, the volume may be smaller and have voids which are not reachable by certain orientations or with certain loads.

Determination of the work volume or envelope is, in general, difficult to represent explicitly by geometric equations [26]. Two issues arise when determining the required workspaces:

1. Given the robotic structure, that is, a table for the robotic joint parameters, what is the geometric structure of the workspace?
2. Given a desired geometric description of a workspace for robotic applications, what is the necessary robotic structure?

While there is not a general solution to the above problems, a method has been suggested to rank given robots according to a workspace performance index [14].

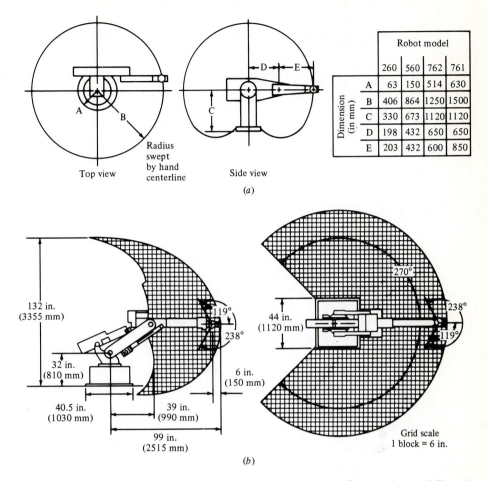

	Robot model			
	260	560	762	761
A	63	150	514	630
B	406	864	1250	1500
C	330	673	1120	1120
D	198	432	650	650
E	203	432	600	850

Dimension (in mm)

Top view — Radius swept by hand centerline — Side view

(a)

132 in. (3355 mm)

32 in. (810 mm)

40.5 in. (1030 mm)

39 in. (990 mm)

99 in. (2515 mm)

119°

238°

44 in. (1120 mm)

6 in. (150 mm)

270°

238°

119°

Grid scale
1 block = 6 in.

(b)

FIGURE 10.19 (a) The PUMA workspace (Unimation/Westinghouse); (b) Milacron T³-726 workspace (Milacron).

The Lee-Yang theorem says that, for a given manipulator, the ratio between the volume of the workspace and the cube of the total length of the extended robot arm is a constant. That is,

$$\frac{V}{(l_1 + l_2 + l_3 + \ldots + l_{2n})^3} = \frac{V}{L^3} = \text{constant} \qquad (10.3)$$

where l_n is the length of link n. This index gives a ranking of the percentages of the actual volume of reach to the maximum volume achievable. The derivation of this formula is beyond the scope of this text, but Table 10.2 gives the performance indices of some industrial robots. The maximum index, VI, is attained using a highly articulated robot of maximum radius L (total length of links) such that the workspace is a sphere of radius L. A normalized performance index, NVI, can be defined based on the maximum possible VI to rank actual robots [26].

TABLE 10.2
Workspace performance indices for selected manipulators

Robot	Degrees of freedom	L (in.)	V (in.³)	VI	NVI
PUMA	6	51.0	183986	1.39	0.331
CMT3	6	101.0	2295223	2.23	0.532
Pana-Robo	5	59.84	355937	1.66	0.397
AID-800	5	59.02	267350	1.30	0.310
CT-V30	5	94.49	1299251	1.54	0.368

10.6   APPLICATIONS

As shown earlier in Table 10.1, welding accounted for approximately 36% of U.S. sales applications in 1985, with assembly accounting for about 11%. By 1995 it is predicted that assembly applications will increase to 26%, with welding dropping to 19%. The other large growth increase for robot application is predicted to be inspection.

Regardless of rankings, the application spectrum can be as broad as the analyst's imagination. This section will discuss a few application areas briefly in order to demonstrate some current robot capabilities. The reader who needs a more in-depth coverage should get a copy of one of the books devoted just to robotics. Excellent coverage can be found in *Robotics: A Manager's Guide* [17].

10.6.1   Welding

Welding is a repetitive, tedious activity that is often accomplished under poor environmental conditions: hot and cramped. Because of the attributes of a human welder, the arc time (the percentage of time a weld is being accomplished) may be only 20–30%. Robot manufacturers indicate increases to 70–80% with robot welding.

Because welding has been such a fertile area for robot application, many robot programming languages include commands for simplifying the coding of welding applications. Typical of these languages is RAIL.

The largest number of welding applications to date has been for spot welding, a pick-and-place type of operation. In addition, arc welding has been accomplished in many other robot applications. *Spot welding* uses an electric current to join two pieces of metal together in a desired point location.

Arc welding provides a continuous seam weld to join two parts and requires a feeding wire to be melted for the seam. Such a system is shown schematically in Figure 10.20. A photograph of a similar system is shown in Figure 10.21.

An example of *adaptive* welding is a car body on an assembly line that needs silicon bronze welding. The location, width, and depth of the seam are variable. A vision system locates the seam and corrects the robot motion. In addition, the

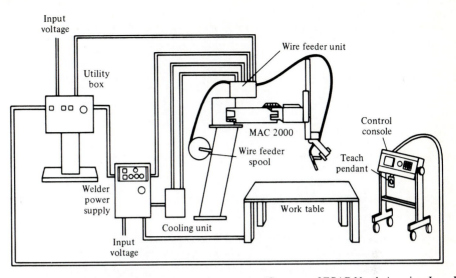

FIGURE 10.20   Composition of a MAC 2000 arc welding robot system. (Courtesy of ESAB North America, Inc., Fort Collins, Colo.)

FIGURE 10.21   An ESAB welding robot. (Courtesy of ESAB, Fort Collins, Colo.)

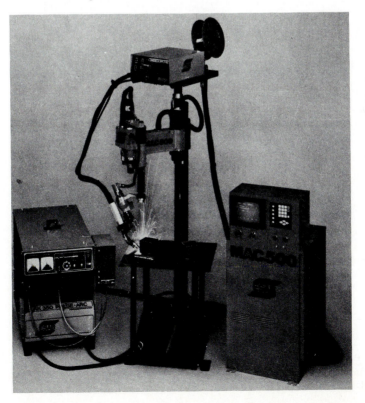

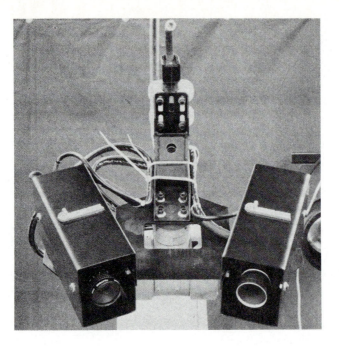

FIGURE 10.22   An Automatix Robotic Welding Vision System. (From *Partracker—Vision-Guided Robots*, Automa-tex, Billerica, Mass., © 1984, with permission.)

seam is gauged visually, and correct weld parameters (speed, voltage, weave, etc.) are automatically selected to optimize welding of the seam geometry. Figure 10.22 shows a vision system used in welding operations.

Stonecipher [27] has compiled a list of factors the user should consider when evaluating particular robot welding capabilities:*

1. The weld gun supplied by a number of robotic vendors has a tendency to deviate, and this can lead to miswelding or overshooting of the material.
2. The robotic unit may have to vary the voltage and amperage settings in order to handle varying material thicknesses and viewing.
3. The welding gun speed before and after it gets to a particular weld point may vary.
4. The robotic welder must correctly orient the weld gun and position it at the particular spot to be welded. In the initial analysis, a range of velocity in inches per minute must be defined.

---

* Permission granted by the Hayden Book Company to use this list from Stonecipher, Ken: *Industrial Robotics: A Handbook of Automated Systems Design,* © 1985 by the Hayden Book Co., Hasbrouk Heights, N.J.

5. The ability of an arc welding robot unit to weld in a wide variety of patterns will increase any future flexibility for a particular welding work station.

6. The robotic welder should be able to perform multi-pass weldings close to the original seam.

7. The robotic unit should also be able to vary the width of a particular weld seam by utilizing a short "weave" function of weld patterns. This involves the ability of a robotic unit to move a weld gun from side to side so that a wide weld width is accomplished.

10.6.2 Spray Painting

Spray painting is a task that is environmentally undersirable for a human operator, but it is quite difficult to get the uniform coating so often required in many products. The requirement for expensive fume elimination can often be eliminated with robot spray painting.

As with robots used in welding applications, considerable dexterity is often needed to allow painting to be accomplished in locations that are difficult to get at. Also, because of the possible hazardous effects of the paint on the robot's linkages and control equipment, a protective cover system may be necessary.

10.6.3 Materials Handling

Materials handling refers to the movement of parts, tools, materials, and equipment in a facility. The cost of materials handling can be extreme, especially when the materials are easily damaged, environmentally a problem, bulky, or have to be oriented correctly.

A manufacturing axiom for many years has called attention to the fact that materials handling in manufacturing is nonproductive and should be eliminated as much as possible. To offset this, a part is sometimes operated upon while it is being moved, such as in an assembly line.

Tompkins and White [30] state that materials handling accounts for some 10% to 80% of a facility's cost, and that robots are used to perform in-process materials handing under a variety of environmental conditions. It is suggested that a robot be considered when:

- The rate of handling is about 15 or less pieces per minute,
- The load weight is less than 1000 lb.
- The orientation of the objects to be grasped is consistent.

Actually, with vision systems, the last suggestion is not mandatory, although ensuring a consistent orientation through mechanical and/or palletizing means may well be cheaper than incorporating a vision system.

One advantage for grouping parts, and the equipment for making those parts, into manufacturing cells is the resultant reduction in materials handling. A

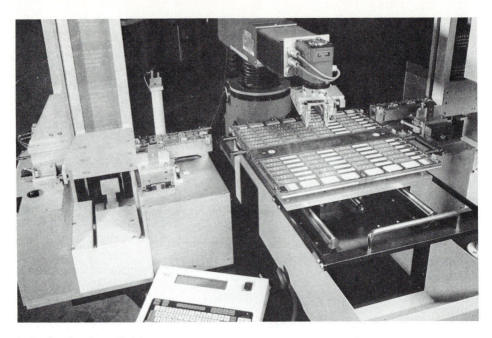

FIGURE 10.23 An Intelmatic robot palletizing system. (Courtesy of Intelmatic, Haywood, Calif.)

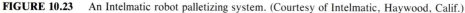

robot might move parts from a conveyor to an incoming parts area, move parts to the machine, move from the machine to inspection, and move from inspection to an outgoing area or conveyor.

Another lucrative opportunity for robots in materials handling is in palletizing and stacking operations. These are labor-intensive operations requiring strength and dexterity. Robots can be programmed to stack or palletize in a minimum amount of time. Figure 10.23 shows an Intelmatec system for palletizing wire-bonded lead frames in prestaging for molding. These are utilized in the electronics industry.

10.6.4 Assembly

Assembly of components has long been a task thought most appropriate for humans to perform. After all, components must be aligned and sometimes intelligently fitted to mating components. Assembly tasks occupy 53% of time and 22% of labor costs in manufacturing a product. Assembly requires skill in both selecting and orienting parts and in dealing with fitting together mating components, sometimes using serpentine motions to align the parts. Joining parts into a composite involves welding, screwing, glueing, or fitting parts together. Welding has been discussed in Section 10.6.1. This section covers the other joining operations.

Robots are good at moving from place to place, but they need special help to grab parts in the correct orientation for assembly and also for performing peg-in-hole-type mating, especially if the fit between the peg and the hole is tight.

Automated assembly involves the use of robots to select and assemble components, but it also must be matched with a system of designing components that are assemblable by robots. Boothroyd [1] has proposed a standard procedure to evaluate the assemblability of objects for automated assembly. Two important parameters are the symmetry of the components and the number of components in the assembly. Boothroyd's plan reduces the number of parts so that assembly operations are reduced and either requires the components to be totally symmetric, so that the orientation is not critical to the matching of mating parts, or instead requires them to be *very* asymmetric, so that the asymmetry is obvious to the robot without extensive examination of the small features of a part. The use of Boothroyd's scheme or something similar prepares the design for easy assembly by a robot.

Typically, a pick-and-place robot can grab parts from bins as long as the bins present the parts for easy orientation. For years, IBM assembled portions of the Selectric typewriter at their Lexington, Kentucky, plant using robots. The assembly required that gears be placed on hubs and that adjacent gears be meshed accurately. A human assembling such parts would first place one gear on a hub and then slide the other gear on the adjacent hub while turning the gear teeth until they aligned and the gears meshed. The robotic approach to this problem is to place the first gear, gear A, and then with straight-line motion and no rotation, slide gear B onto the adjacent hub. The chances of the gear teeth lining up are less than 50%. If they do not align, gear B is removed, rotated an angle θ around its axis, and tried again. The retrial is performed repeatedly, until either the gears mesh or it is determined that the gears or hub locations must be defective. A simple example like this shows that robotic assembly is not always straightforward and, in fact, we sometimes take for granted the human intelligence required in even simply assembly tasks.

Another complicated assembly problem is the peg-in-hole problem (Figure 10.24). If the peg and hole are a rather close fit, the peg may become angled during assembly and bind in the hole, requiring disassembly and realignment of the peg and hole. Since this is a common problem, researchers have developed what is called "remote center compliance" grippers, which kinematically move the point about which the gripper rotates to align with the center of the end of the peg. Then when the controller commands the gripper to reorient the peg, the center of its far end does not move, since it is the center of rotation and the peg can be rotated to align with the hole. This arrangement reduces the probability that the peg will bind in the hole by simulating the process of *pulling* the peg in by its end. The controller of the manipulator applies insertion forces only when there is no resistance to motion. If the gripper were to push when the parts were not aligned, binding would result and the force required to separate the components might be very large.

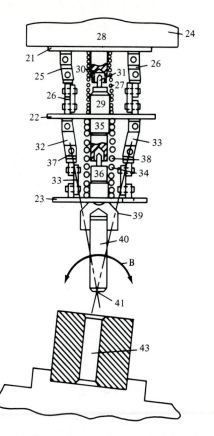

FIGURE 10.24   An end effector for inserting pegs in holes. (Reproduced with permission from Ref. 13.)

Because of the differences in motion capabilities between humans and robots and also because of the intelligence and adaptability differences, a product designed for manual assembly cannot be assembled robotically without redesign. There are two sources for products with potential for robot assembly [22]:

1. New products or redesigned versions of existing but outdated products
2. Existing products that are deemed to have an expanding market and, consequently, increased sales

Owen [22] presents 15 design rules to be observed when designing or analyzing a part for robotic assembly. They are summarized briefly below:

1. Make sure that all components are necessary. Sometimes components may be replaced by one composite part.
2. Construct a precedence diagram to check the sequence of assembly. This

often highlights redundant assembly movements and in what ways it is possible to assemble the parts incorrectly.

3. Make sure that all parts are defined to the robot exactly as they are made.

4. Minimize component variation. If there are similar items in the assembly, see if they can be made identically. Robots have a difficult time separating similar parts.

5. Use symmetric components when possible. This reduces the need for the robot to orient the components.

6. If a part must be asymmetric, make it extremely asymmetric. This allows easier handling by the robot.

7. Minimize the number of components. Each separate part must have its own feeding and handling equipment.

8. Present parts to assembly benches at a known and consistent rate.

9. Design components for unidirectional assembly.

10. Maximize the number of assembly functions for each component to minimize the number of parts in the assembly. Make components do double duty. For example, clamp wires between mating halves of an assembled plug to eliminate the need to use clamps or screws to hold the wires.

11. Design with the robot in mind. Stick with simple pick-and-place motions.

12. Avoid unnecessary reorientation of the assembly during the assembly process.

13. If it is necessary to move a partially completed assembly, the assembly should be sound enough to move without extra jigs.

14. Make subassemblies as large as possible. The farther up the assembly chain each subassembly travels before final insertion, the more money will be saved.

15. Be aware of the power of the robot and grippers and avoid damage to delicate components by making the subassemblies strong in themselves.

Other items to consider when building a robotic assembly station, besides the design of the assembled components, include balancing a robot assembly line, consideration of gripper designs and material feeders, and the inclusion of sensors (vision and touch) in the robots' peripheral equipment.

Manual assembly lines need not be precisely balanced. Human workers can deviate from a mean assembly time and adapt to various assembly rates. Not so with a robot. Robots are precise instruments and perform consistently once programmed and started. Idle time at a station can be minimized by adjusting the robot's pace to coordinate with the functions and times of other robots in the assembly process. The speed at which the tasks are performed at each station is a global, not a local, concern now that robots staff each station and must coordinate their task times. Timing the tasks becomes a system problem because all robots in the system must receive and send parts on a rigid schedule to optimize the assembly process.

Gripper designs can include just the end effector or the possibility of multiple-gripper robot configurations for multihand assembly. We will first discuss single grippers and then explore optional multigripper designs.

An assembly gripper should have some sense of the presence or absence of an object. The IBM 7565 gripper performs a simple open/close operation but contains sensors for both detection of an object and also the amount of force on the jaws. The light source/sensor in opposing jaws (Figure 10.25) permits the robot to sense the presence of an object between its jaws before it grips the object. In assembly, if the material feeder is empty, this sensor can detect that condition by moving to a position where it expects to have an object between its jaws and alert the system to reload the feeder. In addition to knowing merely whether some object is between its jaws, it is also convenient and sometimes necessary to detect if the *correct* object is in place. Given that the robot knows the jaw spread at any time and that the pinch force can be calculated, the robot can sense a dimension of the part by gripping it until the pinch force increases. This technique can be used to see if the component is in the correct orientation to be picked up and assembled. This also allows for exception (error) handling by triggering abort procedures if the jaws grip too tightly on an object. Consider a robot packing eggs in crates. The force sensors, if calibrated correctly and of the required sensitivity, can allow the robot to pick up an egg without breaking it. Similar applications exist in circuit-board assembly and in assembly of sensitive instruments. Although

FIGURE 10.25 IBM 7565 gripper with sensors: (*a*) tactile sensing features; (*b*) light-emitting diode for optical sensing. (Courtesy of International Business Machines Corp., Yorktown, N.Y.)

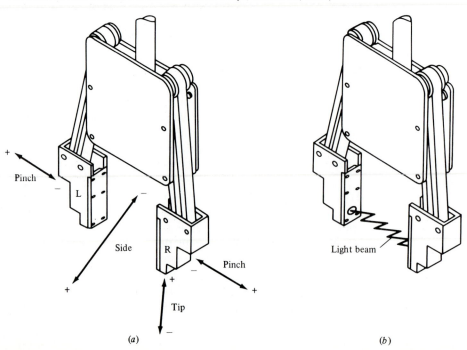

(*a*)   (*b*)

most grippers do not have this sophisticated sensitivity, new gripper designs are being investigated which have hundreds of force sensors on each of the jaw plates to allow the robot to detect not only the force, but also the shape of the imprint of the object being grasped.

A gripper's function is, by definition, to grasp an object. Various types of gripping strategies exist [extracted from 22]. Clamping can be done:

- *Mechanically*, as with fingers or jaws or even anthropomorphically shaped hands
- *Magnetically*, as long as the objects being clamped are metallic and not destroyed by magnetic fields
- *By vacuum*, as long as vibrational shock when the object jumps to the gripper is not a problem
- *By piercing*, which punctures the component, as in a clothing application
- *By adhesion*, using sticky tape or another adhesive

A larger approach to optimizing the assembly robot to replicate a consistent and dextrous human is the multiple-robot work station or multiple-gripper robot in which the grippers work in harmony as would two human hands. An example is shown in Figure 10.26. These designs are especially valuable if multiple components must be assembled simultaneously. There may be problems, however, in

FIGURE 10.26   A two-arm robot. (Courtesy of Prutes Ltd. and PA Technology.)

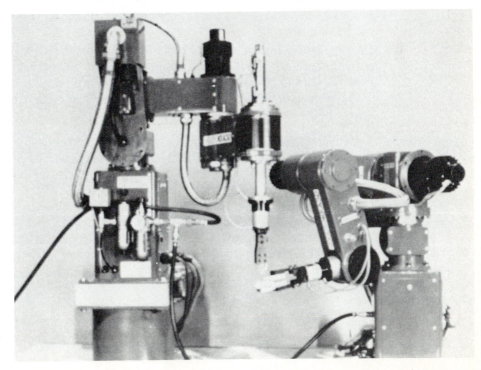

coordinating the accuracy and repeatability of the two arms. Getting two moving grippers to move to a common point introduces twice as much inaccuracy as getting one gripper to move to a fixed point. For fine-motion assembly, two-arm designs may lack the required precision.

Finally, robot arms may possess the capability to change end effectors to suit the job. A robot application to disassemble nuclear reactors at Sandia National Laboratories uses a PUMA robot with interchangeable end effectors. At various times in the process the end effector is a pneumatic wrench, a jaw gripper, or a screwdriver. An example of such a multiple-use end effector is shown in Figure 10.27. Note that the gripper base contains electricity and air power to suit the various end effectors.

Material feeders are an important part of the assembly work station. It is crucial that parts are presented to the gripper in a manner that allows the parts to be grasped. Some robots may be more adept (and more expensive) at picking up randomly oriented parts, but most inexpensive pick-and-place robots require a fixed-part orientation. Material feeders can be classified in three categories [22]:

1. *Bowl feeders* use vibratory action to move items from the bottom of the bowl up an internal ramp, being positioned using gravity devices along the way. Only correctly oriented parts reach the top where the robot can grasp them. Bowl feeders are one of the most successful methods of presenting parts to a robot.

2. *Autoscrewdrivers* feed screws, nuts, rivets, and studs using an air/vacuum system to hold the part. In some cases all of the parts are prealigned coaxially in a long hole inside the autoscrewdriver and the components are presented at the tip one at a time.

3. *Gravity feeders* consist of ramps or tubes which hold the parts and are used for parts that are not conducive to vibratory feeding. As the items are removed from the bottom of the ramp, other parts slide down into place to be grasped. This device assumes that the parts have been preloaded by a human or other device in the correct orientation.

Of course, the final evaluation of a robot assembly system must be economic justification. If no financial benefit is obtained over manual assembly, the system is obviously undesirable. Parameters for which robotic assembly seem to help include: reduced labor costs, improved scrap rates because of consistent assembly practice, and reductions in total work time. Areas of increased cost associated with robotic assembly include capital costs, maintenance, and possibly a short estimated lifetime of the robot. As technology produces faster, smarter, and, supposedly, cheaper robots in the future, robotic assembly will become even more commonplace.

10.6.5   Inspection

General-purpose robots can perform limited inspection routines. For example, a simple manipulator may be able to detect the presence or absence of a component

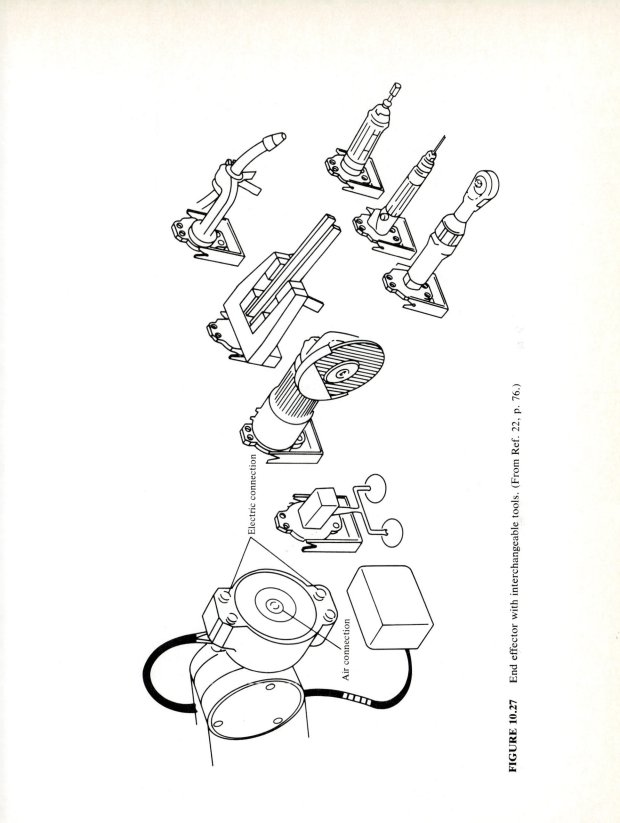

Electric connection

Air connection

FIGURE 10.27 End effector with interchangeable tools. (From Ref. 22, p. 76.)

in an assembly by moving to the suspected location and using a touch or optical sensor. Specialized robots can perform more rigorous inspection procedures. For example, a coordinate measuring machine (CMM) which has the capacity for programmed motion can be used to check the dimensional correctness of a part. CMMs typically have a repeatability measure of ±0.0001 in. based on the accuracy of the position transducers in the CMM. A CMM is usually a cartesian robot, which means that the first three axes of motion are orthogonal translations. The wrist then allows for the roll, pitch, and yaw motions required for accurate measurement sensing. The CMM end effector is a probe which is touch-sensitive and registers its location upon contact with an object. CMMs can be programmed to inspect parts by defining the particular points to be measured and defining the path from one point to another. A computer on board the CMM allows the measurements to be converted into part dimensions by establishing local part origins and measuring relative distances from reference surfaces and points.

It is impossible to mention robotic inspection without at least briefly discussing robot vision. The purpose of attaching a vision sensor is to give the robot a global view of its workspace and to allow the robot to identify parts for either inspection or assembly and determine the quality of the parts. The vision systems usually consist of multiple TV cameras capable of digitizing, storing, and processing the information in the scene. The camera is typically attached to a frame grabber, a processor which stores the image in terms of picture elements or pixels. The camera delivers the image through a digital sensor which divides the scene into a pixel array, usually of a resolution greater than or equal to 256×256 pixels. High-resolution cameras have resolutions of 1024×1024 pixels and are capable of seeing finer detail in the scene and making more accurate analysis of the objects in the scene.

The cameras can be mounted on the robot itself as in a self-guided vehicle or separately. The manipulator processor must at all times be able to calculate the position of the end effector with respect to the objects in the scene. In order to gain depth-of-field information, multiple cameras are required or the ability to move one camera and take multiple views of the scene. This location information allows the manipulator to gain information about where the objects are in the workspace and to determine exactly how to approach to grasp the objects. It also lets the robot know if any obstacles lie in its path. Many types of cameras are available, which can be classified by the type of electronics used to sense the image. Current sensing methods are listed here without explanation just to give the reader an idea of the types of technology to pursue in further reading [26]:

1. Image orthicon tubes
2. Vidicon tubes
3. Plumbicon tubes
4. Iconoscope tubes
5. Image dissector tubes

6. Charge-coupled devices (CCDs)

7. Charge-injected devices (CIDs)

Sensing the image means computing certain parameters to try to identify the objects in the camera's field of view. Since the image consists of an array of pixel information, it is possible to scan the array numerically in order to compute which pixels comprise the objects and which pixels are a part of the background. The results of this analysis can provide a description of the object's edges, center of area, orientation, size, similarity to other objects, and features of the object such as slenderness, roundness, etc.

The above-mentioned features can be calculated as follows. This analysis assumes a picture made up of individual pixels as shown in Figure 10.28. The vertical and horizontal axes represent the pixels in the camera and consequently the scene registered on the frame grabber. Each pixel can be either a 0 (black) or a 1 (white); that is,

$$I(x, y) = 1 \qquad \text{for white pixels}$$

$$I(x, y) = 0 \qquad \text{for all black pixels}$$

The total area of the black image can be calculated by the double integral

$$A = \int_{a_1}^{a_2} \int_{b_1}^{b_2} I(x, y) \, dx \, dy$$

The center of the area (x_a, y_a) can be calculated similarly as

$$\int_{a_1}^{a_2} \int_{b_1}^{b_2} sg(x - x_a) I(x, y) \, dx \, dy = 0$$

and

$$\int_{a_1}^{a_2} \int_{b_1}^{b_2} sg(y - y_a) I(x, y) \, dx \, dy = 0$$

FIGURE 10.28   Binary image bit map.

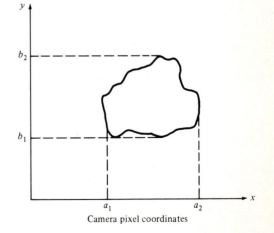

Camera pixel coordinates

Once the features of the image have been calculated, it can be classified or identified by comparing the features against the features for all known objects that might match the image. This template-matching process can be initiated by teaching the system the templates by having the system look at known objects, calculate the features, and store the results under the identifier of the object's name. For example, a nut may be taught to the vision system for later identification of other nuts by grabbing a picture of the nut, calculating the size, center of area, etc., and storing these results with the name "nut." Subsequent pictures displaying a nut can then be classified as such by comparing these parameters to the stored nut parameters. Problems can occur, however, especially if objects are viewed from different angles or other objects overlap the object of interest.

Inspection of objects using a vision system can be done as long as it is realized that the limit of accuracy is determined by the resolution of the camera and the size of the object on the image. For example, if an object which is 10 in. long occupies an area of 512 pixels in length, then the maximum differential of measurement of this length of any distance on the object is limited to 10/512 in., or about 0.019 in.

A better way to use a vision system for inspection is to use the image and the resultant feature information to classify the part and then use that information to select the correct inspection routine for the particular part. Then let the robot or CMM use its high-precision measuring capabilities to locate and confirm reference locations on the object.

10.7 ROBOT ACCURACIES AND REPEATABILITIES

Independent of the range of motion of the manipulator, it is important to discuss the concepts of *accuracy* and *repeatability* of the end effector. Manufacturers usually refer to these robot values when specifying the robot's performance. Two examples will serve to define these terms.

There are several ways to teach a robot to move to a point. One method leads the robot either by a handle near its gripper or by a calculator-type teach control box to move to a specific location and orientation. Once at its taught location, the joint angles and displacements are stored in memory so that the robot can return to that position by itself at some later time. The measure of the robot's precision in returning to the same location time after time is termed *repeatability* [2].

If, instead of teaching the manipulator using a teaching pendant or handle, the user specifies a particular *x, y, z* coordinate and roll, pitch, yaw orientation, then the controller must convert these numbers into joint angles using the matrix approach discussed earlier. The robot's precision with which to move to a computed point is termed the *accuracy* of the manipulator. While the repeatability of most robots is quite good, the accuracy is usually much worse. The reason for poor accuracy comes from the calculation of joint parameters from the matrices.

Round-off of real numbers, other problems with matrix calculations, and the resolution of the driving motors cause much of the inaccuracy [26].

The above definitions are working definitions of accuracy and repeatability. Robot accuracy has not yet been uniformly defined. The International Standards Organization and the National Institute of Standards and Technology (formerly NBS) are proceeding toward definitions but without so far reaching agreement.

Repeatability is a function of number of cycles, location in workspace, temperature, speed/acceleration, payload, etc. Accuracy (Figure 10.29) is a function of many variables, which may be classified into the following categories:

- Environmental factors: temperature, humidity, electrical noise
- Parametric factors: link lengths, structural compliance, friction, hysterisis and backlash
- Measurement factors: computation errors, round-off, control errors
- Computational factors: path computation errors, round-off errors
- Application factors: installation errors, workpiece coordinate system definition errors

These categories may not be independent of each other. For example, temperature may affect link length, electronics drift, friction, etc. Since accuracy depends on some variables that are beyond the control of the manipulator manufacturers, they generally do not publish accuracy specifications, but absolute inaccuracies have been reported in the range from 5 mm to 10 mm. Manufacturers do give

FIGURE 10.29 Absolute robot accuracy. (From Ref. 4, p. 3.)

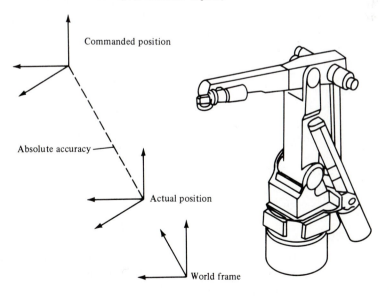

numbers on repeatability. Typical repeatability of high-quality industrial robots is on the order of ± 0.002 to ± 0.010 in.

Day [4] discusses each of the above problems and their individual contributions. Day also mentions the following ways that robot manufacturers are addressing the improvement of robot accuracy:

1. Calibration methods can improve accuracy, but unless the complete workspace is calibrated, this technique usually improves the accuracy only in the region of calibration.
2. Open-loop methods correct the kinematic equation parameters for improved calculation of joint angles. Drive-train errors can be minimized through the application of correction factors to the controller.
3. Closed-loop methods use sensors to detect errors and feed them back to the controller for real-time correction of inaccuracies.

10.8   ECONOMIC JUSTIFICATION OF ROBOTS

It was not uncommon in the "early days" of robots for a company to purchase a robot for evaluation purposes in order to determine the potential technical benefit that might accrue if such a robot were to be implemented in the company's production processes. Many universities benefited from donated robots as a result of such evaluations. Now that the capabilities of robots are well founded, it is doubtful that a company will purchase a robot strictly for evaluation. Robots with known capabilities will be considered for specific production tasks. No company should invest in a robot (or any other piece of equipment), however, without a thorough *economic analysis* of the decision.

An *economic* justification conjures up possible *direct cost* savings, where direct costs are those costs charged specifically to a *particular* product that is manufactured. Labor that is ticketed to the item being produced would be direct labor, while materials required for the product would be direct materials. Other labor and material costs, such as those that occur in the personnel department, would be a part of the plant's *indirect* costs. Indirect costs may far exceed direct costs in a company's operation, and the effect on indirect costs of the purchase of a robot should certainly be considered.

Fleischer [10] discusses the fact that robots generally perform no functions that cannot otherwise be performed by combinations of human workers, machines, and devices. As a consequence, the economic aspect is fundamental to the user's decision to acquire the equipment.

10.8.1   Life-Cycle Costs

The costs that should be considered when evaluating a robot's potential—say, in replacing humans in a repetitive task—are very broad in scope. In many evaluations only a few of these cost are considered.

Fleischer [10] indicates that economic consequences may be grouped into three broad categories, as follows:*

1. Plant and equipment
    1.1 The robot(s), including sensors and interlocks
        1.1.1 Initial cost
        1.1.2 Service life (not a cost)
        1.1.3 Residual value (net salvage value) at the end of the service life
    1.2 Associated tooling
    1.3 Spare parts
    1.4 Property taxes
    1.5 Insurance (property only)
    1.6 Energy requirements
    1.7 Tax consequences
        1.7.1 Investment credit
        1.7.2 Tax savings due to depreciation
        1.7.3 Gains (loss) on disposal
    1.8 Space requirements
    1.9 Installation (including rearrangement of existing facilities)
    1.10 Safety equipment (protective clothing, etc.)
    1.11 Programming
    1.12 Modification of existing equipment to ensure compatability with robot(s)
2. Operation and maintenance
    2.1 Operating labor
        2.1.1 Salaries/wages
        2.1.2 Fringe benefits (costs to employer)
    2.2 Maintenance labor (for periodic maintenance)
        2.2.1 Salaries/wages
        2.2.2 Fringe benefits (costs to employer)
    2.3 Direct cost of injuries and illness (hospitalization, medical care, etc.)
    2.4 Absenteeism (cost of lost productivity)
        2.4.1 Illness
        2.4.2 Feigned illness
        2.4.3 Injury
    2.5 Training
    2.6 Supervision
    2.7 Insurance (personnel only)
    2.8 Overtime (not included in 2.1 and 2.2 above)

---

* This material was extracted with permission from Fleischer, G. A.: "A Generalized Methodology for Assessing the Economic Consequences of Acquiring Robots for Repetitive Operations," in *Robotics and Industrial Engineering Selected Readings,* vol. II, published by Industrial Engineering and Management Press, Norcross, Ga., © 1986.

        2.8.1 Operating labor

        2.8.2 Maintenance labor

    2.9 Labor turnover

        2.9.1 Termination

        2.9.2 Recruitment

        2.9.3 Training

    2.10 Retooling and setup costs for batch processing

    2.11 Maintenance tools and supplies

    2.12 Documentation (operation and maintenance)

    2.13 Costs of interrupted production not included in 2.10, especially downtime

3. Product

    3.1 Required changes in product design

    3.2 Raw material requirements

    3.3 In-process inventory

    3.4 Effects of production rate on:

        3.4.1 Other plant activities

        3.4.2 Shipping schedules

    3.5 Defective (substandard) product

        3.5.1 Scrap rate (not a cost)

        3.5.2 Net cost of handling and reworking defective product

        3.5.3 Costs due to undetected defective product released to customer (e.g., loss of goodwill, responding to customers' complaints, replacing returned products)

In addition to the above, certain assumptions are required to complete discounted-cash-flow economic analysis. These include:

    4.1 Income tax rates

        4.1.1 Federal

        4.1.2 State

        4.1.3 Local

    4.2 Engineering (and consulting) costs not included above

    4.3 Cost of capital (to be used as the discount rate)

A computer program will now be used to demonstrate how some of the cost factors play a role in the economic evaluation of a robot against an alternative approach.

10.8.2 A Computer Program for Robot Evaluation

Sullivan and Liu [28] published a 255-line BASICA program that assists the user in making financial decisions relative to robot adoption versus manual operation in repetitive production. It would not take too much imagination to use the program in evaluating different robot possibilities for a new operation. Since the program is only 255 lines long, it is not too time-consuming for the reader to type in the program. Since the program demonstrates typical costs needed in such an evaluation, an example will be presented.

A printout of the interactive run is given in Figure 10.30. The first question concerns the expected useful life of the robot. Since the program takes income taxes into consideration, it is necessary (in 1990) to depreciate over at least a 5-

FIGURE 10.30   Economic analysis using a computer program.

```
run
THIS PROGRAM IS INTENDED FOR THE AFTER-TAX ECONOMIC ANALYSIS OF INDUSTRIAL
ROBOTS.  END-OF-YEAR CASH FLOW CONVENTION IS ASSUMED.  CASH OUTFLOWS (COSTS)
ARE ENTERED AS NEGATIVE NUMBERS AND CASH INFLOWS (SAVINGS) ARE ENTERED AS
POSITIVE NUMBERS.  THE INTEREST RATE, INFLATION RATE, AND EFFECTIVE INCOME
TAX RATE ARE ENTERED AS DECIMALS.

HOW LONG IS THE STUDY PERIOD (USEFUL LIFE OF THE ROBOT), (N)? 5

THE DIRECT LABOR SAVINGS INCLUDE THE FOLLOWING COMPONENTS:
     1. PLANNED PRODUCTION QUANTITY EACH YEAR, (P)
     2. OUT-OF-POCKET COST PER HOUR FOR MANUAL LABOR, (CM)
     3. AVAILABILITY OF MANUAL LABOR, (AM)
     4. CYCLE TIME (MIN./PC.) FOR MANUAL OPERATION, (TM)
     5. OUT-OF-POCKET COST PER HOUR FOR ROBOT APPLICATION.  (CR)
     6. AVAILABILITY OF ROBOT APPLICATION.  (AR)
     7. CYCLE TIME (MIN./PC.) FOR ROBOT APPLICATION.  (TR)
─────────────────────────────────────────────────────────────────────────
OUT-OF-POCKET COST PER HOUR FOR MANUAL LABOR IS ? 22.5
OUT-OF-POCKET COST PER HOUR FOR ROBOT IS ? 9.75
AVAILABILITY OF MANUAL LABOR IS ? .8
AVAILABILITY OF ROBOT IS  ? .98
CYCLE TIME (MIN./PC.) FOR MANUAL OPERATION IS ? 18
CYCLE TIME (MIN./PC.) FOR ROBOT OPERATION IS ? 16.5

THE PRODUCTION CAPACITY FOR EACH YEAR IS DETERMINED BY:
     1. NUMBER OF SHIFTS/DAY
     2. WORK-HOURS/SHIFT
     3. WORKING DAYS/YEAR

DO YOU HAVE SAME PRODUCTION CAPACITY FOR ALL PERIODS (Y OR N)? y

INPUT NUMBER OF SHIFTS/DAY? 2
INPUT WORK-HOURS/SHIFT?  8
INPUT WORKING DAYS/YEAR?  300

YOUR PRODUCTION CAPACITY FOR THE STUDY PERIODS ARE:

                   CAPACITY
    YEAR           (HR/YR)
    ────           ───────
      1              4800
      2              4800
      3              4800
      4              4800
      5              4800

WHAT ARE THE PLANNED PRODUCTION QUANTITIES DURING THE STUDY PERIOD:
YEAR 1 ? 10000
YEAR 2 ? 10500
YEAR 3 ? 10500
YEAR 4 ? 10000
YEAR 5 ? 9500
```

THE INDIRECT LABOR SAVINGS OR COST ARE:
YEAR 1 ? −8000
YEAR 2 ? −7500
YEAR 3 ? −6000
YEAR 4 ? −6000
YEAR 5 ? −6000

THE EXTRA ANNUAL MAINTENANCE SAVINGS OR COST DURING STUDY PERIOD ARE:
YEAR 1 ? −15000
YEAR 2 ? −15500
YEAR 3 ? −16000
YEAR 4 ? −16000
YEAR 5 ? −17000

THE EXTRA ANNUAL SAVINGS IN MATERIALS DURING STUDY PERIOD ARE:
YEAR 1 ? 21000
YEAR 2 ? 21000
YEAR 3 ? 22000
YEAR 4 ? 22000
YEAR 5 ? 20000

OTHER SAVINGS OR COST ASSOCIATED WITH ROBOT VS. MANUAL DURING THE STUDY
PERIOD ARE:
YEAR 1 ? 0
YEAR 2 ? 0
YEAR 3 ? 0
YEAR 4 ? 0
YEAR 5 ? 0

AFTER-TAX MINIMUM ATTRACTIVE RATE OF RETURN (MARR) (BETWEEN 0.0 AND 1.0) =? .14

EFFECTIVE INCOME TAX RATE (BETWEEN 0.0 AND 1.0) =? .28

ANNUAL INFLATION RATE =? .05

INVESTMENT TAX CREDIT (BETWEEN 0.0 AND 1.0) =? 0

CAPITALIZABLE COST OF ROBOT SYSTEM =? −132000

ACRS RECOVERY PERIOD =? 5
ACRS PERCENTAGE OR DEPRECIATION RATE IN
YEAR 1 ? .2
YEAR 2 ? .2
YEAR 3 ? .2
YEAR 4 ? .2
YEAR 5 ? .2

DO THE ACRS PERCENTAGES SUM UP TO 1 (Y/N)? y

ESTIMATED INFLATED MARKET VALUE (SALVAGE VALUE) OF THE ROBOT SYSTEM AT END
OF YEAR 5 ? 0

CASHFLOW AND DEPRECIATION

YEAR	D.L. SAVINGS	I.L. SAVINGS	MAINT.	MATLS.	OTHER SAVINGS
1	57015	−8000	−15000	21000	0
2	59866	−7500	−15500	21000	0
3	59866	−6000	−16000	22000	0
4	57015	−6000	−16000	22000	0
5	54165	−6000	−17000	20000	0

FIGURE 10.30   (continued)

PRESS ANY KEY TO CONTINUE!

YEAR	BTCF	DEPRE- CIATION	TAXABLE INCOME	INCOME TAXES	ATCF
1	55015	−26400	28615	−8012	47003
2	57866	−26400	31466	−8811	49056
3	59866	−26400	33466	−9371	50496
4	57015	−26400	30615	−8572	48443
5	51165	−26400	24765	−6934	44230

SUMMARY : PW(14 %) AFTER TAXES = 52847.77
AW(14 %) AFTER TAXES = 15393.69

FIGURE 10.30   (continued)

year life. Obviously, operational costs could be given as 0 for later years if the robot's expected life is less than 5 years.

Information needed to determine direct labor savings (positive or negative) through robot implementation is requested next. Hourly costs for running the robot and manual operation are requested, so good accounting data for the manual operation is needed, as are good estimates for the robot operation. Availability of manual labor refers to the percentage of time a worker would actually be productive. The industrial engineering department should have allowance factors for rest breaks. The robot availability is a function of maintenance/reliability.

The cycle times for producing a unit of product by manual means or the robot are input next. It is often assumed that the robot will always be faster than a worker, but for certain operations this might not be the case.

Standard data regarding expected production capacity and product quantities are input next to allow direct labor savings to be determined.

Estimates of *annual* savings (costs) for other factors are then input. The one that might be difficult to estimate is *indirect* labor savings. It would be expected that these would increase due to programming needs, but other indirect costs are possible.

Now, we come to information that probably needs to be discussed a little. The minimum attractive rate of return (MARR) is what management would specify as the investment return they would like to see if money was invested elsewhere. Obviously, management does not want a breakeven situation on investment in any capital acquisition. The effective income tax rate has to be obtained from the accounting department and will include state and local taxes in addition to federal taxes.

The expected annual inflation rate affects cash flows, including taxes. In some time envelopes it might be desirable to put this in as an annual variable. The program we are looking at assumes a constant inflation rate over the life of the robot.

The program was written before investment tax credits were eliminated by the Internal Revenue Service; for the present, at least, this is set at 0.

The capitalizable robot cost is the capital acquisition cost of the robot, including permanent tooling.

The Internal Revenue Service has discontinued the ACRS (Accelerated Cost Recovery System) and, at the present, allows straight-line depreciation over at least a 5-year life.

It should be mentioned that all costs and incomes are assumed to be *present-day* equivalents; inflation has *not* yet been considered. This is just as well, since most of us have a difficult time thinking in terms of present-day dollars, let alone future values!

Now, finally, the program gives us something to mull over—results. The first tableau summarizes direct labor (DL), indirect labor (IL), maintenance, materials, and other savings. None have yet been adjusted for inflation.

The second tableau of results gives before-tax cash flow (BTCF), depreciation (capitalizable robot cost multiplied by the straight-line depreciation factor), taxable income (BTCF + depreciation), income taxes (taxable income multiplied by effective income tax rate), and after-tax cash flow (ATCF), which is (BTCF − income taxes). The ATCF figures look terrific: at least $44,000 benefit per year! However, if we do not want to be fired, we need to realize two things. First, inflation and the MARR (14%) have not been included. Second, the $132,000 capitalizable cost of the robot has not been included, except for tax determination.

So, the real meat is the *summary* data:

```
PW (14%) AFTER TAXES = 52847.77
AW (14%) AFTER TAXES = 15393.69
```

The *present worth* (PW) of all costs/incomes over 5 years assuming a 5% inflation rate and a 14% MARR is a *positive* cash flow of $52,847.77. On an *annual* basis (AW), this is $15,393.69. Is this sufficient to justify the investment risk? This is up to management to decide. Also, these figures can be compared against another potential robot's figures.

Now, how did the program come up with the PW (14%) and AW (14%) values?

INFLATION EFFECT. The BTCF values given in the program's tableaus have to be increased due to inflation. For example, the BTCF at year 1 (BTCF$_1$) would really be $(55015) * (1.05) = 57766$. BTCF$_2$, with 2 years of inflation, would be

$$\underbrace{\underbrace{(57866)*(1.05)}_{\text{Year 1}}*(1.05),}_{\text{Year 2}}$$

or

$$(57866)*(1.05)^2$$

Depreciation is *not* affected by inflation, and so the computer's tableau can be recomputed as shown in Table 10.3. Year 0 is included in the table to allow the

TABLE 10.3
Cash flows corrected for inflation

End of year	BTCF (with inflation)	Depreciation	Taxable income	Income taxes	ATCF (after inflation)
0	−132000	—	—	—	−132000
1	$55015*(1.05)^1$ = 57766	−26400	31366	−8782	48984
2	$57866*(1.05)^2$ = 63797	−26400	37397	−10471	53326
3	$59886*(1.05)^3$ = 69302	−26400	42902	−12013	57289
4	$57015*(1.05)^4$ = 69302	−26400	42902	−12013	57289
5	$51165*(1.05)^5$ = 65301	−26400	38901	−10892	54409

$132,000 capital payment to be included, thus showing *all* required data. The ATCF values are now *future* values assuming 5% inflation per year.

MARR EFFECT. To get a present-worth value of the six ATCF values requires us to consider the 14% desired rate of return. Consider the year 1 ATCF of $48,984 in Table 10.3. A 14% return would require that the present worth would be less than:

$$PW = (48,984)*(1.14)^{-1} = \$42,968$$

The other values would be as given in Table 10.4. The difference in PW (14%) from the computer value (52846 versus 52847.77) is due to calculator round-off error.

Finally, the *equal annual worth* (AW) is found as follows:

$$\underbrace{AW(1.14)^{-1}}_{\text{Year 1}} + \underbrace{AW(1.14)^{-2}}_{\text{Year 2}} + \underbrace{AW(1.14)^{-3}}_{\text{Year 3}} + \underbrace{AW(1.14)^{-4}}_{\text{Year 4}} + \underbrace{AW(1.14)^{-5}}_{\text{Year 5}} = \underbrace{52847.77}_{\text{PW(14\%)}}$$

TABLE 10.4
Calculation of PW (14%)

End of year	ATCF (after inflation)	PW (14%)		
0	−132000	$(-132000)(1.14)^0$	=	−132000
1	48984	$(48984)(1.14)^{-1}$	=	42968
2	53326	$(53326)(1.14)^{-2}$	=	41033
3	57289	$(57289)(1.14)^{-3}$	=	38668
4	57289	$(57289)(1.14)^{-4}$	=	33919
5	54409	$(54409)(1.14)^{-5}$	=	28258
		Total PW (14%)	=	52846

Multiplying both sides of the equation by 1.14 gives

$$AW + AW(1.14)^{-1} + AW(1.14)^{-2} + AW(1.14)^{-3} + AW(1.14)^{-4} = (52847.77)(1.14)$$

Subtracting the second equation from the first gives

$$AW[1 - (1.14)^{-5}] = (52847.77)(1.14 - 1.0)$$

and

$$AW = \frac{(52847.77)(0.14)}{[1 - (1.14)^{-5}]}$$

It would be found that AW (14%) is $15,393.69 per year as given in the program. Also, we have developed a standard engineering economy factor that says:

$$AW = \frac{(PW)(i)}{[1 - (1 + i)^{-n}]}$$

where AW = equal annual amounts
    PW = present worth amount
       i = required interest rate

One last comment concerning the program: It allows a sensitivity analysis to be performed on many of the costs/incomes. This is demonstrated in Figure 10.31, where the robot hourly out-of-pocket cost was changed from $9.75 to $13.50. The $3.75-per-hour increase affected the AW (14%) by a 57% reduction. Obviously, this shows that good estimates are mandatory if the feasibility study is to be beneficial.

10.8.3   Indirect Savings

An interesting case was mentioned by Potter [25] in which a project had a direct labor savings of $75,000, which was too small an amount with which to justify the project. Subsequent analysis of indirect savings resulted in additional savings of $151,253, an amount that made the project economically justifiable.

One of the problems with indirect savings is they are not usually as quantified as are direct costs. A little investigation will allow estimates, or even actual values, to be ascertained.

What, then, are some of the potential indirect savings? Potter offers several possibilities:*

1. *Inventory control.* Work in process, frequently called WIP, is production that has not been completed and shipped. Money invested in manufacturing cannot

---

* Extracted with permission from Potter, R. D.: ''Analyze Direct Savings in Justifying Robots,'' in *Robotics and Industrial Engineering Selected Readings*, vol. II, Industrial Engineering and Management Press, Institute of Industrial Engineers, Norcross, Ga., © 1986.

WOULD YOU LIKE TO TRY ANOTHER RUN (Y OR N)? Y
WOULD YOU LIKE TO HAVE A SENSITIVITY ANALYSIS (S) RUN (CHANGE ONE PARAMETER AT A
TIME) OR A COMPLETE NEW (N) RUN? PLEASE ENTER AS (S OR N)? S
CHANGES IN STUDY PERIOD (N) REQUIRE A NEW PRODUCTION PLAN TO BE ENTERED, AND IT
IS EQUIVALENT TO A COMPLETE NEW RUN.
DO YOU WANT TO CHANGE THE STUDY PERIOD (Y OR N)? N
THE FOLLOWING PARAMETERS ARE ALLOWED TO CHANGE:
    1. OUT-OF-POCKET COST PER HOUR FOR MANUAL LABOR
    2. OUT-OF-POCKET COST PER HOUR FOR ROBOT APPLICATION
    3. AVAILABILITY OF MANUAL LABOR
    4. AVAILABILITY OF ROBOT
    5. CYCLE TIME (MIN./PC.) FOR MANUAL OPERATION
    6. CYCLE TIME (MIN./PC.) FOR ROBOT OPERATION
    7. EXTRA ANNUAL MAINTENANCE SAVINGS OR COST
    8. EXTRA ANNUAL SAVINGS IN MATERIALS
    9. OTHER SAVINGS OR COST ASSOCIATED WITH ROBOT VS. MANUAL
   10. AFTER-TAX MINIMUM ATTRACTIVE RATE OF RETURN (MARR)
   11. EFFECTIVE INCOME TAX RATE
   12. ANNUAL INFLATION RATE
   13. INVESTMENT TAX CREDIT
   14. CAPITALIZABLE COST OF ROBOT SYSTEM
   15. ESTIMATED MARKET VALUE (SALVAGE VALUE) OF THE ROBOT SYSTEM

WHICH ONE DO YOU WANT TO CHANGE (1–15)? 2
THE NEW OUT-OF-POCKET COST PER HOUR FOR ROBOT APPLICATION IS ? 13.5
       CASHFLOW AND DEPRECIATION

YEAR	D.L. SAVINGS	I.L. SAVINGS	MAINT.	MATLS.	OTHER SAVINGS
1	46492	−8000	−15000	21000	0
2	48817	−7500	−15500	21000	0
3	48817	−6000	−16000	22000	0
4	46492	−6000	−16000	22000	0
5	44168	−6000	−17000	20000	0

PRESS ANY KEY TO CONTINUE!

YEAR	BTCF	DEPRE-CIATION	TAXABLE INCOME	INCOME TAXES	ATCF
1	44492	−26400	18092	−5066	39426
2	46817	−26400	20417	−5717	41100
3	48817	−26400	22417	−6277	42540
4	46492	−26400	20092	−5626	40866
5	41168	−26400	14768	−4135	37033

    SUMMARY : PW(14 %) AFTER TAXES = 22680.81
            AW(14 %) AFTER TAXES = 6606.548

WOULD YOU LIKE TO TRY ANOTHER RUN (Y OR N)? N
OK

FIGURE 10.31   Sensitivity analysis using a computer program.

be captured until the product has been shipped to the customer. As a result, capital is lost due to the interest charges on that money at the minimum attractive rate of return (MARR) as well as due to the floor space that is utilized by the in-process items. A robot in, say, cell-type manufacturing, can take the part through several operations with considerable reductions in inventory needs. In a similar vein, the need for protective safety stocks can be reduced when production becomes more deterministic with automated operations than with the more stochastic manual operation.

2. *Quality improvements.* As Potter points out, this can be a pretty subjective category, but it has a significant effect on sales and incomes. For example, Potter cites the automotive production use of robots in spot welding and painting with the associated public awareness of quality improvement.

3. *Scrap and rework reduction.* Many inspection operations are incorporated directly into a robot's cycle, and thus scrap and rework can be reduced considerably. Also, the *consistency* of robotic operations such as spray painting often eliminates the need for scrap or rework. Rework, of course, requires direct labor, and so a cost can easily be obtained for this savings.

4. *Human and safety factors.* Humans, but not robots, must be buffered from adverse conditions. The robot does not fatigue as does a laborer and does not require retraining because a robot quits to go to a new job.

5. *Floor-space savings.* Often, a robot has a larger space intrusion than does its human counterpart. If this is the case, then this factor needs to be included in the evaluation equation. As mentioned earlier, space savings may be realized through in-process inventory reductions.

6. *Flexible operation.* If a robotics operation is contrasted to a "fixed automation" situation, then the ability of the robot to be adjustable to different products and different technologies should be considered. If two approaches are economically of the same value, then flexibility, for example, could well be a logical tie-breaker.

OTHER NONECONOMIC FACTORS. The ICAM Robotics Application Guide [29] indicates that most robot gains are realized through the robot's consistent operation: increased productivity by maintaining a constant rate of production over an extended period of time with overall cycle time reductions as well as fewer bad parts to be scrapped as well as less material waste. However, the guide also recognizes *noneconomic* benefits for robotic applications. There may be operations performed in high-temperature environments (or low-) or where toxic fumes are dangerous, where a human operator should not be allowed to work. A robot may not be the obvious substitute, but, as the guide sagely suggests, the *economic* justification of the robot against other nonhuman alternatives should still be evaluated. Another noneconomic factor mentioned by the ICAM guide is management direction. This is the case where a manager directs that come what may, a robot will be used in a particular application. This may result from a misdirected idea that status will be gained through having a robot in the empire. This type of justification should be avoided if at all possible.

Hopefully, this section has pointed out that robots should not be recommended in a "willy-nilly" type of atmosphere. As with any capital requisition, a thorough economic evaluation should be made, and the item should be acquired only if the analysis so dictates. To assist the analyst, many good engineering economics texts are available, *typical* of which are those by DeGarmo et al. [5], Fabrycky and Thuesen, [8], and Grant et al. [11].

10.9 ROBOT PROGRAMMING LANGUAGES

There are many examples of good programming languages in the field of robotics. Needless to say, not all can be discussed here. A good robotics language should be user-friendly, have English-language (or the user's-language) statements that reflect what have to be accomplished, allow for expansion/flexibility, have "teach" capability so that specific coordinate locations do not have to be input by the user, and have the ability to allow ease of interface with the surrounding environment. This last characteristic refers to the need for the robot to know when external acts occur: a part arrives on a conveyor or a machining operation is complete, for example.

Languages go from very low-level to expert system-oriented languages. For overview purposes, this section will be divided into discussions of

- VAL II—a language that has its roots in other languages
- AML/X—a language evolved uniquely for robot assistance
- Off-line programming and *simulation* with respect to robotics

Only the VAL II and AML/X languages will be treated in sufficient depth to allow example programs to be generated. The remaining topics will be covered in summary form to round out the discussion.

10.9.1 VAL II*

The VAL II language is a computer-based control system and language designed specifically for use with Unimation/Westinghouse's industrial robots. Typical of these is the PUMA 560 shown in Figure 10.32. VAL II will be exemplified in terms of the PUMA 560.

The PUMA 560 is a six-axis robot which has a computer-servo control system and has accuracy stipulated at 7.5 thousandths of an inch with repeatability of 4 thousandths of an inch.

The six axes were shown in Figure 10.8. Joints 1 through 5 obviously follow anthropomorphic characteristics. Joint 6, the flange, allows manipulation about a particular wrist position (bolt turning, for example). It is possible, under manual

---

* This material is summarized with permission from *User's Guide to VAL II, The Unimation Robot Programming and Control System*, Preliminary Draft #3, Unimation/Westinghouse, Danbury, Conn., April 1983.

FIGURE 10.32   Unimation's PUMA 560 robot. (Courtesy of Unimation/Westinghouse, Danbury, Conn.)

control, to manipulate any of the six joints independently from the other five joints.

Two ways to define the conventional cartesian *x, y,* and *z* coordinate system are of benefit to the user. The *world* coordinate system is shown in Figure 10.33. Moving the gripper vertically in the *z* direction will cause several joints to move in order to allow smooth vertical motion. Most manufacturing functions will be concerned with tool/gripper motions. For example, once a drill is positioned correctly for drilling, the actual drilling direction will be in a straight line that passes through the drill. The *tool* coordinate system, shown in Figure 10.34, allows the *z* axes to be redefined to a position vertical to the flange. This allows simplification of "teaching" functions, as will be demonstrated shortly.

MANUAL AND COMPUTER MODE. The hand-held pendant shown in Figure 10.35 can be used to control the robot under *manual* control. Using this pendant, the operator can choose the coordinate system desired and can move the gripper/tool to any desired location. Pressing the JOINT, WORLD, or TOOL button automatically puts the control system into manual mode. Another button, labeled COMP, puts the robot into computer mode to allow current coordinates of the joints to be stored in memory or to allow a controlling program to be developed, tested, and implemented. In addition to allowing coordinate-mode switching and manual positioning, the pendant has switches and buttons to allow speed-of-motion adjustment and "panic" stopping.

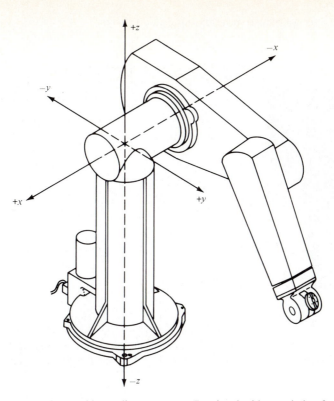

FIGURE 10.33   PUMA 560 world coordinate system. (Reprinted with permission from *PUMA Mark II Robot 500 Series Equipment and Programming Manual*, 398 Pl, Unimation/Westinghouse, Danbury, Conn., 1983.)

FIGURE 10.34   PUMA 560 tool coordinate system. (Reprinted with permission from *PUMA Mark II Robot 500 Series Equipment and Programming Manual*, 398 Pl, Unimation/Westinghouse, Danbury, Conn., 1983.)

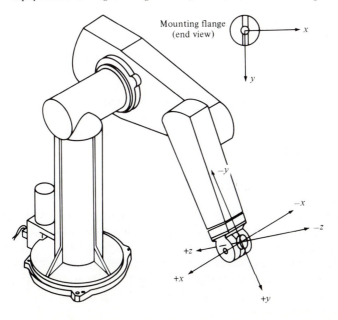

FIGURE 10.35 · "Teach" pendant for PUMA 560 robot. (Reprinted with permission from *PUMA Mark II Robot 500 Series Equipment and Programming Manual*, 398 Pl, Unimation/Westinghouse, Danbury, Conn., 1983.)

"TEACHING" THE ROBOT. One way to teach locations is as follows:

1. Press JOINT button. This allows each joint to be moved manually, independently of the other five joints. Either WORLD or TOOL could have been chosen instead.
2. Using the pendant's control buttons, move the tool/gripper to the desired location.
3. Press COMP for computer mode. Go to the robot's computer terminal and type:

`HERE (location name desired)`

For example, if we are going to use LOCA as the location name in the computer program, we would type

`HERE LOCA`

Any program that requests the tool/gripper to go to LOCA will now cause the correct move. It should be apparent that this type of language structure is meaningful to the user.

Any number of locations can be taught in the manner just presented, switching between manual and computer modes.

A SIMPLE PROGRAM. Suppose we want to drill 16 holes according to the pattern shown in Figure 10.36. The pendant procedure can be used to teach the 16 locations, but this would be quite time-consuming and using the same program in different robot installations would require all points to be taught at each location. VAL II allows location adjustment under computer control.

The program in Figure 10.36 allows all holes to be drilled given just one location, called STA at the bottom right-hand corner of the diagram. Actually, two programs are required, since one will be a subroutine.

Let us input the main program first. We go into computer mode and type

EDIT MC

Assuming that memory does not already contain a program called MC, we can start inputting at line 1. (The line numbers are inserted automatically, as in BASICA).

The MC program should be clear, with two possible exceptions. First, the SPEED command allows robot motion speeds to be adjusted under computer control. SPEED 20 says to run at 20% of normal speed. Second, nested loops are

FIGURE 10.36   Simple VAL II example.

MAIN PROGRAM (MC)	SUBROUTINE (DRILL)
1. K = 0	1. XM = 20 * I
2. SPEED 20	2. YM = 20 * J
3. FOR I = 1 TO 4	3. MOVE SHIFT(STA BY XM,YM,0)
4. FOR J = 1 TO 4	4. DEPART −20
5. K = K + 1	5. SPEED 10
6. CALL DRILL	6. DEPART 20
7. END	7. TYPE/B,"COMPLETED HOLE",/I2,K
8. END	8. SPEED 20
9. TYPE "DRILLING COMPLETE"	

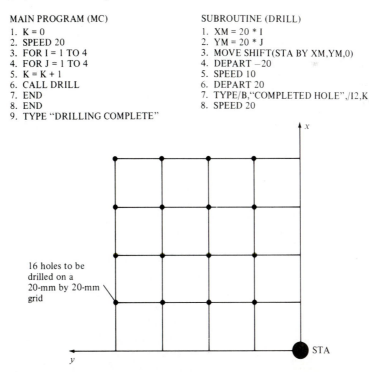

16 holes to be drilled on a 20-mm by 20-mm grid

accomplished within the following construct:

```
┌ FOR    I   = 1 to 4
│               .
│        ┌ FOR    J = 1 to 4
│        │               .
│        │               .
│        │               .
│        └ END
└ END
```

An END statement is *not* used to denote the end of a program; rather, in this case, it is used to signify the end of a loop. After 16 holes are drilled, the message DRILLING COMPLETE will be printed on the screen.

Now we type EDIT DRILL to start input of the subroutine. The first time the subroutine is called, I and J are both 1, and so XM and YM will each have initial values of 20 (statements 1 and 2).

Statement 3 requires some explanation. The command

MOVE ⟨location specified⟩

will cause the gripper/tool to go to the location specified (assuming that it has been taught). The command

SHIFT (⟨location specified⟩ by X', Y', Z')

will add the coordinate values specified by X', Y', Z' to the X, Y, Z coordinate values of the specified location, giving new coordinates of (X + X'), (Y + Y'), and (Z + Z').

MOVE SHIFT(STA BY XM,YM,0) will move the gripper/tool to the first drill location (AA). (It should be mentioned that it is assumed that all coordinate values are in millimeters.)

Line 4 of the subroutine says to move *down* the z axis a distance of 20 mm, which in effect allows drilling. It is assumed that the drill is rotating at the correct speed for the material to be worked on.

Line 5 inserts a speed change to 50% of that when moving from location to location. Line 7 retracts the drill. Finally, the last subroutine line says to beep (signified by B), type COMPLETED HOLE followed by the value of K (hole counter) in the Fortran I2 format. So we will get 17 lines of print in the screen:

```
COMPLETED HOLE 1
COMPLETED HOLE 2
          .
          .
          .

COMPLETED HOLE 15
COMPLETED HOLE 16
DRILLING COMPLETE
```

Now, how do we run the program? After STA has been defined and programs MC and DRILL have been typed in, we just type

EXECUTE MC

and that is it. If we type

EXECUTE MC, 100

then the program will run 100 times. A negative number will run the program indefinitely until the system is stopped manually. Also, if the program is to be run more than once, then some delay is needed between runs to allow whatever is being drilled to be changed. Inserting DELAY (seconds) will allow such a delay. For example,

DELAY 3

will delay the next step operating for 3 s.

The VAL program structure seems to be founded on a combination of BASICA and Fortran with considerable robot user-friendliness added in. Figure 10.37 summarizes some of the VAL II commands discussed and adds VAL II's version of some typical BASIC and Fortran commands.

COMMENTS ON LOCATIONS. As shown earlier, the PUMA 560 has six joints. Three are expressed in x, y, and z millimeter values. The other three joints are expressed in angles. When we accomplish a HERE PART command, the coordinates stored might be

37.75,360.25,−51.31, −160.321,31.218,−25.311

x, y, z in mm   Rotation in angles

Joints 1, 2, 3   Joints 4, 5, 6

It is possible to input the coordinate values manually, but it is readily apparent that a "teach" operation will almost always be the way to handle them.

A clever trick to allow program simplification is through compound transformations, where locations can be *relative* to other locations. As an example, suppose that an assembly has a base on which is attached a frame. Also, a ring is mounted on a corner of the frame. This relationship is summarized in Figure 10.38. If the base is to be moved to several assembly stations, it will be helpful for the ring and assembly locations to be input *relative* to the base location. This means that only the *base* location has to be retaught when it is to be moved.

The "teaching" sequence might be as follows, with the end effector being moved to the correct location before typing the HERE command:

```
HERE BASE            (teach BASE location)
HERE BASE:ASSY       (teach ASSY location)
HERE BASE:ASSY:RING  (teach RING location)
```

Command structure	Example ((END) is end of structure, not of program)	

I. Typical single commands

MOVE (location)	MOVE LOC1	Move end effector in smooth path to LOC1)
MOVES (location)	MOVES LOC1	Move end effector to LOC1 in straight line
APPRO (location),(distance)	APPRO LOC1,50	Approach LOC1 but stop 50 mm away in z-direction)
APPROS (location),(distance)	APPROS LOC1,50	Same as APPRO but in a straight line
DEPART (distance)	DEPART 50	Depart from current location to a point 50 mm away in z-direction
DRIVE (joint),(degree or distance),(speed)	DRIVE 6,35,10	Drive joint 6 a positive 35° at a speed of 10
HERE (location)	HERE LOC1	Gives name LOC1 to current position—can be a program command as well as being used in the teach mode)
SET (location 1)=(location 2)	SET TEMP=LOC5	TEMP now has the same coordinates as LOC5
OPENI		Opens the bang-bang gripper
CLOSEI		Closes the bang-bang gripper
GOTO (statement #)	GOTO 50	Sends program control to command preceded by number 50
CALL (subroutine)	CALL SR1	Transfers control to program SR1; return is to command following the CALL statement
DELAY (seconds)	DELAY 1	No action for 1 s

II. Other typical commands

A. DO
   (group of steps)
   UNTIL (logical expression)

```
K=0
I=0
DO
I=I+1
K=K+(I*I)
UNTIL I=10
```
Sum I-squared for I=1 to 10, then continue

B. FOR (real variable)=(initial)
     TO (final) (STEP) (increment)

```
X=0
FOR J=1 TO 10 STEP 2
X=X+(J*J)
END
```
Sum J-squared for J=1,3,5,7,9,then continue

C. IF (logical expression) THEN
   (first group of steps)
   (ELSE
     (second group of steps))
   END

```
IF K>5
J=10
TYPE "J SET TO 10"
ELSE
J=12
TYPE "J SET TO 12"
END
```
If K is greater than 5, print "J SET TO 10"; otherwise print "J SET TO 12"

D. WHILE (logical expression) DO
   (group of steps)
   END

```
K=0
WHILE K<5 DO
K=K+1
TYPE 14,K
END
```
Print: 1, 2, 3, 4
Using I4
format then
continue

E. MOVE SHIFT(LOCATION BY dx,dy,dz)

[Tool will move to the LOCATION's coordinates modified by the incremental changes specified]

```
XM=3
MOVE SHIFT(STA BY XM,5,3)
```

[If STA has coordinates 10.5,3.2,4.7,then tool will move to position (3+10.5,5+3.2,3+4.7)

FIGURE 10.37 Some VAL commands.

F. K=5
TYPE /B,"FINISH DRILL",/I2,K
[the (B) will cause a BEEP followed
by printing of (FINISH DRILL)
followed by the number K being printed
with I2 format).

G. TIMERs

There are 15 timers, TIMER 1, TIMER 2, etc.

Typical statements:

a) TIMER 3=0
b) IF TIMER(10) < 5 GOTO 15
c) WAIT TIMER(12)=>(TIMER(6)+5.0)

5 second delay:

TIME=TIMER(3)
WAIT TIMER(3)=((TIME+5.0)
(continue)

Another delay:
TIMER 1 = 0
10 IF TIMER(1)<5 GOTO 10
(continue)

FIGURE 10.37   (continued)

If we have a program command

$$\text{MOVE BASE: ASSY: RING}$$

relative relative
to   to

then we move to RING.

$$\text{MOVE BASE: ASSY}$$

will move to ASSY.

Since this is somewhat cumbersome, a SET command will simplify the program writing. For example,

$$\text{SET LOC1 = BASE: ASSY: RING}$$

FIGURE 10.38   Assembly example for compound transformations.

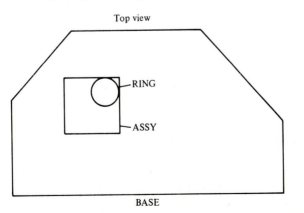

Top view

will move the end effector to RING each time there is the command

MOVE LOC1

in the program.

ANOTHER "TEACH" METHODOLOGY. If a particular path is required for the end effector to travel—say, in paint spraying—it can be accomplished using the record (REC) button on the pendant. To start teaching, TEACH PAINT [   ] is typed into the CRT. The end effector is positioned to each location desired and the REC button is pressed before moving to the next location. Each successive location is stored automatically in the array as PAINT[2], PAINT[3], and so on. If we had typed TEACH PAINT 1, then the locations would have been PAINT1, PAINT2, PAINT3. A path and subsequent program might be as given in Figure 10.39. The teaching could be accomplished by an expert painter. A similar example could have seam welding, for example.

VAL II AND CONTROL COMPUTER FUNCTIONS. Suppose that we have a robot application as shown in Figure 10.40. Work is palletized in the load station and depalletized in the unload area. Parts go in sequence through station 1 to station 2, but the processing times are a function of the numerical control program and are not known to the robot. If a part is set up in the load area, then the robot is to move it to station 1, assuming that station 1 is free. Similar moves are from station 1 to station 2 and from station 2 to the unload area. The neutral location is where the robot goes after each move—possibly an inefficient move. The station z locations are 40 cm above the load and unload area z coordinates.

FIGURE 10.39   Use of arrays in teaching a path.

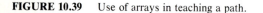
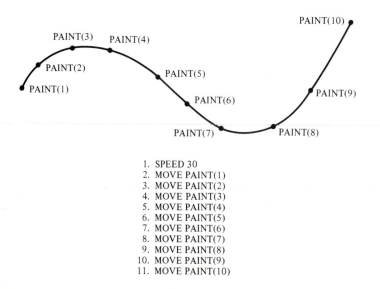

1. SPEED 30
2. MOVE PAINT(1)
3. MOVE PAINT(2)
4. MOVE PAINT(3)
5. MOVE PAINT(4)
6. MOVE PAINT(5)
7. MOVE PAINT(6)
8. MOVE PAINT(7)
9. MOVE PAINT(8)
10. MOVE PAINT(9)
11. MOVE PAINT(10)

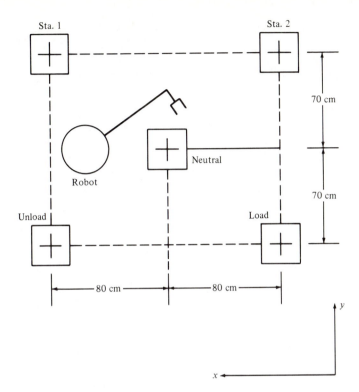

FIGURE 10.40 Layout for cell programming example.

A system of switches sends digital input information to the VAL II programs in the same manner that process input/output was discussed in Chapter 8. The information is maintained as given below:

Variable	Status
S1B	Has station 1 a part?
	(0 = No, 1 = Yes)
S1	Is part in station 1 complete?
	(0 = Yes, 1 = No)
S2 and S2B	Same as above except station 2.
LD	Status of load area.
	(0 = no part; 1 = part waiting)
UNL	Status of unload area.
	(0 = area available; 1 = not available)

Figure 10.41 gives a possible VAL II program for the workcenter. The MAIN program determines if a move can be made, and the subroutine accomplishes that move. For example purposes, no SPEED information is given. How-

```
        Program main                                    Comments
 1. 50  FLAG = 0                                         Indicator to tell if move is to be made–0 is no and 1 is yes
 2. 60  IF (S1B==1) AND (S1==0) AND (S2B==0) THEN        Is a finished part in Sta. 1, and is Sta. 2 available?
 3.       SET SOURCE=SHIFT(LOAD BY 1600,1400,400)        Pickup point is set to Sta. 1
 4.       SET DESTIN=SHIFT(LOAD BY 0,1400,400)           Drop point is set to Sta. 2
 5.       FLAG=1                                         Flag indicates a move is to be made
 6.     ELSE
 7.       IF (S2B==1) AND (S2==0) AND (UNL==0) THEN      Is Sta. 2 item finished and unload slot available?
 8.         SET SOURCE=SHIFT(LOAD BY 0,1400,400)         Pickup point is set to Sta. 2
 9.         SET DESTIN=SHIFT(LOAD BY 1600,0,0)           Drop point is set to unload area
10.         FLAG=1
11.       ELSE
12.         IF (LD==1) AND (S1B==0) THEN                 Is part at load area, and is Sta. 1 available?
13.           SET SOURCE=LOAD                            Pickup point is load area
14.           SET DESTIN=SHIFT(LOAD BY 1600,1400,        Drop point is set to Sta. 1
                 400)
15.           FLAG=1
16.         ELSE
17.         END
18.       END
19.     END
20.     IF FLAG=0 GOTO 60                                Loop until a move is required
21.     CALL MOVIT                                       Go to subroutine if a move is needed
22.     GOTO 50                                          After move, loop until another move

        Subroutine movit
 1.     APPRO SOURCE, 50                                 Approach to 50 mm of pickup point
 2.     OPENI                                            Open gripper
 3.     MOVE SOURCE                                      Move to part
 4.     CLOSEI                                           Close gripper
 5.     DEPART 50                                        Go to 50 mm from pickup point
 6.     APPRO DESTIN, 50                                 Approach to 50 mm of drop-off point
 7.     MOVE DESTIN                                      Move to drop-off point
 8.     OPENI                                            Open gripper
 9.     DEPART –50                                       Move 50 mm from drop-off point
10.     MOVE SHIFT(LOAD BY 800,700,200)                  Move to neutral location

        Status indicators
        S1B = HAS STATION 1 A PART? (0 = NO AND 1 = YES)
        S2B = HAS STATION 2 A PART? (0 = NO AND 1 = YES)
        S1  = STATION 1 PART STATUS (0 = COMPLETED AND 1 = NOT COMPLETED)
        S2  = STATION 2 PART STATUS (0 = COMPLETED AND 1 = NOT COMPLETED)
        UNL= STATUS OF UNLOAD AREA (0 = NO PART IN AREA AND 1 = A PART IN AREA)
        LD = STATUS OF LOAD AREA   (0 = NO PART WAITING AND 1 = A PART IS WAITING)
```

FIGURE 10.41 Manufacturing cell programming example.

ever, in all probability, the APPROACH speeds would be considerably faster than the MOVE speeds.

It is certainly not the intent to make the reader an expert in VAL II programming, but a comment should be given relative to the IF…THEN…ELSE structure found in lines 2 through 19 of the MAIN program. Nesting is accomplished in just the same way as in BASICA. An END statement is required for each nest, though, and these are grouped at the bottom of all the nested statements. The last END (line 19) relates to the IF-THEN-ELSE that starts at line 2. The second sequence, starting at line 7, completes at line 18, and the inner grouping goes from line 12 to line 17. A "quirk" of the system is seen in lines 7 and 20. A simple IF statement using an equality is programmed with a single equality sign. An IF statement followed by THEN requires two equality signs.

Two points are of interest from this example: (1) Real scheduling problems would be far more complex, and (2) how would the process information be available to the robot?

First, consider the scheduling situation. We saw in Chapter 7 that scheduling is a nontrivial component of the factory. It can affect whether a profit is made or not. If we had a two-station cell as shown in the example, parts could probably go from station 1 to station 2 or vice versa as well as to only one of the two stations. Also, most cells would have more machines/inspection stations than just two. If the processing times are available to the robot, then what moves should be made when more than one is feasible? Where should the end effector stay after a move in order to minimize travel time? These are nontrivial questions, and brute-force programming will probably not be optimum. One possible solution is to use an expert system, where rules can be stipulated with which the scheduling system will try to optimize some criterion, such as maximizing equipment processing time.

Now, how about the process input/output? VAL II allows a *process* control program to run concurrently with the *motion* control program. The process control program can input/output digital and analog data, and the input information is available to the motion control program as with the status indicators shown in the sample program (Figure 10.41). The process program is now allowed to accomplish robot motions.

Priority interrupt capability (see Chapter 8) is available to VAL II users through the REACT command:

```
REACT <signal>, <program>, {<priority>}
```

The <signal> refers to external binary data, and the REACT command is looking for a change of ''0'' to ''1'' or ''1'' to ''0'' (16 such signals may be monitored simultaneously). If a change occurs, then program control is switched to <program> in the same manner as a CALL<program>. An optional priority may be set such that the *program reaction* is queued until the priority is greater than the running program's priority. Priorities may be changed under program control.

SUMMARY. We have just seen an introduction to VAL II. With its priority interrupt and process control capabilities, it is an extremely versatile language enabling relatively simple programs to effect complex motion control.

10.9.2 AML/X Programming Language

The AML/X language is a derivative of the AML language developed by IBM in the early 1980s specifically for the IBM 7565 cartesian robot system. AML, which stands for A Manufacturing Language, is different from VAL in the sense that AML is meant to be a general programming language with constructs available for nonrobotics programming. It is also object-oriented. The overall goal of AML was to develop a language for manufacturing and computer-aided design. The revision of AML that resulted in AML/X has extended the language to be more useful in robotics, machine vision, and computer-aided design in general.

In the area of robot programming, it has been found that specification of robot motion represents only a small proportion of the total code. Other major requirements of a robot language are the handling of input/output for auxiliary devices such as sensors, cameras, or other robots, operator interfaces, calibration and setup routines, bookkeeping, access to manufacturing data bases, etc. Because robotic applications do rely heavily on geometric calculations (as do CAD systems), means must be provided for expressing these conveniently. Also, powerful exception handling must be considered part of a good robot language. For example, if the robot encounters an obstacle or unsafe condition, certain actions must be taken to prevent damage to the robot or environment. In addition, AML/X provides improved ways to promote better integration of robot and vision programs. Incredibly large amounts of data must be dealt with in vision systems; camera resolutions of 512×512 pixels each of which having 8 bits of information create 2 Mbits of data with every frame of picture captured. Each of these pixels must be stored in a quick-access and compact-memory-storage implementation. In computer-aided design, it is important that the language support layering, user interaction, and facilities for accurate numerical computation and support of abstract data types for complex geometric modeling applications.

In short, AML/X is a general-purpose language with special constructs for the above applications. It contains the standard types of data objects such as integers, real numbers, strings, bits, booleans, symbols, and pointers. But it also has the unique feature of the aggregate data type. An *aggregate* is a collection of objects that can be treated as one. An aggregate is similar to an array in Fortran, but the elements of an aggregate need not be the same type. Once an aggregate is created, it must remain the same size, but the types of the objects in the aggregate can change. A simple example aggregate is:

$$
\text{m: NEW} < <1, 2, 3, 4>,
$$
$$
<5, 6, 7, 8>,
$$
$$
<9, 10, 11, 12>>;
$$

which is an aggregate of three elements, each of which has four objects. Referencing the individual elements is somewhat different from using an array. Notice that an aggregate may have aggregates as members. Subscripts can be used to reference individual elements or aggregates, where the subscripts are listed in order of the level of aggregate. For example, $m(i,j)$ refers to the jth element of the ith element. That is, $m(3,2) = 10$. Since m has three elements which we can consider matrix rows, we can also refer to each row individually by specifying just the aggregate i as $m(i)$, where i is either 1, 2, or 3. $m(2)$ would be the aggregate $<5,6,7,8>$. The data can be also be referenced in selected row or column subsets as follows:

m(<1,3>) is the first and third rows.

m(,<4,1>) is the fourth and first columns (the absent first index means take the complete aggregate, and the 4 and 1 are the fourth and first elements in each subaggregate).

m(<1,2>,<1,2>) is the upper left 2 × 2 matrix (take the first and second elements in the first and second aggregates).

The power of aggregates can be shown through examples of matrix algebra and exception-handling routines. A transformation matrix or **A** matrix for robotic kinematics analysis can be represented as a four-member aggregate, each member being a four-object aggregate; for example,

```
m:  NEW < <1, 2, 3, 4>,
          <5, 6, 7, 8>,
          <9, 10, 11, 12>,
          <13, 14, 15, 16>>;
```

Once defined, the matrix elements can be accessed using subscripts and also, using a construct called *operator mapping*, the matrix elements can be changed according to operators with succinct commands. For example, scalar multiplication can be performed by the statement:

```
n = s op m;
```

where m is the 4 × 4 matrix, op is an operator from the set (+,−,*,div,**), and s is a scalar object. The elements of two matrices may be operated on together by the command:

```
n = m op p;
```

where p is an aggregate of the same size as m. This should not be construed as a matrix multiplication, but as the corresponding elements of each aggregate operated on by op.

Aggregates can be used as a "frame" construct to group together items which may be used together. For example, in robotics it may be convenient to group the name of a part to be grasped with its orientation vector and the name of a subroutine to be called when an exception occurs, such as a collision with an object or reaching an impossible position with the gripper. The aggregate might be defined as:

```
<part2, <0.547, 0.547, 0.632>, action23>
```

where part2 is the part name, the aggregate <0.547,0.547,0.632> is the orientation vector, and action23 is the subroutine name. This structure lends itself very well to object-oriented programming, where objects are defined as collections of descriptive data.

When an error is detected during program execution, an exception is raised. The exception handler defined for that particular exception specifies the action to be taken. Each exception has a name defined as a symbol datatype and can be raised either by the system in case of a system exception or by the user using the RAISE_EXCEPTION subroutine. The exception routine can either stop the program and execute a CLEANUP routine to close files and perform other shutdown procedures or continue with the program with a flag raised or the problem resolved with alternative action.

AML/X provides for the defining of objects called *classes* in object-oriented programming vernacular. A class is a type of object. For example, a chair may be an object and the class *chair* may contain the definition of properties of a chair that may be important, such as height, material, etc. A class instance is a particular object derived from the class. For example, a class instance of the chair would be a chair with a height of 30 cm and a material of oak/wool. This is analogous in other programming languages to types and instances of types. A class definition can contain methods, that is, special subroutines that deal with that particular class of objects. For example, a method for a chair may be to determine the type of material and that method may contain statements to ask the user and store the material typed in by the user in response to a question. The theory of object-oriented programming is beyond the scope of this text, but the technique is very popular today, especially in programming intelligent systems as in artificial intelligence and expert systems.

Classes and methods can be used to define robot paths and exceptions during path traversal, parts to be gripped by the object, and subsequent motion after gripping. Object-oriented programming enables the programmer to group data logically into classes which may be accessed at once. This speeds program execution and provides a structure for the data which encourages organization and a logical approach to problem solving.

As an example of the use of classes and methods and, in fact, AML/X, let us look at operators for handling three-element vectors, (x, y, z). Vectors such as this arise when defining points in space and relative motion vectors such as MOVE(3.5,5,1.2), which means move to a new location by moving 3.5 in the x, 5 in the y, and 1.2 in the z. This is a relative move, because the x, y, and z represent changes in motion, not absolute points in space. We can define the class *vector*, which has three arguments as follows:

```
vector:CLASS(xx DEFAULT 0.0, yy DEFAULT 0.0, zz DEFAULT 0.0)
   IVARS
     x: NEW REAL (xx);
     y: NEW REAL (yy);
     z: NEW REAL (zz);
   END;
```

Note that the three input arguments xx, yy, and zz, if not specified, have default values. NEW and REAL describe the type of variable as a new decimal value. We can continue to define operations involving vectors, for example, dot product and scaling and the cross product. The dot product and scaling method follows:

```
$*:  METHOD(v)
     SELECT(?v)
        CASE vector THEN RETURN(x*v.x+y*v.y+z*v.z)
        OTHERWISE RETURN (vector(x*v,y*v,z*v)
     END;
  END;
```

This method takes in an object v, and if v is a vector, returns the dot (or inner) product, and if v is a scalar, returns a vector with each of the components multiplied by the scalar. Additional methods could be defined for the translation, rotation, and scaling of matrices, etc. And other methods and classes could be defined for computing the gripper location and orientation from the joint angles and the reciprocal.

We could spend much more time describing AML/X, but an example from Ref. 20 may be the best way to give an overall view of the language used in a robot assembly situation. The goal of this program is to find a parts tray which has a fixture attached to it holding a box and cover to be assembled. In this case we can define an object called the tray which contains the box and its cover. The cover and box have grasp points defined in their own coordinate systems, and we can assume the functions grasp_object, move_object, release_object have been defined elsewhere, either as system calls or as user-defined subroutines. The object class tray is initially empty and is filled by the method locate_object by somehow locating the tray on the conveyor or assembly table using possibly a vision system or algorithmic search routine by moving the gripper to find some reference points. Then the cover and box classes are defined with respect to the tray by the transformations given as . . . in parentheses. The cov_grasp is defined with respect to the cover coordinate system, and the box_top and box_grasp points are defined with respect to the box coordinate system. The program is very short, but realize that the routines called with arguments may also need to be defined in a particular system. Exceptions (error handling) have been omitted. It has been estimated that error handling alone occasionally accounts for more than 50% of a robotics program.

```
tray:      NEW frame();
cover:     NEW frame(tray,trans(...));
cov_grasp: NEW frame(cover,trans(...));
box:       NEW frame(tray,trans(...));
box_top:   NEW frame(box,trans(...));
box_grasp: NEW frame(box,trans(...));

tray = locate_object(DEFAULT, ...);  ## no a priori information
cover = locate_object(cover, ...);   ## locate cover better
box = locate_object(box, ...);       ## locate box better
```

```
grasp_object(cover_grasp, ...);      ## grasp the cover
move_object(cover,box_top, ...);     ## move it
release_object( ... );               ## let go
grasp_object(box_grasp, ...);
move_object( ...);
```

The definition of grasp_object is a method and is sketched below:

```
grasp_object: SUBR(grasp_frame, ... )
    ...
    move_robot(grasp_frame,... );
    close_gripper( ... );
    (grasp_frame.rigid_ancestor()),affix_to(robot);
    ...
    END;
```

The functions move_robot and close_gripper may be system subroutines.

AML/X has been in use at IBM since 1985 in the research lab. The language is interpreted instead of compiled, but it gives reasonably good performance. Work on a compiler has begun because of some slowness of the system in large-scale CAD applications. In addition to robotic applications, AML/X has been used to write a user front-end for a geometric modeling system. The classes and methods are a natural way to express solid geometric models (see Chapter 2). A windowing package for AML/X was reported in 1987 to increase the user-interface applications [15].

10.9.3 Off-Line Programming and Simulation

In some ways, it is surprising that a special section concerning off-line programming for robots should be considered necessary in an introduction to robotics. When that close kin of robotics, numerical control, was discussed in Chapter 9, there was no special section that discussed off-line programming. Morris [19] observed the unique difference by pointing out that numerical control started out with off-line programming and is moving toward on-line systems, while robotics started out with on-line programming and is moving toward the off-line approach.

Before one is tempted to deride the robot industry for its emphasis on on-line programming, it should be realized that there are many reasons why some task programming has to be on-line. The biggest is probably the fact that many tasks include a large number of variables to be controlled, such as torch attitude, contact-tube-to-work distance, and so on in arc welding [19].

The advantage of off-line programming is the elimination of production downtime when the robot has to be programmed. Typical programming time for a robot application is 1 to 2 weeks [19].

Even though some 90% of robots are controlled in an on-line (teach) manner [3], there is no doubt that significant benefits will accrue when a larger percentage will be off-line programming. In fact, one report predicts that nearly a third of all U.S. robots will be programmed in an off-line fashion by 1995 [9].

The U.S. Air Force's ICAM operation developed off-line programming procedures for a Cincinnati Milacron robot in the early 1980s. One advantage of this development is the fact that the language developed, MCL (Machine Control Language), was based on the numerical control programming language APT, which for years had allowed for off-line programming. A 1985 handbook indicates that as of the writing there were no general-purpose off-line programming systems available commercially [21]. Eight systems that are in various stages of development are discussed. Morris [19] discusses some advances being made in off-line programming: Cincinnati Milacron's Robot Off-Line Programming System [ROPS] and a Michigan Technological University development calls GRIPPS. GRIPPS stands for Graphical Off-Line Robotics Planning and Programming System and incorporates facilities for work cell development as well as path programming and verification.

One innovation that off-line programming should incorporate is a *simulation* capability to allow robot motions to be verified prior to implementation.

There are many types of simulations, physical, analog, and digital being the common three. A physical simulation models the true situation with a physical (iconic) model. A case in point would be an oil-refinery piping model to determine possible problems prior to implementation. Physical simulations have been accomplished for manufacturing cells and flexible manufacturing systems, but they have definite accuracy problems if complex robot positioning is involved [32].

DIGITAL SIMULATION. *Digital* computer simulation models the real system using graphical representations. Such systems might be wire frame representations of anthropomorphic robots that test the feasibility of getting to specific coordinates as well as judge the ability to sidestep interfering objects within the work area.

Other digital simulations are oriented to evaluation of the operating characteristics of the entire work facility. Such a system can assist the design of the facility in such decisions as the following [6]:

1. Work station design
   a. Select the type of machine and equipment.
   b. Decide on the number of work stations, including machine load/unload stations, inspection units, and cleaning stations.
   c. Evolve the best work station layout.
2. Materials handling system design
   a. Choose the best workpiece/tool transporters.
   b. Decide on the number of transporters.
   c. Investigate transporter characteristics, such as moving speed, loading capacity, etc.

   d. Decide workpiece routing and related policies (priorities, etc.).

   e. Determine the size and number of storage buffers, as well as their locations.

3. Control strategy development

   a. Work-order scheduling and dispatching.

   b. Investigate reactions to disruptions, such as system component failure, maintenance schemes, unexpected interruptions, etc.

4. Supplementary considerations

   a. Number of fixtures/pallets to maintain.

   b. Tool management procedures.

In order to allow such decisions as these to be made, a digital simulation has to have information relative to the probability distributions for occurrences. For example, what are the failure characteristics of a robot or a conveyor? The simulation allows the modeled system to operate with respect to time and usually much faster than real time, so that a week's production run might be simulated in less than an hour.

An example of digital simulation would be the modeling of a manufacturing work cell such as pictured in Figure 10.42 [6].

While this type of simulation will *not* allow for robot dexterity verification, it will allow evaluation of potential problems that might occur in operation of the manufacturing cell. Statistics are maintained on such variables as average processing time, buffer storage levels, machine failure times, and so on. The effect of

FIGURE 10.42   Graphics screen representation of manufacturing cell. (From Ref. 6.)

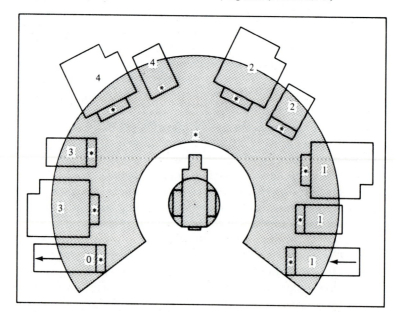

changing cell parameters can be evaluated *prior* to implementation of such changes. In addition, an animation option allows the analyst to *see* the effects of system changes projected on the graphics screen.

The research project that evolved the procedures just described took an existing simulation language and developed a user-friendly preprocessor so that manufacturing cells can be simulated by analysts with no previous simulation experience. The commercial simulation package used is called SIMAN (SIMulation for ANalysis) [24]. Certain features are included for modeling materials handling devices that make SIMAN especially suitable for *manufacturing* simulations.

10.10 SUMMARY

The field of robotics is in many ways an extension of machine tool numerical control. As with numerical control, robotics has to be a key to many computer-aided manufacturing functions. Typical of many tasks for which the inclusion of a robot is almost mandatory are those that have to be accomplished in hazardous environments as well as those with repetitive and monotonous work requirements.

The application spectrum for robots will undergo an interesting change over the next decade, with emphasis moving somewhat from welding to inspection and assembly. This does not mean that welding applications will decrease in number; rather, the *percentage of total applications* is expected to decrease but the total number of applications will see a significant increase.

It is of paramount importance that the potential robot purchaser perform (or have performed) a thorough economic evaluation before deciding if a robot really is a feasible and desirable alternative. Even if a robot is deemed beneficial, an economic and capability evaluation should be accomplished to pick the best robot from among the myriad available. Even with hazardous operations or dull, repetitive tasks, it should be realized that there may be alternatives to robots that might be more economical.

Finally, this chapter began with a somewhat gloomy aura brought about by the U.S. robot industry's early emphasis on large, nonflexible behemoths. The emphasis for the future will undoubtedly be on relatively small, inexpensive, electric-powered robots that are sufficiently dexterous to allow handling a wide variety of tasks. This flexibility should not be based solely on the robot linkage configuration. Rather, a major component will be the programming language, which must include user-friendly, task-related statements. Expert system capability will undoubtedly offer a significant benefit in optimization of robot operations, such as in part movement scheduling. Off-line programming capabilities will improve drastically, and an emphasis on the use of digital simulation for checking out these off-line programs will become more prevalent.

In spite of the gloomy comments presented earlier, there is no doubt that the robot, especially with its enhanced characteristics, will become firmly entrenched in computer-aided manufacturing.

EXERCISES

1. Review a *current* article to determine the robot applications to date in the United States by percentage, in the manner given in Table 10.1. Discuss any significant trends, if any, that are apparent.

2. Within your particular field of interest, specify a potential application for a robot. Define the tasks that the robot will have to accomplish and, based on these tasks, suggest the logical configuration for the robot (cylindrical, spherical, etc.) as well as any special instrumentation and end-effector requirements that need to be considered.

3. The introduction to this chapter gave some rather gloomy statistics relative to robot manufacturing in the United States. Review a *current* article that is more positive concerning the robot industry, and indicate any changes that seem to have occurred since the *Time* article [16] that indicate improvement.

4. Review the literature for articles that discuss *assembly* robots. What characteristics are required for an assembly robot that might not be needed for a welding robot?

5. Two robots are to be considered for a particular application. Pertinent data for these robots are as follows:

	Robot A	**Robot B**
Capital cost	$163,000	$138,000
Out-of-pocket cost per hour	$8.90	$9.05
Cycle time per part	14 min/piece	14.2 min/piece
Availability	0.95%	0.92%
Part cycle time	14 min/piece	14.2 min/piece
Income per part	$16.00	$16.00
Material cost per part	$6.50	$6.50

The life of the robot is assumed to be 5 years, with yearly variable costs (indirect, maintenance) as follows:

	Robot A	**Robot B**
Year 1	$5000	$5800
Year 2	$3500	$5200
Year 3	$3500	$5200
Year 4	$4000	$6000
Year 5	$5000	$6500

The plant will produce to the robot's capacity, since demand is high. The plant will work 24 hours per day and 340 days per year. Determine before-tax cash flows (BTCF) for each of the five years for each robot. Obviously, minimum annual rate of return and inflation rate will not be considered, since they have not been given.

6. Using your Exercise 5 results, determine after-tax cash flow (ATCF) assuming an effective tax rate of 31%. Assume 5-year, straight-line depreciation.

7. Assume that inflation over the next 5 years is estimated to be 5%, 6%, 8%, 6%, and 5% respectively. Correct your Exercise 6 results to account for inflation. For robot A only, evaluate the effect of using an average inflation rate (6%) for each of the five years.

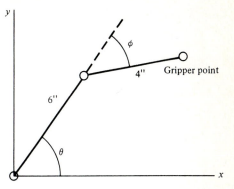

FIGURE 10.43

8. Determine the present worth of both robot investments in the previous exercise, given that management requires a minimum annual rate of return (MARR) of 14%. Which investment seems the best? The present worth is PW (17%).

9. From your results in Exercise 8, determine the annual worth of the two investments, AW (17%).

10. Two programming languages were discussed in this chapter. Review a *third* language by perusing your library's publications. A recent robotics text might be your best source.

11. Consider the VAL II example given in Figure 10.36. Plot the x–y path that the robot will take in moving from hole to hole; assume that the end effector starts at STA and, after drilling the 16 holes, comes back to STA. How optimum does this travel seem?

12. Redo the VAL II program given in Figure 10.36 to minimize the distance traveled assuming the start and finish locations as given in Exercise 11. Be sure to include the results of this assumption in your program. Also, after each 16 holes are drilled, incorporate a 25-s delay to allow a new part to be inserted in the fixture.

13. Write the matrix transformations to describe the gripper point location in terms of the joint angles for the robot shown in Figure 10.43. If the gripper is at the point (8.0,2.0), what are ϕ and θ?

14. A robot is to be used to place a pin in a hole. The hole has a location from a fiduciary pin A as shown in Figure 10.44. The hole is dimensioned and toleranced as shown. The

FIGURE 10.44

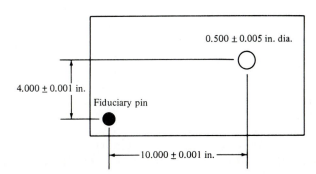

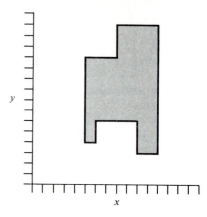

FIGURE 10.45

pin has a diameter of 0.490 ± 0.001 in. What is the repeatability required? What is the required precision?

15. Describe the robot workspace in Exercise 13.

16. Take apart a mechanical pencil and propose an assembly plan to be used with robot assembly. Plan the sequence. Use Owen's 15 guidelines to evaluate the pencil design for robot assembly. Propose part feeders and a precedence for assembly. What would the gripper(s) look like?

17. Calculate the area and centroid of the pixel pattern for an object viewed by a robot vision camera as shown in Figure 10.45. How would the vision system identify the part? What if two parts have the same area and centroid?

REFERENCES AND SUGGESTED READING

1. Boothroyd, G.: *Design for Assembly Manual*, University of Massachusetts, Amherst, Mass., 1985.
2. Craig, John J.: *Introduction to Robotics Mechanics and Control*, Addison-Wesley, Reading, Mass., 1986.
3. Critchlow, Arthur J.: *Introduction to Robotics,* Macmillan, New York, 1985.
4. Day, Chiaa P.: "Robotic Accuracy Issues and Methods of Improvement," *Robotics Today*, vol. 1, no. 1, Spring 1988.
5. DeGarmo, E. Paul, William G. Sullivan, and John R. Canada: *Engineering Economy*, 7th ed., Macmillan, New York, 1984.
6. Du, Wen-tzong: "Development and Evaluation of a Graphical Planning/Layout Pre-Processor for Flexible Manufacturing Systems," Unpublished Master's Report in Industrial Engineering, Arizona State University, Tempe, 1986.
7. Engleberger, Joseph F.: *Robotics in Practice*, AMACOM (A Division of American Management Associations), New York, 1980.
8. Fabrycky, W. J., and G. J. Thuesen: *Economic Decision Analysis*, 2d ed., Prentice-Hall, Englewood Cliffs, N.J., 1980.
9. Farnum, Gregory T.: "Industrial Robots—The Next 10 Years," *Manufacturing Engineering*, December 1985.
10. Fleischer, G. A.: "A Generalized Methodology for Assessing the Economic Consequences for Acquiring Robots for Repetitive Operations," in *Robotics and Industrial Engineering Selected*

Readings, vol. II, Industrial Engineering and Management Press, Institute of Industrial Engineers, Norcross, Ga., 1986.

11. Grant, Eugene L., W. Grant Ireson, and Richard S. Leavenworth: *Principles of Engineering Economy*, 7th ed., John Wiley, New York, 1982.

12. Groover, M. P., M. Weiss, R. N. Nagel, and N. G. Odrey: *Industrial Robotics: Technology, Programming, and Applications*, McGraw-Hill, New York, 1986.

13. Lammineur, P., and O. Cornillie: *Industrial Robots*, Pergamon Press, Elmsford, N.Y., 1984.

14. Lee, T. W., and D. C. H. Yang: "On the Evaluation of Manipulator Workspace," *Trans. ASME, Journal of Mechanisms, Transmissions and Automation in Design,* vol. 105, no. 3, pp. 70–77, March 1983.

15. Lieber, Derek: "A Window Package for AML/X," IBM Watson Research Center, Report #RA 185 (#56116), Yorktown, N.Y., January 7, 1987.

16. "Limping Along to Robot Land," *Time*, July 13, 1987.

17. Maus, R., and R. Allsup: *Robotics: A Manager's Guide*, John Wiley, New York, 1986.

18. Morris, Henry M.: "Profitable Robotic Work Cells Result from Interconnecting the Islands of Automation," *Control Engineering*, May 1985.

19. Morris, Henry M.: "Robot Programming Goes Off-Line," *Control Engineering*, May 1985.

20. Nackman, Lee R., M. A. Levin, R. H. Taylor, W. C. Dietrich, and D. D. Grossman: "AML/X: A Programming Language for Design and Manufacturing," IEEE Fall Joint Computer Conference, Dallas, Tex., November 2–6, 1986.

21. Nof, S. Y., Ed.: *Handbook of Industrial Robotics*, John Wiley, New York, 1985.

22. Owen, Tony: *Assembly with Robots*, Kogan Page, London, 1985.

23. Paul, R. P.: *Robot Manipulators: Mathematical Programming and Control*, MIT Press, Cambridge, Mass., 1981.

24. Pegdon, C. Dennis: *Introduction to SIMAN*, Systems Modelling Corporation, State College, Pa., 1982.

25. Potter, Ronald D.: "Analyze Indirect Savings in Justifying Robots," in *Robotics and Industrial Engineering Selected Readings*, vol. II, Industrial Engineering and Management Press, Institute of Industrial Engineers, Norcross, Ga., 1986.

26. Shahinpoor, Mohsen: *A Robot Engineering Textbook*, Harper & Row, New York, 1987.

27. Stonecipher, Ken: *Industrial Robotics: A Handbook of Automated Systems Design*, Hayden Book Co., Hasbrouk Heights, N.J., 1985.

28. Sullivan, W., and Ming Liu: "Economic Analysis of a Proposed Robot," in *Robotics and Industrial Engineering Selected Readings*, vol. II, Industrial Engineering and Management Press, Institute of Industrial Engineers, Norcross, Ga., 1986.

29. Toepperwein, L. L., M. T. Blackmon, et al.: ICAM Robotics Application Guide (RAG), Technical Report AFWAL-TR-80-4042, vol. II, Air Force Wright Aeronautical Laboratories, Wright Patterson Air Force Base, Dayton, Ohio, April 1980.

30. Tompkins, James A., and John A. White: *Facilities Planning*, John Wiley, New York, 1984.

31. *User's Guide to VAL II, The Unimation Robot Programming and Control System*, Preliminary Draft #3, Unimation/Westinghouse, Danbury, Conn., April 1983.

32. Viswanathan, K.: "The Physical Simulation of a Three-Machine Flexible Manufacturing System," Unpublished Master's Report in Industrial Engineering, Arizona State University, Tempe, 1983.

CHAPTER
11

MEASUREMENT, ANALYSIS, AND ACTUATION

Control only occurs when there is an interaction between Measurement, Decision, and Actuation.

Control Engineering [2]

Many topics have been discussed in this text that have somewhat blithely assumed that information can be obtained automatically from the process being controlled or relayed automatically to the process. Typical situations included the following:

- *Computer control* of a process requires analog input/outputs as well as digital input/outputs. The analog and digital information has to be available somehow from the process or receivable from the computer.

- *Programmable (logic) controllers* (PLCs) require information processing similar to that in computer control. Information availability was assumed when PLCs were discussed.

- *Numerical control* for machine tools assumes that specified design paths can be realized, at least within the tolerance constraints. It has been assumed that the tool/part position relationships can be obtained automatically in some fashion.

- *Adaptive control* of machine tool operations assumes that conditions can be sensed which will trigger adaptive reactions. For example, are the part dimensions during machining according to planned operations? This implies that a way is available to measure part dimensions and it is possible to correct for unexpected results.

- A *shop-floor control system* assumes that on-line information can be gathered in real time on which timely decisions can be based. Similarly, it is assumed that the timely decisions can be translated into corrective physical motions by such equipment as machine tools, materials handling devices, robots, automatic warehousing facilities, and so on.

- *Manufacturing cell* operation assumes that operations within the cell are sequenced correctly: For example, a part will not be removed from a machine tool chuck by a robot until machining is completed and the chuck is released.

All these situations assume that control follows the sequence of (1) sense current conditions in a process, (2) analyze those conditions, and (3) effect conditions through changes to the process. In turn, this assumes that the conditions can be detected and the information can be transmitted to the analysis device. Further, once analysis has been concluded, it is assumed that corrective information can be transmitted to the process and that this information can be utilized in correcting the process operation.

The objective of this chapter is to answer some questions as to how conditions may be detected in a process and how changes can be effected to a physical process. This will be accomplished primarily through discussions of equipment for sensing, measuring, analyzing, and actuating.

11.1   INTEGRATIVE ROLE IN CAD/CAM

Since this chapter is somewhat ancillary to previous chapters, it is moot to say that instrumentation for gathering process information and actuators that effect changes in a process based on transmitted information have an integrative role in the CAD/CAM process. As an example of the need for sensors and actuators in the CAM environment, consider Figure 11.1.

Quoting from Bollinger and Duffie [13] relative to this figure:

An automatic milling machine with a loading-unloading robot relies on diverse sensors, actuators, and displays. On the machine tool, dc motors (1) provide movement on the *x, y,* and *z* axes; tachometers (2) sense the speeds of the axis motors; resolvers (3) sense axis-motor shaft position; an ac motor (4) drives the tool spindle; and limit switches (5) sense when the milling table is approaching its maximum allowable bounds and thus prevent overtravel. A stepping motor (6) positions the tool changer so that the spindle can accept a new tool at the appropriate moment, and a tactile probe (7) measures the dimensions of the workpiece at each machining step. In the machine-control unit, servo amplifiers (8) regulate the machine drives, a computer (9) exercises overall control, and a display (10) keeps a human supervisor informed of the machine status. On the robot, hydraulic servo valves (11) actuate the arm, optical encoders (12) sense the position of the arm, a pneumatic control valve (13)

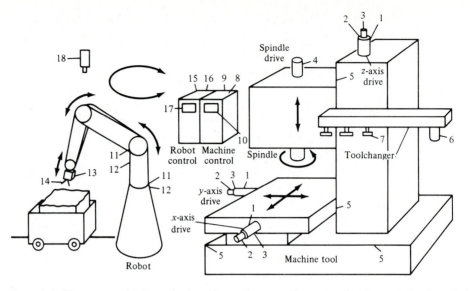

FIGURE 11.1   Examples of instrumentation in a robot/machine tool system. (Reproduced with permission from Ref. 13, © 1983 IEEE.)

actuates the robot's gripper, and a tactile sensor (14) measures the gripper force. The robot control contains servo amplifiers (15), a computer (16), and a display (17). Overhead, a TV camera (18) identifies parts and guides the robot.

It is appropriate to mention that lack of available sensors and actuators has historically been the stumbling block in many automation projects. In a paper production process, for example, an experienced operator will "feel" the paper with his or her hands as the paper moves into the roll. How can this inductive process be captured by instrumentation? Similarly, how can a machine tool adapt to the fact that a cutting tool is becoming dull? Further, once this is accomplished, how can the cutting speeds be adjusted fast enough to correct for the dullness? These are just a couple of cases where instrumentation was at one time a major restriction on automating a process. Subsequent sections of this chapter will introduce some existing sensor, analysis, and actuator capabilities. New instrumentation equipment is being introduced daily that will greatly further the facilitation of CAD/CAM operations.

11.2   KEY DEFINITIONS

Many terms specific to automatic control will be defined as they arise. A few key definitions are given at this time to prepare the reader for the specific material.

Actuation is the control function that allows a process variable to be changed by some other energy input. For example, a flow valve might be rotated with an electrical input to an actuator.

Analysis, in this chapter, will be the determination of whether the process being controlled is conforming to operation standards. If not, analysis will compute corrections needed.

Annunciator will refer to instruments that can measure a process variable, compare it to a desired set point, and alarm if necessary.

Bang-bang control refers to control action that is two-position rather than multiposition. A robot gripper that can be only fully open or fully closed conforms to this type of control.

Electromechanical relays are actuators that open or close relay contacts upon receiving an electrical signal.

Encoded media refers to an identification process whereby the information is coded in some fashion. The most common types are bar codes and magnetic stripe coding.

Limit switches are on/off switches that are actuated automatically by some object that hits a mechanical arm, cam, or other protrusion.

Measurement, in this chapter, will refer to determining process values automatically. Typical values include pressure, temperature, number of parts, and so on.

Photoelectric switches are on/off switches that change state as a result of some light emission characteristic.

PID controllers use some combination of proportion, integral, and derivative aspects of process error to automatically correct for that error.

Programmable limit switches are switching instruments that can have a limit switch programmed to open and close at specific time intervals for fixed-cycle operations. A drum timer is somewhat analogous to this type of switch.

Proximity switches are actuated by the *presence* of an object rather than by actual contact with the object. Typical proximity switches use inductive, capacitive, or magnetic properties for this purpose.

SCADA is an acronym for *s*upervisory *c*ontrol *a*nd *d*ata *a*cquisition system. As such, a SCADA can enable any computer to have real-time, process control capability.

Sensing is the determination of process conditions by automatic means.

Servomotor is a DC electrical motor that uses a control winding to allow the motor to automatically correct for errors in specified position.

Stepping motors are DC motors whose speed and position are accurately controlled by input *pulse* signals. As such they are highly amenable to communication from a digital device.

Transducers are instruments that will receive information from a process in one manner (position, pressure, velocity) and output that information in another form, usually an electrical signal.

11.3 SENSING AND MEASURING

Whatever the problem, no solution or corrective action should be attempted until the problem has been fully defined and understood. This is a key to the so-called systems approach to complex problem solutions where the problems have interacting components [11]. Computer-aided manufacturing systems are obviously extremely complex and have many interacting components. It is imperative that there be a clear understanding of current conditions before one attempts an analysis of those conditions, with the analysis followed by subsequent control (actuation) action.

As a consequence, an introduction as to how information can be obtained automatically from a physical process will now be presented. Technology with regard to sensing is expanding at a tremendous rate. Only a *brief* overview will be possible in this section. The reader interested in a more thorough discussion should consult the literature for texts which cover subsets of the information.

Sensing and measurement will be broken down into four categories: (1) object detection, (2) object identification, (3) condition detection, and (4) sensing for machines and robots. The latter category could be incorporated into the first three categories, but it should be of interest to see some of the advances being realized specifically in the fields of robotics and machines.

11.3.1   Object Detection

The simplest form of object detection is the mechanical-contact limit switch, which has been available for decades. A part coming down a conveyor might hit the arm of a limit switch, thus closing a set of contacts. In turn, closing the contacts allows current to flow, indicating to a computer or other device (light, annunciator, programmable controller, etc.) that a part is available. The limit switch closure might also shut off the power to the conveyor, through a programmable controller, allowing the part to be worked on.

Technology has come a long way since the mechanical limit switch was developed. For example, *proximity* switches measure the *presence* of an object without physical contact being made. *Photoelectric* switches, a form of proximity switch, also have wide application potential. Trends in the purchase of these three limit switches are shown in Figure 11.2. In 1976, photoelectric and limit switches had 91% of the market. In 1985, proximity switches accounted for 43% [26].

LIMIT SWITCHES. A conventional light switch is construed as a *manual* switch. A *limit* switch has the same on/off characteristics but changes position *automatically* when an object forces closure of the switch contacts. There are many ways in which this can be accomplished. The limit switch may be *pressure*-sensitive, so that an object on a conveyor will close the contacts just by its own weight. The majority of limit switches have a lever arm that the object can hit, thus causing closure. An object might have a magnet attached that causes a contact to rise and close when the object passes over it. A model train in an automation laboratory at Arizona State University has position sensed in just this manner.

Limit switches can either be normally open (NO) or normally closed (NC) as well as having multiple-pole capability. Figure 11.3 shows the possibilities. A NO switch has continuity when pressed, while a NC switch opens when actuated. A

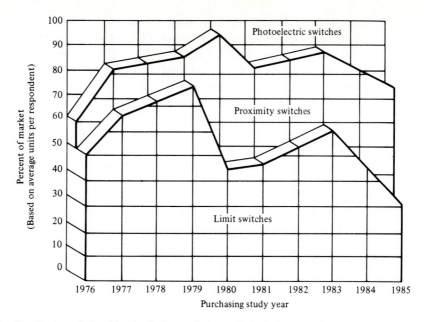

FIGURE 11.2 Purchasing relationships for limit, proximity, and photoelectric switches. (Reprinted with permission from Ref. 26, copyright © Cahners Publishing Co., 1987.)

FIGURE 11.3 Three limit-switch configurations: (*a*) single-pole, normally open limit switch; (*b*) single-pole, normally closed limit switch; (*c*) double-pole, normally open limit switch.

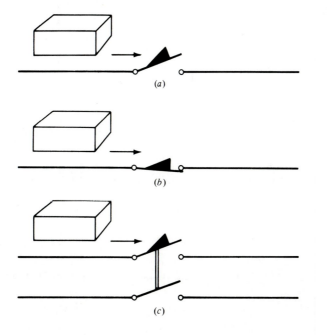

single-pole switch allows one circuit to be closed or opened upon switch contact, whereas an *n*-pole switch allows *n* poles to be opened or closed. In general, a four-pole limit switch is about the maximum usually encountered.

Two potential problems with mechanical limit switches are possible mechanical failure and relatively slow speed of operation. *Control Engineering* magazine reported on a *photoelectric* microsensor which had switching speeds up to 3000 times faster than conventional switches [7].

PHOTOELECTRIC SENSORS. A typical photoelectric sensor will provide a voltage—say, 10 V DC, if a light beam is not interrupted by some object. If the beam is interrupted, then no voltage is provided. Thus the voltage is the equivalent of an on/off switch.

Three possible modes for photoelectric sensing are *through scanning, retroreflective scanning,* and *diffuse* (proximity) *scanning* [18]. These three modes are shown in Figure 11.4. An obvious disadvantage of through scanning is the need to position the emitter and reflector directly across from each other. The second two modes do not have this problem.

Typical applications for the three modes of sensing are given in Figure 11.5. In these particular examples, a fiberoptic cable is utilized for transmitting the light source as well as receiving the light reflection.

FIGURE 11.4   Three modes of photoelectric scanning: (*a*) through scanning; (*b*) retroreflective scanning; (*c*) diffuse scanning. (From Ref. 18, courtesy of Hyde Park Electronics.)

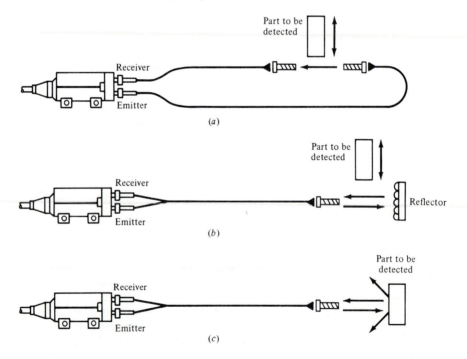

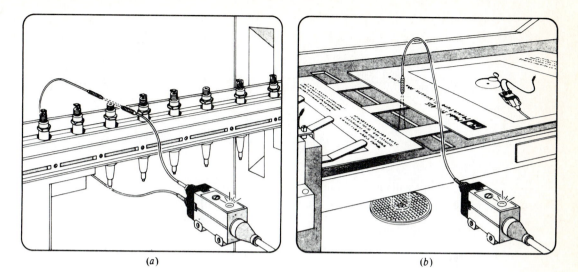

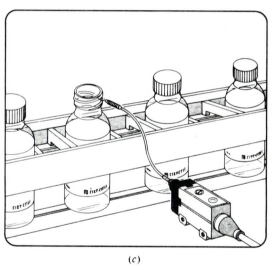

FIGURE 11.5   Typical photoelectric sensing applications. (*a*) Small parts detection: The through-scan mode can be used to detect small parts where the parts can be passed between the two fiberoptic cables. (*b*) Paper count: The retroreflective-scan mode can be used to count any opaque object that can be placed between the fiberoptic cable and a reflector. (*c*) Cap detection: The diffuse mode can be used to detect any object that has the ability to reflect light, such as a bottle cap, a label, or any other reflective surface. (From Ref. 18, courtesy of Hyde Park Electronics.)

An important characteristic to be noted when considering photoelectric sensing is *hysteresis*, which is the percent of the received light beam that must be broken to cause an "on" or "off" condition [18]. A typical hysteresis has 10% blocked for an "on" condition and 95% unblocked for an "off" state. This is shown in Figure 11.6. When the part blocks 10% of the through beam, the sensor

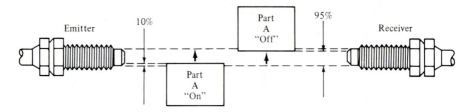

FIGURE 11.6   Hysteresis representation. (From Ref. 18, courtesy of Hyde Park Electronics.)

output will change state. When the part has moved a sufficient distance to unblock 95% of the light beam, the sensor output will revert back to its original state. This prevents the possibility of multiple sensings if the part vibrates while moving through the beam.

The capability of photoelectric sensing equipment is affected by both the ambient lighting as well as by workplace environmental conditions. Such conditions might include dirt or other materials affecting the emission and reflection of light. There are other methods of object detection that are similar in philosophy to photoelectrics but that overcome some of the problems just mentioned. One such way is through proximity sensing.

PROXIMITY SENSING. In some ways, all photoelectric sensing can be considered proximity sensing, because nothing physically touches the part when it is detected. To differentiate from photoelectric devices, *proximity detectors* are electrical or electronic sensors that respond to the presence of a material [32]. The major categories are inductive, magnetic, and capacitive, with the inductive and magnetic sensors requiring that the detected material be metallic. Cars at traffic lights, as an example, are frequently detected by proximity sensors buried just below the surface of the road.

Inductance and capacitive sensors effect switch changes on the basis that a correctly designed electrical device can be made to undergo an inductive change or a capacity change when an object is in close proximity. Seippel [32] indicates that inductive sensors can be designed for steel, chrome-nickel, stainless, brass, aluminum, and copper parts, whereas capacitive sensors can detect such diverse materials as steel, water, wood, glass, and plastics. Magnetic detectors are based on the fact that a current is induced in a coil located in a magnetic field if the magnetic flux changes due to a ferrous object moving into close proximity to the device [32].

Other categories of proximity detectors are available that affect switch on/off position. An exciting example uses an ultrahigh-frequency acoustic transmitter and tuned receiver for part sensing [19]. In a manner similar to the photoelectric through or retroreflective scanner, an object is detected when it interrupts a sound beam or reflects the sound beam from the transmitter to the receiver. Some application examples are shown in Figure 11.7.

FIGURE 11.7   Examples of ultrasound proximity sensor applications: (*a*) messy conditions; (*b*) dusty conditions; (*c*) level sensing through opaque can. (Courtesy of Hyde Park Electronics, Dayton, Ohio.)

So far, we have discussed only object *detection*, a far cry from object *identification*. Detection requires only a binary presence/no presence determination. Identification requires more detailed information to be gathered automatically. This is the topic of the next section.

11.3.2   Object Identification

In any type of manufacturing system it is mandatory that the status of all components of the system be known. This includes where parts, materials, and components are located, what machines and people are busy, the status of the materials handling equipment, and so on. In an automated facility it is desirable to be able to monitor the identification of moving materials and parts. The most common means of automatic identification is through bar codes, though many other approaches offer advantages under certain conditions. We will introduce the concept of encoded media for identification (bar codes, magnetic stripes) and then summarize some other possibilities.

BAR CODES. The person who has not seen bar code identification used in stores is rare indeed. The Universal Product Code (UPC), used in retail stores, is a standard 12-digit code. Five of the digits represent the manufacturer and five the item being scanned. The other two digits are used for identification of the type of number system being decoded (standard supermarket item, for example) as well as a parity digit to determine correctness of reading. The first half of the digits are represented by codes in the pattern of a light bar followed by a dark bar followed by a light bar followed by a dark. Different bar widths allow for many character combinations. The second half of the digits are formed by a dark bar followed by a light bar, etc., in the reverse sequence of the first digits. This allows backward scanning detection, with the decoder then reversing the sequence of digits. Figure 11.8 shows these pattern arrangements. A problem with the UPC is that it handles

FIGURE 11.8 Typical UPC representation.

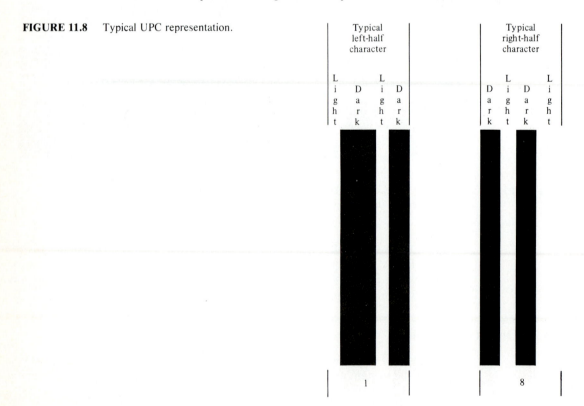

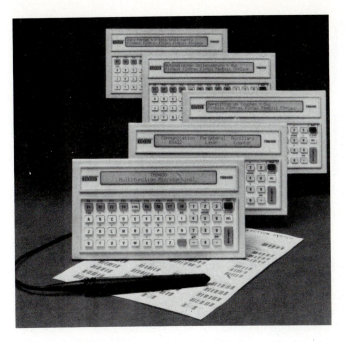

FIGURE 11.9   Bar code terminal. (Courtest of Burr-Brown, Tucson, Arizona.)

only *numeric* data. A bar code reader is shown in Figure 11.9. This particular reader can handle several different bar code standards, decoding the stripe without being told the particular standard (i.e., Code 39).

There are many other possible code schemes, but Military Standard 1189 specifies the type of coding (Code 39) used by the Department of Defense [12]. Code 39 provides for 44 characters including the letters A through Z; because of its alphanumeric capability, Code 39 is very effective for manufacturing applications. The "39" comes about because 3 of the 9 bars (light and dark) which form a character are wide, the rest being narrow.

Bar code labels are easy to produce. Figure 11.10 shows Code 39 labels that were generated by a personal computer program [36]. Such labels are ideal for inventory identification, part tracking throughout the manufacturing cycle, and other types of fixed-information gathering. Bar code does not have to be on a label. Tools, for example, have had the code etched on their surface to allow for tool tracking. Techniques have been developed that allow bar codes to be molded on rubber tires so that identification can be readily ascertained. Holographic scanners allow reading "around corners," so that parts need not be oriented perpendicular to the reader as they feed down a processing line. The list goes on and on.

Even though the Department of Defense has chosen Code 39 for their applications, many other systems exist and at least one more is being proposed. According to Ribeiro [29], a new Japanese company, the Calra Company, has been

ABCDE–123

CAD/CAM

FIGURE 11.10 Bar codes generated by personal computer program.

established to develop the code structure shown in Figure 11.11. Supposedly, the new code will be able to handle printing errors and eliminate the need for the high-precision printing required with conventional systems. The "Calra" code is supposed to be able to handle an error up to 0.75 mm, as contrasted to one of 0.01 mm with conventional systems. Also, if two rows are stacked with eight squares per character, the new code will handle 2^8, or 256, characters.

A problem with bar code has been the fact that the code cannot be read if the bars become covered with certain substances, though infrared scanners are used to read codes that are coated with a black substance to prevent secrecy violations through photoreproduction of the codes [35]. One way to generally offset the problem of dirty environments is to use magnetic stripe–encoded information.

MAGNETIC STRIPES. Information can be coded on a magnetic stripe in much the same way that bars represent information on a bar code label, since the light and dark bars are just a form of binary coding.

Worker identification data is often coded on a magnetic stripe that is an integral component of the worker's badge.

Figure 11.12 shows a machine operator wanding magnetic stripe information into a computer. Such information might include the fact that a job is complete,

FIGURE 11.11 The Calra code. (Courtesy of *Manufacturing Engineering*—see Ref. 29.)

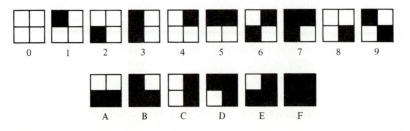

FIGURE 11.12   Magnetic stripe scanning station. (Courtesy of Garrett Turbine Engine Company, Phoenix, Arizona.)

the number of parts produced, the part number, the operator's identification number, and so on. However, it should be pointed out that the same scanning station could have been set up with bar code information.

A possible advantage of magnetic striping, as mentioned earlier, is that the information can still be read even if the stripe is contaminated with dirt or cutting oil. A disadvantage of magnetic striping is the fact that the reader has to contact the stripe in order to recall the information.

OTHER IDENTIFICATION TECHNIQUES. While bar code labels and magnetic stripes are very effective on the shop floor, there are many shop situations that require more information to be gathered about a product that can be realistically handled with encoded media. For instance, with automobiles being assembled to order in many plants, a lot of information is needed to indicate the options required for a particular assembly. Radiofrequency (RF) devices are used in many cases. An RF device, often called a *transponder*, is fixed to the chassis of the car. The "chip" in such a device can handle hundreds of bytes of information. A radio signal at specific assembly stations causes the transponder to emit information which can be "understood" by a local receiver. The transponder can be coated with grease and it will still work. The potential in any assembly operation is readily apparent. Some smart transponders have read/write capability, thus allowing support of local decision making [5].

A process that is similar to the RF device is surface acoustic waves (SAW). Object identification is triggered by a radar-type signal, which can be transmitted over greater distances than in RF systems.

Another form of automatic identification is optical character reading (OCR). English-type characters form the information, which the OCR reader can "read." High-speed mail sorting by the U.S. Postal Service is accomplished in mail processing centers using OCR readers. The potential extension to manufacturing information determination is obvious.

There are also many other means for object identification, such as vision systems and voice recognition systems. With vision systems, TV cameras read alphanumeric data and transmit the information to digital converter. OCR data can be read with such devices, as can conventionally typed characters [1]. Voice recognition systems have potential where a person's arms and hands are utilized in some function that is not conducive to reporting information. Such an application might be the inspection of parts by an operator who has to make measurements on the part.

As has been seen, object identification can be accomplished in a myriad of

TABLE 11.1
Comparison of some object identification approaches

Technology	Read and controller price*	Price of label or tag	Read distance	Line of sight required
Bar code	$500–$7,500	Very low	Contact to 10 ft	Yes[†]
Magnetic stripe	$500–$2,000	Low	Contact	Yes
Optical character recognition	$500–$2,000	Very low	Contact	Yes
Radiofrequency and surface acoustical wave	$700–$2,500	High	RF: <30 ft SAW: <5 ft	No
Machine vision	$28,000–$50,000	None to low	Depends on lens system Nominal: <10 ft	Yes
Voice recognition	$25,000–$35,000	N/A	N/A	N/A

* Range in cost depends on such features as functionality, microprocessor intelligence, and input/output capabilities.

[†] Holographic scanners can read "around corners."

[§] Infrared readers can read bar code when covered.

Source: Reprinted from Ref. 5 by permission of *Modern Materials Handing,* copyright © 1985 by Cahners Publishing Company, Div. of Reed Publishing USA.

ways. Which is the best way to handle a particular application? Table 11.1 shows typical comparisons of characteristics that might aid this decision.

11.3.3 Measurement of Conditions

So far in this section we have considered how the presence of something, such as a part, might be detected and identified. This information is not sufficient in many cases to allow realistic control to be maintained. Detecting a dull tool in a machining operation is mandatory if correction for tool wear is to be implemented. Knowing *how* dull will enhance the correction. There are times when it is desirable to know how fast something is moving, at what temperature is a particular bearing, and other variable conditions.

Automatic detection of conditions is not only an extremely broad application area, it is in many cases a limitation to complete automation of a process. There are, however, many standard ways to measure certain conditions automatically.

TRANSDUCERS. A *transducer* inputs one form of energy or characteristic and has as an output a form of energy or characteristic that is different from the input. The most common output is a voltage; the digital-to-analog conversion process discussed in Chapter 8 is such a transducer situation. So is analog-to-digital conversion, which does not have voltage as an output.

Seippel [32] gives a very comprehensive introduction to the field of transduc-

Can read if label or tag is covered	Durable lable or tag	Read laser-etch code	Human-readable label or tag	Reencodable label or tag	Code reading speed	Handheld reader available
No[§]	No	Yes	No	No	Fast	Yes
Yes	No	No	No	Yes	Moderate	Yes
No	No	Yes	Yes	No	Moderate	Yes
Yes	Yes	No	No	RF: Yes SAW: No	Fast	No
No	Yes	Yes	Reads both OCR and typed characters	N/A	Moderate	N/A
No	Human dependent	No	Yes	N/A	Moderate	N/A

ers and lists 18 parameters that are commonly measured with transducers: linear distance, velocity, and acceleration; angular displacement, velocity, and acceleration; force; torque; vibration—displacement, velocity, acceleration, and torsional; sound; pressure; flow; temperature; viscosity; and humidity.

With each measurement parameter, Seippel gives the transduction principle. For example:

Linear displacement	1. Strain gauge
	2. Linear potentiometer
	3. Linear (digital) encoder
Angular displacement	1. Strain gauge
	2. Shaft encoder
	3. Capacitor
Pressure	1. Bellows—potentiometer
	2. Diaphragm—strain gauge
Temperature	1. Thermocouple
	2. Optical pyrometer

The reader who is interested in a rather quick introduction to the underlying principles of transducers will benefit from reading the book by Bannister and Whitehead [10]. They show that a large percentage of transducers can be explained in terms of only a small number of basic principles. Some of these include the following:

- *Electromechanical transducers* use mechanical movement of some form to realize an electrical output.
- *Resistance transducers* use potentiometers to measure rotary or linear position.
- *Capacitive transducers* rely on the fact that a percentage change in the separation between two parallel plates produces a corresponding change in capacitance.
- *Piezoelectric transducers* use the property of certain crystals that an electric charge is generated when the crystal is deformed.
- *Photovoltaic transducers* (solar cells) convert light energy to electrical energy; no external power supply is required.

As one example of a transducer, consider the shaft encoder. Shaft encoders are found in some machine tool situations such as rotary table positioning, as well as in robot positioning situations [21]. A shaft encoder transduces angular position to binary high and low voltage outputs as shown in Figure 11.13. If mechanical probes conduct a current when a dark area is sensed and do not conduct when in a light area, then the binary codes for the encoder shown in Figure 11.13a can be evolved as shown in Table 11.2. The codes assume that the outside ring is the lowest-order binary position. For the four concentric rings we have evolved 16

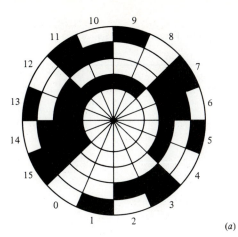
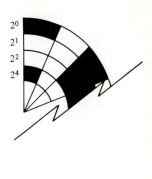

(a)

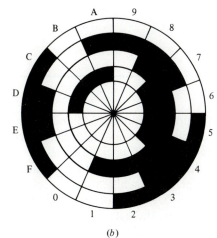

(b)

FIGURE 11.13   Possible shaft encoder configurations: (*a*) example code (adapted with permission from Ref. 21); (*b*) Gray code.

(2^4) different values, giving an angular resolution of 22.5°. Obviously, the more rings, the better the accuracy. The first code has the problem of having more than one change in conducting/nonconducting position for adjacent characters (i.e., 15 to 0). Sensing position at these adjacent points can give rise to potentially large errors. The Gray code eliminates this problem [21]. A 4-bit Gray code sequence is given in Table 11.2, with the equivalent shaft encoder configuration given in Figure 11.13*b*.

11.3.4   Machine Tool Sensing

Excellent coverage of all facets of machine tools is found in Manfred Weck's four-volume *Handbook of Machine Tools* [37]. The third volume of this unique work is

TABLE 11.2
Shaft encoder bit sequences (for Figures 11.13a and 11.13b)

4-Bit sequence	Koren code	Gray code
0000	0	0
0001	1	F
0010	2	9
0011	3	E
0100	4	1
0101	5	2
0110	6	8
0111	7	3
1000	8	B
1001	9	C
1010	A	A
1011	B	D
1100	C	6
1101	D	5
1110	E	7
1111	F	4

devoted entirely to automation and controls. A considerable portion of the third volume discusses sensors and transducers used in gathering the necessary information for control of machine tools.

Readers interested in the interface of CAD with CAM cannot expect to become expert in machine tool controls; in fact, one generally has to emphasize either CAD or CAM in choosing a career path, because of the complexities of both fields. However, the user should realize that the modern machine tool is a very intricate and complex piece of machinery that incorporates many facets of sensor and transducer technology in the information-gathering step. A few examples of the complex sensing necessary within automatic machine tools will now be given. The reader who needs in-depth information should consult Weck's handbook.

PART MEASUREMENT. When one considers the conditions surrounding high-speed machining, it is remarkable that design tolerances of a thousandth of an inch or better can be maintained. It should be realized that design tolerances refer to points in space, not the ability of axis drives to maintain axial tolerances. Conditions affecting tolerance maintenance include high temperatures, tool wear, adverse environmental conditions, variations in metal structure, gear slippage, and other mechanical problems.

One way to offset some of these problems is to enclose the work area fully and then flood coolant to try to maintain optimum temperature conditions and eliminate some of the environmental characteristics. However, tools still wear and parts still need to be measured in order to allow corrective actions to be taken if needed.

Some of the ways to handle production gauging are given in Sheffield's history of measurement [33]. One of the most common on-line methods is to use a *probe* which emits an electrical signal proportional to the part's surface distance from a reference point. The probe is very successful in helping to maintain machining requirements for rotational (turned) parts. Mechanical feelers, similar to probes, can be used to sense if a tool is broken [37].

Coordinate measuring machines (CMMs) are often incorporated into a flexible manufacturing system as a separate station. The CMM, under computer control, can measure an extremely wide variety of part configurations.

Optical gauging utilizes light reflection or light modification as it relates to the variable being measured. Types of systems discussed by Sheffield [33] include the following:

- *Scanning laser:* A laser beam scans the measurement area at a constant speed, and the object being measured interrupts the beam for a time which is proportional to the diameter or thickness of the part being measured. Resolutions of 10 millionths of an inch are possible.

- *Linear array:* Parallel light beams are emitted from one side of the object to be measured to a photo optical diode array that is mounted on the object's opposite side. Diameters are measured by the number of array elements that are blocked. Sheffield indicates that resolutions of 50 millionths of an inch, or better, are feasible.

- *TV camera:* A TV camera is used in the digitizing of the part, and the result is compared to a stored image. Dimensions can be measured, and part orientation and feature presence can be checked. Some exploratory work is being accomplished with cameras that can fit in a tool-changer mechanism. The camera can be brought to the part in the same manner as a tool, and part characteristics can then be verified automatically.

TOOL WEAR. The sensors accomplishing part measurements can compare the information obtained with expected results, and a computer inference program could well determine the presence of tool wear. A more common approach, as with many sensing applications, is through *indirect measurement*. Information that is relatively easy to obtain can be interpreted to indirectly estimate hard-to-measure conditions. A dull tool, for example, will put a strain on the drive motors of the machine. The increase in amperage can then be interpreted in terms of tool wear, and speeds and feeds can be adjusted to compensate if the tool wear is not too drastic.

AXIAL MOTION. If spacial tolerances are to be achieved, it is logical that axial tolerances have to be maintained. This means, in turn, that axial position has to be measured. Optical encoders, discussed earlier, can be utilized for this purpose. Another method, suggested by Koren [21], is the *inductosyn*. The inductosyn can be obtained for rotary (lead screw) or linear (table position) applications.

SEQUENCE OF FUNCTIONS. There are many occasions where specific sequences of motions are required in machine tool operation. A case in point is automatic tool changing. Once the tool magazine is positioned with the correct tool in the unload position, the tool is removed from the magazine, transported to the work area, inserted in the tool holder, and set to the correct depth. Sensing of the tool at each sequence point can be by limit switches (a very common sensor for any control function). The results of this sensing can then be transmitted to a programmable controller for actuation.

TOOL IDENTIFICATION. It is very common for cutting tools in a machining center to be coded so that they can be automatically identified in the tool-changing mechanism. Some flexible manufacturing systems which have lot sizes of one utilize computer programs that identify what tools are needed based on the part arriving. If the tools are coded in a fashion that allows the correct tool to be identified in the changer mechanism, then the tools do not have to be placed in the mechanism in a specific, predetermined order.

Special binary codes are utilized by some manufacturers, whereas others utilize etched bar codes. It is possible to utilize a similar form of part coding that can trigger the computer to automatically set programmable fixtures (manipulated by hydraulic controls); thus fixturing and tool selection can be evolved automatically based on what part arrives in the system.

Now that we have seen some of the sensing needs for machine tools, similar overview comments will be made concerning robots.

11.3.5   Robot Sensing

As is the case with machine tool sensing, there is a tremendous amount of literature dealing with robotics; not surprisingly, in view of the analogy between some robots and human beings, a large portion of the robot literature relates to sensing. In fact, Pugh [28] has edited a two-volume work devoted strictly to robot sensors. Volume 1 covers vision, while Volume 2 is centered on tactile and nonvision sensing.

Whatever the type of robot and the function it performs, a robot has to have *position*-sensing capability. As with machine tool position, this can be accomplished with a variety of sensors, including *encoders*. To demonstrate some of the special sensing needs of robots, this section will concentrate on the human sensing analogies: feel, hear, and see.

TACTILE SENSING. The human hand is a wonderful mechanism. It can pick up an egg without breaking it, crack it on a countertop, and then peel the shell while leaving the inside intact. Making a robot hand emulate this simple task requires considerable tactile skill—skill that relates to the sense of *touch*. It should be mentioned, though, that robots have been used to break eggs for a baking process.

Many robot grippers are capable only of going from fully open to fully closed, and vice versa. Pneumatic, two-position controllers are excellent for this type of hand operation. Two-position hand operation, however, does *not* represent tactile sensing.

As we saw when sensors were discussed earlier in this chapter, the simplest form of presence detector, or touch, is the on/off switch. A problem with this type of sensing is that the robot arm can have an error in orientation after the switch closes due to overshoot or the inability of the switch to allow orientation sensing [27].

Stonecipher [34] indicates that there are two approaches to tactile sensing: (1) skin sensing and (2) "haptic" perception. The latter relates to *indirect* measurement by interpreting what the joints are doing. Groover et al. [17] give the example of using a force-sensing wrist in the haptic category.

A single-force sensor can be utilized as a touch sensor—say, with a strain gauge. Jayawani [in 28] claims that tactile sensing is the *continuous* sensing of forces in an array, thus approximating skinlike properties. A variety of sensing devices have been developed that can be set in an array pattern to allow variable-force sensing (as contrasted to an array of microswitches or light diodes as discussed earlier in relation to part measurement). Groover et al. [17] discuss an 8×8 matrix of conductive elastomer pads as one such array. The electrical resistance of each pad changes in direct relation to the pad deflection, which is itself proportional to force.

Research is continuing with regard to tactile sensing. Multiple-finger "hands" have been experimented with. In fact, it has been suggested that robot fingers will eventually be equipped with high-precision tactile sensors to produce a digital image of manipulated objects [16]. The aim is to approach human tactile sensing ability.

VOICE RECOGNITION. Computer voice recognition has progressed greatly in recent years. The "hearing" function is accomplished basically by the human voice first being converted to a voltage signal. Next it is passed through an analog-to-digital conversion, and lastly, the digital representation is compared with a library of prestored voice inputs. Obviously, voice recognition could be used in the programming of the robot and for possible emergency commands such as "Stop."

For strictly *control* functions, voice recognition will have limited application. It does not take a great deal of imagination, however, to see the potential for *sound* recognition. For example, consider a manufacturing cell operation. An experienced operator is often responsible for all the automatic machines in the cell: setup, tear down, monitoring. Operators can *sense* if problems are occurring through sound or smell sensing. Sound-sensing capability may well be automated in the future. Similarly, the presence of parts may well be accompanied by an appropriate sound. Golf ball sound sensors have been developed to assist blind people who play that sport. Why not similar concepts in the manufacturing process?

VISION SENSING. The technological requirements for vision sensing are extremely complex and are beyond the scope of this text. However, introductory comments may be beneficial for the CAD/CAM analyst.

Vision is utilized in robot applications for part recognition, sensing a part's orientation on a conveyor, inspection, and so on. In many cases, other approaches might be better and cheaper. For example, part orientation can be accomplished through a vendor requirement that a part be positioned on a pallet so that the orientation will be as required when the conveyor comes down the line. Similarly, mechanical orienting devices have been around for years that have ensured that parts have the correct orientation in automated production lines. Be that as it may, many production systems will benefit from vision systems. In fact, it has been predicted that over 25% of all robots sold in the United States in the early 1990s will have some vision capability [34].

The steps in vision sensing are (1) obtaining the pictorial image, (2) digitizing the image, and (3) processing and analyzing the digitized image with a computer.

The image is usually captured with a camera, several varieties of which have appeared on the market. A common type is the vidicon tube, which puts out a continuous voltage signal for each image line processed [15, 17]. A digital conversion process then generates the digital representation of the image for storage in a computer. Processing now has to be accomplished to account for lost or unclear information.

The cost of vision systems has been of some concern. Groover et al. [17] suggest that such systems need to cost in the area of $30,000 or less if they are to be viable. Unfortunately, it is becoming more and more difficult to cost-justify robot systems in manufacturing today. Stonecipher [34] sums up the situation succinctly:

> In time, the country's research efforts will no doubt yield solutions to some of the problems which have long plagued industry, particularly those associated with computer vision. Robots will have to be able to provide better accuracy and additional speed. But right now, speed plus accuracy is hard to obtain at any price, even though a low price is required for the system to sell.*

11.4 ANALYSIS

In any control situation, the data input from the process being controlled has to undergo analysis to determine if information *needs* to be transmitted back to the process to correct for problems that are occurring or are expected to occur.

The material in this book has been oriented to CAD/CAM in discrete-item production. In any manufacturing plant, however, there are many needs for *continuous* process control. For a machine tool, continuous control would relate to

---

* Reproduced with the permission of the publisher, Howard W. Sams & Co., Indianapolis, Ind., from *Industrial Robotics*, by Ken Stonecipher, Copyright 1985.

the movement of the tool cutter in a continuous path or maintenance of a desired flow of coolant fluid. The majority of motor drive controls fall into the continuous control spectrum. An exception is stepping motors, which require a continuous pulse stream to operate. Because the analyst involved in CAD/CAM is likely to run into continuous control applications, particularly proportional-integral-derivative (PID) controllers, the first subsection on analysis will *introduce* continuous control concepts. Then, some comments will be made regarding digital computer and programmable controller roles relative to analysis for control.

11.4.1 PID Control

An advantage of using a digital computer in the analysis phase of control is that decision logic can be programmed so that control can be accomplished in much the same way as an expert operator reacts to disturbances in the system being controlled. The computer has an advantage over an operator in being able to respond to such changes much more quickly.

Using a cell control computer to determine the correct tool position in an NC continuous path operation or for many local control applications is not logical. Rather, local controllers react in a feedback mode to changes from desired conditions. This eliminates the need for the cell computer to be burdened by many control functions that could well prevent it from responding to critical conditions. A widely used control scheme in such autonomous *continuous* controllers is called proportional-integral-derivative control, often abbreviated PID. Such control capability frequently is found in programmable controllers, which can be located where required in a factory. Also, manufacturing cell controllers may have *interface devices* which have PID capability. Such devices will be discussed immediately following this section.

The complex PID mathematics will be avoided by presenting basic continuous control approaches in a sequential manner: (1) two-position, (2) proportional, (3) integral, and (4) derivative control. The introduction is given at this time because PID controllers have enjoyed successful widespread use in manufacturing for many years and are significant tools in CAM applications.

TWO-POSITION CONTROL. We are all familiar with a central heating system where the heat is turned on if the temperature falls below the thermostat setting and cuts off when the temperature is at the setting. This is called two-position control. If the heating system worked exactly as just described, then the heater would be continually turning on and off in very short time intervals. Not only would this be annoying to those in the heated environment, it would be damaging to the equipment. A *differential gap* is used to space out the on/off cycles.

The differential gap is analogous to hysteresis, which was discussed earlier relative to photoelectric controllers. For temperature control we might set the differential gap between 70 and 72°F. If the controller is turned off at the upper range of the differential gap, 72°F, then it will not go on again until the differential gap has been transversed to 70°F. The same is true for the on-to-off range.

The differential gap in a thermostat is usually set by adjusting a magnet's position with a screwdriver. The magnet holds the heater control contacts closed for a length of time that is proportional to the distance between the magnet and the contacts. The closer the magnet is to the contacts, the wider is the differential gap.

The two-position control with differential gap response is exemplified in Figure 11.14. A problem with such control action is that a specific setpoint cannot in general be maintained exactly. PID is a way to reduce this problem.

PROPORTIONAL CONTROL. The control of level of a liquid process—say, a swimming pool—is frequently accomplished by using a device where the control function is *proportional* to the deviation of the level from some specified setpoint. A swimming pool might have a floating ball, similar to the floating ball in a toilet, that is linked to a water-entry valve that opens an amount proportional to the deviation from the desired pool level. If the level drops at a rapid rate, the controller will attempt to fill the pool at a rapid rate. The heating system discussed with two-position control would have to have a way to input heat at a rate proportional to the deviation from a setpoint if the controller were to be proportional.

If ΔY is the process deviation and ΔX is the controller deviation, possibly from a closed position, the proportional action can be expressed by

$$\Delta X = G_p\, \Delta Y \qquad (11.1)$$

where G_p is a proportionality factor.

Obviously, ΔX and ΔY in Eq. (11.1) can be specified in terms of actual movement (perhaps feet for ΔY and inches for ΔX). A more common way is to express a proportional controller in terms of its *proportional band*:

$$\text{PB} = \left[\frac{\Delta X_m}{G_p\, \Delta Y_m}\right] (100\%) \qquad (11.2)$$

FIGURE 11.14   Example of two-position temperature control.

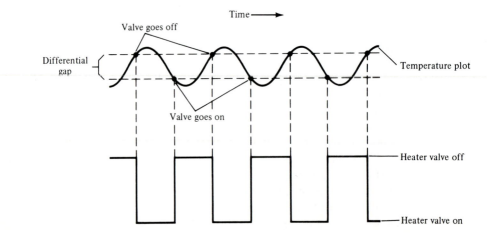

where ΔY_m is the maximum possible process deviation, and ΔX_m is the maximum possible controller deviation. The proportional band, PB, gives the percentage of process deviation required to effect 100% of controller change. For example, if ΔY_m is 1 ft and ΔX_m is 0.5 in., then a controller with G_p of 1 in./ft will have a PB of 50%.

The lower the PB percent, the more responsive the controller will be to a unit change in the process. In fact, if the PB is turned to 0%, the controller approaches a two-position controller. One problem with a proportional controller, however, is the fact that it will not stabilize at a desired setpoint while a sustained disturbance is in effect. The offset is called a *proportional offset*.

The intent of this section is not to transform the reader into a control engineer; rather, it is to introduce continuous control concepts that often arise in relation to computer-aided manufacturing. But since proportional control is fairly straightforward, it might be interesting to see how a proportional controller reacts to changes in the proportional band. To do this, a simple BASICA program was written as given in Figure 11.15.

The user inputs the proportional band percent and setpoint percent. The latter, logically, is the setting the process is desired to maintain. Next, a steady-state controller percent is input. This is the valve opening, or other controlled action, that will maintain the process at the specified setpoint if no process disturbance is encountered. A controller effect factor percent follows. To simulate the process control situation we have to specify how the control action will affect the process. The controller effect percent—for example, 5%—says that for every

FIGURE 11.15   BASICA program to demonstrate proportional control action.

```
10   DIM C(31),P(31),TP(31)
20   FMT$="##                    ###.###              ###.###               ###.###"
30   FMT1$="     ###.###            ###.###            ###.###"
40   INPUT "INPUT PROPORTIONAL BAND PERCENT"; PB
50   INPUT "INPUT SETPOINT PERCENT"; P(0)
60   INPUT "INPUT STEADY STATE CONTROLLER PERCENT"; C(0)
70   INPUT "CONTROLLER EFFECT FACTOR PERCENT"; CFAC
80   INPUT "INPUT STEP DISTURBANCE";DIS
90   TP(0)=P(0)
100  PRINT " TIME PERIOD   PROCESS PERCENT   CONTROLLER PERCENT   TRUE PROCESS"
110  FOR T=0 TO 30
120  IF T=0 THEN TP=P(0): GOTO 210
130  P(T)=TP(T-1)
140  X=(P(T)-P(0))*(100/PB)
150  C(T)=C(0)-X
160  IF C(T)>100 THEN C(T)=100
170  IF C(T)<0    THEN C(T)=0
180  TP(T)=TP(T-1)+(C(T)-C(T-1))*(CFAC)
190  PRINT USING FMT$; T,P(T),C(T),TP(T)
200  GOTO 260
210  PRINT " 0 (BEFORE DIST)";
220  PRINT USING FMT1$; P(T),C(T),TP(T)
230  TP(T)=P(T)+DIS
240  PRINT " 0 (AFTER DIST)";
250  PRINT USING FMT1$; P(T),C(T),TP(T)
260  NEXT T
270  END
```

percent of controller value change we will get 5% process change (a huge amount). Finally, a sustained step disturbance to the process is input to the program.

Figure 11.16 shows four results from running the program. The format for the output is from LaJoy [23] and warrants a brief explanation. We are approximating *continuous* control with *discrete* time periods. In run one, at time 0 (after

FIGURE 11.16   Effects of proportional band on controller.

```
RUN
INPUT PROPORTIONAL BAND PERCENT? 40
INPUT SETPOINT PERCENT? 40
INPUT STEADY STATE CONTROLLER PERCENT? 60
CONTROLLER EFFECT FACTOR PERCENT? .02
INPUT STEP DISTURBANCE? 1
  TIME PERIOD     PROCESS PERCENT     CONTROLLER PERCENT     TRUE PROCESS
0 (BEFORE DIST)       40.000               60.000              40.000
0 ( AFTER DIST)       40.000               60.000              41.000
1                     41.000               57.500              40.950
2                     40.950               57.625              40.953
3                     40.953               57.619              40.952
4                     40.952               57.619              40.952
5                     40.952               57.619              40.952
```

Run 1
— Proportional band = 40%
  Steady-state offset = 0.952

```
RUN
INPUT PROPORTIONAL BAND PERCENT? 30
INPUT SETPOINT PERCENT? 40
INPUT STEADY STATE CONTROLLER PERCENT? 60
CONTROLLER EFFECT FACTOR PERCENT? .02
INPUT STEP DISTURBANCE? 1
  TIME PERIOD     PROCESS PERCENT     CONTROLLER PERCENT     TRUE PROCESS
0 (BEFORE DIST)       40.000               60.000              40.000
0 ( AFTER DIST)       40.000               60.000              41.000
1                     41.000               56.667              40.933
2                     40.933               56.889              40.938
3                     40.938               56.874              40.937
4                     40.937               56.875              40.938
5                     40.938               56.875              40.938
```

Run 2
— Proportional band = 30%
  Steady-state offset = 0.938

```
RUN
INPUT PROPORTIONAL BAND PERCENT? 100
INPUT SETPOINT PERCENT ? 40
INPUT STEADY STATE CONTROLLER PERCENT? 60
CONTROLLER EFFECT FACTOR PERCENT? .02
INPUT STEP DISTURBANCE? 1
  TIME PERIOD     PROCESS PERCENT     CONTROLLER PERCENT     TRUE PROCESS
0 (BEFORE DIST)       40.000               60.000              40.000
0 ( AFTER DIST)       40.000               60.000              41.000
1                     41.000               59.000              40.980
2                     40.980               59.020              40.980
3                     40.980               59.020              40.980
```

Run 3
— Proportional band = 100%
  Steady-state offset = 0.980

```
RUN
INPUT PROPORTIONAL BAND PERCENT? .01
INPUT SETPOINT PERCENT? 40
INPUT STEADY STATE CONTROLLER PERCENT? 60
CONTROLLER EFFECT FACTOR PERCENT? .02
INPUT STEP DISTURBANCE? 1
  TIME PERIOD     PROCESS PERCENT     CONTROLLER PERCENT     TRUE PROCESS
0 (BEFORE DIST)       40.000               60.000              40.000
0 ( AFTER DIST)       40.000               60.000              41.000
1                     41.000                0.000              39.800
2                     39.800              100.000              41.800
3                     41.800                0.000              39.800
4                     39.800              100.000              41.800
```

Run 4
Proportional band = 0.01%
Controller acts as two-position device

the disturbance), the true process indicates the effect of the disturbance. The process percent does not reflect this until the next time period because of the time lag in measurement capability. Eventually, both process percent and true process will stabilize at the same steady-state value.

Now, remembering that we have a *sustained* step disturbance in the process, it can be seen for run 1 that an offset from the setpoint of 0.952 eventually occurs. This offset decreases as the proportional band decreases and increases as the PB increases. The fourth run shows a problem: A very low PB causes the process to oscillate, with the controller going from fully open to fully closed condition (bang-bang or two-position control). If sustained disturbances are expected for any length of time, it is apparent that the offset is undesirable.

INTEGRAL CONTROL. Using the same symbology as we used for the proportional controller, an integral controller follows

$$\Delta X = G_i \int_0^t \Delta Y \, dt \tag{11.3}$$

where G_i is an integral constant.

The controller (X) has a rate of motion that is proportional to the process deviation, ΔY. Because the control action usually is used to *reset* the offset found with proportional controllers, this type of control is often called reset control. A PI controller will combine Eqs. (11.1) and (11.3):

$$\Delta X = G_p \, \Delta Y + G_i \int_0^t \Delta Y \, dt \tag{11.4}$$

and G_p and G_i can be adjusted on the controller for the sensitivity of proportional control desired as well as the ability to reset the proportional offset. Cassell [14] indicates that PI controllers are probably the most common type of continuous process controllers in use.

DERIVATIVE CONTROL. Derivative control follows

$$\Delta X = G_d \left[\frac{dy}{dt} \right] \tag{11.5}$$

where G_d is the derivative tuning factor.

The controller changes according to the *rate of change* of the process error (deviation from setpoint), and so is often called *rate* control. Cassell [14] says that this allows for rapid response to a *desired* change in setpoint. Because it is very sensitive to random changes in the process, it is only utilized with one or more of the previous types of control. A complete PID controller has

$$\Delta X = G_p \, \Delta Y + G_i \int_0^t \Delta Y \, dt + G_d \left[\frac{dy}{dt} \right] \tag{11.6}$$

Adjustment of the three coefficients dictates the end effect of the three forms of control. The relationship between proportional offset by a proportional controller and how PI and PID attempt to correct for the offset is depicted in Figure 11.17.

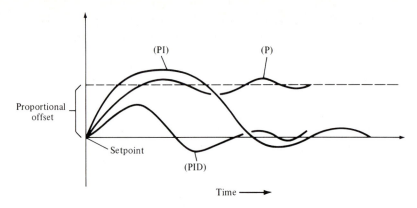

FIGURE 11.17   Comparison of relative controller actions. (From Ref. 14, with permission.)

11.4.2   Analysis for Control by Programmable Controllers and Computers

If we were to consider all the different analysis approaches utilized in digital computer control, this text would become many volumes in length. The task with programmable (logic) controllers (PLC) would not be so immense, since there is a generic characteristic about analysis with PLCs. This section will therefore first consider PLC analysis and then conclude with some generic comments on computer control. It will be assumed that the reader has perused Chapter 8 on computer control, which discussed programmable controllers.

PROGRAMMABLE CONTROLLERS. Johnson [20]* indicates that there are five general areas for PLC application: (1) sequence control, (2) motion control, (3) process control, (4) data management, and (5) communications.

Sequence control is usually handled by boolean evaluations of the status of combinations of digital inputs, with digital outputs being set dependent on the boolean analysis. This was the analysis approach presented in Chapter 8 when PLCs were discussed. The concept of a ladder diagram being the medium for input programming for such logic was also introduced at that time. Counters and timers are incorporated into the boolean logic. Sequencing of a materials handling operation would fall into the sequence control function. Similarly, sequencing of traffic at a traffic light is a sequencing function handled by a PLC. A left-turn arrow might turn on, for example, if three or more cars arrive in the left-turn lane during a red-light cycle. The PLC can sense the number of cars and the fact that they arrived during a red-light cycle, and trigger the left-turn arrow to be on.

Most PLCs have bit-sensing capability as well as register-shifting capability. Bit sensing refers to the ability of testing for 0 or 1 in bit register positions.

---

* Adapted from Ref. 20 by courtesy of Marcel Dekker, Inc.

Register-shifting capability means that the bit patterns contained in a register can be shifted to the right or left. For example, if a 12-bit register contains the values

| 1 | 0 | 1 | 1 | 0 | 1 | 1 | 0 | 1 | 0 | 1 | 1 |

a logical left shift, four places, would give

| 0 | 1 | 1 | 0 | 1 | 0 | 1 | 1 | 0 | 0 | 0 | 0 |

Using these types of commands it is possible to make the bits represent the status of many products in a production line. *Matrices* of register values extends this capability from just the number of bits that comprise a single register.

Wilhelm [38] offers an interesting example using the PLC's capability for acting on a matrix of bit values. A conveyor moves parts through a sequence of four test stations. If a product fails any test, it is so classified and sorted at the end of the conveyor. If a product fails at a test station which has other test stations following, then the product is not tested at the subsequent stations. A PLC can handle the sequencing of parts into test according to the specifications and can keep track of many parts at one time using bit handling.

Motion control, the second of five application areas, refers to linear or rotary motion control. Johnson [20] mentions that PLCs are used for motion control in many machine tool applications, assembly operations, and even robot motions. Special-function I/O boards are incorporated into the PLC for this purpose. Moore [25] shows that some PLCs have the capability to drive the control position, velocity, and torque of DC motors. PID, discussed earlier, is often utilized in the control board to correct for position error.

The third area of capability, process control, utilizes PID control methodologies to provide error reduction in many process applications. PLCs that have such PID capability often allow input of setpoints, offsets, and tuning parameters through a CRT interactive menu process.

The fourth application area for PLCs is data management. The merging of digital computer capability with PLC characteristics has allowed data management to become a reality in only the last few years. In fact, it has been suggested that such merging of capabilities might well portend the death knell for PLCs in the not-too-distant future. If one argues that a PLC is *not* a special-purpose computer, then this argument is undoubtedly true. The need for special-purpose computers to handle sequencing as well as motion and process control will be mandatory for years to come in order to prevent the functional overloading of the computer that integrates the operation of many devices. A manufacturing cell controller falls into this category.

Johnson [20] discusses the data management function as being the capability to compare machine or process data with reference conditions with results possibly being sent to a computer for further analysis or report generation.

Lastly, the communication application thrust for PLCs refers to the ability of the PLC to operate in a manufacturing local area network. A cell controller has to communicate through such a network to all the programmable devices under its control. PLCs are used to take some of the real-time sequencing load from the cell controller. For example, when a part reaches a pickup point on a conveyor, a PLC can notify the robot to pick up the part. The PLC has to be able to notify the robot that the part is available. The PLC might prevent indexing of the conveyor until the part has been picked up. In addition to communicating with devices in the same hierarchy level, the PLC must be able to notify the higher-level computer (cell controller) of significant events (part batch completion, conveyor failure, and so on).

The limitation for PLC analysis has historically centered on the PLC's *cyclic* operation. For example, in a sequencing operation it might be mandatory to check *all* boolean conditions within a millisecond in order to maintain control. This eliminates the ability of the PLC to do management reporting and other time-consuming tasks. It also indicates the need for PID and motion control modules. The use of distributed control—say, several PLCs reporting to one supervisory computer—can eliminate many of the inherent PLC restrictions.

DIGITAL COMPUTER CONTROL. All of the analysis functions discussed for programmable controllers could be handled by a digital computer with process I/O capability; after all, the PLC is just a special-purpose computer. The advent of the microprocessor generated an almost unlimited spectrum of affordable computer control applications. The disadvantage that the control computer has as contrasted to the PLC is the relative difficulty in programming due to the generic characteristics of the computer. Typical programming examples for control computer applications include some given by Auslander and Sagues [4]: (1) DC motor control and testing, (2) position control with a stepping motor, (3) temperature control, (4) blending process control, (5) automatic weighing, and (6) automatic cutting. Computers are used for such diverse applications as rapid transit control, shipboard control, landing airplanes on an aircraft carrier, supervising complete manufacturing operations, and many others. The list is mind-boggling.

Probably the main advantage of the control computer over the PLC in control, other than the size of the problem it is able to handle and the ability to operate outside the cyclic constraint of the PLC, is its ability to incorporate *real* decision-making capabilities into its operation. The evolution of expert systems, often misguidedly called artificial intelligence, is a case in point. The optimum sequencing of parts within a cell by a computer might be an impossible problem to program with conventional abilities. Should the robot wait until one part is complete after delivering the part to a work area? Which part should be moved next if several are waiting? An expert system will allow the programming of rules and constraints regarding what the robot should do and not do, as well as time constraints on these activities. The expert system will then allow robot action according to satisfaction of those rules and constraints.

Expert systems derived their name from the principle that the information upon which the expert system would operate was obtained from *experts* in the particular field. There are many cases where multivariable process control cannot be well defined in terms of mathematical equations. For example, the clinker output of a cement kiln is dependent on many things: kiln rotation speed, fuel characteristics, load in the kiln, ambient conditions such as rain or snow on the kiln, draft characteristics, and so on. Early successful cement kiln control was obtained by programming what an expert operator would do when specific problems occurred in the clinker production. In many ways, those early 1960 systems were brute-force forerunners of today's expert systems. Only a generic-type computer could really handle such analysis.

As a final comment, the ability to handle a wide spectrum of mathematics is a positive point for the computer in control. For example, control of catalytic crackers in petroleum refineries would not be feasible if the computer could not handle a linear programming algorithm. The cement kiln could not be effectively controlled without the computer's simultaneous linear equation solution capability.

To repeat a comment made when discussing PLCs: The broad capability of the control computer does not nullify the benefits offered by programmable controllers, or by small, single-purpose microprocessors. A single control computer cannot effectively control *all* the controllable items in a manufacturing cell. So much control information would have to be input and output in short increments of time that the computer would be rendered essentially immobile. The only way to handle the problem, at least with today's technological capabilities, is with a distributed/hierarchical manufacturing computer structure. This allows all control components, computers or programmable controllers, to best utilize their unique analysis capabilities.

Now that an introduction to analysis has been given, the next section will discuss typical actuation equipment to allow the process to be physically set according to the analysis.

11.5   ACTUATION

Following sensing and analysis, the final step in control is actuation. *Actuation*, simply put, is just the reverse of transduction. A value from the analysis device—say, an electrical signal from a computer—is transformed to a desired on/off condition or required motor velocity, and so on. A plethora of devices are available for actuation, and this section will discuss a few of these: on/off switches, timing of on/off conditions, alarm and annunciation, and motor drives.

All sorts of kinematic linkages could be included with the actuation devices. Vibratory orienting equipment which orients small parts for mechanized assembly operations, for example, utilizes linkages to effect the required motion. The spectrum of applications as well as the clever linkages that enable the applications is very broad and therefore will not be covered.

11.5.1   On/Off Switches

Almost all processes could be controlled with just on/off digital outputs. Motor speeds can be stepped up by sequentially controlling on/off switches. For a large process this would not be very efficient, but there is no doubt that on/off switches form a major grouping of actuators.

A very common form of controller switch is the electromechanical switch, illustrated schematically in Figure 11.18. The magnetic coil is energized by the controller, which causes the contact to be pulled down and closed. The spring allows the contacts to open once power is removed from the coil. The switch pictured in Figure 11.18 is normally open. Closing the contacts allows current to flow, and opening the contacts prevents current flow. Electromechanical relays can be purchased that will handle a variety of coil amperage conditions as well as many contact amperage levels. Also, they can be obtained in a variety of contact configurations such as a combination of normally open and normally closed.

Solenoids, another form of electromagnetic actuation, can be used for linear or rotary motion (frequently, on/off electromagnetic switches are classed as solenoid switches or solenoid relays). The rotary action switches might include a ratchet action so that each energizing of the coil indexes the switch a certain amount—say, 2° [24]. The old telephone switching circuits used this capability.

TIMING ON/OFF CONDITIONS. When timing of on/off conditions is required on a cyclic basis such that the on/off changes are to be effected at specific times within the cycle, a sequence timer might be incorporated. Figure 11.19 shows a modern sequence timer, usually called a programmable limit switch. The switch shown has a motor that runs at 1800 rpm. A digital encoder reads angular position of the motor shaft. This angular position is compared to desired programmed setpoints, and outputs are switched on or off as desired at the setpoints. The device shown

FIGURE 11.18   Schematic of electromechanical relay.

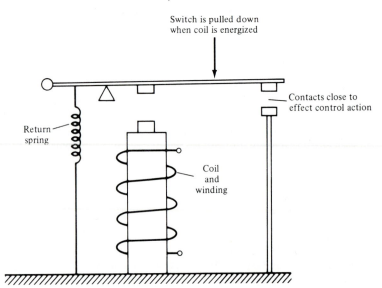

FIGURE 11.19   Modular programmable limit switch. (Courtesy of the Autotech Corporation, Carol Stream, Illinois.)

has 48 output channels (on/off switches), and the minimum angular resolution is 0.36°.

11.5.2   Alarms and Annunciators

There are many places in factory operations where autonomous annunciation/alarm operation is utilized. Basically, this is an on/off type of action. An annunciation device monitors a variable—say, pressure in a particular process. If a setpoint value is reached, an alarm is triggered—a light, horn, printer message, etc. Some annunciators will also shut down the process. So a typical annunciator might encompass sensing, analysis (compare setpoint to variable), and control (alarm and/or cut process off).

A variety of monitoring switches can be utilized for such variables as temperature and pressure. The switch is closed when a predetermined setting is reached. The simplest may well be the familiar bimetal temperature switch, which has two dissimilar metals banded together. The metals expand at different rates, thus allowing deformation to close a contact when a specific temperature is reached.

The programmable controller can be used as an annunciator/alarming device, but more often local monitors will be used as cost-effective devices. Many will be in a complex machine tool with output being to an operator panel. Annunciators and alarm monitors are available in single variable control loops as well as multiple alarm capability devices that can handle up to 24 input channels [6]. Most are modular, allowing for ease of expansion.

11.5.3   Motor Drives

The required motion for control will often require motors to be driven at variable speeds. Machine tool and robot positioning require extremely close control of the drive speeds. A wide variety of motor drives are available to the user, depending on the application. Electrical drive motors can be classified as stepping motors, AC drive motors, servomotors, and DC motors. Other power sources for motors may be hydraulic or pneumatic.

STEPPING MOTORS. Motors whose motion is controlled by a series of input DC pulses are called stepping motors. Because they are not continuous in the sense of AC and DC motors, they have been utilized successfully in digital computer control since the motor accepts digital (pulse) information. Figure 11.20a shows a typical stepping motor, and Figure 11.20b shows the translator used to allow the pulses to be correctly sequenced so that the motor can run either clockwise or counterclockwise.

Figure 11.21 shows a simple stepping motor schematic with eight stator

FIGURE 11.20   (a) Stepping motor and (b) translator. (Courtesy of Crouzet Controls, Schaumberg, Illinois.)

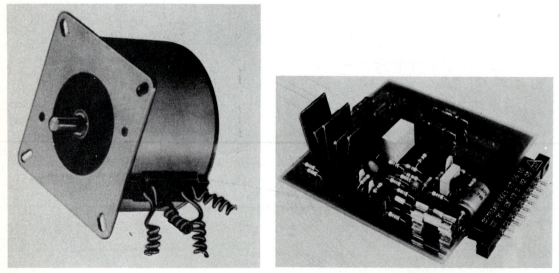

(a)   (b)

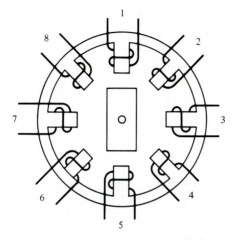

FIGURE 11.21   Simple stepping motor structure (Reproduced from Ref. 14, with permission.)

poles where each pole has a separate winding [14]. To drive in a clockwise direction, pole 1 is energized. Next, pole 2 is energized and pole 1 is deenergized. This sequential process causes the motor to turn. The need for pulse switching in the translator for the stator pole-energizing sequence is apparent.

Industrial applications requiring linear motion can take advantage of a linear motor system which operates on the same principle as a rotary hybrid stepping motor.* Such a system is shown schematically and pictorially in Figure 11.22. The moving element is called a forcer, and the stationary part is called a platen. The forcer consists of two electromagnets driven by the two phases of a pulse width-modulated microstepping drive. The forcer also contains a permanent magnet that has an electromagnet aligned at each pole, as shown in the figure. All of the active components (permanent magnet, coils, and bearings) are combined into the forcer so that the platen, as a passive element, can be provided in various lengths, as shown in Figure 11.22b.

The flux path of the permanent magnet passes through one of the electromagnets, across the air gap between the forcer and the platen, along the platen, back across the air gap, and through the other electromagnet to complete the magnetic circuit. If neither electromagnet is energized, the permanent magnet flux is divided equally at the four poles of the electromagnets. When one of the electromagnets is switched on, the current in the winding directs the total flux due to both magnets through the one pole common to both. For example, if the energized electromagnet is at the north pole of the permanent magnet (see Figure 11.22a), then the flux will pass through its north pole into the platen. No flux will pass

---

* Information on linear stepping motors was furnished by the Compumotor Division of the Parker Hannifin Corporation, Petaluma, California.

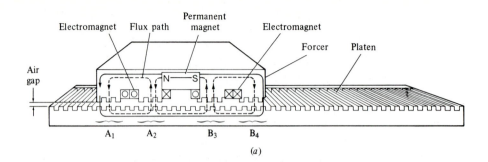

(a)

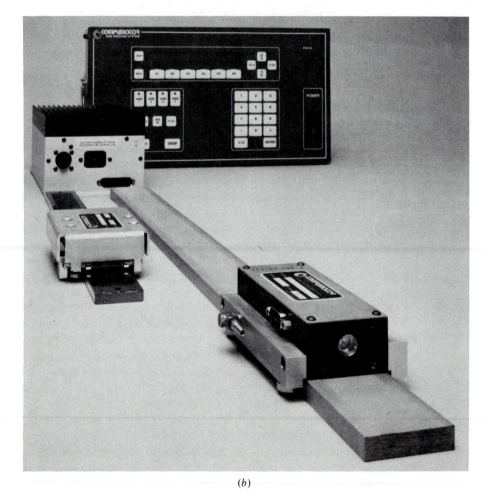

(b)

FIGURE 11.22 Linear stepping motor: (a) principle of operation; (b) typical equipment. (Courtesy of the Compumotor Division of the Parker Hannifin Corp., Petaluma, California.)

through the south pole of the electromagnet. The flux through the one pole generates a force parallel to the platen, which pulls the forcer along the platen until the teeth of that pole are aligned with the platen teeth. Switching current from one magnet to the other moves the forcer one full step along the platen. Changing the direction of the current each time an electromagnet is energized changes the orientation of the poles. In this manner, the poles are aligned with the platen in sequence (A2, B3, A1, B4) and the forcer continues in the initial direction. Reversing the sequence changes the direction of the forcer. A full step for the linear motor is equal to one-fourth the platen tooth pitch.

An application of linear motors is shown in Figure 11.23. Linear motors are used to turn a stop/start process into a continuous-flow process on a metal-forming machine. A flat plate is fed off a spool of raw material. The plate is formed by a series of synchronized rollers. The motors are then linked at the indexer electrically to assure that they all turn at the appropriate speed in synchrony with the rest of the rollers. This technique allows for repeatable, consistent forming of the metal into a U-shaped channel. A rotary encoder is used to provide velocity information to an indexer. The indexer accelerates the linear motor to match the speed of the channel. This prevents bending of the channel and binding of the saw. At the same time, the indexer positions the linear motor (i.e., the saw blade) at the precise spot on the channel to give the desired length. Two more linear motors are used to move the saw and the beam it rides on into and out of the cut. Once the cut cycle is completed, the linear motor carrying the saw is decelerated and stopped,

FIGURE 11.23   Linear stepping motor application. (Courtesy of the Compumotor Division of the Parker Hannifin Corp., Petaluma, California.)

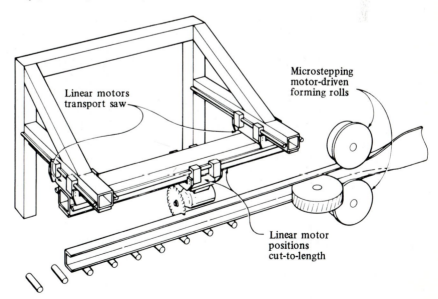

The direction is reversed and the saw is returned to the starting position at high speed.

Stepping motors are very accurate for low loads and so can often be used in open-loop situations. However, their limitation is the low load requirement.

OTHER ELECTRICAL MOTOR DRIVES. AC motors are available in all segments of our industrialized society. They have the advantage of being brushless which, according to Koren [21], eliminates a maintenance problem that accrues with DC motors. Koren also shows that when an AC motor has to have *controlled* speed, it has to have a costly inverter to allow frequency variation which, in turn, controls the motor speed. AC motors are found in the myriad constant-speed operations needed on the factory floor, including conveyors, pumps, and so on.

Direct motor drives, on the other hand, are controlled by the voltage level applied to the motor. Much control is effected by DC motors, especially small-to-medium numerical control machines [21].

A *servomotor* is designed in such a way that the amplitude of the current applied to one of two windings is a function of an error signal. As a result, the motor speed increases as the error increases and decreases as the error becomes smaller. Cassell [14] indicates that servomotors deliver maximum torque at lowest speeds which helps to accelerate the load quickly when the error signal is high or when the motor is starting. Cassell also shows that the same design characteristics help to hold the load in position. Asfahl [3] indicates that this principle is very important for NC machines and robots, since they must be able to hold the latest position until a new command comes from the control unit. When the error signal is zero, the motor has reached its current position, and the motor automatically stops.

An interesting question that often arises concerns whether to use a stepping motor or a servomotor where either is applicable. One company has a policy of recommending stepping motors in such cases, since they will generally be more cost-effective [9]. As the number of steps per revolution increases, the incremental stepping motor approaches the characteristics of a servomotor, especially if encoders are utilized for feedback control. Bailey [8] discusses circuitry that enhances the traditional 200 steps per revolution to a rumored 50,000 steps (in increments of 200), which gives a potential 0.018° of resolution. A drive, also mentioned by Bailey, operates step motors at 20,000 steps per second. The advances and possibilities are certainly interesting!

11.5.4 Hydraulic and Pneumatic Actuation

Not all actuators are electrical, of course. Linear and rotary motion can be effected by hydraulic or pneumatic power. Both require a cylinder to be linked to the object being moved in order to obtain the motion. Hydraulic power is used for heavy load requirements and, as such, it is used in some machine tool situations that are not amenable to stepper or DC drives. Similarly, some robotic applications require hydraulic power. A couple of problems accrue from hydraulics: The system tends to be noisy and also potentially dirty. A third possible problem in this era of expensive factory square footage is that space requirements tend to be relatively large.

Pneumatic actuation tends to be bang-bang, or two-position, due to difficulty in controlling position with air cylinders. Robot grippers which are used for pick-and-place operation with no force constraints will often use pneumatic power because it is cost-effective. Some robot manufacturers have attempted to develop medium-load pneumatic-drive robots in order to achieve significant cost savings. Other than accuracy problems, a major limitation in itself, such robots tend to have considerable vibration problems. There is no doubt that the future for the clean and accurate stepper and servomotor actuator is very bright.

11.6 COMPUTER COMMUNICATION WITH SENSORS AND ACTUATORS—SCADA

Chapter 8 discussed concepts of computer control where the computer has the ability to input and output information from and to a process. The concepts of digital and analog input/output were presented with that material. To conclude this current chapter's emphasis on sensor and actuator hardware, we will return to the topic of computer communication with the process. Specifically, comments will be made on the efficient interfacing of the computer with a *supervisory con-trol and data acquisition* system, commonly referred to as a SCADA [22].

Historically, stand-alone SCADA systems were developed for the process industries to enable local data logging— say, in a petroleum process. The SCADA philosophy has broadened somewhat to allow SCADA units to be interfaced with *any* computer to remove much of the input/output burden from the computer.

As an example, the IBM PC control computer pictured earlier in Chapter 8 is just a personal computer with a process interface board installed. The board has assembly programs permanently entered in the board memory that make it simple for the user to communicate with the process. The multiplexing required for information input/output is handled in this manner, as is the possibility of bringing in several analog values with one command request.

A new breed of SCADA is exemplified by Burr-Brown's SCADAR [30, 31]. The base board that forms the heart of the SCADAR is shown in Figure 11.24. Any digital computing device that has ASCII capability can function with this system. The electronic base board contains all the information necessary to oper-ate a small control station: power supply, microprocessor, memory, dual com-munication channels, EPROM* resident software, and field I/O [31]. Field input/output includes digital I/O, analog I/O, and pulse input. Pulse input can be used to count input transactions or to measure input signal frequency.

---

* EPROM stands for *e*raseable *p*rogrammable *r*ead-*o*nly *m*emory. It has reliable program retention capability combined with the ability to be erased and reprogrammed many times [14].

FIGURE 11.24   SCADAR base board (enclosure case in background). (Courtesy of Burr-Brown, Tucson, Arizona.)

FIGURE 11.25   SCADAR series 10 program flow. (From Ref. 31, with permission of Burr-Brown Corporation.)

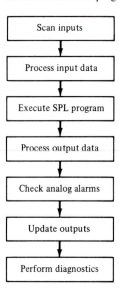

The SCADAR flow is as depicted in Figure 11.25. SPL is the control programming language designed for use in the SCADAR series 10. SPL can read input date, change output data, and perform mathematical or logical operations [31]. It will be apparent to the reader that the flow in Figure 11.25 summarizes the functions that have been discussed in this chapter: input sensor information, analyze that information, output to actuator, and alarm as needed. Operation of these functions by the SCADAR is triggered by the host computer, which can obtain information for the SCADAR when needed.

The point to be made with regard to CAM is that the host computer might well be a cell controller. The SCADAR device is a possible interface between the cell controller and the devices on the shop floor: sensors, actuators, machine controls, programmable controllers, and so on. The cell control computer can rely on the interface device to perform the real-time input/output functions, thus relieving time for the cell control computer to do the things it does best. Typical of these functions are cell scheduling, downloading NC program information, and communicating with the manufacturing mainframe computer.

11.7 SUMMARY

This chapter has centered on the control methodology process, where process information is first captured, this information is then analyzed, and, based on the analysis, corrective information is furnished back to the process.

The need for this chapter came about due to assumptions in earlier material that equipment was available to allow the capture of the process information as well as the actuation of the process through output information.

As a result, basic concepts of sensors, transducers, and actuators were presented. Also, to round out earlier analysis material, an introduction to proportional-integral-derivative control (PID) was given. Lastly, the SCADA approach to generic, local input/output interfaces was given, primarily as a possible way to handle the manufacturing cell computer I/O interface.

EXERCISES

This chapter gave background information on sensors, control, and actuation to fill in the gaps in previous material where it was assumed that the reader was familiar with the subject. As a result, the exercises suggested will be brief in number and, for the most part, nonquantitative.

1. Consider a four-way intersection where the traffic lights are controlled automatically. If two or more cars appear in the left-turn lane, they will be given access by a left-turn arrow. Minimum times are set in a programmable controller for timing the normal red- and green-light sequences. However, if three or more cars wait at a red light, the controller can switch from red to green earlier than expected. Comment on the sensing, actuation, and analysis capability that will be required to allow the lights to be controlled.

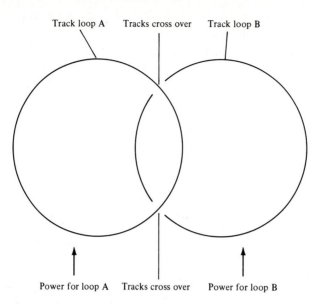

Track loop A   Tracks cross over   Track loop B

Power for loop A   Tracks cross over   Power for loop B

FIGURE 11.26   Track arrangement for Exercise 2.

2. Suppose that you want to control switching for a model train. The track arrangement is as shown in Figure 11.26. You are going to have one train run on Track A and a second train on Track B; both will travel counterclockwise. Suggest sensing and actuation equipment that will enable a control computer to start the trains and stop one of the trains if a collision is imminent. Flow-chart the logic that would be programmed into the computer to allow the collision avoidance system to work.

3. Why is the UPC bar code standard not practical for manufacturing tool identification schemes?

4. The digital outputs of a control computer are usually "high" and "low" voltages corresponding to the "1" and "0" possibilities. These voltages are 5 V DC and 0 V. Suggest hardware that would allow a 110 V AC motor to be started and stopped by one of the outputs.

5. Suppose that you need a shaft encoder that has an angular resolution of at least 1°. How many concentric circles would be required if an encoder similar to the one in Figure 11.13a is desired?

6. Consider the "Calra" alternative to conventional bar code striping. Instead of the 16-character code shown in Figure 11.11 we have a 256-character code formed with a two-by-four pattern of squares per character. Assume that the first characters are the digits 0 through 9 followed by A through Z (the other 220 characters are to be defined). Draw the "Calra" representation for Y 3 B F 6.

7. Review an article in the literature that presents a unique application for code sensing. Magazines that might be of help include *Industrial Engineering, Manufacturing Engineering, Production Engineering*, and *Control Engineering*. Write a 500-word summary of the article, giving special emphasis to the benefits of the system.

8. Review an article in the literature that discusses recent advances in robot tactile sensing. Develop a 500-word review of the article that stresses the advantages of tactile sensing.

9. Review a current article that gives an example of vision application in manufacturing. Indicate in your review the cost of the system as well as the unique benefits that have accrued since adoption of the vision system.

10. It has been suggested that a robot will never (a dangerous word) be successful in playing tennis against a human opponent. From an instrumentation point of view, suggest why two robots playing tennis might be realistic.

REFERENCES AND SUGGESTED READING

1. "6 of the Newest Information Techniques," *Modern Materials Handling*, August 5, 1983.
2. "About the Cover . . . ," *Control Engineering*, September 1984.
3. Asfahl, C. Ray: *Robots and Manufacturing Automation*, John Wiley, New York, 1985.
4. Auslander, David M., and Paul Sagues: *Microprocessors for Measurement and Control*, Osborne/McGraw-Hill, Berkeley, Calif., 1981.
5. "Automatic Identification Posts Major New Gains," *Modern Materials Handing*, November 1985.
6. Babb, Michael: "Alarms and Annunciators: Keeping the Lid on in Industrial Control," *Control Engineering*, February 1986.
7. Bailey, S. J.: "High Speed Production Lines Benefit from Latest Object Motion Detectors," *Control Engineering*, September 1985.
8. Bailey, S. J.: "Lessening the Gap between Incremental and Continuous Control," *Control Engineering*, May 1987.
9. Bailey, S. J.: "Servo vs Stepper—Motion Control Design Decision with Dynamic Overtones," *Control Engineering*, May 1987.
10. Bannister, B. R., and D. G. Whitehead: *Transducers and Interfacing*, Van Nostrand (UK), Wokingham, Berkshire, England 1986.
11. Bedworth, David, and James E. Bailey: *Integrated Production Control Systems*, 2d ed., John Wiley, New York, 1987.
12. Berry, Gayle, and Glenn Dunlap: Shop Floor Information System, Report #AFWAL-TR-82-4147, Volume III, AFWAL/MLTC, Wright-Patterson Air Force Base, Dayton, Ohio, October 1982.
13. Bollinger, John G., and Neil A. Duffie: "Sensors and Actuators," *IEEE Spectrum*, May 1983.
14. Cassell, Douglas A.: *Microcomputers and Modern Control Engineering*, Reston Publishing Company, Reston, Va., 1983.
15. Critchlow, Arthur J.: *Introduction to Robotics*, Macmillan, New York, 1985.
16. Edson, Joseph: "Giving Human Hands a Human Touch," *High Technology*, September 1985.
17. Groover, Mikell P., M. Weiss, Roger N. Nagel, and Nicholas G. Odrey: *Industrial Robotics*, McGraw-Hill, New York, 1986.
18. *Hyde Park Sensing and Control Solutions*, Marketing Brochure #61835KMBA, Hyde Park Electronics, Dayton, Ohio, 1986.
19. *Hyde Park Sensing and Control Solutions*, Marketing Brochure #PP-05-86/1-A, Hyde Park Electronics, Dayton, Ohio, 1986.
20. Johnson, David G.: *Programmable Controllers for Factory Automation*, Marcel Dekker, New York, 1987.
21. Koren, Yoram: *Computer Control of Manufacturing Systems*, McGraw-Hill, New York, 1983.
22. Laduzinsky, Alan J.: "Would SCADA by Any Other Name Still be the Same?," *Control Engineering*, February 1986.

23. LaJoy, Millard H.: *Industrial Automatic Controls*, Prentice-Hall, New York, 1
24. Lenk, John D.: *Handbook of Controls and Instrumentation*, Prentice Hall, I
    N.J., 1980.
25. Moore, J. A.: *Digital Control Devices*, Prentice-Hall, Englewood Cliffs, N.J. (Ir
    of America), 1986.
26. Morris, Henry M.: "Object Detection—The First Level," *Control Engineering*
27. Osborne, David M.: *Robots: An Introduction to Basic Concepts and Applicatio*
    Tech, Detroit, Mich., 1983.
28. Pugh, Alan, Ed.: *Robot Sensors*, IFS Publications, Kempston, England, 1986.
29. Ribeiro, Jorge: "New Code to Replace Conventional Bar Codes?," *Manufactu*
    June 1987.
30. *SCADAR Series 10 Hardware Manual*, Document #MA51921, Burr-Brown Cor
    Arizona, May 5, 1985.
31. *SCADAR Series 10 Software Manual*, Burr-Brown Corporation, Tucson, Arizo
32. Seippel, Robert G.: *Transducers, Sensors and Detectors*, Reston Publishing (
    1983.
33. Sheffield Measurement Division, *66 Centuries of Measurement*, 3d ed., Sheffie
    1984.
34. Stonecipher, Ken: *Industrial Robotics: A Handbook of Automated Systems Desi*
    Co., Hasbrouk Heights, N.J., 1985.
35. *SuperCard*, Advertising Brochure of Identatronics, Elk Grove Village, Ill., 19
36. Todd, James L.: "Program Constructed for On-Demand Bar Code Printing," *In*
    ing, September 1986.
37. Weck, Manfred: *Handbook of Machine Tools*, John Wiley, New York, 1984.
38. Wilhelm, Robert E., Jr.: *Programmable Controller Handbook*, Hayden Boo
    Heights, N.J., 1985.

COMPUTER-
INTEGRATED
MANUFACTURING

There is nothing more difficult to carry out, nor more dangerous to handle, than to initiate a new order of things.

Machiavelli (1469–1527)

Up to this point, many components of automated design and manufacturing systems have been presented. Comments on how the material relates to an "integrated" system have been made in various chapters. Now, instead of talking about somewhat independent systems and associated functions, it is time to discuss integrated systems in the manufacturing firm. Often this is classified as CIM.

In 1973, Joseph Harrington published the initial concepts of CIM in a book entitled *Computer Integrated Manufacturing* [12]. Initially these concepts received little attention. It was not until about 1984 that people began to realize the potential benefits these concepts promised. Since 1984, thousands of articles have been written on this subject.

12.1 A DEFINITION OF CIM

Unfortunately, it has been difficult to define CIM; the many different definitions in the literature attest to this statement. Although in the mid-1980s there was an initial burst of enthusiasm regarding the potential of CIM, this enthusiasm waned as firms were slow to approve the large expenditures necessary to support a CIM

599

plan. However, in the 1990s, with the emergence of a better understanding of the meaning of CIM, firms are showing renewed interest. In retrospect, this is reasonable: Management cannot be expected to approve large expenditures for something they do not understand.

Some of the difficulties in arriving at a consensus definition for CIM can be attributed to the words "integration" and "manufacturing." Ronald Reimink says that "Integration is one of the most overused, abused and misunderstood words in the technical world, and yet it is a uniquely important factor in system success" [28]. This word alone makes it difficult to define CIM.

Now consider the word "manufacturing." When this word is considered by itself, we can arrive at a consensus as to what it means. Unfortunately, CIM is broader than just manufacturing. Figure 12.1, which shows the major elements of a manufacturing business, depicts the scope of CIM. In this context, manufacturing is defined as the collection of physical and intellectual activities associated with designing and making tangible, movable items of value, either by hand or through the use of machinery.

Although there is not one definition of CIM that is universally accepted, as we understand this subject better, a consensus is emerging. Computer-integrated manufacturing is a management philosophy in which the functions of design and manufacturing are rationalized and coordinated using computer, communication, and information technologies. According to Mize and Palmer, "rationalized" in this context means:

> The entire [manufacturing] system, from product definition and raw material acquisition to the disposition of the final product, is carefully analyzed such that every operation and element can be designed to contribute in the most efficient and effective way to the achievement of clearly enunciated goals of the enterprise [22].

FIGURE 12.1 Major elements of a manufacturing business.

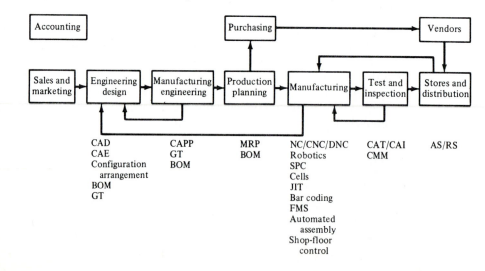

TABLE 12.1
Potential benefits of CIM

Improved customer service
Improved quality
Shorter time to market with new products
Shorter flow time
Shorter vendor lead times
Reduced inventory levels
Improved schedule performance
Greater flexibility and responsiveness
Improved competitiveness
Lower total cost
Greater long-term profitability
Shorter customer lead times
Increase in manufacturing productivity
Decrease in work-in-process inventory

It should be noted that CIM is not a specific technology that can be purchased. In the mid-1980s, some firms tried to buy CIM from other firms that were "selling" this product. This led to disappointment and bad publicity for this emerging philosophy. Rather, CIM is a strategic goal that a firm strives to achieve over time. The definition of a goal is something that is continually strived for but never attained, such as the goal of becoming a better student. You can always improve, so this is a goal. The term "integrated" should possibly be changed to "integrative," because in a CIM environment we are involved in a continually evolving integrative process.

CIM is an important philosophy; Table 12.1 lists the benefits some firms have realized from successfully implementing this philosophy in portions of their business. Therefore, the difficulties that have been experienced in attempting to define and implement these concepts should not deter us from learning the fundamentals necessary to be successful.

In the remainder of this chapter we will discuss the technology issues that are essential in a CIM environment. The importance of some of these was not well understood until recently. Undoubtedly, as our conceptual understanding of CIM improves, new technology issues will be added to this list. Also in this chapter, computer network fundamentals will be explained. Network technology is maturing to the point that it will have a very significant impact in the 1990s. In addition, some guidelines will be given for developing and implementing a successful CIM strategy. Finally, a CIM example will be presented.

12.2 KEY DEFINITIONS

Baseband Digital bits are sent on the network by raising and lowering a voltage [18].
Broadband Any cable network using analog transmission [35].
Bus A communication path with access capability for multiple nodes [18].

Bus network   A network in which transmission and receiving stations share a communications path. All stations can communicate with each other without the intervention of an intermediate station [18].

Carrier-sense multiple access with collision detection (CSMA/CD)   A LAN which senses when another station is transmitting, permits two or more stations to start transmitting concurrently, and resolves collisions using random re-transmissions [35].

Collision   When two or more stations transmit concurrently on a LAN [35].

Computer-integrated manufacturing   A management philosophy in which the functions of design and manufacturing are rationalized and coordinated using computer, communication, and information technologies.

Configuration management   The definition and communication of the form and function of a completed entity and the control and incorporation of changes to that entity throughout its life cycle [5].

Ethernet   A networking standard that uses baseband signaling in a bus topology that connects all stations on the network as peers and uses the CSMA/CD access method. Each station receives all messages placed on the network [35].

Head end   The device, in a broadband LAN, that translates between frequencies of the forward and return channels [18].

Local area network (LAN)   A nonpublic data network for direct communications between data stations, such as programmable machines and computers [18].

Manufacturing Automation Protocol (MAP)   A broadband, token-bus network protocol based on the Open Systems Interconnection (OSI) reference model that will accommodate a broad range of manufacturing environments.

Open Systems Interconnection (OSI) reference model   An International Standards Organization (ISO) standard that specifies the conceptual structure of systems that are to communicate with each other.

Protocol   A set of rules that defines how two or more devices engage in data communication.

Technical and Office Protocol (TOP)   A set of protocols defined by Boeing for a LAN designed for the office environment.

Token   A unique string of bits that serves as a control signal in a network.

Token-bus network   A bus network in which a token-passing procedure is used.

Token-ring network   A ring network in which a token-passing procedure is used.

12.3 TECHNOLOGY ISSUES

In the mid-1980s, as the concept of CIM began to become more widely known, there was the initial rush of euphoria that often accompanies a major new idea or technique that appears to promise significant benefits. As the euphoria subsided, many people began to question the envisioned benefits. This process of examination brought the realization that CIM is difficult to define; consequently, many different definitions were in use. As a result, there is still some confusion about what this concept actually means.

Today, a consensus is emerging as to the meaning of CIM. With this understanding, we can proceed to develop an understanding of what technologies are necessary to implement these concepts. Table 1.1 (Chapter 1) lists many of the functions included in CIM. At first glance the number of functions is somewhat overwhelming, and the relationships of various functions are not very clear.

Another way to begin to understand the technologies involved is to evaluate the sequence of functions performed in a typical discrete manufacturing firm (see Table 12.2) for possible integration opportunities. Companies that use CAD often develop several models (sometime five or more) of a part. These models have different uses, such as stress and thermal analysis, aerodynamic studies, process planning, tooling design, NC program development and verification, and coordinate measuring machine programming. Different types of models may be involved, such as solid, wire frame, 2-D, and 3-D models. Usually, more than one type of CAD system is used.

12.3.1 One-Model Concept

Ideally, only one part model will be created. This model would contain all the part information needed during the life cycle of the part, from design through field support. The PDES concept, which was discussed in Chapter 6, is a major step toward achieving this goal; however, this standard is some years away from being fully developed and accepted. Today, design engineering has difficulty in transferring just the graphic portion of a design between dissimilar CAD systems. IGES,

TABLE 12.2
Sequence of functions performed in a manufacturing firm

Design assemblies and perform tolerance analyses on those assemblies.

Prepare engineering drawings of assemblies, individual parts, tooling, fixtures, and manufacturing facilities.

Create analytical models of parts for structural and thermal analysis.

Calculate weights, volumes, centers of gravity, and costs of manufacturing.

Classify existing parts according to shape, function, and the processes by which they are manufactured, and retrieve these parts from the data library on demand.

Prepare parts lists.

Prepare process plans for individual part manufacture and assembly.

Program NC tools for machining complete parts.

Program tools for bending and punching sheet metal parts.

Program the movement of robots in work cells.

Draw isometric sketches of parts and assemblies for use in process planning sheets and technical manuals.

Prepare inspection programs, including programs for coordinate-measuring machines (CMMs).

Control the effect of part design changes on assemblies and their manufacture.

which was discussed in Chapter 6, is normally used. Also, the different types of CAD applications, which often correspond to a different CAD model of the same part, may require the graphic data to be in a different format. This discussion did not address the problems of transferring a model between different firms that may utilize different data management techniques as well as different CAD systems. As a result, just the seemingly simple task of interfacing dissimilar CAD systems or model applications presents some serious technology limitations.

Representation of a part in graphic form on a CAD system is an evolving science. A complex part surface might be represented (approximated) using a Bezier, B-spline, or nonuniform rational B-spline (NURBS) function. None of these mathematical functions will handle all types of surfaces satisfactorily. Some CAD systems do not provide all of these functions; and, where the same function is provided on dissimilar systems, the implementation may vary. As a result, systems utilizing apparently similar functions for modeling surfaces may be operationally incompatible.

Companies have tried to mandate a solution to these difficult integration problems by buying only one brand of CAD system. However, this has not provided a satisfactory solution, because no one system is superior in all aspects; moreover, if one were superior today, it might not be superior in the future as the technology evolves. In addition, since most companies purchase some parts from other companies, there is a need to transfer part models from one company to another. For the defense industry, CALS (Computer Aided Acquisition and Logistics Support) demands that in the future all contractors will be required to provide data through an electronic representation [19]. Consequently, mandating one type of graphics system is not prudent management.

As CIM has become better understood, an important integration principle has emerged. We have realized that ideally only one part model should be maintained for each design release. This model should contain any information required by any functional group, from design engineering to field support. When this philosophy is applied, the way parts are designed and manufactured changes. Some of these differences are reflected in the following ways: errors from using an incorrect model are eliminated; assemblies fit together because they have been modeled as an assembly; and concurrent engineering and manufacturing are facilitated. When paper drawings and 2-D graphics systems are used, engineering designs are often incomplete and contain errors. A 3-D part model, however, provides another perspective and permits a more accurate evaluation of assemblies; consequently, many design errors can be eliminated. Also, realizing that the design of a complex part is normally an evolutionary process, concurrent engineering and manufacturing will be facilitated because the model can be easily accessed. This access can be controlled so that the correct engineering release of the design is always utilized. As a result, design and manufacturing lead times are reduced, productivity increases, and quality improves.

At this point, the concept of one model is far from reality. In practice, we may never achieve this ideal. However, this concept aids us in defining research thrusts. PDES is a good illustration of this process.

12.3.2 Configuration Management

Some have conjectured that as much as 90% of part costs are related to documentation. Boznak [5] reports that uncontrolled design accounted for 27% of product cost for one firm that he analyzed. The information associated with the one-model concept is a type of documentation, and it is obvious that enormous amounts of data are involved. Initially capturing this data is only a portion of the effort, because significant effort must be expended to maintain the integrity of this data. With this understanding comes the realization that configuration management is an extremely important function within a firm. Unfortunately, most firms devote few resources to this function; as a result, configuration management-related activities are performed as time permits. This is surprising given an understanding of the costs involved. The concept of one model and the implied requirements for data management reinforce the importance of configuration management.

Configuration management is a powerful yet subtle concept. The fact that it is subtle may be one reason that this subject has not received the emphasis it deserves. In fact, in the early euphoria of CIM, configuration management was not even addressed. As our understanding of CIM has improved, however, configuration management came to the forefront as a necessary technology.

Since configuration management is a subtle concept, we should begin developing a better understanding of this concept with a definition. *Configuration management* is the definition and communication of the form and function of a completed entity and the control and incorporation of changes to that entity throughout its life cycle [5]. Figure 12.2 depicts the phase in the life cycle of a product. The objective of configuration management is first to describe clearly what end result is desired and then to control changes that are made in the means to obtain this end result.

To better understand the concept of configuration management, Boznak [5] segments it into four principles: definition, communication, control, and incorporation. If an entity is not clearly defined, it will be difficult to understand and manage. An entity can be an engineering design, a software program, a strategy, or a product. The definition should identify and bound the important parameters of the entity. The functions that control these parameters should also be identified. Whenever any of these parameters change, all functions that might be affected by these changes must be notified.

Communication is vital in any management endeavor. This is especially true in configuration management, where communication can occur in many forms,

FIGURE 12.2   Phases in the life cycle of a product.

such as specifications, drawings, parts lists, rework instructions, computer listings, 3-D part models, or electronic mail. Every time a document is created there is an opportunity for effective communication where all relevant parties are informed and understand the content and implications of the changes. When this is done before changes are made, management can function proactively and control the processes.

Disruptions have a negative impact on an organization. A change to a part design will usually create a disruption. If these disruptions are unplanned, the negative impact will be more significant. The objective of configuration control is to plan and introduce changes in a manner that minimizes the effects on functional cost and scheduling. Also, change control will ensure that changes are not only an engineering decision; instead, the changes will also be evaluated in terms of the impact on costs, schedules, and return on investment. In some engineering-dominated firms, engineers receive recognition based on the number of changes made to the product. In this environment, without control, the number of changes made without concern for costs or schedules will have a major negative impact on the profitability of a firm.

Because of the importance of change control, this function should be an integral part of management. Instead, in many firms, this function is delegated to a change control board that has little authority. If the importance of configuration management were well understood, this would not happen.

The initial step in change control is to screen all requests as being essential or nonessential. Nonessential requests should not be approved. Essential requests should then be screened again into those that can be scheduled and those that cannot. A request can be scheduled if the change is for producibility or product improvement. Urgent requests, such as for safety, should be made immediately; therefore, they should not be scheduled for a more convenient time. After a request has passed the screening process, the change must be thoroughly reviewed by all functional organizations that might be affected. Corrective actions should be defined jointly by these organizations before the change is approved. This joint review should specify when the change becomes effective, usually denoted by an effectivity date or serial number. When corrective actions are not performed jointly, additional change may be generated. A study by a major aerospace company concluded that change requests generated 1.7 additional changes when the corrective actions were not defined jointly [5].

In addition to effectivity, document changes must be synchronized. Table 12.3 lists some of the documents that might be affected by one change. As product traceability requirements increase, the associated documentation demands increase significantly. For instance, there are very strict traceability requirements associated with the production of a jet propulsion engine. Configuration management in this type of environment is expensive and extremely important.

Product design is an evolutionary process. Even the best-designed products will require some changes during their life cycle. However, changes must be managed in order to minimize disruption to the functional organizations of a firm; otherwise, costs will be much higher than necessary.

TABLE 12.3
Some documents affected by an engineering design change

Engineering	Sales/marketing	Manufacturing	Product support
Drawings	Purchase orders	Schedules	Operation manuals
Schematics	Contracts	BOM	Maintenance manuals
Specifications	Option catalog	Processes	
BOM	Pricing data	Equipment	
Software	Quotes	Costs	
		NC programs	

Incorporation, the fourth principle of configuration management, means implementing the corrective action as specified by an approved change request. Usually there are many changes, possibly thousands in a large firm, in process at any one time. Consequently, management of even a relatively simple change request can be a complex task. This complexity is compounded by customer order expediting, machine maintenance, or many other unplanned events that occur in a company.

12.3.3 Data Base Management Systems

Because of the large amounts of data associated with the concepts of a one-part model and configuration management, data base management is an important consideration. Data encountered in an engineering/manufacturing environment can be classified into four basic types: (1) product data, which consists of graphic, text, and numeric data; (2) production data, which describes how the parts are to be manufactured; (3) operational data, which describes the events of production, such as schedules and lot sizes; and (4) resource data, which describes the resources involved in production, such as machines and tools. Other types of data are encountered in a firm, such as financial and marketing data.

Most data management systems in use today were developed for business applications characterized by text and numeric data stored in relatively homogeneous records. Table 12.4 is a relative comparison of business and engineering/manufacturing data base characteristics. Virtually no data base management systems are available that combine CIM requirements with business needs.

The magnitude of the data management problem is easier to comprehend when the tasks listed in Table 12.2 are to be performed. Note the many ways that data needs to be associated if it is to be managed efficiently (stored, retrieved, modified, and reported). In addition, the data of interest may be in a distributed system at many locations. Also, there will be an enormous amount of data that will further complicate operational tasks such as backup, transaction traceability, fault tolerance, and security. Compare this with the current environment in a typical manufacturing firm, where difficulties are encountered in managing data in one data base so that inventory and bill-of-material records are accurate enough to

TABLE 12.4
Comparison of business and engineering/manufacturing data base characteristics

	Business	Engineering/ manufacturing
Record type	Few	Many
Relationships	Simple	Complex
Queries	Simple	Complex
Analysis	Rare	Normal
Updates	Short	Long
Records	Few	Many

Source: Ref. 8.

support MRP. However, on the positive side, extensive research is being performed in the area of data base management systems that will support a CIM environment. Some initial systems are now being marketed; however, none has been used long enough to be considered a mature product.

Most data base management systems arrange data in one of three ways: hierarchical, network, or relational (see Figure 12.3). In a *hierarchical structure*, data records are related in a treelike manner. Starting at the root of the tree, each record (parent) has a one-to-many relationship to its branches (children). A parent record can have several children, but a child record can have only one parent record.

In a *network structure*, each child record can have more than one parent record. Thus, there can be a many-to-many relationship among data. Hierarchical and network-type data bases require that all data relations be predefined and embedded in the structure. Consequently, data access is defined when the structure is defined. These types of data bases are suited for environments requiring high transaction rates and limited access paths.

In a *relational data base*, data is stored as a collection of tables composed of rows and columns. Rows in the tables need not be in any special order; therefore, it is easy to add data to a relational data base. Data relationships are established by queries; data is located through row-by-row searches; consequently, data access is more flexible. This type of data base is appropriate when many unanticipated queries might be made. However, transaction rates will usually be slower than would be experienced with hierarchical or network data base structures.

A few data base management systems provide all three structures, since no one of the three types of data base structures is best for all circumstances. Research is being performed to find a more robust structure. Some promising results have been obtained with an object-oriented structure in which data and related attributes are organized as a group (an object) with standardized inputs and outputs [16]. Objects are manipulated by operations. Anything can be described as an object, and objects can be associated with one another. An object can be made up

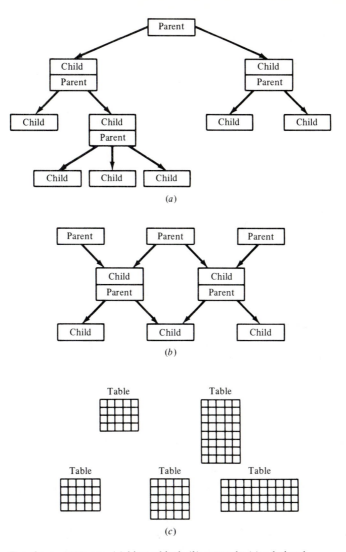

FIGURE 12.3 Data base structures: (*a*) hierarchical; (*b*) network; (*c*) relational.

of multiple types of data. The user can develop object classes depending on application requirements. The applications would be under control of the data base management system which would provide an unchanging user interface. This type of structure appears promising; however, it is still in the developmental stage.

The data base management system ideal for a CIM environment is not available. Table 12.5 list some of the desirable features. The needs are much clearer than they were 5 years ago, and progress is being made toward a data base management system that will facilitate the development of a CIM environment.

TABLE 12.5
Desirable features for a CIM data base
management system

Data integration across heterogeneous computer systems
Transaction processing
Real-time interactions
Distributed processing and data
Multiple views of data
Levels of security, data integrity
Multiuser networks across dispersed work environments
Geometry management
Versioning
Backup and recovery
Good user interface and knowledge-base support
Materials management integration
Quality assurance integration
Configuration management

Source: Ref. 1.

12.3.4   Networking

As islands of automation proliferated in factories, users demanded communication networks to connect these islands. In response, the major computer manufacturers initially developed proprietary network architectures in an attempt to meet the need for information sharing between disparate systems (mainframes, minicomputers, and microcomputers). These architectures established the interconnection rules and specifications (interfaces, cabling, and protocols) to link a particular vendor's systems. Usually, these network architectures supported only the computers, peripherals, and related networking devices of a particular vendor. These incompatibilities were amplified when some vendors would not release all of their proprietary specifications to other vendors and users, and would not assist customers in connecting systems from other vendors.

As computer technology matured, the growing need for communication between computers from different vendors was costing user organizations billions of dollars in duplicated system expense, large technical staffs to support different vendor systems and associated networks, and lost productivity. Finally, because the growth in information system markets was being threatened by the incompatibilities of network architectures, the major computer manufacturers and communication service suppliers reached an agreement in the latter 1970s to develop a new standard within the framework of the International Standards Organization (often referred to as ISO). ISO's charter is to develop a new, open, and standard architecture for computer-to-computer communications, and then define standard protocol specifications that will enable vendors to implement this architecture [17]. The new standard that has been evolving since then is called the Open Systems Interconnection (OSI) reference model.

The complexity of the hardware and software existing in the latter 1970s at many firms, and the rapidly evolving related technologies, made satisfying this charter a very difficult task. As a result, over a decade later, OSI is not yet a reality. However, significant progress has been made. This is exemplified by the increasing number of new networking product announcements that conform to defined portions of the OSI specifications. The mid- to later 1990s should see OSI become an international standard. OSI has evolved as a seven-layer structure. These layers will be described later in this chapter.

Unfortunately, the term ''open'' has different meanings to various users and vendors. Some vendors' interpretation of the meaning of an open system is too narrow. To them, an open system has the ability to send data back and forth between nodes of a heterogeneous computer environment. From a user viewpoint this interpretation is not broad enough. Instead, accessibility by the user must also be included. The end user should have the ability to interact with facilities such as software systems, data management, and hardware interchangeability. A system should have the ability both to transport application software and to have applications cooperate. Open systems should literally ''open'' the system to the end user.

Baron [2] describes three stages through which open-system architectures will evolve as integration progresses: connected, cooperative, and coherent. The first stage, *connected integration*, allows for the physical and logical linkage of islands of automation. Communications is the predominant technology utilized. Rules in the form of communication data protocols specify how data is exchanged. Heterogeneous systems can be linked; however, flexibility is severely limited by the level of compatibility between any two communication protocols. This technology is exemplified by routers, gateways, interface hardware, and physical communication media.

Cooperative integration permits heterogeneous systems to execute applications that are designed to interact with a limited number of other applications. This stage builds on the capabilities of connected integration. There must be facilities for consistent data representation and management so that different machines and software packages can cooperate. More general data consistency methods must be developed so that information can be retrieved from multiple machines. This capability will permit more generic application cooperation between machines, which will result in lower support and maintenance costs. Also, portability of applications within architectures becomes feasible. Some of the concepts represented by this stage will be discussed in Section 12.3.5.

According to Baron [2], *coherent integration* is the ultimate objective. End users can interact with the system without knowing where the information resides. Operations will be simplified when end-user information analysts, not systems analysts, control the system.

The specifics of Baron's broader notion of open systems could be debated. However, his basic premise that the meaning of the term ''open system'' will change as integration progresses is valid. Open systems require more than the ability to communicate with dissimilar systems, because the end user must be able to integrate several computing resources easily.

FACTORY COMMUNICATION REQUIREMENTS. Industrial network requirements are different from those of an office. The environment of a typical shop floor is much more hostile. Equipment is subject to electromagnetic interference, dirt, voltage fluctuations, and temperature variations. Communications are often made to support a process, such as transferring an NC program to a CNC mill. Also, timing of communications may be critical, especially when a real-time process control operation is involved. In addition, reliability is extremely important, because downtime on production equipment can be very expensive.

The type of computer application and where it resides in the factory network hierarchical structure (see Figure 12.4) usually characterizes the communication network requirements. Table 12.6 lists some communications attributes for typical computer applications at various levels in Figure 12.4. It can be seen that communication requirements for the shop floor are very diverse. Consequently, designing the best factory network is not an easy task. This is further complicated by the lack of standards and the existence of many competing proprietary products. Some network fundamentals dealing with cabling and network structures will be presented in Section 12.4.

FIGURE 12.4   Factory network hierarchical structure [1].

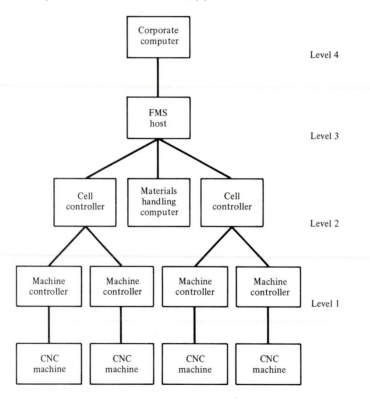

TABLE 12.6
Communication attributes of a factory network

Hierarchical Structure	Level 1	Level 2	Level 3	Level 4
Traffic pattern	Very frequent, almost continuous, and of very short durations from 1 bit to a few bytes	Can have bursts of traffic, be less frequent, and be of both short and long durations from 1 bit to 100,000 bytes	Can have bursts of traffic, be less frequent, and be of mostly long durations (approximately 100,000 bytes)	Mostly of very long durations and randomly distributed
Data rate	From less than 300 bits/s to more than 50 kbits/s	Mostly low-speed asynchronous communications via an RS-232 line at 4800–9600 bits/s or via an IEEE-488 bus at a higher speed	Both high-speed synchronous computer-to-computer communications and low-speed, asynchronous computer-to-terminal communications	Mostly high-speed, synchronous computer-to-computer communications at approximately 50 kbits/s and above
Time critical	Very time critical	Time critical, as communications involved include sequencing and coordination	Less time critical, but still important in that production may be delayed	Not time critical to the operation of the FMS
Data base requirements	Limited to simple performance histories, diagnostic information, adaptive control data, and instructions	Store NC programs, machine status, job status, operations status, tool status, and QC and production scheduling information	Store parts design data, process plans, operations data, and data on the status of machines, robots, tools, and parts, as well as QC data, tool management data, and NC programs	Store all data for a traditional MIS
Distance covered	Very short—less than several meters in the proximity of a machine	Short—less than several hundred meters in the proximity of a factory building	Short—less than several thousand meters in the proximity of factory premises (within a factory building or a cluster of factory buildings)	Short—less than several thousand meters in the proximity of factory premises
Allowable error rate	Zero	Zero	Zero	Zero
Typical communication	Interprocess communications in a distributed processor system Communications in a servo control loop Simple machine input/output	Communications acknowledgment Simple inquiry Notification of an activity's completion Transfer of information, NC programs Simple instructions	Transfer of files, NC programs, executive programs, and instructions Simple inquiry Communications acknowledgment	Data entry File transfer Batch/RJE (remote job entry) Simple inquiry response Data access

Source: Ref. 1.

12.3.5 Distributed Data Base Systems

As the capabilities of networking, microcomputing, and data base management system technologies evolve, distributed data base systems are becoming a reality. Because networking technology is maturing at such a rapid rate, the 1990s may be characterized as the distributed data base systems decade. It is important to distinguish between a networking system which provides remote data base access and a distributed data base system. In a networking system, the user may operate on data at one or more remote sites simultaneously. The user is aware that the data are distributed, and work is performed by following procedures that facilitate utilization of this data. In a true distributed data base system, the data-manipulation user is not aware that the data is distributed.

Stonebraker [34] lists seven features of distributed data base systems:

1. *Location transparency:* A user can submit a query that accesses distributed objects without having to know where the objects are.
2. *Performance transparency:* A query can be submitted from any node, and it will run with comparable performance. A distributed optimizer determines the best way to execute any distributed command.
3. *Copy transparency:* The system supports the optional existence of multiple copies of data base objects. Consequently, if a site is down, users can still access data base objects by obtaining a copy from another site.
4. *Transaction transparency:* A user can run an arbitrary transaction that updates data at a number of sites.
5. *Fragment transparency:* The distributed data base management system allows a user to segment a relation into multiple pieces and place them at multiple sites according to certain distributed criteria.
6. *Scheme change transparency:* Users who add or delete a data base object from a distributed data base need to make the change only once, to the distributed dictionary.
7. *Local data base management transparency:* The distributed data base system must be able to provide its services without regard for the local data base systems that are actually managing local data.

No distributed data base management system on the market has all of these features. However, significant work is being done to address the shortcomings. So, over the next few years, systems will evolve to the point that most if not all of these features will be available.

12.3.6 Management of Technology

Up to this point we have been addressing technologies that must be developed further before the potential of CIM can be fully realized. An important related issue is the management of technology. During the last two decades, U.S. firms have not done a good job applying the new research developments. Management receives most of the blame.

For more than one reason, Digital Equipment Corporation (DEC) has a major interest in CIM. Using these concepts, DEC has achieved some real successes in many of their plants. Also, DEC's products represent a major share of CIM-related sales. So it is not surprising that DEC conducted a study to identify the issues that are blocking the implementation of CIM. The following list of issues represents the results of that study [30]:

1. Lack of understanding or knowledge of CIM
2. Lack of management support or commitment
3. Resistance to change
4. Cost
5. Turf protection
6. Job security
7. Training, retraining, education
8. Organization restructuring

This list is interesting from at least two aspects. First, the omission of technology limitations is surprising, and second, these issues are either directly or indirectly related to management.

In addition, some dramatic changes are occurring in accepted management practices. For instance, the traditional hierarchical organization that has been in vogue for decades is now being replaced by flatter organizations. The resulting organizations improve communication and facilitate teamwork; as a consequence, costs go down, quality improves, and productivity increases.

These observations lead to the conclusion that management of technology must be addressed as we strive to achieve the full potential to be realized from CIM.

12.3.7   Other Emerging Issues

Hardware-related technology, such as computers, CAD terminals, and networks, is evolving faster than can be fully implemented and utilized. Nonhardware technology, such as software, human-computer interfacing, and data base management, is proving much tougher to overcome than anticipated. In the mid-1980s, many articles were written about "lights-out factories." Today, this terminology is seldom used, because we realize that people will be an important component of manufacturing environments in the foreseeable future.

Emphasis is shifting to software issues, such as:

1. How do we write and maintain the millions of lines of code that CIM systems require?

2. As technology changes, how do we convert the many existing programs to utilize the capabilities of this technology?

3. How do we maintain the data integrity of the large, distributed data bases?

4. How do we manage large, distributed data bases?

Important people-related questions are also being asked:

1. What are the best human-computer interfaces?

2. What organization structures are best?

3. How do we manage in a multiple-discipline team environment?

4. How do we cope in environments where change occurs at an increasing rate?

These lists do not address all of the known emerging issues, but they are indicative of the types of issues that require substantial work. Most of these issues are very complex, meaning that finding solutions will be very difficult. Also, as we progress, other issues will emerge that will be equally difficult to resolve.

12.4 FUNDAMENTALS OF NETWORKING

12.4.1 Networking Concepts

Two types of signaling methods are predominately used in a LAN network: baseband and broadband. A third type based on fiberoptics, which uses light instead of electricity, appears to be very promising. This latter technology is just emerging. The primary applications have been to connect different LANs at widely separated locations.

In baseband applications, digital bits are sent on the network by raising and lowering a voltage. A typical baseband system uses a specified base voltage to signify an "off" (0) bit and then varies the voltage to a specified level for a specified period of time to signify an "on" bit. Transmission occurs millions of times per second. If cable lengths are too long, signal distortion can occur; consequently, cable length restrictions must be followed [17].

Broadband systems use radiofrequency signals to transmit multiple network signals simultaneously. The network information is sent on channels with separate receive and transmit frequencies. The large total frequency bandwidth of a broadband system is divided into intervals of 50 kHz for voice and 6 MHz for video. Consequently, several baseband networks, such as Ethernet, can be implemented on the same broadband cable along with radio and TV transmissions. In fact, broadband technology is used by cable TV. Tapping into a broadband cable requires a network interface card (NIC). An NIC creates a radiofrequency signal and performs the required protocol functions of baseband. The integrity of the network can be impaired by the frequency drift of any node's transmitter. Consequently, a broadband network must be "tuned" from time to time. A broadband network is often used as a high-bandwidth bridge between less expensive baseband networks.

Network topology deals with how the nodes on a network are physically connected and how they logically interact. The major topologies are star, linear bus, and ring (see Figure 12.5). In a star network, a dedicated cable connects all stations to a central point, which is usually the server. Since each cable to the server is unique (none are shared), the impact of a cable or NIC fault is limited to a single station. Also, no complex protocol is needed to control sharing of the physical connections. The major disadvantage of a star topology is that large amounts of cable may be required. A PBX telephone system, in which all phones are connected to a central switchboard, is a good example of a star network.

A linear bus network consists of a single length of cable (called a bus). An Ethernet LAN uses a bus topology. All stations are connected with a stub to this bus, and the end of the bus are terminated (not connected to each other). The length of a stub is limited; for example, Ethernet's limit is 6 ft if a thin-wire coaxial cable is used [17]. Any station on the network can transmit a message to any other

FIGURE 12.5   Network topologies: (*a*) star network; (*b*) ring network; (*c*) bus network.

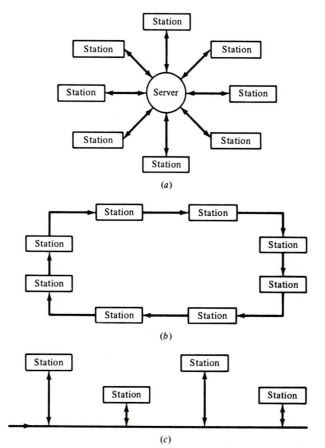

station on that network; consequently, each station must inspect the address of all messages. One advantage of a bus network is wiring simplicity. However, any break in the bus may cause the entire network to fail. Also, the location of the break may be difficult to identify.

In a ring network, the stations are connected in a series with the cable ends connected together to form a ring. All messages pass through all stations. When a message arrives at a station, the message's address is checked, and if that station is the intended recipient, the message is stored for use.

Stations may be physically wired in one topology and logically interact in another topology. The IBM token-ring network is an example of this type situation (see Figure 12.6). The stations are connected to hubs in a star arrangement. But the hubs are connected in a wire ring, and all devices are logically configured as a ring.

When several stations are connected to a network, contention for access will occur when more than one station tries to send a message at the same time. Consequently, network access is an important consideration in network design. Two basic methods are used: deterministic (token passing) and contention (carrier-sense multiple access with collision detection, which is often referred as CSMA/CD). In the deterministic approach, access is available only to the station in possession of a "token" (a unique string of bits that serves as a control signal) that is continuously passed among the stations. Therefore, a station wishing to gain access to the network must wait until it can take possession of the token. Most token-ring networks have rules (protocols) that ensure that every station will

FIGURE 12.6   Token-ring network.

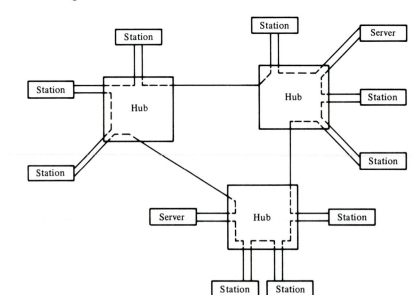

have an opportunity to send a message. Because the delivery time of a message can be controlled, token-based networks are chosen for real-time manufacturing applications. MAP uses a deterministic access method.

In a contention-based system, each station wanting to transmit a message must "listen" to make sure that no other station is sending a message [17]. When the network is accessible, a message can be sent. When two or more stations transmit simultaneously, a collision is detected and each station immediately terminates the transmission of data. All transmitting stations must wait a random length of time before attempting to retransmit. This process continues until the contention is eliminated. The scheduling of retransmissions is determined by a process called truncated binary exponential back-off. After each repeated transmission failure, the station doubles the mean value of the random delay. As a network becomes busier, stations dynamically adjust to longer times between transmission attempts. Data are transmitted in packets of $8n$ bits, where $46 \le n \le 1500$. The minimum size of $46*8$ bits ensures that collisions are detected while transmission is in progress [17]. Short messages are padded with binary zeros, and long messages are segmented into several packets. This type of network is often used in office environments, where typical traffic occurs as large blocks of data with no critical timing for delivery. Ethernet is a contention-based network.

Raw data transmission speed capability of a network is an important performance characteristic. The speed varies considerably. One million bits per second (Mbps) is relatively slow. Ethernet specifies 10 Mbps [17]. These speeds will increase as the technology evolves.

Network installation costs are influenced by the type of cable used. CSMA/CD uses two types of coaxial cable (coax) and twisted pairs. The maximum distance between the farthest nodes on a network is called the geographic span [17]. Thick coax is used for long runs (up to 1000 m); thin coax may be used up to 305 m. These distances can be expanded with devices called repeaters.

Token-passing bus networks use several types of 75-Ω coax. The fastest implementations use semirigid cable TV-type cable. Token-ring networks use twisted-pair cables: shielded and unshielded (ordinary telephone wire). The latter should be used only in environments with low electromagnetic interference.

In general, network technology is evolving toward a media-independent implementation of the major types of networks. The trend is toward fiberoptic media for the following reasons:

1. Immunity from electromagnetic interference
2. Small size and weight
3. Nonelectrical property reduces risk in hazardous environments
4. Small size and weight (a single fiber can replace a 300-pair telephone cable)
5. Low signal loss

At this time, fiberoptical technology is difficult to work with and expensive. However, as the technology evolves, this should become the preferred cable media.

12.4.2   OSI Fundamentals

It was previously noted that many vendors developed proprietary computer system architectures. Networks were a very important part of these architectures. Since no network standards existed, many types of networks were created, no one being the best in all aspects. Efforts were established to develop some standards for the better networks. As a result, the Institute of Electrical and Electronic Engineers (IEEE), American National Standards Institute (ANSI), and representatives from many companies, over a period of several years, have written a family of specifications for LAN hardware [17].

In 1981, the IEEE agreed with the ISO that all standards should be in accordance with the OSI reference model. This is an assortment of specifications which comprise the OSI repertoire of standards. Implementors choose which specifications to build into a product. For instance, there are several types of medium access control, such as token and CSCA/MD. The implementor must choose one from among those specified in the OSI repertoire.

Communicating between dissimilar systems is not a simple process. Specifications must address this complexity and accommodate changes as technology evolves. This was one reason why the architects decided to segment the specifications into groups called *layers*. In this way, changes could occur within a layer without harming compatibility of the adjoining layers, provided the layer interfaces were maintained. A seven-layer model evolved (see Table 12.7) in which messages would pass down, starting at layer 7, through the stack of layers, utilizing particular specifications at each layer. After passing through these layers, a message would be transmitted over a cable to the address of the recipient. At that point the message would ascend through a corresponding stack of layers to the recipient.

Layer 1 (the physical link) specifies what type of cable is used. This layer addresses actual physical attachment to the network and the physical form of the communication. The physical link layer is implemented by hardware. Using a telephone call as an analogous example, this layer corresponds to the phone-cable connection to the phone jack, the phone jack, and the type of dialing (rotary or touch-tone).

TABLE 12.7
OSI Reference Model

Layer	Name	Function
7	Application	Selects appropriate service for applications
6	Presentation	Provides code conversion and data reformatting
5	Session	Coordinated interaction between end-application processes
4	Transport	Provides end-to-end data integrity and quality of service
3	Network	Switches and routes data
2	Data link	Transfers units of data to other end of physical link
1	Physical	Transmits onto network

Layer 2 (the data link) packages the data into packets and places these packets onto the cable. Upon receiving data, if something is wrong, this layer notifies layer 4 (the transport layer) of the error. In a telephone call, this might correspond to picking up the receiver, which notifies the central office that a call is coming, and then hanging up after the call is completed. Layer 2 is implemented by a combination of hardware and software. The remaining five upper layers are implemented entirely by software, such as the Manufacturing Automation Protocol (MAP). As noted previously, a given product may address several layers; however, to comply with the OSI model, a vendor must provide compatibility with the layers that are not addressed.

The IEEE standards that specify the two lowest layers are [17]:

1. Standard 802.2, data or logical link control
2. Standard 802.3, carrier-sense multiple access with collision detection (CSMA/CD) bus LANs, such as Ethernet
3. Standard 802.4, token-passing bus LANs, such as MAP
4. Standard 802.5, token-passing ring LANs, such as IBM's token-ring network.

An 802.6 standard is being developed for metropolitan area networks, and Standard 802.7 will cover broadband networks.

Layers above layer 3 (the network layer) assume that a message has arrived at the designated station. The network layer is responsible for translating logical addresses into physical addresses and for picking the best network route, if more than one is available. Non-OSI networks or machines are often interfaced to an OSI LAN utilizing this layer and devices such as protocol converters or gateways. In the phone-call example, this layer corresponds to utilizing the phone number dialed to select a route to the appropriate phone.

Layer 4 (the transport layer) ensure that data passed up to it is sent on successfully. It establishes and maintains the connection and has responsibility for demanding a retransmission if the data did not arrive error-free. This corresponds to the phone ringing and busy signals sent back to the caller. An answer from the called party would confirm a successful connection.

The session layer (layer 5) decides when to turn a communication between two stations on or off. This layer monitors the communication, and, if an electrical disruption breaks the connection, attempts to reestablish the connection before the station at the other end is aware of a problem.

The presentation layer (layer 6) is responsible for translating commands from layer 7 into syntax common to the network. Also, commands received from the network are translated back to a syntax that is comprehensible to layer 7.

The application layer (layer 7) provides utility functions such as file transfer or electronic mail which a program or station can utilize to communicate with other systems on the network. For example, a LAN network may have a shared printer. Layer 7 could provide access to that printer through some command such as LAN PRINT.

As the number of protocol specifications grows at each layer, choosing the "right" one might appear to be an overwhelming task. Certain combinations of specifications will most likely be used together. Figure 12.7 depicts the message handling system (MHS)/X.400 protocol stack, which is commonly referred to as an electronic mail protocol. The top three layers comprise the MHS/X.400 protocol. In this case, this is called the application-layer protocol. Layer 4 is a robust transport protocol designated by TP-4, and layer 3 contains the Internet Protocol (IP), which provides routing and relaying at the network layer between multiple subnetworks. Layers 1 and 2 comprise the Ethernet specifications.

Two other protocol stacks that have been finalized are: File Transfer, Access and Management (FTAM), and Virtual Terminal Protocol (VTP). FTAM is used for file-transfer applications, and VTP is designed to enable a diversity of terminal types to access a common software application. Other protocol stacks will emerge over time.

12.4.3   MAP/TOP Fundamentals

In 1982, General Motors mandated that any computerized device installed in its factories would communicate via the Manufacturing Automation Protocol (MAP) [25]. The objective was to establish one set of LAN protocols for communications between intelligent devices, such as computer-controlled machine tools, engineering work stations, process controllers, factory-floor terminals, and control rooms.

MAP is a broadband, token-bus network protocol based on the OSI model that will accommodate a broad range of manufacturing environments. Layers 1 and 2 use the IEEE 802.4 specification. The MAP User's Committee has defined the top five layers. Table 12.8 shows how MAP 3.0 relates to the OSI reference model [36]. MAP 2.1 was the first version of the protocol to have products supporting it. However, this was soon replaced by Version 2.2, which has now been replaced by Version 3.0. This latest version differs from Version 2.1 primarily in

FIGURE 12.7   Message handling system/X.400 protocol stack.

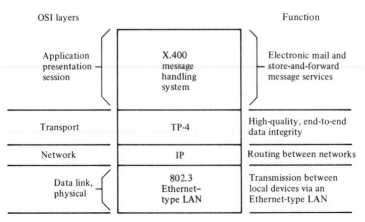

TABLE 12.8
Comparison of MAP 3.0 and the OSI reference model

MAP 3.0	OSI reference model
• Directory services • Network management • Association control service element (ACSE) • Manufacturing message standard (MMS) • File-access transfer method (FTAM)	Application layer
OSI presentation layer	Presentation layer
OSI session layer	Session layer
OSI transport class 4	Transport layer
OSI connectionless network service (final system to intermediate system)	Network layer
• IEEE 802.2 token bus	Data link layer
• IEEE 802.4 10-Mbps broadband	Physical layer
• IEEE 802.4 5-Mbps carrier band	

Source: Ref. 13.

the file-transfer, real-time messaging and presentation protocols, and in the network-management and directory-services capabilities at the upper levels [27]. There are some differences at the lower levels. Because migrating from MAP Versions 2.1 and 2.2 to Version 3.0 is difficult, there is some established user reluctance to support the latest version. To encourage support, the specifications for Version 3.0 were frozen for 6 years. Some companies are waiting on products from large firms, such as IBM and DEC, before committing to installing MAP 3.0. The most important thing that MAP offers is OSI compliance.

As was noted earlier, office LANs have different performance requirements from factory LANs. Baseband systems have been very successful in the office, they are easier to install and maintain, and they are less expensive. Consequently, in 1985, Boeing proposed the TOP (Technical and Office Protocol) as a supplement to MAP. A MAP/TOP User's Group was formed to propose and approve the specifications for the combined protocols. In fact, TOP is also Version 3.0; Version 2.0 was skipped to maintain consistency with MAP version numbers.

TOP and MAP networks are compatible. A core set of protocols, layers 3, 4, 5, 6, and parts of 7, are common to both (see Table 12.9). At layer 7 they diverge; TOP specifies application protocols that address requirements of the office instead of factory-floor requirements. They also diverge at the two lowest levels, where TOP specifies the less expensive, less deterministic 802.3 baseband media and media-access technique in place of the 802.4 broadband that MAP uses [26].

Like MAP, more and more vendors are announcing products that support TOP. There appears to be a growing commitment to TOP. Ultimately, the success of TOP and MAP will depend on whether different vendors' protocol implementa-

TABLE 12.9
Comparison of MAP and TOP

	MAP	TOP
Layer 7		OSI ACSE
		OSI FTAM
		OSI directory services
		OSI network management
	Manufacturing message standard (MMS)	X.400 electronic mail
		OSI office document protocols
Layer 6		OSI presentation layer
Layer 5		OSI session layer
Layer 4		OSI transport class 4
Layer 3		OSI connectionless network service
Layer 2	IEEE 802.4 token bus	IEEE 802.3 CSMA/CD
Layer 1	IEEE 802.4 broadband 10 Mbps	IEEE 802.3 baseband 10 Mbps
	IEEE 802.4 carrier band 5 Mbps	IEEE 802.5 token ring

Source: Ref. 26.

tions communicate correctly and reliably with each other. Three types of testing are required: conformance, interoperability, and functionality. Conformance testing determines that an implementation conforms to specifications. Interoperability testing ensures that two or more conformant implementations interoperate. Functionality testing ensures that interoperable implementations can do useful work.

FIGURE 12.8 Example MAP/TOP network.

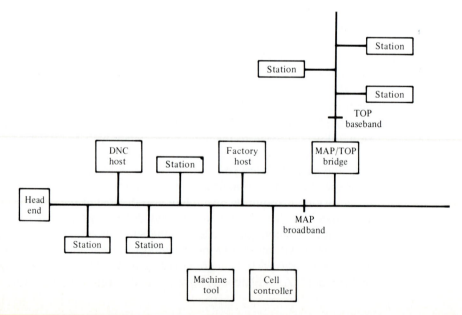

Organizations are developing methods to address these three types of tests. Significant strides have been made toward eliminating the many different protocols used in the office and factory in the early 1980s.

12.4.4 Example MAP/TOP Network

Figure 12.8 illustrates a network that provides a means for integrating the engineering and business environment with the factory floor using a MAP-based broadband backbone. Connected to the backbone is a baseband TOP network.

A gateway device is used to connect two diverse networks. This device performs the appropriate address conversions and protocol changes. A bridge is also used to connect the MAP backbone to the TOP baseband. This type of device is used to connect two LANs that use the same logical link-control procedure but use different medium-access control procedures.

12.5 DEVELOPING A SUCCESSFUL CIM STRATEGY

12.5.1 Guidelines

Global competitive pressures are placing an emphasis on manufacturing and technology management. CAD, CAM, and CIM concepts promise lower cost, higher quality, and shorter times for product development. However, plans should not be made naively, because there are limits to what can be realistically expected.

Given limits on what can be expected from CAD, CAM, and CIM and the desire to develop a CIM strategy, where should we start? Table 12.10 lists advanced manufacturing concepts and technologies by company function [21]. At first glance, the list is somewhat overwhelming. In the mid-1980s, when CIM was seized upon by some firms as the answer to most competitive manufacturing problems, vast amounts of money were spent on automation in order to reduce direct labor costs. Figure 12.9 shows that direct labor accounts for only 10% of the

FIGURE 12.9 Product costs by major category.

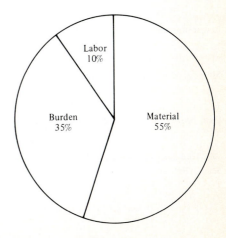

TABLE 12.10
Advanced concepts and technologies by company function

Company function	Applicable advanced concepts and technologies
Strategic business planning	Flexible manufacturing strategies/automation/design for differentiation/ rapid order completion/high-manufacturing-velocity strategies
Sales and quotations	Interactive, integrated data bases/data communication links to field and customer locations
Product design	CAD—design/CAE—engineering analysis/simulation and animation/ design for manufacturability, for automated handling and assembly, and to cost targets/group technology—design retrieval/engineering change control
Process design	CAD—equipment and process design/simulation models/cellular and flexible machining systems
Capacity analysis and facility design	Facility layout—analysis algorithms/simulation models/family-of-parts analysis/computer-aided facilities planning
Process planning and tool design	Computer-aided process planning/group technology/classification and coding/time standards data base
Production planning and scheduling	Master production scheduling/production smoothing/MRP/MRP II/OPT/ JIT/Kan-ban/scheduling algorithms/simulation/plantwide monitoring systems
Purchasing and vendor management	Automated purchasing/electronic links to vendors/automated follow-up
Order entry and order processing	Demand management/forecasting/integrated data bases/on-line, real-time order tracking
Shop-floor management	Dispatching—handling controls/automated identification systems—bar coding, etc./automated input of performance data as order progresses through shop/data integrity audit procedures—cycle counting, etc./ preventive maintenance scheduling/advanced concepts in performance measurement
Fabrication	NC/CNC/DNC/programmable controllers/automated processes/auto- mated tool changing/machine and tool monitoring/robotics/quick-change setups
Assembly	Automated assembly/robotic assembly/automated testing by assembly stages
Inspection and test, QA	Computer-aided inspection and testing/machine monitoring and diagnos- tic systems/statistical process control/machine vision/adaptive controls/ inspection "on the fly"
Materials handling and storage	Automated storage and retrieval systems/automated guided vehicles/ automated conveyors/robot handling/part transfer systems-hard automa- tion
Data processing and factory communications	MAP/TOP, other communication protocol standards/factory networks/ postprocessors and preprocessors/paperless control and reporting/elec- tronic process plans, instructions to operators/integrated data bases

Source: Ref. 21.

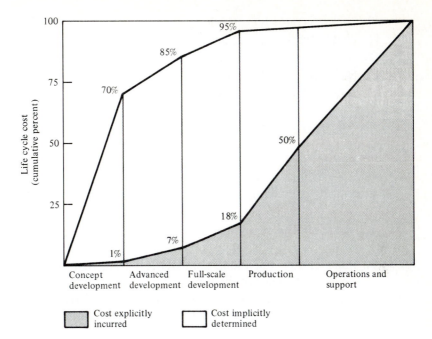

FIGURE 12.10   Determination of costs during product life cycle [38].

cost of sales. From this figure, it is obvious that we should concentrate on burden and materials costs.

It is useful to view product costs from the perspective of when product costs are determined. Figure 12.10 shows when costs are determined and when they are actually incurred in the life cycle of a product. This perspective also verifies that reducing direct labor costs does not provide the most potential for reducing costs.

Indirect labor costs offer significant opportunities for product cost reduction; consequently, CIM projects comprising these functions promise the most potential benefits. Pursuing this reasoning, it is important to identify the indirect labor tasks in a typical discrete-parts manufacturing firm. A new part design or an engineering change to an existing part can result in the following actions:

1. Sketches and drawing retrieval
2. Engineering analysis
3. Testing and simulation
4. Layout and checking
5. Engineering review
6. New drawings
7. Manufacturing modeling
8. Process plans
9. New molds, dies, tools, and fixtures

10. New programs for machining
11. New inspection procedures
12. Quality assurance and testing
13. New assembly instructions
14. New field and service manuals

Many people from several departments will be involved in completing these tasks, and the cycle time will be weeks if not months.

At this point, we have a good idea where efforts should be concentrated. However, before a CIM strategic plan is developed, the following guidelines for successfully completing a CIM project should be understood:

1. Define CIM in terms that are understandable to all, especially top management.
2. Make CIM part of a long-term strategic plan. Also, CIM goals must support the strategic goals of the firm.
3. Top management must sponsor the efforts, and top management must remain involved because success cannot be achieved without a committed management team.
4. Plan from the top. Design a comprehensive system from the beginning; otherwise, the system pieces may not fit together. A CIM plan represents a multi-year effort; consequently, the CIM plan must be robust enough to accommodate changes resulting from rapidly evolving technologies.
5. Implement from the bottom in manageable-sized projects. Select a particular product line which will serve as a focus for developing a CIM business strategy.
6. Develop for the long term so that each project supports the manufacturing process you want to be in place when the CIM plan is realized.
7. Simplify, plan for integration, automate, and then integrate. Too often implementors and management expend substantial money on automating a process that is not well understood. The result is usually a failed project.
8. Emphasize the basics, such as understanding the process so that it can be controlled within known limits, ensuring discipline in following accepted procedures, and stressing the importance of data integrity.
9. Place emphasis on projects that address indirect labor, engineering, and management functions. The greatest benefits will be realized in these areas.
10. Design engineering is an important area, because parts must be designed for manufacturability.
11. Emphasize information flow. Information as well as parts should flow smoothly with a minimum number of steps.
12. Strive to "pay as you go." Many long-term projects fail because no payback

is received in short-term intervals. As a result, management support wanes as pressing demands for resources receive higher priority.

13. Emphasize education of management and the workforce. Change is always disruptive. Education alleviates concern and can attract support.

14. Assign the most capable people to the program. A CIM program results in many cultural changes; consequently, the chances of failure are high. Capable people can make the difference.

15. Communicate, communicate, and communicate. Most perceived problems can be solved with open communication.

Following these guidelines will significantly improve the chances of achieving a successful CIM program.

12.5.2 CIM Example

The Society for Manufacturing Engineers (SME) annually sponsors a conference called AUTOFACT. At this conference in 1989, a CIM demonstration, Partnership in Integration, was exhibited [33]. Approximately 25 companies, ranging from computer system vendors to consultants, worked together to simulate how a fictitious, medium-size company, Huron Manufacturing, exploits CIM.

The demonstration centered around an order for an out-of-production part from an owner of a riding lawn mower. The flexibility of a CIM environment permits making the part cost-effectively without disrupting production schedules. The exhibit illustrated dramatically how several dissimilar computer systems can be integrated using existing technologies. Figures 12.11 through 12.16 depict the networked systems. The following description of the demonstration will be presented in present tense as a case example.

FIGURE 12.11   Retail outlet network [33].

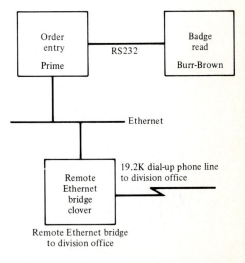

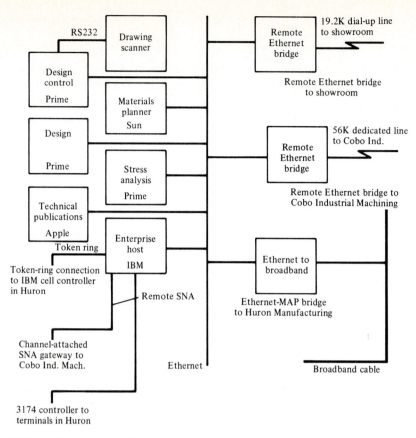

FIGURE 12.12 Division office network [33].

FIGURE 12.13 Cobo Industrial Machining network #1 [33].

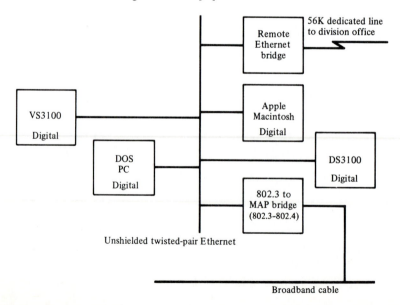

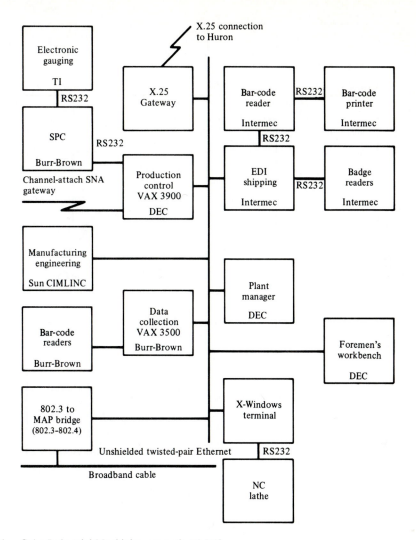

FIGURE 12.14 Cobo Industrial Machining network #2 [33].

Huron Manufacturing produces and markets lawn and garden tractors and service parts through the following interrelated operations:

1. A retail outlet that demonstrates, sells, orders, and services the product line
2. A division office that handles strategic planning, master production schedules, product design, and engineering change management
3. A main manufacturing plant, Huron Manufacturing, that is responsible for all manufacturing engineering, inventory control, component production, and final assembly
4. Cobo Industrial Machining Company, which is Huron's subsidiary job shop.

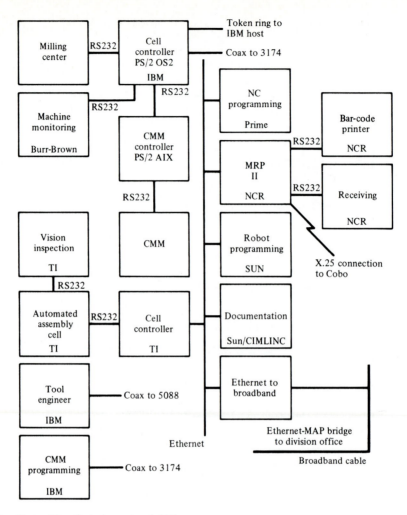

FIGURE 12.15   Huron Manufacturing network [33].

These facilities were simulated in one large exhibit booth, although in reality they could be thousands of miles apart, connected via networks.

The demonstration begins when a customer enters the retail outlet to place an order for a mower deck part on an older-model tractor. The customer does not know the part number, so the dealer uses an on-line CAD part catalog to retrieve the part geometry. The system is capable of displaying the mower deck, after which the user can explode it into components to identify the part that is needed.

The part is available only as an assembly, which the dealer does not have in stock. Using the on-line catalog, the dealer orders four of the assemblies from the factory, because there are several tractors of this model in the area. The order information, including part number, sales order number, quantity, required deliv-

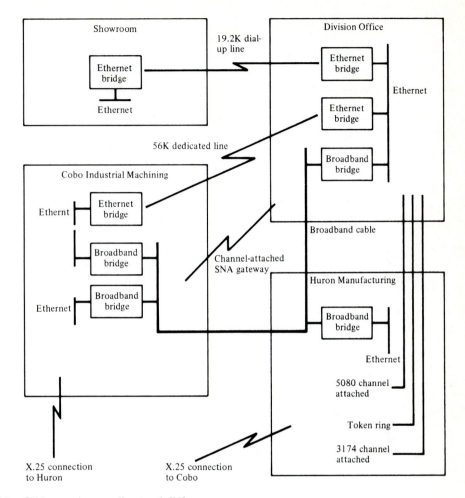

FIGURE 12.16 CIM example: overall network [33].

ery date, and customer number, is sent over a network to an IBM host computer which resides at the division office.

After the order information is received at the division office, the inventory system is checked for part availability. The assembly is out of stock. The order is then forwarded over an Ethernet network to a materials planning system that resides on a Sun microcomputer. At that time, the production planner will review the part assembly bill-of-materials (BOM). The BOM indicates that two of the assembly's components are not available: a traction clutch bellcrank and one bushing.

The material plan data base reveals that both of those parts were produced by subcontractors that are no longer in business. The planner then decides to have the subsidiary job shop make the bushing and the main plant make the traction clutch bellcrank. At that point the planner sends an electronic mail message to the shop analyst's DEC VS3100 work station at Cobo Industrial Machining to see if it is possible to schedule the bushing. The production planner also sends an interoffice electronic message to a design engineer, who works with a Prime computer, asking the engineer to begin preparing the necessary engineering data for both parts.

The design engineer reviews the request and discovers that there is no engineering data on the bushing and the only bellcrank information is a drawing. One possible source of similar bushing data is the group technology data base that resides on the Prime computer. A query into this data base reveals that there is a similar bushing, but it has a slightly larger outside diameter. The design engineer modifies the similar part so it has the proper dimensions and then produces the necessary drawings. The new bushing data are then released and stored in the design control system, which resides on another Prime computer.

By this time, the blueprint for the bellcrank has been retrieved from storage. The engineer uses a large document scanner to quickly capture the drawing in electronic form. Using this as a reference, a 3-D model is created. The engineer notices that the part has square holes for carriage bolts. However, manufacturing engineering has switched to round holes. So the part is redesigned and the new design requires stress analysis, which requires a program residing on another Prime computer. After passing all tests and analysis, the modified part design receives a revision-level number. The new components replace the existing ones in the assembly, which also gets a revision-level change. The design and analysis data are sent to the design control system.

Using electronic mail, the design engineer notifies the production planner of the new assembly number. In addition, the new assembly geometry is sent from the design control system to the IBM host computer, where copies of all part geometry are managed and stored in a consolidated design file.

The new assembly requires updating part catalogs and service manuals with correct engineering data. This is done with a technical publications system that runs on a Macintosh. The original assembly documentation is extracted from the design control system and revised. The revised documentation is sent to the design control system.

Now the production planner releases the order using an MRP II (see Chapter 7) system which resides on the IBM host computer. Huron Manufacturing's current production schedule is maintained in this system. The planner then retrieves the rough-cut capacity plan and material requirements for both parts from the MRP II system on the IBM. These data are combined with other data from MRP II and shop-floor control systems and loaded into a simulation-based finite-capacity sequencing and scheduling package that resides on the Sun microcomputer. Using this package, the production planner evaluates alternative production schedules that will not disrupt current manufacturing flow. After a feasible schedule has been selected, the revised production schedule, including the special order, is sent by the IBM host computer to the main plant's NCR manufacturing production control system (another MRP II system).

At Cobo Industrial Machining, a DEC VAX 3900 retrieves the new master schedule for the bushing and its IGES drawing file from the IBM host computer using a channel-attached SNA gateway and VAXlink software.

Cobo's shop analyst, using a DEC VS3100 system, receives an electronic mail message from Huron's production planner requesting permission to schedule the bushing into the job shop's production. The analyst retrieves the old routings, bill-of-materials (BOM), and tooling data from the VAX 3900. The BOM specifies a bar stock that is not in inventory; however, there is a bar stock close to the specified size in inventory. The analyst retrieves the bushing geometry from the VAX 3900 and proceeds to change the part specifications to fit the available raw stock. Next Huron's planner is notified by electronic mail that an engineering change approval is requested. The planner obtains the approval and notifies the analyst. Now, the new geometry is sent to the VAX 3900.

Because Cobo is at full production, the plant manager uses electronic mail over Ethernet to request an analysis of the bushing order's effect on the existing production schedule. An analyst, using a MAC operating system and an electronic spreadsheet, accesses the data base on the VAX 3900 to do a cost analysis. The same data base is accessed by another analyst using a DS3100 system with a UNIX operating system to do a schedule-and-labor analysis. The results are merged by the assistant plant manager using a DS316 personal computer running a DOS operating system. The assistant plant manager adds his recommendation and sends the report to the plant manager.

Cobo's plant manager reviews an electronic version of the report, decides that the shop can run the bushing order, and notifies the analyst to proceed. The analyst sends the news to Huron's production planner by electronic mail.

The production control system residing on the VAX 3900 requests the IBM host at the division office to download the released order for the bushings. The production control system then does the necessary materials allocation, operation scheduling, and workcenter dispatching, as well as directing a printer to produce a bar-coded shop traveler.

Manufacturing engineering now must develop an NC program and manufacturing documentation. The BOM and the IGES geometry file are transferred from the VAX 3900 to manufacturing engineering's Sun microcomputer. The IGES geometry is converted to CIMLINC format, which is a format utilized by CIMLINC systems. Using CIMLINC software, an NC program for the bushing is prepared, verified, and postprocessed. In addition, the necessary documentation for the machine operator is prepared using CIMLINC software. The NC program and manufacturing documentation are sent from the Sun system to the X-Windows terminal, and a notification of completion is sent to the VAX 3900.

The production schedule, manufacturing documentation, and inspection data are then transferred from the VAX 3900 to the X-Windows shop-floor terminal. The machine operator reviews the on-line documentation to determine how to make the bushing, including what tooling to use. The operator then requests a transfer of the NC program from the Sun microcomputer to her X-Windows terminal. The shop-floor terminal now loads the program into the CNC milling machine.

After the machining has been completed, the operator uses a Texas Instruments (TI) electronic gauging system to inspect the bushing. This data is sent to a Burr-Brown gauge interface and statistical process control (SPC) system. The SPC system passes the analysis to the VAX 3900, which downloads the SPC data to the X-Windows shop-floor terminal for the operator.

The operator also does a bar-code labor-reporting transaction using a Burr-Brown bar-code reader connected to a VAX 3500. Software on this VAX validates the labor transaction interactively and then sends the data to the VAX 3900.

At any time, the job-shop foreman, using the foremen's workbench, can perform on-line queries against the management information system that resides on the VAX 3900 to analyze the schedule and track production.

When the machine operator reports that the lot has been completed, she closes the order using the bar-code reader. The VAX 3500 sends a "close order" transaction to the VAX 3900, which communicates job completion information to the MRP system residing on the IBM host at the Huron division office.

Finished parts, in groups of four, are taken to shipping in tote boxes. Bar-coded shipping labels (containing part number, quantity, and serial number) are produced by an Intermec printer. The labels are attached to the tote boxes. The shipping clerk scans the bar codes, and the Intermec system passes the data to the VAX 3900 to complete a pack/ship transaction.

The bushings are delivered to Huron Manufacturing in small just-in-time (JIT) batches. Notification of parts shipment comes from electronic data interchange (EDI) between the Cobo VAX 3900 and the Huron plant's NCR MRP II system using an X.25 connection. X.25 is an international standard for packet switching. The EDI Intermac sends shipping information to the VAX 3900, which then notifies the NCR MRP II that the parts have shipped.

When the bushing arrives at the main plant, a receiving clerk scans the bar code on the tote box with an NCR data-acquisition terminal connected to the NCR MRP II system. This logs the parts into inventory. The NCR system receives the order for the service part from the division office's IBM host computer and verifies that it is a valid assembly. The production schedule is checked to determine where to insert the order without disrupting the plant. The order is then sent to manufacturing engineering.

Manufacturing engineering prepares the robot assembly program, NC machining program, computer-aided process plans, and manufacturing documentation for the bellcrank and final assembly.

Assembly data, including work-order number, parent part number, component part number, quantity, description, date in, date out, and part type, are sent from the NCR MRP II system to a Sun microcomputer where documentation will be prepared. The assembly engineer requests the part geometry (3-D solid model in IGES format) and tooling geometry (3-D wire frame model in IGES format) from the design control system located on the Prime computer at the division office. After this information is received, the assembly engineer transmits assembly instructions to a robot programmer at a Sun computer.

The robot programmer prepares a robot simulation program that is postprocessed to produce the robot program. When complete, an indicator (including work-order number, part number, and ready flag) is sent to the NCR system.

A tool engineer, working at an IBM 5080 graphics system, receives part geometry (3-D solid model) over a coax cable connected to the IBM host computer. This engineer selects the appropriate tooling from a library and stores the data in the consolidated design file on the IBM host computer so that other manufacturing engineers can use it.

In another area, a quality assurance (QA) engineer is using an IBM graphics terminal connected by a coax cable to the IBM host computer. The QA engineer selects the bellcrank part geometry (3-D surface model) stored in the IBM host. Using special software, coordinate measuring machine (CMM) instructions are generated and downloaded to the IBM cell controller on the shop floor.

Upon receipt of the order, the NCR system transfers the bellcrank production data to the Prime CAD system where the NC programming will be performed. Here, a manufacturing engineer uses the bellcrank geometry (3-D solid model) to generate an NC APT program for the CNC milling machine. The program is verified using visual simulation and is postprocessed. This same engineer also requests the tooling geometry (3-D wire frame) and associated data from the IBM host computer to prepare a fixture layout. A process plan is also prepared. Then the NC programs, data, and documentation are stored on the IBM host, and a completion flag is sent to the NCR system.

At this point the NCR system checks the status of the IBM cell controller on the plant floor. If the cell is operational, the NCR sends the production schedule to this controller.

The cell controller requests the NC program and manufacturing documentation from the IBM host. The cell controller display also shows the manufacturing documentation to the machine operator. The controller downloads the NC program to the CNC machine. The Burr-Brown machine monitoring system and status information from the CNC machine are continuously interrogated by the cell controller. The machine monitor tells the cell controller when the job is finished.

An operator takes the finished part and initiates the inspection process on the CMM. Inspection data from the CMM is transferred to the CMM controller and compared to the part specifications. A pass/fail message is sent to the IBM cell controller for display.

When the cell controller notifies the NCR MRP II system that machining is finished, the NCR checks the status of a TI assembly cell controller. If the cell is operational, the NCR then sends the production schedule for the part assembly to the TI controller.

As pallets are loaded onto the conveyor, the TI cell operator enters into the TI cell controller the identification of which parts are on each pallet. The TI cell controller notifies the TI programmable logic controller (PLC) of pallet contents. The robot controller commands the TI PLC to present parts to the robot, which then builds the traction clutch bellcrank assembly by using a nutrunner, an end-effector tool changer, and materials handling conveyors. The TI PLC will index the conveyor to the pick-up position and stop it when the bar-code scanner reads the bar code for the correct component and sends a signal to the PLC.

The robot delivers the completed assembly to the TI vision system and receives back pass/fail dimensional quality data. Once inspected, the robot moves the assembly to the proper (good/bad) bin. The robot controller provides cycle complete, quality, and production data to the TI cell controller. The cell controller displays quality charts for the operator and sends quality and ''order complete'' data to the NCR MRP II system.

The NCR system also creates a bar-code shipping label using a bar-code printer. The label is put on the shipping box, updated documentation is placed inside, and the parts are ready to go!

In the exhibit at AUTOFACT 89, parts were actually made and the transactions described above occurred. It was a paperless, integrated environment. This demonstrated the state of the art. Seeing this technology at work is exciting.

12.6   CHAPTER AND TEXT SUMMARY

This text has presented what the authors feel are the important concepts in CAD and CAM. CIM was discussed from a pragmatic perspective. Some related concepts were omitted because of space limitations. It should be obvious that some technologies are maturing to the point that the next decade will be an exciting era in manufacturing. We have much more to learn and develop. However, much has been learned, and sufficient technology is available to allow a careful progression toward achieving CIM.

EXERCISES

1. There is a growing awareness that people-related issues are very important considerations in the implementation of a CIM environment. Using current references, write a report on this subject.
2. Using current references, write a report on the current status of the OSI reference model for networking.
3. Write a report describing a successful implementation of CIM. If possible, include a discussion of benefits, reasons for success, and effort involved (costs, time, and resources).
4. Write a report describing an engineering data management system that is commercially available.
5. Write a report on the computer network or networks that are utilized at your educational institution.

REFERENCES AND SUGGESTED READING

1. Ang, Cheng Leong: "Planning Factory Data Communications Systems," *CIM Technology*, August 1987, pp. 39–44.
2. Baron, R. C.: "Picking the Varied Fruit of Open Systems," *Manufacturing Engineering*, May 1988, pp. 52–54.

3. Barr, John R.: "Connectivity in the Factory," *Unix Review*, June 1987, pp. 33–42.

4. Behringer, Catherine A.: "Steering a Course with MAP," *Manufacturing Engineering*, September 1986, pp. 49–52.

5. Boznak, Rudolph: "Reducing Hidden Cost—A New Look at Configuration Management," *1989 International Industrial Engineering Conference Proceedings*, Toronto, Canada, pp. 47–52.

6. Brenner, Aaron: "OSI Model Update," *LAN Magazine*, June 1987, pp. 48–51.

7. Chang, Chao-Hwa: "A New Perspective on Realization of Computer Integrated Manufacturing," *Manufacturing Review*, June 1989, pp. 82–90.

8. "Data Base Management: Gateway to CIM," *American Machinist & Automated Manufacturing*, October 1987, pp. 81–88.

9. Date, C. J.: "Twelve Rules for a Distributed Data Base," *Computerworld*, June 8, 1987, pp. 75–81.

10. Gayman, David: "CAD/CAM Opens Up," *Manufacturing Engineer*, February 1986.

11. Hales, H. Lee: "Producibility and Integration: A Winning Combination," *CIM Technology*, August 1987, pp. 14–18.

12. Harrington, Joseph: *Computer Integrated Manufacturing*, Krieger Publishing, Malabar, Fla., 1973.

13. Hodges, Parker: "Manufacturing Automation's Problems," *Datamation*, November 15, 1989, pp. 32–38.

14. Hurt, James: "A Taxonomy of CAD/CAE Systems," *Manufacturing Review*, September 1989, pp. 170–178.

15. IBM Corp.: *Computer Integrated Manufacturing: An IBM Perspective*, G361-0004-0, IBM, Rye Brook, N.Y., 1987.

16. Kim, Won: "Defining Object Databases Anew," *Datamation*, April 1, 1990, pp. 22–30.

17. Krumrey, Art, and John Kolman: "LAN Hardware Standards," *PC Tech Journal*, June 1987, pp. 55–68.

18. Lefkon, Richard: "OSI Becoming Required Knowledge for LAN Users," *Management Information Week*, January 12, 1987, pp. 27–29.

19. McGrath, Michael F.: "CALS: A Strategy for Change," *CIM Technology*, May 1987, pp. 15–19.

20. Mize, Joe H.: "CIM—A Perspective for the Future of IE's," *1987 IIE Integrated Systems Conference Proceedings*, Nashville, Tenn., pp. 3–5.

21. Mize, Joe H.: "Success Factors for Advanced Manufacturing Systems," *1987 Institute of Industrial Engineers Conference Proceedings*, Washington, D.C., pp. 546–551.

22. Mize, Joe H., and Glenn Palmer: "Some Fundamentals of Integrated Manufacturing," *1989 International Industrial Engineering Conference Proceedings*, Toronto, Canada, pp. 575–580.

23. Moad, Jeff: "The Software Revolution," *Datamation*, February 15, 1990, pp. 22–30.

24. Moad, Jeff: "The New Agenda for Open Systems," *Datamation*, April 1, 1990, pp. 22–30.

25. "Putting MAP to Work," *American Machinist & Automated Manufacturing*, January 1986, pp. 75–79.

26. Rauch-Hindin, Wendy: "Revamped MAP and TOP Mean Business," *Mini-Micro Systems*, November 1986, pp. 95–109.

27. Rauch-Hindin, Wendy: "Users Impose Standards on Interfaces, Protocols," *Mini-Micro Systems*, January 1988, pp. 65–72.

28. Reimink, R. L.: "Integrating AGV's with Automated Manufacturing," *SAE Technical Paper Series*, International Congress and Exposition, Detroit, Mich., February 1986.

29. Rouse, Nancy E.: "Managing Engineering Databases," *Machine Design*, September 10, 1987, pp. 108–112.

30. Savage, Charles: "The Human Side of CIM," DEC World Show, Digital Equipment Corporation, Boston, Mass., 1987.

31. Schonberger, Richard J.: "Frugal Manufacturing," *Harvard Business Review*, September–October 1987, pp. 95–100.

32. Secula, Lawerance: "How to 802.3: Guidelines for Implementing Thin Ethernet," *LAN Magazine*, August 1987, pp. 54–62.

33. Society of Manufacturing Engineers: "Partnership for Integration Technical Application," *AUTOFACT*, 1989.

34. Stonebraker, Michael: "The Distributed Database Decade," *Datamation*, September 15, 1989, pp. 38–39.

35. Tannenbaum, Andrew S.: *Computer Networks*, Prentice-Hall, Englewood Cliffs, N.J., 1988.

36. Voelcker, John: "Helping Computers Communicate," *IEEE Spectrum*, March 1986, pp. 61–70.

37. Williamson, Mickey: "In Pursuit of Integration," *Computerworld*, July 6, 1987, pp. 51–56.

38. Winner, Robert L.: "Cross Functional Management," National Technological University, November 14–15, 1989.

39. Wolfe, Philip M., and F. Stan Settles: "Computer Integrated Manufacturing from a Pragmatic Perspective," *1989 International Industrial Engineering Conference Proceedings*, Toronto, Canada, pp. 592–597.

NAME INDEX

Allen, D. K., 291
Allen, G., 70
Allsup, R., 478, 553
Anderberg, M. R., 231
Anderson, D. C., 282, 292
Ang, C. L., 638
Armstrong, G. T., 69
Asfahl, C. R., 408, 592, 597
Ashton, J. E., 353
Auslander, D. M., 584, 597

Babb, M., 597
Bailey, J. E., 10, 20, 352, 597
Bailey, S. J., 592, 597
Bairstow, J., 20
Baker, M. P., 132
Bancroft, C. E., 175
Bannister, B. R., 570, 597
Baron, R. C., 611, 638
Barr, J. R., 639
Baumgart, B. G., 69
Bedworth, D. D., 10, 20, 352, 408, 597
Behringer, C. A., 639
Bell, A. C., 176
Berenji, H. R., 291
Berry, G., 597
Berry, W. L., 353
Bertoline, G., 132
Besant, C. B., 4, 20
Bezanson, L., 408
Bezier, P., 69
Billatos, S. B., 175
Blackmon, M. T., 553
Blauth, R., 20
Bollinger, J. G., 555, 597
Boothroyd, G., 150, 175, 507, 552
Boyse, J. W., 69
Boznak, R., 605, 639
Braid, I. C., 45, 69
Brenner, A., 639
Bridenstine, D. R., 175
Brown, C. W., 352
Brown, K. H., 352

Buker, D. W., 348, 352
Burbidge, J. L., 184, 231
Burch, J. G., 292
Byrne, D., 175

Canada, J. R., 552
Canden, D., 231
Carey, G. C., 69
Cassell, D. A., 361, 408, 581, 592, 597
Chang, C-H., 639
Chang, G. J., 70
Chang, T. C., 69, 231, 292
Chavez, P., 70
Chevalier, P. W., 292
Childs, J. J., 459, 461, 468, 469, 475
Choi, B. K., 292
Clark, K. E., 352
Clausing, D., 175
Cleary, C. M, 352
Cornillie, O., 553
Craig, J. J., 552
Cralley, W., 175
Crestin, J. P., 70
Critchlow, A. J., 552, 597
Crosby, P. B., 334, 352
Crow, K., 175
Curtin, F. T., 20

Darbyshire, I., 292
Date, C. J., 639
Davidian, R., 175
Davies, B. J., 292
Davis, F., 175
Davis, R. P., 69
Day, C. P., 518, 552
DeGarmo, E. P., 552
Deming, W. E., 296, 334, 335, 350
de Pennington, A., 69
Descotte, Y., 292
Dewhurst, P., 150, 175
Diebold, J., 296
Dietrich, W. C., 553

641

Dillman, R., 133
Disa, R., 292
Draper, A., 20, 476
Du, W-T., 552
Duffie, N. A., 555, 597
Dunlap, G., 597
Dunn, M. S., 293

Eckert, R. L., 231
Edson, J., 597
Encarnacao, J., 71, 132
Engleberger, J. F., 477, 552
Erickson, P. E., 70
Euler, L., 52
Eversheim, W., 292

Fabrycky, W. J., 552
Farin, G., 69
Farnum, G. T., 552
Fleischer, G. A., 518, 552
Foley, J., 132
Ford, H., 296
Fu, K. S., 293
Fuchs, H., 292

Gayman, D. J., 464, 475, 639
Gilchrist, J. E., 69
Glassner, A. S., 132
Goss, L., 132
Gossard, D. C., 176
Grant, E. L., 553
Granville, C. S., 292
Graves, G. R., 217, 218, 231
Grayer, A. R., 69
Grayson, T. J., 181
Greene, T. J., 352
Groover, M. P., 20, 132, 231, 352, 382, 408,
    452, 475, 478, 553, 575, 576, 597
Grossman, D. D., 553
Grudnitski, G., 292

Hales, H. L., 639
Hallquist, J., 132
Ham, I., 181, 231
Harrington, J., 599, 639
Hathaway, H. K., 231
Hax, A., 217, 218
Hayes, R. H., 353
Hayes-Roth, F., 132
Hearn, D., 132

Heidenreich, P., 175
Henderson, M. R., 70, 175, 282, 292
Higgins, W., 408
Hillyard, R. C., 69
Hinson, R., 486
Hirleman, E. D., 175
Hitomi, K., 181, 231
Hodges, P., 639
Hordeski, M. F., 21, 70, 132
Hornberger, L., 175
Huebner, K. H., 133
Hurt, J., 639
Hyer, N. L., 231

Ireson, W. G., 553
Ishii, K., 175
Iwata, K., 292

Jalubowski, R., 292
Jared, G. E., 70
Johnson, D. G., 583, 597
Jones, A. T., 20, 353
Juran, J., 334, 335

Kakazu, Y., 292
Kelly, J. C., 293
Kennedy, M. E., 176
Khoshnevis, B., 291
Kieffer, P., 293
Kim, W., 639
King, B. E., 353
Kinney, H. D., Jr., 353
Kolman, J., 639
Koren, Y., 425, 475, 573, 592, 597
Korn, G., 408
Krajewski, L. J., 350, 353
Kral, I. H., 476
Krumrey, A., 639
Kumar, P., 231
Kung, H. K., 293, 294
Kung, J. S., 284, 285, 293
Kyprianov, L. K., 293

Laduzinski, A. J., 597
LaJoy, M. H., 580, 598
Lammineur, P., 553
Latombe, J. C., 292
Leavenworth, R. S., 553
Ledley, R. S., 354, 408
Lee, T. W., 501, 553
Lee, Y. C., 293

Lefkon, R., 639
Lenat, D. B., 132
Lenk, J. D., 598
Levin, M. A., 553
Li, R-K., 242, 270, 293, 410, 476
Lieber, D., 553
Lin, L., 293
Liou, M., 175
Liu, C. R., 293
Liu, D., 293
Liu, M., 520, 553
Logan, F. A., 293
Luggen, W., 425, 476

Machover, C., 4, 20
Mann, T., 177
Mann, W. S., 293
Mantyla, M., 70
Maus, R., 478, 553
McAuley, J., 206, 207, 231
McGinnis, L. F., 353
McGrath, M. F., 639
McLean, C. R., 353
McNeill, B. W., 175
McWaters, J. F., 70
Meal, H., 217, 218, 231
Mellichamp, D., 408
Meriwaki, T., 292
Miller, J. H., 70
Mitchell, L. D., 133
Mitrofanov, S. P., 180, 227, 231
Mize, J. H., 600, 639
Moad, J., 639
Montgomery, D. C., 330, 353
Moore, J. A., 583, 598
Morris, H. M., 546, 547, 553, 598
Mortensen, M. E., 24, 70
Musti, S., 70, 175

Nackman, L. R., 553
Nagel, R. N., 553, 597
Nevins, J. L., 134, 175
Niebel, B. W., 20, 476
Nilson, E. N., 293
Nof, S. Y., 553
Norman, R., 176

Oba, F., 292
Odrey, N. G., 553, 597
Okino, N., 292
Opitz, H. A., 231

Orlicky, J., 353
O'Rourke, J. T., 1
Osborne, D. M., 598
Owen, T., 508, 553

Palmer, G., 600, 639
Pao, Y. C., 133
Paul, R. P., 553
Pease, W., 476
Pegden, C. D., 553
Perlotto, K., 293
Petterson, O., 408
Pflederer, S. J., 293
Phillips, L., 292, 293
Pickett, M. 69
Potter, R. D., 526, 553
Pugh, A., 574, 598
Puszta, J., 476

Quatse, J. T., 403, 408

Ralston, A., 476
Ranky, P. G., 133
Rauch-Hindin, W., 639
Razdan, A., 70
Reimink, R. L., 600, 639
Rembold, U., 133, 408
Requicha, A. A. G., 70
Ribeiro, J., 565, 597
Rirok, J., 293
Ritzman, L. P., 353
Rogan, E., 175
Rony, P., 408
Rouse, N. E., 639
Rowe, G. W., 176
Runcimen, C., 176

Sack, C. F., 293
Sackman, H., 353
Sagues, P., 584, 597
Salva, M., 476
Sata, T., 70
Savage, C., 176, 639
Schaffer, G. H., 231
Schlechtendahl, E. G., 132
Schonberger, R. J., 353, 639
Schwartz, S. J., 293
Secula, L., 639
Seifoddini, H. S., 211, 231
Seippel, R. G., 562, 570, 598
Seth, M. K., 408

Settles, F. S., 640
Shahinpoor, M., 553
Shigley, J. E., 75, 133
Shreve, M. T., 293
Singh, C. K., 231
Sink, D. S., 334, 335, 353
Skinner, W., 349, 353
Smith, A., 295
Smith, P. R., 291
Sneath, P. H., 206, 232
Sokal, R. R., 210, 232
Soloja, V. B., 178, 180, 232
Srinrasan, R., 293
Staley, S. M., 293
Steinberg, H., 70
Steudel, H. J., 293
Stoll, H. W., 176
Stonebraker, M., 614, 640
Stonecipher, K., 504, 553, 575, 576, 598
Stoner, D. L., 353
Strater, F. R., 292
Stroud, I. A., 69
Subrin, R., 293
Suh, N., 176
Sullivan, W. G., 520, 552, 553
Sutherland, I., 4
Swift, K., 176

Taguchi, G., 158, 159, 175
Tannenbaum, A. S., 640
Tapadia, R. K., 70
Taylor, F. W., 180, 296
Taylor, R. H., 553
Thornton, E. A., 133
Thuesen, G. J., 552
Tice, K. J., 353
Todd, J. L., 598
Toepperwein, L. L., 553
Tompkins, J. A., 353, 505, 553
Tulkoff, J., 293

Ursoevic, S. M., 232

Vail, P. S., 408
VanDam, A. D., 132
Veraldi, L., 139
Viswanathan, K., 553
Voelcker, H. A., 70
Voelcker, J., 640
Vollman, T. E., 353

Wade, O. R., 269, 293
Waldman, H., 293
Waterman, D. A., 132
Weck, M., 464, 476, 571, 572, 598
Weinstein, J. B., 408
Weiss, M., 553, 597
Wheelwright, S. C., 353
White, J. A., 505, 553
Whitehead, D. G., 570, 597
Whitney, D. E., 134, 175, 176
Whyback, D. C., 353
Wiemer, R., 416
Wilhelm, R. E., Jr., 583, 598
Williamson, M., 640
Wilson, C. W., 176
Winner, R. L., 640
Wolfe, P. M., 294, 640
Wong, D. S., 353
Woo, T. C., 70, 294
Wright, J., 293
Wysk, R. A., 20, 69, 231, 292, 294, 476

Yang, D. C. H., 501, 553

Zdeblick, W. J., 294
Zimmers, E., 132, 231, 352
Zons, K. H., 292

SUBJECT INDEX

Acceptance sampling, 329
Accuracy, robot, 516–518
  absolute, 517
  categories of, 517
  ways to improve, 518
Actuation in control:
  alarms and annunciators, 587–588
  definition of, 557
  electromechanical, 586
  hydraulic, 592–593
  motor drives, 588–592
  on/off switches, 585
  pneumatic, 592–593
Adaptive control, NC, 414, 466
Adaptive welding, robotic, 482
Aggregate planning, 15
Aggregate production plan, 299, 308
Alarms, 587–588
Analog input/output (I/O), definition of, 359
Analog to digital (A/D) conversion:
  definition of, 359
  procedure, 369–371
  programming concepts, 377
  resolution of convertor, 369
Analysis in CAD, 116–128
  data structure for, 121
  finite-element analysis, 123–125
  kinematics/dynamics analysis, 125–128
  mass properties calculation, 124
Analysis in control, 576–585
  digital computer control, 584–585
  PID control, 577–581
  programmable controller, 582–584
Annunciator, 557, 587
Anthropomorphic robot configuration, 481
APT (automatically programmed tool), 414, 425–440
Arc welding, 482
Articulated robot configuration, 481
Artificial intelligence (AI):
  in CAD, 128
  in process planning, 288–289
AS/RS (automatic storage/retrieval system), 9
Assemblability, 137

Assembly by robot, 506–512
  analyzing part for, 508–509
  material feeder assist, 512
Assembly evaluation method, 152
Assembly robot, 481, 490
Attribute (polycode) code, 178, 189
Automated drafting/documentation, 77
Automation, islands of, 2
Average linkage clustering algorithm, 178, 210–211
Axes in NC, 418
Axiomatic design, 137, 146–147
Axioms for design optimization, 146–147

Bang-bang control, 557
Bar coding, 564–566
Baseband, 601
Bezier curves, 39–40
  first-order continuity, 40
  illustration of, 40
  zero-order continuity, 40
Bezier surfaces, 43
Bicubic surfaces, 42–44
  Bezier form, 43
  B-spline form, 44
  Hermite form, 43
  networked, 42
Bill of materials (BOM), 299, 301
  example of, 302
Binary image map, 515
Biological classification, 178
Bit (binary digit), 359
Bit-mapped screen, 93
Blending functions, curve, 39
Boolean operations, two- and three-dimensional, 51
Boolean operator, 24
Bottleneck machine, 178
Boundary evaluator, 50
Boundary representation (BREP), 24, 47
  part interpretation with, 65–66
  solid modeler storage data base, 54–61
BREP (*see* Boundary representation)
Broadband network, 601

B-spline curves, 40–42
  examples of, 41
  first- and second-order continuity, 41
BUILD modeler, 45
Business plan, definition of, 295
Bus network, 601, 602
BYTE, definition of, 368

CAD (*see* Computer-aided design)
CAD/CAM:
  evolution of, 3
  operations flow, 13
  role in integrating by:
    computer control, 357–359
    group technology, 182–183
    integrative manufacturing, 300–303
    measurement, analyis, actuation, 555–556
    numerical control, 410–414
    process planning, 236–237
    robotics, 479–481
  system, 22
CADD (*see* Computer-aided drawing and
    drafting)
CAE (*see* Computer-aided engineering)
Calra code, 566
CAM (*see* Computer-aided manufacturing)
CAMI automated process planning, 239, 240,
    242–247
  data elements in, 246
  flow diagram for, 243
  operation plan data elements, 250
Canned cycle, CNC, 414, 458–461
Capacity planning, 299, 308, 317–322
  example of, 321
  rough-cut, 300, 314–315
CAPP (*see* Computer-aided process planning)
CAT (*see* Computer-aided testing)
Cathode-ray tube (CRT), 72
Causal model in forecasting, 311
Cell decomposition (CD), 46, 48
  illustration of, 49
Cell manufacturing, 7, 178, 299, 339–341
  control system for, 344
  functions and control, 8
Chain code in group technology, 189
CIM (*see* Computer-integrated manufacturing)
Classification in group technology, 179
Classification of parts, definition of, 186
CL (cutter location) data, 414, 462
CLDATA (*see* CL data)
Closed-loop control, NC, 464–465
Cluster analysis, 179
CMPP (*see* Computer-managed process
    planning)

CNC (*see* Computer numerical control)
Code 39 bar code, 565
CODE group technology code, 193, 194
Coding of parts, definition of, 186
Component-machine chart, 185
Computer-aided analysis, 72
Computer-aided design (CAD), 71–133
  analysis in, 116–128
  data base, 95
  definition of, 3, 16, 72
  evolution of, 3
  geometry, 95–97
  integrated, 128–131
  module categories, 77
  qualities of good, 130
Computer-aided drawing and drafting (CADD),
    3, 72, 96
Computer-aided engineering (CAE), 12
Computer-aided inspection (CAI), 331
Computer-aided manufacturing (CAM):
  applications, 297
  evolution of, 4
Computer-Aided Manufacturing—International
    (CAMI), 239
Computer-aided process planning (CAPP), 179
  computer-managed process planning (CMPP),
    233–294
  generative, 235, 239–242
  role in CAD/CAM integration, 236–237
  variant, 236, 238–239
Computer-aided testing (CAT), 333
Computer control, 354–408
  in analysis, 584–585
  background of, 361–362
  multiple machine, 7
  multitasking in, 384–388
  priority interrupts in, 382–384
  production line, 7
  programming concepts:
    analog input, 377–378
    analog output, 378–380
    digital input, 374–376
    digital output, 376–377
  role in CAD/CAM, 357–359
  timing in, 381–382
Computer graphics, 3
  and the part model, 98–116
    interactive, 98–100
    three-dimensional, 108–109
      composite transformations, 110–111
      transformation, 109
    two-dimensional, 101–108
      transformations, 103–104
        combining, 105

Computer graphics (*Cont.*)
    coordinate systems, 106–108
    homogeneous coordinates, 104
    scaling, 104
Computer-integrated manufacturing (CIM), 3,
    599–640
  definition of, 599–601
  management of technology in, 614–615
  networks in, 610–613
  operational flow, 16
  scope of, 10–15
  strategy development for, 625–629
  structure of, 11
  technology issues, 602–616
Computerized servo robot, 490
Computer-managed process planning (CMPP),
    235, 242, 262–268
  feature capabilities, 268
  feature specification methodology, 267
  process decision model, 264
  process planning function, 268
  system overview, 263
Computer numerical control (CNC), 7, 414,
    422, 457–461
  advantages over conventional NC, 461
  benefits through, 458
  canned cycles for, 458–461
Computers, categories of, 361
Concurrent design, 76
  steps in, 143–145
Concurrent engineering, 134–176
  driving forces behind, 138–141
  engineering tools summary, 169–173
  goals of, 143
  meaning of, 141–145
  schemes for, 145–169
  typical domains, 141
Configuration management in CIM, 602, 605–
    607
  principles of, 605–606
Constraints, design, 73
Constructive solid geometry (CSG), 24, 47, 49
  in feature-based design, 62–65
  solid modeler storage data base, 54–61
    data structure, 57
    object storage, 58
    tree format, 57
Continuous cutting in NC (contouring), 414,
    466
Continuous production process, 1, 359
Control computer, 7
  definition of, 359
  generalized schematic for, 355
Control points for curve shape, 37

Controls, in NC, 464–467
  adaptive control, 466
  closed-loop control, 464–465
  interpolation, 467
  open-loop control, 464–465
  point-to-point, 466
Conversion count in A/D, 359
Coordinate measuring machine (CMM), 331,
    514
Coordinate systems in computer graphics, 106–
    108
Correlation coefficient, 305
Criteria of design, 73
CSG (*see* Constructive solid geometry)
Cubic curves, 36–42
  Bezier, 39
  B-spline, 40
  Hermite, 37
  storage of, 42
Cursor, 73
Cutter offset, NC, 454
Cycle time in production rate, 4

Data base:
  BREP, 54, 55, 57, 58
  CAD, 95
  CSG, 54, 55, 57
  solid modeler storage, 54
Data base management system, 607–610
  distributed, 614
  hierarchical structure, 608
  network structure, 608
  relational, 608
Data glove, 88
Data handling with microcomputers, 368–373
Data transfer standards, 66–67
  initial graphics exchange standard (IGES), 66
  product definition exchange specification
    (PDES), 66
DCLASS (*see* Design and classification infor-
    mation retrieval)
Decision table, 235, 249
  example of, 253
  for part family process plans, 254
Decision tree, 235
  example for DCLASS, 259
  example for process plan, 255
Decision variables, 179
Delphi forecasting technique, 310
Dendogram, 179, 209
Departments in manufacturing, 298
Derivative control in PID, 581

Design:
  applying computer to, 76–77
  axiomatic, 137, 146–147
  axioms for, 146–147
    corollaries to, 147
  concurrent, 76
  cycle costs, 140
  definition of, 73
  engineering analysis in, 77
  process, 74–77
    phases in, 75
  review in CAD, 77
  traditional, 135
  variables, 73
Design and classification information retrieval
    (DCLASS), 235, 240, 249–262
  basis of, 196
  example of, 194
  feature complexity code, 198
  group technology code, 193, 196–200
  logic tree, 197, 199
  material code logic tree, 201
  part code segments, 196
  precision class code, 200
  size code, 200
Design for assembly, 150–158
Design for manufacturability (DFM), 137
Design for manufacturing:
  guidelines, 148–149
  method and application, 172
  methodology comparisons, 171
  tools and capabilities, 170
Design science, 137, 149
Detection of objects, automatic, 558–
    563
Dials in CAD input, 86
Digital I/O, definition of, 359
Digital-to-analog (D/A) conversion:
  definition of, 359
  procedure, 371–373
  programmable gain in, 378
  programming concepts, 378
  resistive circuit for, 372
Direct numerical control (*see* Distributed nu-
    merical control)
Discrete production process, 1, 2, 359
Dispatching, 17
Distributed data base system, 614
Distributed numerical control, 414, 422, 461–
    464
  CLDATA in, 462
  retrofitting for, 464
  simplified schematic for, 462
DNC (*see* Distributed numerical control)

Economic justification of robots, 518–528
  indirect savings, 526
  inflation effect, 524–526
  life-cycle costs, 518–520
    factors in, 519–520
Economic order quantity (EOQ), 316
Edge, in solid modeling, 24
  list, 35
  loops, 58
Electrical drive power, robot, 488
Encoded media, 557
End effector, robot, 481
  motions of, 487
Engineering work station, 5, 73
EOQ (*see* Economic order quantity)
Equation:
  explicit, 36
  implicit, 36
  parametric, 37
Equipment control system, 345
Ethernet, 602
Euler:
  equation, example of, 53
  formula, 52
  operator construction, 54
  operators, 52, 53
Evaluation in CAD, 77
Expert systems, 73
  architecture, 129
  in CAD, 129
  inference engine, 129
  knowledge base, 129
Explicit equation, 36

Face, in solid modeling, 24, 32
Facility control systems, 342
Facility design using GT, 205–211
Failure-mode evaluation analysis (FMEA), 137,
    168–169
Feature-based design, 62–65
Feature recognition, 281–287
  case study, 62
  research examples, 281
Features modeler, 282
  example for, 283–287
Fiberoptic media, 619
Finite-element analysis (FEA), 73, 123–125
Flexible manufacturing system (FMS), 8, 339
FLOPS (floating-point operations per second),
    78
FMS (*see* Flexible manufacturing system)
Forecasting future events, 309–312
  Delphi technique, 310
  long-term, 308

Forecasting future events (*Cont.*)
  moving-average technique, 311
  seasonal variations, 311
  time-series model, 311
Form features, 23, 24
Forward transformation, in robot motion, 497–498
Frame buffers, 73
Function factory layout, 179

Gain, programmable, 378
Gantt chart, 325
Generative process planning, 235, 239–242
  commercial systems, 241
Geometric entities, 30–44
  point, track, surface, 58
Geometric Modeling Applications Interface Program (GMAP), 235
Geometry, 21–24
  CAD, 95
  definition of, 25
  object, 23
  wire frame, 27–28
GMAP (*see* Geometric Modeling Applications Interface Program)
Gray code, 571
Gripper, robot, 481
  strategies for, 511
Group factory layout, 179
Group technology, 4, 137, 177–232
  advantages through, 226–227
  attribute code, 178
  chain code, 189
  classification and coding in, 186–193
  code selection, 190–192
  coding systems, 193
  definition of, 178, 179
  design benefits from, 221–223
  developing a unique code, 192
  economic modeling for, 214
  economics of, 221
  hierarchical code, 179, 189
  history of, 180–182
  hybrid code, 179, 187
  information content of a code, 195
  managerial benefits through, 225–226
  manufacturing benefits through, 223–225
  monocode, 179, 187
  objectives for, 190
  part family development methods, 183–186
  polycode, 178, 187
  role in CAD/CAM integration, 182–183
  selecting a code in, 190–192

Hardware in computer-aided design, 78–95
  input/output devices, 80–95
Hardwired control, NC, 415, 420
Hermite curves, 37–39
  cubic, 38
  geometric form, 39
  illustration of, 39
Hermite surfaces, 43
Hierarchical computer system, 357
  definition of, 359
  representative diagram of, 358
Hierarchical group technology code, 179, 187–189
Hierarchical manufacturing control, 341–346
  structure for, 343
Hierarchical tree, 235
Homogeneous coordinates, 104
Hybrid code in group technology, 187, 189
Hydraulic drive power, robot, 487
Hysteresis in sensors, 561, 563

Icon, 85
Identification of objects, automatic, 563–569
  bar code, 564–566
  magnetic stripe, 566–567
  optical character reading (OCR), 568
  radio frequency (RF), 567
  surface acoustic waves (SAW), 568
IGES (*see* Initial Graphics Exchange Standard)
Image generation, 113–116
  ray tracing in, 113
  rendering in, 113
Image storing, 92–95
  display processor unit (DPU), 93–94
  frame buffer, 92
Implicit equation, 36
Inference engine, 129
Information management system, 345
Initial Graphics Exchange Specification (IGES), 66, 67, 235
  annotation entities, 278
  geometric entities, 278
  structure entities, 279
Input devices in computer graphics, 80–88
  categories of functionality, 80
  characteristics summary, 87
Inspection, by robot, 512–516
Integral control in PID, 581
Integrative manufacturing, 295–353
  role in CAD/CAM integration, 300–303
Interactive computer graphics, 98–100
Interlacing, 73, 94
Interpolation, in NC control, 415, 467
Islands of automation, 2

JIT (*see* Just-in-time)
Job-shop production, 1, 2
Joystick, 81, 82
Just-in-time (JIT), 346–349
  benefits of, 347
  philosophy behind, 299
  principles to follow in, 348

Kernel, 22
Keyboards, 86
Kinematic link chains, 491–492
Kinematics, 73
Kinetics analysis, 73
Knowledge base, 129

Ladder diagram, definition of, 359
  programming with, 391–395
LAN (*see* Local area network)
Lee-Yang theorem, 501
Liaison sequence in assembly, 138
Light pen, 81
Linear regression, 299
  procedure in, 304–306
Line factory layout, 179, 206
Line in design, 31
  representation formats, 32
Link geometries in robot motion, 492–493
Local area network (LAN), 78, 602
Logic tree, 179
Long-term forecasting, 308
Lot-for-lot order sizing, 317

Machine component chart, 179, 207
Machine control unit (MCU), 416
Machine tool sensing, 571–574
  axial motion, 573
  in sequencing functions, 574
  part measurement, 572–573
  tool identification, 574
  tool wear, 573
Machining plan, in NC, 426, 440–451
Machining specifications, in NC, 426, 451–454
Macro, in NC programming, 415, 442–443
Magnetic stripe identification, 567
Mainframe computer (maxicomputer), 78, 354
  definition of, 360
Management of technology, 614
Manufacturing, categories of, 1
Manufacturing automation protocol (MAP), 602, 622–625
Manufacturing cell, 7
  example of, 211–214
Manufacturing costs:
  direct labor, 214

Manufacturing costs (*Cont.*):
  direct materials, 214
  indirect materials, 216
  tooling, 218–221
Manufacturing engineering, 303–307
Manufacturing operations, responsibilities of, 298
Manufacturing planning and control, 295–353, 336–339
Manufacturing resources planning (MRP-II), definition, 299
MAP (*see* Manufacturing automation protocol)
MAPICS (Manufacturing Accounting and Production Information Control System), 336–339
Masking, in data I/O, 360
Mass production, 1, 2
Master production schedule, 13, 312–314
Material requirements planning (MRP), 13, 299, 308, 315–317
  benefits of, 320
Materials handling, by robot, 505–506
Mathematical models, 73
Measurement, analysis and actuation, 554–598
  integrative role in CAD/CAM, 555–556
  situations requiring, 554–555
MICLASS group technology code, 193, 200–205
Microcomputers in control, 354
Minicomputer:
  definition of, 78
  work stations, 5, 73
Monocode in group technology, 179, 187
Mouse input device, 83–85
MRP (*see* Material requirements planning)
MRP-II (*see* Manufacturing resources planning)
Multitasking systems, 384–388
  state-space diagram for, 385
  typical CALLS in, 386–387

Natural quadrics, 49
NC (*see* Numerical control)
Network architecture example, 345
Networking in CIM, 616–625
  concepts in, 616–619
  Ethernet, 602
  factory communication requirements, 612
  local area network (LAN), 602
  MAP, 602, 622–625
  OSI reference model, 620–622
  token ring, 618
  TOP, 602, 622–625
  topologies for, 617
Number systems, 362–368

Numerical control, 6, 7, 409–476
  advantages of, 468
  axes in, 418
  CNC, 457–461
  controls in, 464–467
  definition of, 415
  DNC, 461–464
  equipment examples, 423–425
  operational sequence, 415–420
    analyst's role, 416
    operator's role, 416
  programming, 425–457
  role in CAD/CAM integration, 410–414
  types of systems, 418–420
NURBS, 44

Object instancing and specification, 102
One-model concept in CIM, 603–604
Open-loop control in NC, 464–465
Operation code (op code), 235, 250
  for DCLASS, 258
Operation plan (op plan), 235
Operations management, 14
OPITZ group technology code, 194, 195
Optical character reading (OCR), 568
Optimization in design, 74
Order release, 299, 322
Orthographic drawing, 26
OSI reference model, 602, 620–622
Output devices for computer graphics, 89–92

Parallel design, 74, 76
Parametric equation, 37
Part definition in NC, 415, 426–440
Part effectivities, definition of, 299
Part family, definition of, 179
Part family matrix, 235, 244
Part grouping:
  by geometric shape, 183
  by manufacturing process, 184
Part representation, 24
PDDI (*see* Product definition data interface)
PDES (*see* Product definition exchange specifi-
    cation)
Performance variables in design, 74
Pick-and-place robot, 481, 489
PID (proportional/derivative/integral) control,
    557, 577–581
  derivative control, 581
  integral control, 581
  proportional control, 578
  two-position control, 577
Pixel, 74, 92
Planning horizon, definition of, 299

Plasma screen, 92
PLC (*see* Programmable controller)
Plotter, 89, 90
Pneumatic drive power, robot, 488
Point geometry entity, 58
Point-to-point machining, 415, 440–442
  controls for, 466
Polygon, 25
  planar, 30
  modeling, 33–36
Polygonal model, 31
Polyhedra, multiple, 52
Polyhedron, 25, 52
Port, communication, 360
Postprocessing, NC, 415
Printer, 90
Priority interrupt, 360, 382–384
Priority sequencing rules, 327
Process decision model (PDM), 264
Process plan, 25, 179
Process planning, 17, 233–294
  approaches to, 237–242
  artificial intelligence in, 288–289
  automated (CAPP), 239–268
  benefits of, 234
  criteria for selecting CAPP, 275
  generative, 235, 239–242
  information flow in, 237
  manual approach, 238
  role in CAD/CAM integration, 236–237
  tolerance charting for, 271–275
  variant, 236, 238–239
Producibility, 138
Product data exchange specification (PDES),
    66, 235, 279–280
Product definition data interface (PDDI), 236
Product development team, 138
Production and information flow, 307
Production categories, 1, 2
Production control responsibility, 309
Production flow analysis, 179, 184–186
Production planning and control, definition of,
    298
Product life cycle, 138
Product-structure tree, 302
Programmable (logic) controller (PLC), 388–403
  analysis for control with, 582–584
  definition of, 360
  future of, 401–403
  ladder diagram concepts, 391–395
  languages for, 403
  modular structure of, 389
  network compatability, 403
  operating system, 403

Programmable gain, 378
Programming NC:
  looping, 443–446
  machining plan, 440–451
    contouring, 446–451
    point-to-point, 440–442
  machining specifications, 451–454
  macros, 442–443
  part definition (geometry), 426–440
    circles, 436
    cylinders, 436–438
    lines, 431–434
    patterns, 429–431
    planes, 434–436
    points, 427–429
Programming robots, 529–547
  AML/X, 541–546
  control computer functions, 538–541
  teaching, 532
  tool coordinates, 530
  VAL-II, 529–538
  world coordinates, 530
Projection in CAD, 112–113
Proportional control in PID, 578
Protocol, 602
Proximity sensing, 558, 562–563
Pure primitive instancing (PPI), 46, 47

Quadric surface, 25
Quality assurance, 328–336

Radio frequency (RF) transponder, 567
Radix in number systems, 360
Random-scan device, 74
Raster device, 74
Ray tracing, 113
Real-time control, definition of, 360
Real-time in computer control, 381–388
  multitasking, 384–388
  priority interrupt, 382
Refreshing of computer screen, 74
Register, computer, 361
Rendering, 24
  in image generation, 113
Repeatability, robot, 516
Retrofit of NC machine, 415, 464
Robot:
  accuracy and repeatability, 516–518
  anthropormorphic, 481
  applications, 502–516
  assembly, 481, 490
  characterization of, 482–490
  computerized servo, 481, 490
  definition of, 478

Robot (*Cont.*):
  drive power, 487–490
  economic justification of, 518–529
  end-effector motions, 487
  forward transformation, joint angle solution,
    498–499
  frame of reference, 493, 495–497
  kinematic link chains, 491–492
  link geometries, 492–493
  off-line programming, 546–547
  orientation, 494–495
  pick-and-place, 481, 489
  programming languages, 529–546
  sensing, 481, 490
  vision, 514–516
  workspace description, 499–502
Robotics, 477–553
  definition of, 481
  role in CAD/CAM integration, 479–481
  sales by application, 479
Robot sensing, 574–576
  tactile, 574
  vision, 576
  voice recognition, 575
ROMULUS solid modeler, 282
Rotational part, definition of, 180
Rough-cut capacity planning, 300, 314–315

SCADA (supervisory control and data acquisi-
    tion system), 557, 593–595
Scaling in computer graphics, 104
Scheduling, shop-floor, 324–328
Sensing and measuring in control, 557–576
  machine tool sensing, 571–574
  object detection, 558–563
  object identification, 563–569
  robot sensing, 574–576
Sensing robot, 481, 490
Sequencing rules, priority, 327
Serial design, 74
Serviceability, 138
Servomotor, 557, 592
Shaft encoder, 571
Shop-floor control, 300, 323–328, 343
Shop-floor scheduling, 324–328
Similarity coefficient, definition of, 180
Similarity matrix, 208
Simulation, in robotics, 481, 547–549
  benefits from, 547–548
Single-linkage clustering algorithm (SLCA),
    180, 206–210
  chaining in, 209
SKETCHPAD, 4

Skinner hand, 483
Solid modelers, 44–61
    BUILD modeler, 45
    construction techniques, 46
    Euler formula in, 52
    ROMULUS modeler, 282
    storage data base, 54–61
    tesselated, 30–32
Span of control, 357
Spatial occupancy enumeration (SOE), 46, 48
SPC (see Statistical process control)
Spot welding, 482
Spray painting, robot, 505
Standard for Transfer and Exchange of Product
        Model Data (STEP), 236, 280
Stanford manipulator, 495, 496
Statistical process control (SPC):
    charts in, 329
    definition of, 300
    reasons for use of, 330
STEP (see Standard for Transfer and Exchange
        of Product Model Data)
Stepping motor, 488, 557, 588–592
Storage tube, direct view (DVST), 91
Supercomputer, 78
Surface:
    Bezier, bicubic, Hermite, 42, 43
    models, 28–30
Surface acoustic waves (SAW), 568
Surface geometry entity, 58
Sweeping, 46, 50
Switches for actuation:
    electromechanical relay, 557, 586
    limit, 557, 558–560
    on/off, 585
    photoelectric, 557, 560–562
    programmable limit, 557
    proximity, 558, 562–563
Synthesis in design, 74
Systems approach, 10

Tablet input device, 85
Tactile sensing, robot, 481, 574
Taguchi method, 138
    for robust design, 158–167
Teaching a robot, 532

Teach pendant, 482
Teach robot, 481
Technical and office protocol (TOP), 602, 622–
        625
Tesselated modeler, 30, 32
Threshold level, 208
Threshold value for clustering, 180
Through-put time, manufacturing, 325
Time-series forecasting, 311
Token-ring network, 602, 618
Tolerance chart, 236, 271–275
Tolerance setting in NC, 453
Tool coordinates, robot, 530
TOP (see Technical and office protocol)
Topology, 25
Total quality management (TQM), 300, 334–336
TQM (see Total quality management)
Track geometry entity, 58
Transducers, 557, 569–571
Transfer machines, automatic, 6
Transformations in design, 74
Transponders, 567

Unified life-cycle engineering, 138
UNITE command, 46
Universal product code (UPC), 564–565
User interface, 74

Value engineering, 138, 169
Variant process planning, 236, 238–239
    commercial systems, 240
Vertex, 25, 31
    list data structure, 33, 34
Viewport in computer graphics, 107
Vision sensing, 514, 576
Voice recognition, automatic, 575

Warehousing, intelligent, 14
Welding, robot, 502–505
    evaluating capabilities for, 504
Window, in computer graphics, 107
Wire frame geometry, 27–28
Work station, engineering, 5, 73
Work station control architecture, 344
Work station control system. 344
Workspace, description for robot, 499–502
World coordinates, robot, 530